XXI International Scientific Conference – The Application of Experimental and Numerical Methods in Fluid Mechanics and Energy (AEaNMiFMaE-2018)

MATEC Web of Conferences Volume 168 (2018)

Rajecke Teplice, Slovakia
25 - 27 April 2018

Editor:

Milan Malcho

ISBN: 978-1-5108-6387-3

Printed from e-media with permission by:

Curran Associates, Inc.
57 Morehouse Lane
Red Hook, NY 12571

Some format issues inherent in the e-media version may also appear in this print version.

Printed by Curran Associates, Inc. (2018)

For additional information, please contact EDP Sciences – Web of Conferences at the address below.

EDP Sciences – Web of Conferences
17, Avenue du Hoggar
Parc d'Activité de Courtabœuf
BP 112
F-91944 Les Ulis Cedex A
France

Phone: +33 (0) 1 69 18 75 75
Fax: +33 (0) 1 69 28 84 91

contact-edps@webofconferences.org

Additional copies of this publication are available from:

Curran Associates, Inc.
57 Morehouse Lane
Red Hook, NY 12571 USA
Phone: 845-758-0400
Fax: 845-758-2633
Email: curran@proceedings.com
Web: www.proceedings.com

TABLE OF CONTENTS

Department of Power Engineering
Faculty of Mechanical Engineering
University of Žilina

XXI. International Scientific Conference

THE APPLICATION OF EXPERIMENTAL AND NUMERICAL METHODS IN FLUID MECHANICS AND ENERGY 2018

25.04. - 27.04.2018

Rajecké Teplice – Slovakia

Organizing Committee - Editors and technical reviewers

Ing. Richard Lenhard, PhD., (richard.lenhard@fstroj.uniza.sk)

Ing. Katarína Kaduchová, PhD. (katarina.kaduchova@fstroj.uniza.sk)

Conference Secretariat

Ing. Slavka Jelušová (slavka.jelusova@fstroj.uniza.sk)

Guarantors of the conference

prof. RNDr. Milan Malcho, PhD., ŽU Žilina, SR

prof. Ing. Jozef Jandačka, PhD., ŽU Žilina, SR

Scientific committee and Reviewers

prof. Ing. Mária Čarnogurská, CSc., TU Košice, SR

prof. Ing. Jozef Černecký, CSc., TU Zvolen, SR

prof. Ing. Ladislav Dzurenda, PhD., TU Zvolen, SR

prof. Ing. Jozef Jandačka, PhD., ŽU Žilina, SR

prof. Ing. František Kavička, CSc., VUT Brno, CZ

doc. Ing. Branislav Knížat, CSc., STU Bratislava, SR

prof. Ing. Václav Kopecký, CSc., TU Liberec, CZ

prof. dr. sc. Ladislav Lazič, University of Zagreb, Zagreb Croatia

prof. RNDr. Milan Malcho, PhD., ŽU Žilina, SR

doc. Dr. Ing. Tadeáš Ochodek, VŠB - TU Ostrava, CZ

doc. Ing. Róbert Olšiak, PhD., STU Bratislava, SR

prof. Ing. Jaromír Příhoda, CSc., UT AVČR Praha, CZ

prof. Ing. Miroslav Příhoda, CSc., VŠB - TU Ostrava, CZ

prof. Ing. Miroslav Rimár, CSc., TU Košice - Prešov, SR

prof. Ing. Josef Soukup, CSc., UJEP Ústi nad Labem, CZ

prof. Andrzej Szlek, PhD, DSc, University of Technology, Gliwice, PL

prof. Ing. Pavel Šafařík, CSc., ČVUT Praha, CZ

doc. Ing. Peter Tauš, PhD., TU Košice, SR

prof. Ing. František Urban, PhD., STU Bratislava, SR

prof. Ing. Augustín Varga, CSc., TU Košice, SR

Additional Scientific Reviewers

prof. RNDr. Milan Malcho, PhD., ŽU Žilina, SR

prof. Ing. Jozef Jandačka, PhD., ŽU Žilina, SR

Ing. Richard Lenhard, PhD., ŽU Žilina, SR

Ing. Katarína Kaduchová, PhD. ŽU Žilina, SR

doc. Ing. Radovan Nosek, PhD. ŽU Žilina, SR

doc. Ing. Lukasz Orman, PhD. UT Kielce, PL

Ing. Alexander Čaja, PhD. ŽU Žilina, SR

Ing. Peter Ďurčanský, PhD. ŽU Žilina, SR

Ing. Michal Holubčík,, PhD. ŽU Žilina, SR

Ing. Marek Patsch, PhD. ŽU Žilina, SR

Ing. Peter Pilát, PhD. ŽU Žilina, SR

Ing. Patrik Nemec, PhD. ŽU Žilina, SR

Sponsors

AREKO s.r.o.

KVANT®
PRE VÁS
od roku 1995

TECHSOFT
ENGINEERING

Presentation of the AEaNMiFMaE 2018

Conference opening ceremony, speaker prof. Doc. Jozef Jandačka, PhD.,
Vice-Rector for Development, University of Žilina (Photo: Katarína Kaduchová)

Conference opening ceremony, speaker prof. Doc. Milan Malcho, PhD.,
Head of the Department of Power Engineering, University of Žilina (Photo: Katarína Kaduchová)

Conference participants (Photo: Katarína Kaduchová).

**Preface: Proceedings of the XXI. International Scientific Conference –
The Application of Experimental and
Numerical Methods in Fluid Mechanics and Energy 2018**

(XXI. AEaNMiFMaE-2018)

The International Scientific Conference on Application of Experimental and Numerical Methods in Fluid Mechanics and Energy 2018 (XXI. AEaNMiFMaE-2018) has been scheduled to take place at Hotel DIPLOMAT – Rajecké Teplice, Slovakia, between 25-27 April 2018.

The aim of the international conference is to bring together researchers, designers, experimenters, modeller's and numerical analysts from laboratories, industry and academia working in the field of two-phase and multiphase flow and experimental methods used in the field of fluid mechanics, thermodynamics and energy.

The international conference covers all topics in Measurement and calculation of state variables in the fluid flow, modelling and simulation in fluid mechanics and energy, optimization of flow parameters in thermal and hydraulic machines and systems, the application of the latest knowledge in the field of fluid mechanics and energy in technical practice, flow visualization, measurement of the energy systems and systems, special experimental methods in fluid mechanics and energy, renewable energy, current problems in the energy sector.

The conference should contribute to the establishment closer contacts, presentations on the latest knowledge, exchange of experience in the application of different experimental methods, preparation and implementation experiments, results processing and numerical simulations of experiments.

The scientific committee determined the following conference topics:

- Measurement and calculation of state variables in the fluid flow.
- Modelling and simulation in fluid mechanics and energy.
- Optimization of flow parameters in the thermal and hydraulic machines and systems.
- Application of latest developments in the field of fluid mechanics and energy in technical practice.
- Visualization of flow.
- Measurement of energy systems and systems.
- Special experimental methods in fluid mechanics and energy.
- Renewable energy.
- Current problems in the energy.

We would like thanks:

- The Scientific Committee and Reviewers of AEaNMiFMaE-2018 for their help and support.
- The Session Chairman's for their considerable effort the managing of individual sections.
- The Guarantors of the conference (prof. Ing. Jozef Jandačka, PhD., Vice-Rector for Development, University of Zilina, Slovakia, prof. RNDr. Milan Malcho, PhD., University of Zilina, Slovakia).

Statement of Peer review

In submitting conference proceedings to *Web of Conferences*, the editors of the proceedings certify to the Publisher that

1. They adhere to its **Policy on Publishing Integrity** in order to safeguard good scientific practice in publishing.
2. All articles have been subjected to peer review administered by the proceedings editors.
3. Reviews have been conducted by expert referees, who have been requested to provide unbiased and constructive comments aimed, whenever possible, at improving the work.
4. Proceedings editors have taken all reasonable steps to ensure the quality of the materials they publish and their decision to accept or reject a paper for publication has been based only on the merits of the work and the relevance to the journal.

Title, date and place of the conference

XXI. international scientific conference -
The Application of Experimental and Numerical Methods in Fluid Mechanics and Energy 2018,
25.– 27.04.2018
Wellness Hotel Diplomat
Rajecké Teplice - SLOVAKIA

Proceedings editor(s):

prof. RNDr. Milan Malcho, PhD., ŽU Žilina, SR
prof. Ing. Jozef Jandačka, PhD., ŽU Žilina, SR
Ing. Richard Lenhard, PhD., ŽU Žilina, SR
Ing. Katarína Kaduchová, PhD. ŽU Žilina, SR
doc. Ing. Radovan Nosek, PhD. ŽU Žilina, SR
doc. Ing. Lukasz Orman, PhD. UT Kielce, PL
Ing. Alexander Čaja, PhD. ŽU Žilina, SR
Ing. Peter Ďurčanský, PhD. ŽU Žilina, SR
Ing. Michal Holubčík,, PhD. ŽU Žilina, SR
Ing. Marek Patsch, PhD. ŽU Žilina, SR
Ing. Peter Pilát, PhD. ŽU Žilina, SR
Ing. Patrik Nemec, PhD. ŽU Žilina, SR

Date and editor's signature

20th April 2018

Measurement of Thermal Conductivity of Sludge and Liquid Materials

Mária Čarnogurská[1,*], *Miroslav* Příhoda[2], *Tomáš* Brestovič[1], *Marián* Lázár[1], *Natália* Jasminská[1], and *Jiří* Marek[2], *Dušan* Puškár[1]

[1]Technical University of Košice, Department of Power Engineering, Vysokoškolská 4, 042 00 Košice, Slovakia
[2]VŠB - Technical University of Ostrava, Department of Thermal Engineering, 17. listopadu 15, 708 33 Ostrava-Poruba, Czech Republic

Abstract. The present article describes two various methods of determination of thermal conductivity of semi-solid and liquid materials. One is based on experimental measurements of heat flux through a layer of semi-solid material while using the Heat Flow Plates, and the second one uses Peltier Elements to determine the heat flux density. The first method was used to determine thermal conductivity of sludge taken from natural gas coolers. Its value, identified by measurements, was 0.746 $W \cdot m^{-1} \cdot K^{-1}$. The second method was applied to determine thermal conductivity for sludge taken from natural gas filtration units. Values thereof were ranging from 0.608 $W \cdot m^{-1} \cdot K^{-1}$ (at the mean sludge temperature of 22 °C) to 0.652 $W \cdot m^{-1} \cdot K^{-1}$ (at the sludge temperature of 43 °C).

1 Introduction

In the execution of the cooling or heating processes it is often required to know the value of thermal conductivity λ of semi-solid or liquid materials of unknown composition, particularly in the cases of various heat exchangers, gas coolers, and other equipments the use of which requires identification of the quantity of heat removed/supplied through the heat-transfer surface which is contaminated, or even significantly glazed with such materials. In the analytical heat flow determination, it is necessary to identify several thermal and technical parameters [1-3]. Some of the values of these parameters can be found in physical tables; others must be identified analytically or by experiments. Such parameters include also thermal conductivity of materials. Determination thereof is very complicated, even with substances of known compositions. The present article describes two methods of determination of this coefficient λ for the sludge taken from a natural gas cooler and from a natural gas filter at the KS01 Compressor Station in Veľké Kapušany.

* Corresponding author: maria.carnogurska@tuke.sk

2 Description of the methodology for measuring the thermal conductivity of sludge

Standard methods typically applied to identify thermal conductivity were not applicable to the given consistency; therefore, an original laboratory experimental method for measuring the thermal conductivity for arbitrary semi-solid materials was designed. The principle thereof is based on the elementary relation applicable to the heat flux density q at unidirectional steady heat transfer in a planar board (layer) [4-6], i.e.

$$\lambda = \frac{q \cdot b}{\Delta t} \quad (\mathrm{W \cdot m^{-1} \cdot K^{-1}}), \tag{1}$$

where b is the layer thickness (m),
Δt - difference in temperatures of both layer surfaces (K).

2.1 Measurement using the heat flow plates

Provided that a heat flux density value and the temperature difference may be identified experimentally, then formula (1) can be used for simple identification of the thermal conductivity. Determination of heat flux through a 12.5 mm thick sludge layer taken from a natural gas cooler was carried out while using an experimental device (Figure 1) consisting of the following parts:

- a block with dimensions of 400 x 192 x 85.2 mm made of electrolytic copper,
- an electric heater providing the necessary heat input for temperatures ranging from 45 to 70 °C,
- an aluminium frame with internal dimensions of 220 x 140 x 12.5 mm in which the measured sludge specimen was placed, hermetically enclosed in the stretch foil,
- the Heat Flow Plates Type FQ 90150-1 heat flow sensor,
- a memory system by AHLBORN, ALMEMO 2390-8 type, recording the heat flow,
- the Grant 2020 measuring centre for continuous temperature scanning.

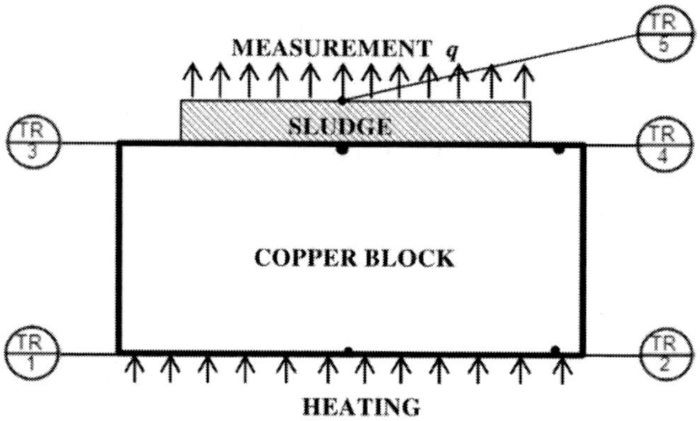

Fig. 1. Scheme of the measuring device.

Prior to the measurement itself, a type of the thermal filed inside the copper block was identified (Figure 2). To achieve stationary temperature distribution in the copper block along its height as well as its width, the block had to be heated from a cold state for the period of approximately 12 hours. An example of temperature curves for t_3 and t_4 on the top side of the copper object, after achieving the stationary state, is presented in Figure 3.

The graph indicates that the temperatures on the top surface under the sludge specimen, as well as next to it, are actually identical. The average difference between them did not exceed 0.07 °C.

After the measurement completion, data files from memories of both centres were exported to a PC, evaluated while using the Excel calculator, and subsequently used to calculate the thermal conductivity.

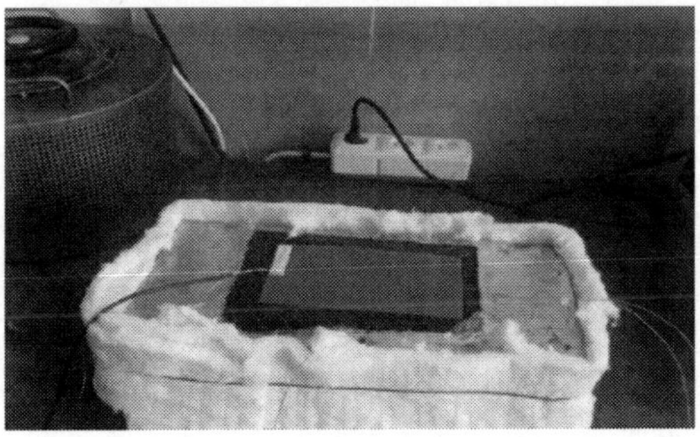

Fig. 2. Experimental device during the measurement.

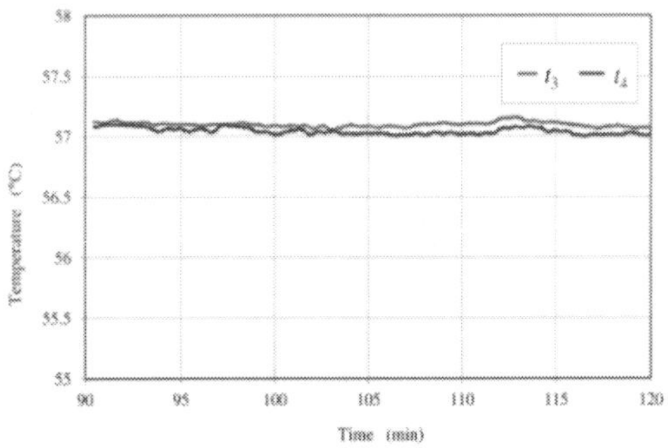

Fig. 3. Copper block top surface temperatures.

In the steady state, the measured values of heat flux q through the sludge layer, within the period of 110 min., were ranging between 201.9 and 232.1 $W \cdot m^{-2}$; the average value represented 218.3 $W \cdot m^{-2}$. Within the same period of time, the differences in temperatures of the lower and the top sludge surface were ranging between 3.6 and 3.8 K; the average temperature difference was 3.657 K. The measured parameters q and Δt, as well as the sludge layer thickness of 0.0125 m, were used to determine the thermal conductivity of sludge while using formula (1). It was observed that the average value thereof during the experiment was $\lambda = 0.746\ W \cdot m^{-1} \cdot K^{-1}$. The thermal conductivity of the sludge with such average value was used in the thermal calculation of the glazed natural gas cooler.

2.2 Measurement with the use of Peltier elements

An alternative to the coefficient λ measurement while using the Heat Flow Plates is the measurement carried out while using the device (Figure 4), comprising the Peltier Elements (PE) of the TEC1-12708 type.

In this case again, when identifying the thermal conductivity, it is necessary to respect the Fourier's Law for stationary unidirectional heat conduction. For the purpose of analytical expression of the thermal conductivity, while applying the Fourier's Law, it is necessary to know a particular value of the heat flow passing through the contact surface into the measured material, the thickness of this material layer, as well as the temperatures on both sides of the material layer. It is difficult to determine a particular value of the heat flow passing into the examined material as well as meeting the requirement of conducting the heat only in one direction.

The measured material is put inside a hollow block (1) the bottom of which consists of the PE (4). The vessel is covered with a lid (2) with the attached PE (3). Both elements are directly contacting the material, whereas the PE (3) heats it up from the top side, and the element (4) cools it from the lower side. This prevents the specimen from "stirring" due to free convection. Elimination of the heat gain on the cold side of the PE (3) is provided by the PE (5) located on the top lid surface. The cold side of this element is maintained at the same temperature as the cold side of the element (3). This temperature is scanned by electronic temperature sensors (9) and (10). Material specimen temperature is measured by sensor (6) located on the hot side of the PE (3) surface, sensor (7) located in the half-height of the material specimen, and sensor (8) located on the cold side of the PE (4).

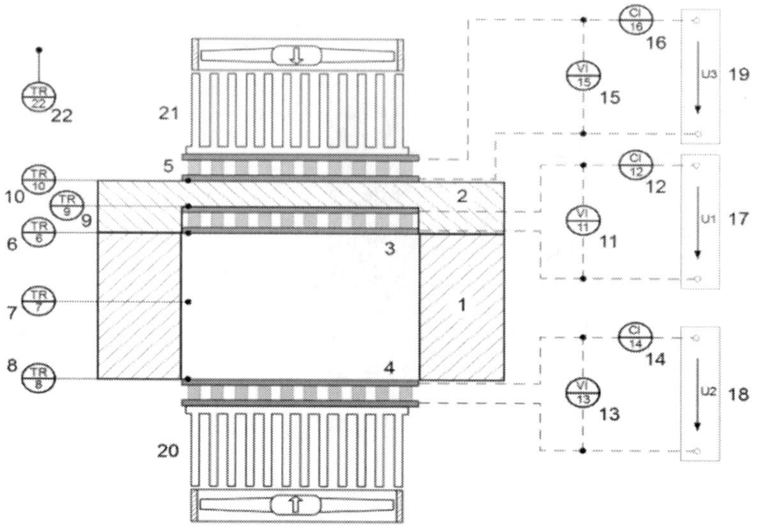

Fig. 4. Device for the measurement of thermal conductivity of materials, based on Peltier Elements.
(1 – hollow block, 2 – lid, 3, 4, 5 – Peltier Elements, 6, 7, 8, 9, 10, 22 – temperature sensors, 11, 13, 15 – voltmeters, 12, 14, 16 – ammeters, 17, 18, 19 – DC power supplies, 20, 21 – coolers)

Unidirectional heat flow along the height of the measured specimen is achieved by maintaining the specimen temperature difference in the vertical direction, which is ensured by PEs (3), (4), and (5). The average temperature of the examined specimen is maintained at the same value as the ambient temperature, measured by sensor (22), ensuring thus the zero total average heat flow from the external surface of the hollow block (1). An option of changing the mean temperature of the examined specimen is not excluded.

Used temperature sensors are of the SMT 160 – 30 type. A prototype of the device for the measurement of thermal conductivity of materials of semi-solid consistency is shown in Figure 5. A hollow block is made of foam glass of the FOAMGLAS PERINSUL type which is a good thermal insulator. Its thermal conductivity is 0.04 $W \cdot m^{-1} \cdot K^{-1}$. The volume of the examined material is 32 ml. A clear distance between the PEs (3) and (4), i.e. the height of the examined specimen, is 20 mm.

Calibration measurements showed that with regard to accuracy of the used temperature sensors, the optimal difference in temperatures of the top and of the lower sides of the measured specimen is approximately 10 °C. The required temperature difference can be achieved by regulating the PE power supply.

Fig. 5. Device prototype.

The thermal conductivity can subsequently be determined by an indirect method, a calculation using the known values of the supply voltage of a DC power supply (17) identified using a voltmeter (11), electric current of the power supply measured by an ammeter (12), temperatures measured by sensors (6, 8), and geometrical parameters of the device, according to formula

$$\lambda = \frac{(U_{11} \cdot I_{12} - P_{str}) \cdot \delta}{(t_6 - t_8) \cdot S} \quad (\mathrm{W \cdot m^{-1} \cdot K^{-1}}) \qquad (2)$$

where λ is thermal conductivity of the examined specimen ($\mathrm{W \cdot m^{-1} \cdot K^{-1}}$), U_{11} – voltage of the DC power supply (V), I_{12} – electric current of the power supply (A), P_{str} – mean loss of heat flow passing vertically through the block walls (W), δ - distance between PEs (3) and (4) (m), t_6 – temperature on the hot side of the PE (3) (°C), t_8 - temperature on the cold side of the PE (4) (°C), S – heat-transfer surface of the PE (m^2).

Heat is removed from the hot side of the PE (4) by an air cooler (20). The same function for element (5) is served by the air cooler (21). The temperature sensor (7) scans the mean temperature of the specimen and controls the stabilisation of the thermal field of the specimen [7]. PE (4) is supplied by the power supply (18), whereas the voltage is measured by the voltmeter (13) and the current by the ammeter (14). PE (5) is supplied by the power supply (19) the voltage of which is measured by the voltmeter (15) and current by the ammeter (16). Figure 6 shows how the device is connected during the measurement.

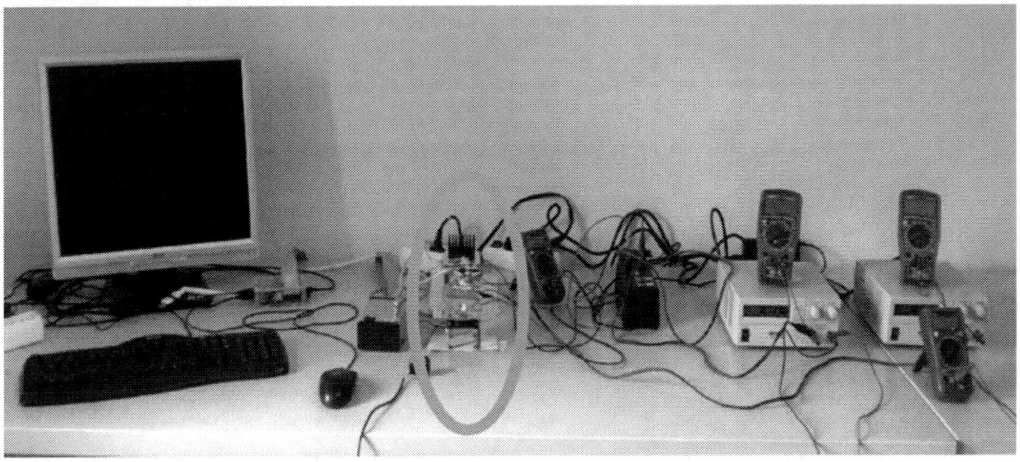

Fig. 6. Device during the measurement.

3 Conclusion

The present article describes two different methods of determination of thermal conductivity of semi-solid as well as liquid materials. One is based on the experimental measurement of heat flux through the layer of semi-solid material, while using a commercially available planar sensor Heat Flow Plates, and the second one uses the PE. As for practical applications, the second method is more complicated but more accurate. It was not possible to experimentally examine both specimens while applying both methods, as they were not obtained at the same time. The sludge from a natural gas cooler was measured two years before the sludge from a filter was obtained, and at that time the device presented in Figure 4 was not yet constructed.

Results of the measurements of the two types of sludge, while applying different procedures, show the deviation of as much as 19 %. This could be explained by different compositions of the two sludge types. Filters are located at the inlet into the compressor station, whereas the coolers are located at the outlet. It is assumable that the sludge taken from coolers is more contaminated with oil that leaks from compressors into the compressed gas.

The value of thermal conductivity of the sludge taken from natural gas coolers was 0.746 $W \cdot m^{-1} \cdot K^{-1}$. The sludge taken from natural gas filtration units was determined for the temperature range from 22 °C to 43 °C. The λ values were ranging from 0.608 $W \cdot m^{-1} \cdot K^{-1}$ to 0.652 $W \cdot m^{-1} \cdot K^{-1}$.

An advantage of measuring the thermal conductivity in a device that uses the PE is the creation of conditions that prevent the formation of free convection in specimens of semi-solid or liquid consistency, with concurrent elimination of heat losses into the surrounding environment, in particular by adjusting the mean temperature of the measured specimen to the ambient temperature.

This paper was elaborated within the tasks related to the execution of the KEGA 003 TUKE – 4/2016, APVV-15-0202, ITM S26220220044 and SP2018/94-FMMI VŠB TUO projects.

References

1. P. Nemec, M. Malcho, Experimental evaluation of cooling efficiency of the high performance cooling device. AIP Conference Proceedings, **1745**, article number 020035 (2016)
2. P. Ďurčanský, Š. Papučík, M. Malcho, Measurement of the performance characteristics of hot air heat exchanger. AIP Conference Proceedings, **1768**, article number 020028 (2016)
3. M. Příhoda, J. Vlček, M. Velička, M. Čarnogurská, R. Pyszko, Heat output of the recuperator for preheating natural gas from a mobile container device. AIP Conference Proceedings, **1768**, article number 020016 (2016)
4. M. Rédr, M. Příhoda, Základy tepelné techniky. SNTL, Praha (1991), in Czech
5. M. Čarnogurská, M. Příhoda, J. Molínek, Determination of deposit thickness in natural gas cooler based on the measurements of gas cooling degree. J. Mech. Sci. Technol. **25,** 11, (2011)
6. B. Knížat, F. Urban, M. Mlkvik, F. Ridzoň, R. Olšiak, Numerical simulation of losses along a natural circulation helium loop. AIP Conference Proceedings, **1745**, article number 020023 (2016)
7. P. Liptai, B. Dolník, M. Pavlík, K. Zbojovský, M. Špes, Check measurements of magnetic flux density: Equipment design and the determination of the confidence interval for EFA 300 measuring devices. Measurement. **111** (2017)

Temperature measurement in the flowing medium

Kamil Sedlák[1,*], and *Lenka* Fialová[1]

[1]Doosan Škoda Power, s.r.o., Experimental research of flow, Tylova 1/57, 301 28 Plzeň, Czech republic

Abstract. The article deals with a brief description of methods of temperature measurements in a flowing water steam. Attention is paid to the measurement of pseudo static temperature by a single sealed thermocouple entering the flowing liquid through the flown-by wall. Then three types of probes for stagnation temperature measurement are shown, whose properties were tested using CFD calculations. The aim was to design a probe of stagnation parameters of described properties which can be used for measuring flow parameters in a real steam turbine. An important factor influencing the construction is not only the safe manipulation of the probe when inserting and removing it from the machine in operation, but also the possibility to traverse the probe along the blade length.

1 Introduction

From the viewpoint of measuring thermodynamic parameters of flowing water steam, it is possible to choose two different and basically correct approaches.

The principle of the first approach is to focus on static parameters measurement and, with the knowledge of velocity field character and the thermodynamic properties of the flowing medium, to calculate the total parameters. This process is well manageable for pressure measurements, but it fails for temperature measurement. Provided the measured temperature was really static, it would be necessary to guarantee the sensor movement in accordance with the flowing medium. In other words, if the static temperature is to be measured, it is necessary for the sensor (thermometer) to move not only at the same velocity, but also in the same direction in the three dimensional space as the flowing medium. From these requirements it is evident that such measurement is not possible in common technical practice.

In this situation the solution could be to measure the temperature by a stationary sensor and, to accept the fact that the measured temperature is neither static nor total but somewhere in between, or to attempt to measure the total temperature and add the so called recovery factor that will be defined later on in the text and which carries, in the standardized form, the information about how far the measured temperature is from the static one.

* Corresponding author: kamil.sedlak@doosan.com

The motivation for this article were some unanswered questions, emerging mainly in relation to the analysis of measurement uncertainties. One of them was how big the error of temperature measurement is when the thermometer does not move with the flowing medium, but the measured temperature is considered static; what values the corrective coefficient (recovery factor) has and if it is possible to measure overall temperature in the flowing steam environment using the experience with measurement in flowing air.

The aim was to design a probe of total parameters which could be used for measurements in steam turbine LP parts. The probe must be constructed to meet the requirements for stagnation parameters measurement. It must be also possible to insert it and remove it from the turbine in operation through the round opening in the turbine casing designed for pneumatic probe measurement.

2 Study description

The influence of thermodynamic heating of the flowing medium depends on thermodynamic properties of the medium, specifically on Poisson constant, and also on Mach number. The following graph shows the influence of both mentioned parameters on the difference between stagnation and static temperature of the ideal gas with selected Poisson constants.

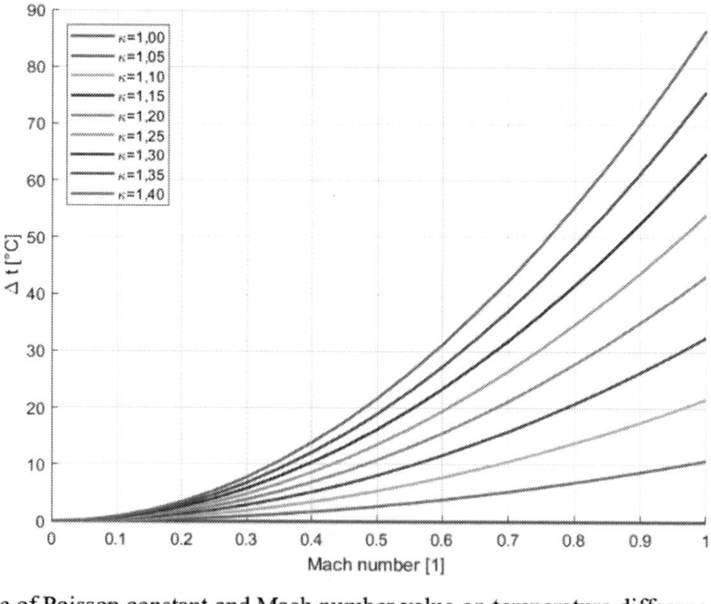

Fig. 1. Influence of Poisson constant and Mach number value on temperature difference.

The difference of both temperatures plotted in the graph above was derived from the equation given below.

$$T/T_0=(1+(\kappa-1)\cdot Ma^2/2) \quad (1)$$

The probe was designed to measure stagnation parameters in the flowing water steam environment for parameters calculated according to industrial formulation IAPWS IF97 and given in the following table.

Poisson constant value is ~1.3, which means that for Mach number 0.4 the difference between stagnation and static temperature is about 10 °C.

Two working media were considered: air, in the graph labelled "air" and ideal gas with water steam parameters given in the previous table, labelled as "ideal gas".

Table 1. Flowing medium parameters.

Quantity	Value	Units
Specific thermal capacity for constant pressure	1 914.91	[J/kg/K]
Poisson constant	1.3	[1]
Dynamic viscosity	$1.454e^{-05}$	[Ns/m^2]
Thermal conductivity	0.0322	[W/m/K]

In this work four geometrical variants of temperature measuring sensors were considered. Variant 0 is a common insulated sealed thermocouple or another cylindrical sensor passing through the flown-by wall and reaching with its active part the flowing medium, see the two following figures.

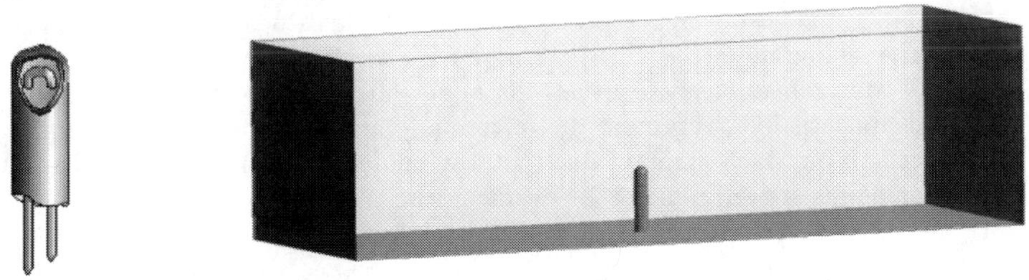

Fig. 2. Sealed thermocouple acc. [6], geometry task variant 0.

The following three figures show the geometry of total parameters probes, variant 1 being the reference probe whose geometry was taken from [5]. Variant 2 and variant 3 are designed solutions where besides the resistance sealed sensor Pt1000 also total pressure take-off is fitted.

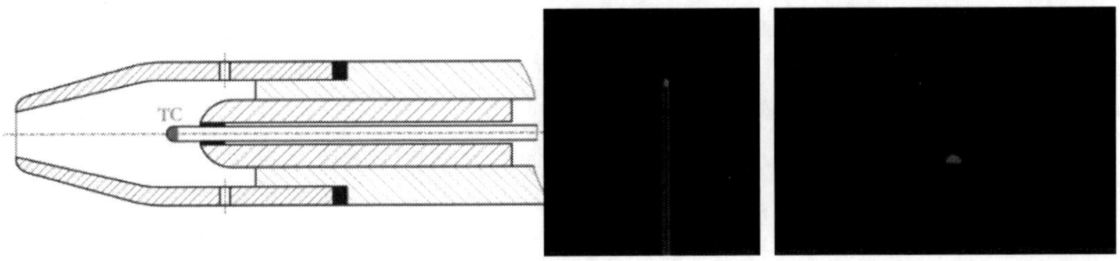

Fig. 3. Reference variant of probe 1 by [5], variant 2 and variant 3.

For comparison of individual probe variants, the so called recovery factor was used that is defined by the following relation.

$$r = (T_{probe}-T)/(T_0-T) \tag{2}$$

It is a dimensionless coefficient that describes the standardized "distance" of measured temperature T_{probe} from static temperature T, where value 1-r is the "distance" of measured temperature from total temperature T_0.

The scope of Mach numbers was from 0.2 to 0.7, for variant 2 the value 0.2 of Mach number was not considered and for variant 3 no results are available for Ma = 0.7.

In all cases several variants of circumferential and radial flow angle of the medium flowing on the probe were calculated.

3 Results

Results are given in the following table where the influence of Mach number on recovery factor of all the above mentioned variants is shown.

Useful results were provided not only by variants of all three probes, but also by the cylindrical sensor, which was labelled as variant 0. This is a common solution for temperature measurement in the flow part of the turbine, where Mach numbers can reach values from 0.2 up to 0.7.

For the lowest Mach number, the recovery factor is 0.375 and with its growth recovery factor keeps dropping down to 0.325. The temperature measured by the sealed cylindrical sensor is nearer the static temperature than the total temperature.

Referential probe characteristics (variant 1) is in the whole range more or less constant and near 1, which means that from the viewpoint of the total temperature measurement quality this is the best designed probe tested within this study. Similar results can be found in the original literature [5].

Out of the remaining two probes the better one seems to be the geometric variant 2, whose characteristic is very near the referential variant, variant 3 characteristics change a lot dependent on Mach number; due to a low number of points it cannot be ruled out that inflection points appear in the probe characteristic.

Table 2. Recovery factor of individual variants for Ma = 0.2 – 0.7.

Variant	Mach number			
	0.2	0.3	0.5	0.7
Variant 0	0.375	0.350	0.340	0.325
Variant 1	0.998	0.999	0.999	0.999
Variant 2	-	0.996	0.993	0.998
Variant 3	0.957	0.970	0.974	-

In the following graph the difference is plotted of total and static temperatures depending on Mach number and Poisson constant. The curves are supplemented by points of temperature difference arising from CFD calculations for both media.

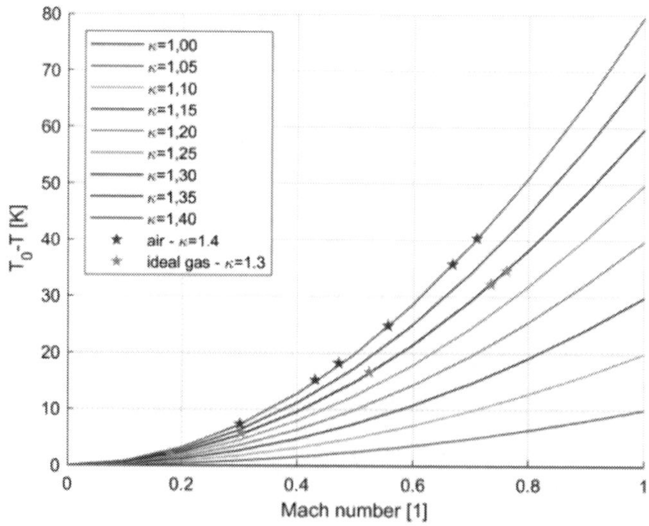

Fig. 4. Influence of Poisson constant and Mach number on temperature differences.

It is evident from the graph that the points match the corresponding curves. It can be thus expected that the error caused by neglecting the influence of thermodynamic heating can be well described by the given relation (2). It means that it is possible to have a relatively large error in defining stagnation or static temperature for the given medium in relation to the Mach number values while neglecting or not knowing the recovery factor.

The error rate is further amplified by the increasing value of Poisson constant. It is necessary to keep in mind that for the real steam the isobaric or isochoric specific thermic capacity is not constant, thus even the iso-entropic exponent κ cannot be considered a constant in the wide range of medium parameters. The value of exponent κ can change dependent on the pressure and temperature in the range from ~1.2 ÷1.7. From the given graphs it is evident that the higher the κ value, the larger error appears for the same Mach number in defining temperature. For $\kappa = 1.7$ the error is 10°C for $Ma \approx 0.3$.

4 Conclusion

Based on the study results the following was obtained:

- Characteristics of sealed cylindrical sensor, or thermocouple or resistance sensor,
- probe design that shows suitable properties for measuring stagnation parameters in flowing water steam in the steam turbine flow part,
- probe labelled as variant 2 fitted on the probe carrier for measurement behind the last stage of steam turbine LP part.

The probe was first tested in the aerodynamic tunnel. Measurement on the work was done in 2017 behind the last stage of LP part of the steam turbine of a large output whose owned and operated by ČEZ. Based on the experience gained the probe geometry was adjusted and fitted with a more accurate sensor for temperature measurement. Measurement on the steam turbine with the second generation probe has not been done yet.

Thanks belong to all involved, Doosan Škoda Power and the owner and operators of the ČEZ block and its employees for their helpfulness and assistance in the preparation and implementation of the measurements.

References

1. R.A. East, J.H. Perry, *A Short Time Response Stagnation Temperature Probe,* (1967)
2. M. Fernelius, S. Gorrell, *Thermocouple Recovery Factor for Temperature Measurements in Turbomachinery Test Facilities*, (2014)
3. A.H. Shapiro, *The Dynamics and Thermodynamics of Compressible Fluid Flow,* (1953)
4. C. Tropea, A. Yarin, J.F. Foss, *Springer Handbook of Experimental Fluid Mechanics,* (2007)
5. E. Winkler, *Stagnation temperature probes for use at high supersonic speeds and elevated temperature,* (1954)
6. http://www.capgo.com/Resources/Temperature/Thermocouple/Thermocouple.html

Theory of Transport of Conteiners in Pipelines

František Klimenda[1,*], *Josef* Soukup[1], and *Lenka* Rychlíková[1]

[1]University of J. E. Purkyně, Faculty of Mechanical Engineering, Institute of Machinery and Energetics, Ústí nad Labem, Czech Republic

Abstract. For transporting materials pneumatic transport in containers is also used. Calculation of this type of transport is quite complicated, especially because of non-stationary and non-isothermal conditions in the pipes. Because of low speeds of the train motion in the pipes, it is possible to consider the phenomenon in the pipes isothermal. The simplified calculation of this type of transport for one-dimensional flow in the pipes is indicated in the article.

1 Introduction

Pneumatic transport of materials in containers is not very widespread. It is most used only in Russia and the states of the former Soviet Union (especially in the Caucasus, where the terrain is inaccessible). Although it is used on such a small scale, it is an interesting and energy-saving mode of transport. This mode of transport in the large-diameter pipeline has been developed to transport raw materials in inaccessible terrains or in challenging climatic conditions. It is possible to imagine its use, for example, for transporting goods in today's long-range gas pipelines. The transport system is presented by pipe tube in which the containers on the wheels are moved in the driving gas (air) stream supplied by the compressor stations. The pressure gradient on the containers is generated by automatically. The energy of the air stream is transferred to the containers.

At low container speeds the pressure gradient is small, for the movement of containers there is no need for great force [1, 4]. By measuring it was found that due to the energy losses and the costs of the piping system, the optimum container speed is between 15 and 30 $km.h^{-1}$. At higher speeds, although energy losses are lower, it is necessary to build a massive transport pipeline with guide rails, which significantly increases costs.

Typically, these systems operate at a compression pressure of 100 – 200 kPa (overpressure). Containers run on wheels at the bottom of the pipeline. The guide wagons also run along the top of the pipeline. The assembly of the entire container train consists of end guides and sealing wagons. Containers with cargo are loaded between these wagons.

The source of air for the drive of the containers are turbochargers, vacuum pumps or ventilators. Depending on the air source and the length of the route, we recognize the systems

- *magistral* – for long-distance traffic, they are always overpressure,
- *industrial* – overpressure,

[*] Corresponding author: frantisek.klimenda@ujep.cz

– vacuum,
– overpressure – vacuum.

Depending on the route, we recognize the systems branched, unbranched, straight and circular. The containers are loaded and unloaded at stations. The stations are done as end (end of route) or as travel (between end stations). The whole system may be one-way, ie. all trains have one loading and one unloading station or multi-directional – can be interpreted on the route or trains run to the branching.

The diameter of the pipeline is dependent on the amount of conveyance, the size of the container, the specific profile of the track and its required throughput. Large pipe diameters are used. More than 1 million tons of material per year are used with two or more piping automated systems.

The container train assembly consists of freight containers, sealing wagons and coupling devices (see Figure 1).

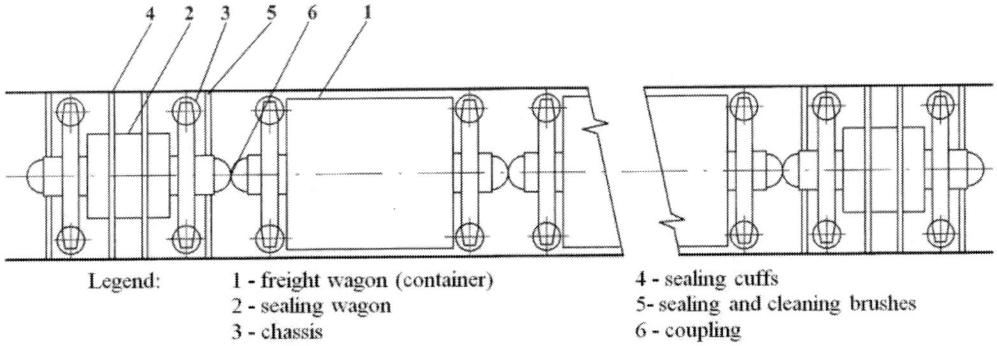

Fig. 1. train assembly – schema [4].

2 Calculation

The movement of the container assembly in the pipeline is described by the usual differential equations II. order. The force acting on the set is a pressure gradient on pneumatic wagons (sealing wagons located at the beginning and end of the container train). The pressure p and the amount of air Q_v characterize the air source. For simplicity, the flow of propellant gas in the pipeline is considered to be one-dimensional [2].

The basic equations describing the one-dimensional flow of gas in a pipeline for continuous (homogeneous) environments are based on the Law of conserving mass (1), momentum (2) and energy (3). The system of differential equations of the one-dimensional flow of gas in the pipeline is characterized by the flow in the axis. We draw from the gas dynamics equation in which all functions represent the values in the center of the cross section and depend only on the position x and time t.

Basic equation [1]

$$\frac{\mathrm{d}}{\mathrm{d}t}\int_V \rho.\mathrm{d}\tau = 0 \tag{1}$$

$$\frac{\mathrm{d}}{\mathrm{d}t}\int_V \rho.w.\mathrm{d}\tau = \oint_\Omega \mathbf{P}_n \mathrm{d}\sigma + \int_V \rho.\mathbf{F}.\mathrm{d}\tau \tag{2}$$

$$\frac{\mathrm{d}}{\mathrm{d}t}\int_V \rho(u + \frac{w^2}{2})\mathrm{d}\tau = \oint_\Omega \mathbf{P}_n w \mathrm{d}\sigma + \int_V \rho.\mathbf{F}.w.\mathrm{d}\tau + \oint_\Omega q_n \mathrm{d}\sigma \tag{3}$$

where $\mathbf{P}_n$ – surface pressure with normal n, $\mathbf{F}$ – mass forces, u – internal energy, w – velocity of the gas (medium), q_n – Heat flow in the normal direction, ρ – gas mass, V – volume, Ω – surface bounding volume.
It is not appropriate to describe the whole theoretical process. After editing and derivation we get the equation of the heat input

$$\rho\frac{\mathrm{d}u}{\mathrm{d}t}=\frac{p}{\rho}\frac{\mathrm{d}\rho}{\mathrm{d}t}+\mathbf{T}\nabla\mathbf{w}+div\mathbf{q} \tag{4}$$

where $\mathbf{T}$ – tensor of tangent pressures (forces), p – pressure.

The calculation involves consideration of changes in enthalpy, internal energy and gas pressure, as well as heat input. For the one-dimensional flow in the pipeline (which we use for these cases in the technological calculations) we modify the equations so that all the variables depend only on the position (x) and the time (t), ie to the shape

$$A(x,t)=\langle A(x,y,z,t)\rangle=\frac{4}{\pi d^2}\int_S A(x,y,z,t)d\sigma \tag{5}$$

where S - sectional area of pipeline, d – pipe diameter, A – function.
After editing the equation, they will have a shape

$$\begin{gathered}\frac{\partial\rho}{\partial t}+\frac{\partial Q}{\partial x}=0 \qquad Q=\rho\cdot w\\ \frac{\partial Q}{\partial t}+\frac{\partial}{\partial x}\left(p+\beta.\rho.w^2\right)=\frac{4}{d}\tau+\rho.F_x\\ \rho\frac{\mathrm{d}u}{\mathrm{d}t}-\frac{p}{\rho}\frac{\mathrm{d}\rho}{\mathrm{d}t}=\langle\mathbf{T}\nabla\mathbf{w}\rangle+\frac{4}{d}q_n+\frac{\partial q_x}{\partial x}\end{gathered} \tag{6}$$

where $Q(x, t)$ – the amount of gas, w – medium gas velocity in pipeline, p – gas pressure in pipeline, τ – the mean tensile stress (gas) on the inner surface of the pipe, β – coefficient of unequal distribution of gas velocity w in the cross section of the pipeline, for turbulent flow $\beta\approx 1$, q_n, q_x – radial and axial heat flux.

Further modifications and simplifications, we obtain 6 equations determining the relationship betweenρ, p, w, Q, T and u

$$\begin{gathered}\frac{\partial\rho}{\partial t}+\frac{\partial Q}{\partial x}=0 \qquad \frac{\partial Q}{\partial t}+\frac{\partial p}{\partial x}=-\lambda\frac{1}{d}\frac{\rho\cdot w\cdot|w|}{2}\\ \rho\frac{\mathrm{d}u}{\mathrm{d}t}-\frac{p}{\rho}\frac{\mathrm{d}Q}{\mathrm{d}t}=\lambda\frac{1}{d}\frac{w^3}{2}+\frac{4.q_n}{d}\\ u=c\cdot v\cdot T+\text{konst.} \qquad p=\rho\cdot R\cdot T \qquad Q=\rho\cdot w\end{gathered} \tag{7}$$

After adjusting and fitting from these equations, we determine the amount of air required to drive the containers in the pipeline. The air is fed behind the container train, the air in front of it, until the next full-pass shutter is led by a passage through the blowers to the next section. For a coherent (homogeneous) environment, we come out of the law on the conservation of matter, energy and movement.

In the solution, we assume that

- the weight, pressure, internal energy and gas temperature change slightly over the pipe cross-section
- the gas velocity on the inner surface is equal to zero
- gas compression is much greater than the flexibility of the pipe walls.

Then we get

$$\frac{\partial Q}{\partial t}+\frac{\partial Q}{\partial x}=0 \tag{8}$$

after editing

$$\frac{\partial Q}{\partial t}+\frac{\partial}{\partial x}\left(p+\beta.\rho.w^2\right)=\frac{4}{d}\tau+\rho.Fx \tag{9}$$

Further simplification can be achieved by hydraulic approach (by conversion to hydraulic flow conditions). When introducing simplification – the surface tension on the inner surface of the pipeline is associated with the average flow parameters.

According to Darcy

$$\frac{4}{d}\tau=-\lambda\frac{1}{d}\frac{\rho.w(w)}{2} \tag{10}$$

where λ - dimensionless coefficient of hydraulic resistance.

This simplification ceases to apply in high speed processes – gas velocity compared to the gas velocity (Mach number)

$$M=\frac{w}{c}<<1 \tag{11}$$

This simplification does not consider the convective transfer of motion in the equation of motion, ie the variable ρw^2 compared to the pressure fluctuations. The fluctuation of pressure according to Zhukovsky is [3]

$$\partial_p=c.\partial Q \tag{12}$$

and fluctuations $\partial\left(\beta\rho w^2\right)\approx\beta.w.\partial Q$

The ratio of fluctuations of the given variables

$$\frac{\partial\left(\beta.\rho.w^2\right)}{\partial_p}\approx\beta\frac{w}{e} \tag{13}$$

The heat flow along the x-axis is much larger than the y-axis flow.

From the equations of the non-stationary isothermal movement of the gas in the pipeline after the modifications we get

$$\frac{\partial p}{\partial t}+c^2\frac{\partial Q}{\partial x}=0 \tag{14}$$

In the quasi-isothermal movement of gas in the pipeline, the equation system is non-linear (c = const.). Linearization of equations allows to find an analytical solution for many tasks of non-stationary motion using rows or operating methods.

The presence of container assemblies in the gas flow is the task of moving the containers in the non-linear pneutage in the case where the gasodynamic equations undergo linearization.

In addition, the complexity of the track profile, the non-linearity of boundary conditions, and the need to monitor multiple container assemblies in the pipeline while moving, leads to solving tasks using numerical methods using computer technology. The description of the solution (ie the integration of equations) will not be addressed in this paper.

For the non-isothermal flow of the equation we modify the shape

$$\frac{\partial p}{\partial t}-\frac{p}{T}\frac{\partial T}{\partial t}+\frac{c^2}{\kappa}\frac{\partial Q}{\partial x}=0 \tag{15}$$

$$\frac{\partial Q}{\partial t}+\frac{\partial p}{\partial x}=\varphi(p,Q,T) \tag{16}$$

$$c_p\rho\frac{\partial T}{\partial t}+c_p.Q\frac{\partial T}{\partial x}-\frac{\partial p}{\partial t}=\varepsilon(T)$$
$$p=\rho\cdot R\cdot T \qquad c^2=\kappa\cdot R\cdot T \tag{17}$$
$$\varphi(p,Q,T)=-\frac{\lambda c^2(T)Q/Q}{2\cdot\kappa\cdot d\cdot p}$$

Sometimes it is necessary (especially for long routes) to solve systems with high density sets. The calculation is complex, it is solved with Dirac functions, double integers, etc. When designing a container pneumatic transport system, we perform basic technological calculations.

System throughput

$$Q_i\le Q_{i\max} \qquad i=(m,V,s,t) \tag{18}$$

quantity of material transported per year Q_r we need

$$m=\frac{Q_r}{q}\ \text{containers} \tag{19}$$

where q – the capacity of one container.

We have n containers in one set, then we have to go through the year

$$v=\frac{m}{n}\ \text{sets} \tag{20}$$

Then the interval of the movement of the container assemblies

$$\tau=\frac{e}{v} \tag{21}$$

The movement of the container sets is characterized by the fact that in the closed part of the length l_i (between the passage closures) there is always only one container set. The throughput, ie, the duration of the stay one of the container set when moving back and forth is given by the relationship

$$\tau_{\min}=\tau_p+\tau_r+\frac{2L}{v_{\max}} \tag{22}$$

where

τ_p, τ_r – time of unloading and time of loading, L – pipe length, v_{max} – maximum mean speed.

In case of two-pipe systems, the duration of the kit is one half in the pipeline and therefore τ_{min} is also half.

3 Conclusion

The diameter of the conveying duct depends on the amount transported, the container size, the specific profile of the track and the required throughput. Large pipe diameters are used.

Several features have been realized, all of which are used to transport building materials, even in hard-to-reach terrain. Typically, the pipeline diameter is between 1.020 - 1.220 m with a pitch up to 3°. The most important is the TDO transport system in St. Petersburg. It works with parameters: pipe diameter 1.220 m, climb max. 2°, full train train weight is 20 t, empty 10.9 t, transmission capacity is 125000 t / year [4].

For horizontal transport there is a pressure drop of about 15 kPa, at a climb of about 45 kPa. The pressure gradient forms sealing wagons.

The research work is supported by the UJEP-SGS-2017-48-003-3, Czech Republic.

References

1. J. Soukup, *Pneumatická doprava odpadů*, Habilitační práce, 2004
2. J. Janalík, *Potrubní hydraulická a pneumatická doprava*, Ostrava, VŠB v Ostrvě, 1984, 209 s.
3. J. Urban, *Pneumatická doprava*, SNTL Praha, 1964
4. J. Soukup, *Nové způsoby odstraňování odpadů (pneumatická doprava)*, Výzkumný ústav výstavby a architektury, Praha, 1990

Controlled outflow of storm water without flooding

Karel Adámek[1,*], *Jan* Kolář, and *Pavel* Peukert

[1]VUTS a.s., Dept. of numerical simulations, Svárovská 619, 46001 Liberec, Czech Republic

Abstract. There are many types of devices used for various purposes, called as vortex valves. The aim of this paper is the design of vortex valves, determined for controlled higher outflows from retention tanks. The paper follows the previous study of smaller sizes of vortex valves. The method of flow numerical simulation allows us to identify the reason of the two-branch operational (resistance) characteristic of the solved valves and the suitable sizes of the main valve dimensions for the given flow rate and water level.

1 Introduction

There are many types of devices, used for different purposes, called vortex valves. The aim of this paper is the design of vortex valves, destined for controlled higher outflows from retention reservoirs. The principle of valve activity is described curiously by Mr. Google and many dealers and producers, for instance, as "the effect is given by flow phenomenon" or "it works on the principle of hydraulic phenomenon". Reality is not so mysterious, the flow inside of such a valve is possible to describe by numerical flow simulation, completed by some essential equations of fluid mechanics, see [1-7] etc.

1.1. The previous model

In the previous work [7], the analysis and solution of vortex valves were made for lower flow rates, therefore, here, it should be repeated the main results, partially used for higher flow rates.

Typical resistance characteristic Δp (Pa) = f (m (kg/s)) of the solved vortex valves is composed from two branches of different gradients and separated by an area of some discontinuity / instability. Typical results of unified dimensions for low and high mass flows are presented in Par. 4.6 below for several combinations of both inlet (tangential) and outlet (axial) orifices. The reason of the two-branch characteristic is explained in [7], too. The preliminary design for higher flow rates is realized as an enlargement of cross sections from the verified solution [7], proportional to the higher mass flow. This hypothesis should be verified on a specific design using the assortment of semi-finished products and of actual vortex valves [8-11]. At first, it will be necessary to enlarge the valve body, to insert the

* Corresponding author: karel.adamek@vuts.cz

enlarged inlet tube. Due to the unified design, it should be suitable to use the same valve body and the mass flow to adjust them by replaceable outlet inserts. As found out in [7], the branch of higher gradient is valid when the valve body is full by water, and the transition between the branches arrives when the water level is on the upper outline of the valve body, approx. For lower water levels, the valve body is not full with water and the inner rotation of water does not arise. It exists a free water level, only, waved by the tangential inlet from below and water is then freely flowing out through the axial outlet without spraying.

It is possible to await that in a larger valve body, the transition between the branches of characteristic begins for a higher pressure than in [7]. However, here assigned and used pressure gradients are similar as in [7], therefore, it is possible to await that the real valve operation could be in the unstable transitional region. In addition, it is necessary to check the global adequacy of the new dimensions.

Note: The aim of this vortex valve solution in not any perfect spraying of outflowing liquid, as for instance in the spraying gun, but the only decreasing of the dynamic effect of the outflowing liquid on the surroundings. The aim is reached for smaller pressures without sprayed outflow, too, with free water level inside the valve body – some part of the flow energy is changed into flow losses inside the valve body (inlet, friction in the inlet tube and during the rotation, the change of the flow direction from peripheral into the axial, flow spraying, back flow of the air from the surrounding, etc.).

Designing the valve, it is necessary to fulfil two contradictory requirements: For a valve of high volume, flow larger inlet and outlet diameters should be designed together with the larger valve body diameter, too. But by this way, the peripheral velocity wo is decreasing and after its second power the centrifugal acceleration

$$ao = wo^2/r \qquad (1)$$

is decreasing, too. And more, through a small volume of the valve body, it cannot flow the required high water flow.

Here, it is alternatively designed a simple prismatic shape of the valve body, where the smaller diameter ensures a quite high peripheral velocity and centrifugal acceleration, too, and the necessary flow cross section is designed in the axial direction, depending on the necessary flow volume (see for instance the Par. 2.3 below).

2 Flow numerical simulation

At first, the above given hypothesis should be verified, i.e. to derive the shape of the new valve for higher flows by a simple increasing of main dimensions from the verified solution [7]. The procedure of the model preparation and solution is identical with [7], therefore, only a short description is made here.

a. Model description

Geometry – as in [7], adapted proportionally to the prescribed flow.

Mesh – model dimensions are various in the range over than 1:100, so in rough mesh some inaccuracy should be taken into account. Never mind, such a possible error is identical for all solved cases and comparable results of many solved cases are obtained quickly and reliably.

Boundary conditions – inlet hydrostatic pressure up to 2.0 [kPa] of the phase 1 (water) is defined in the center of the inlet tangential tube, outlet in the atmosphere 0 kPa is defined on the surfaces of the surroundings, added along the axis of the valve body. In the outlet, there is defined backflow of the phase 2 (air from the surroundings). At the beginning, in all outlet volumes, it is defined phase 2 (air), only, and inside the valve, the phase 1 (water), only.

Note: The boundary condition "inlet pressure" is simply defined as constant in the whole inlet cross section, i.e. as hydrostatic pressure of water column between the water level and the center of inlet. The real difference between the highest and lowest point of the inlet cross

section in [7] is typically ±0.25 kPa - it is neglected here, to simplify the model, without the next connecting channel between the valve inlet and the surrounding free water level. Really measured differences between the simulated and real flows [8] are less than 5 % approx.
Solver: Turbulence model k-ω SST. Used stationary two-phase model "mixture", the unsteady model VOF (volume of flow) needs a much longer solution time, but here it is not necessary to solve details of the flow field [7]. The solution is regarded as converged, when the difference between the inlet and outlet flow is less than 0.1 kg/s.
Results – during several iterative steps, the designed main dimensions are changed to get flow as defined in advance. Illustrative figures of the flow fields are in the text, as necessary. For the text shortening, some scales are left out – in such case, it is in force that red = maximum value and blue = minimum value, identically with the wave lengths in the spectrum of visible light. For a qualitative comparison of individual cases, exact scales are not necessary.

b. Trial prismatic model

The results of this trial valve are added as an illustration, only. Contrary to the next cases, its axis is vertical with outflow up, therefore without a disturbing effect of gravity, as discussed in [7]. Similar case [5] presents the vertical axis with outflow down, when the upper surface of the valve remains dry. With an increasing of the mass flow, the valve main dimensions must be increased, too (body, inlet, and outlet). With increasing the body diameter, the centrifugal acceleration, necessary for a good water spraying, is decreasing with the second power of the body diameter. Therefore, the vaulted valve bodies [7] using semi-finished products [8-11] are not suitable because their outlines are too large. A better solution is a prismatic valve body, where the inlet size is simply changed by a change of the body width (by the cylinder height, in the axial direction) and at a suitable body diameter, the velocity and flow along the periphery do not decrease under the suitable limits. For a given pressure gradient of 16 kPa (1.6 m), the ideal outflow velocity is

$$wid = \sqrt{(2.g.H)} = \sqrt{(2\ .\ 9.81\ .\ 1.6)} = 5.6\ m/s \tag{2}$$

and the theoretic flow of 143.43 kg/s through the prismatic inlet of 0.16x0.16 m. The inlet flow from the simulation is 61.82 kg/s, only, so the flow coefficient is 0.43 (the ratio of « real », here simulated, flow to the ideal one). The next results of the simulation are the air suction of 1.02 kg/s through the front face of the surroundings and the outlet flow of 62.77 kg/s through the cylindrical periphery of the surroundings. Therefore, the error of the mass balance is 0.07 kg/s, i.e. 0.1 % of the inlet flow. It is negligible, therefore, the simulation was finished and evaluated.

For a presentation of the complicated flow field, Figure 1 to Figure 4 present several images of flow field parameters. For a better resolution, different scales are used for the same parameter in different cross sections.

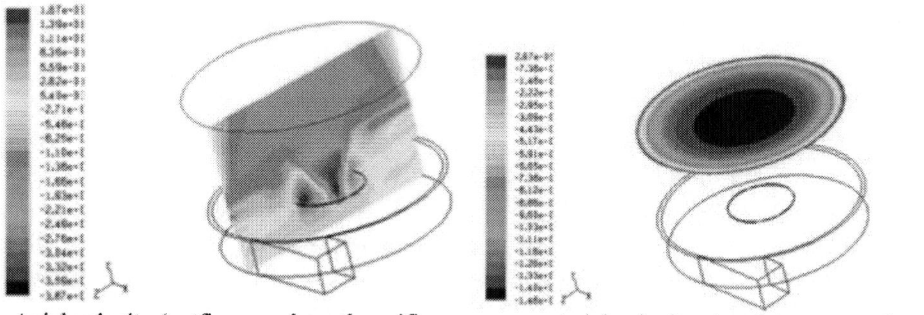

Axial velocity (outflow up along the orifice outline, suction down along axis) | Axial velocity at outlet cross section

Fig. 1. Axial component of velocity.

Figure 1 presents the highest value of the axial component of velocity at the outlet edges, the highest suction inside along the outlet axis. In the area opposite to the tangential inlet, the flow is higher than at the side close to the inlet. At the outlet, the backflow is visible along the axis.

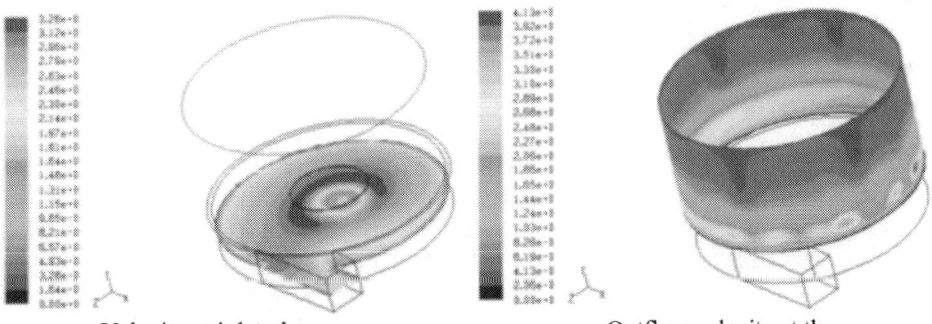

Velocity at inlet plane

Outflow velocity at the periphery of the area

Fig. 2. Inlet and outflow velocity.

Figure 2 left presents the absolute value of the velocity vector in the cross section through the inlet axis. Here, it is approximately evaluated the peripheral velocity of the rotating liquid body as 2.5 m/s at the radius of 0.4 m. In this estimation, some error is expected – it is hardly to estimate, what part of the rotating liquid body to use for the determination of average peripheral velocity and what is the average radius of such a body. Nevertheless, from such an estimation, the centrifugal acceleration is 15.63 m/s^2. Due to the inaccuracy at determination of peripheral velocity and flow radius and also that the gravity affects here in across direction, it supposes that a perfect spraying of the flow will not be created, rather the flow will be spilled out along the whole outlet periphery with some irregularity, due to the one-sided inlet. At the periphery of the outer area (the surroundings, Figure 2 right, there is observable some flow thickness, which is flowing on the body surface out in the direction across to the axis, the most intensive flow is opposite to the inlet. Along the periphery, there are visible some irregularities, feeble points, etc. It could be the error of the simulation due to the coarse mesh or little area of the surroundings – so that to get a shorter time of solution for many solved cases. Similarly, at the upper edge, there are some irregularities, but absolutely small, here, it is hardly any flow.

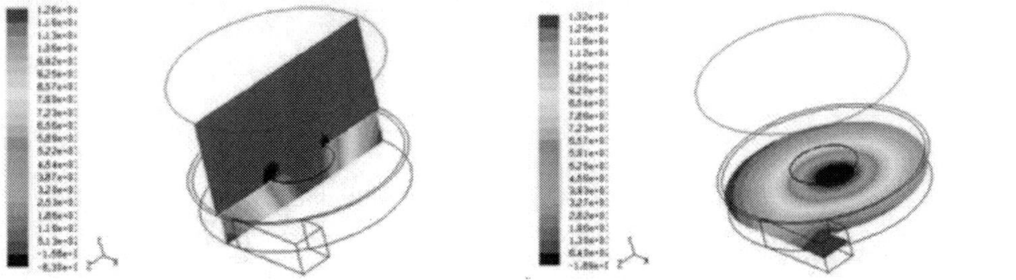

Fig. 3. Pressure – inversed to the velocity (max. in the inlet, min. in the outlet).

Pressure is inverse to the velocity (Figure 3), maximum at the periphery of the valve body, it is decreasing to the axial outlet, where it is going under the atmospheric pressure of the surroundings.

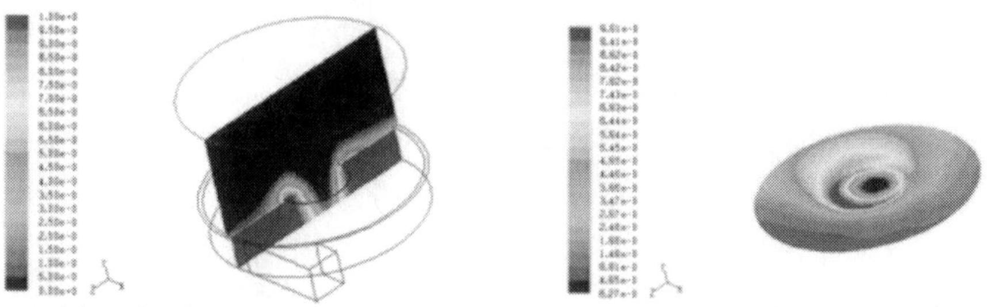

Axial section (water = red, max. inside) Outlet plane – asym. spraying

Fig. 4. Velocity field of water phase.

The field of the water part (phase 1) in Figure 4 presents the water spillage on the body surface after the outflow. The outflow is not symmetrical, facing to the tangential inlet, there is the higher peripheral outlet. The same is visible in the cross section in the outlet plane, too.

3 Modelled solutions

From Par. 2.2, it is visible that the detailed presentation of all flow field parameters of each solved case is very extensive, therefore, here are presented directional fields of both phases (water and air), only, at the outlet from the cylindrical surroundings - through both front and peripheral surfaces. They indicate the quality of the sprayed flow. From the results of four solved cases here are demonstrated two of them, only.

For the case 1 (21.3 kPa, 30 kg/s), the assigned pressure is quite high, therefore, the spraying is very good. In the smaller valve body and at the higher pressure gradient, the image of the outlet directional field is very good, see Figure 5, the deviation of the individual vectors from the radial direction is the consequence of the one-sided tangential inlet, followed by flow rotation.

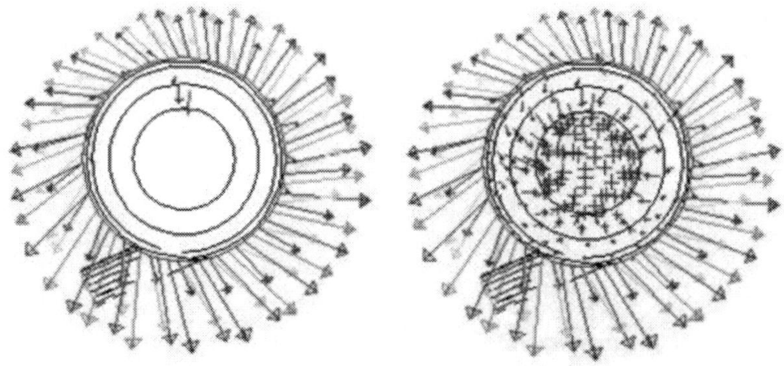

a) phase 1–water is going out through the periphery, only, not through the central part

b) phase 2–air is drawn in through the axial face, and is going out through the periphery, together with water

Fig. 5. Directional fields at the outlet side – case 1.

A large prismatic arrangement of the case 2 with a relative small pressure gradient (15.9 kPa, 105 kg/s). The result in Figure 6 shows the partial spraying, only, through the upper part of the periphery and air suction through both the axial middle part and lower peripheral part.

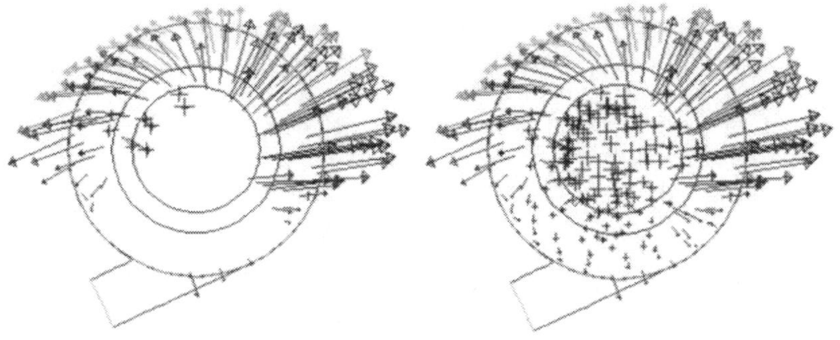

a) phase 1 – water is flowing through the upper part of the periphery, only, it is visible small backflow

b) phase 2 – air is drawn inside through both middle axial part and lower part of the periphery, and going out together with water through the upper part of the periphery

Fig. 6. Directional fields at the outlet side – case 2.

Operating point of the next case 3 (16.3 kPa, 22 kg/s) seems to be similar to the case 1, but the spraying is bad, due to the low pressure – therefore, without Fig. The only remedy is the decreasing of the valve body diameter with the original sizes of the rest parts. By this, the velocity is increasing and the spraying will be better. But the outer diameter decreasing is not possible due to "short-circuit" flow from inlet into outlet, without flow rotation.

The biggest case 4 of prismatic inlet 0.16x0.32 m (15.8 kPa, 125 kg/s). The simulated mass flow corresponds to the defined one, outlet spraying is good, therefore, without next Fig.

Note: The case 4 above was modified by the spiral outer case (designed without detailed calculations). Comparing the results of both cases, it is possible to state that the resulting directional field of the spiral valve body is similar to the cylindrical one. It is interesting that the flow is increased of 20 % approx., at the same boundary conditions. It means that the resistance of such a valve is lower, but it is not good for the intended purpose – to suppress the dynamic effect of the outflow on the surrounding.

4 Conclusions

a. Geometry

For each of 4 solved cases, suitable main geometric parameters of the valve (outlines, inlet and outlet diameters) were found to get the prescribed water flow for a prescribed pressure gradient (hydrostatic pressure of the given water level).

The initial dimensions of the new cases were derived from the former solved and verified cases, using simply the continuity equation on the flow change by a cross section change is the initial inlet parameter, only. But flow resistances during the complicated 3D flow (inlet, rotation, interaction of rotating and entering flow, transition from rotation into axial outlet, spraying, friction, etc.) have important influence. For such case, it is not possible to apply simple formulas from the basic course of fluid mechanics, but a numerical simulation of 3D flow is necessary.

The modelled solution should be verified by experiment – each model is some approximation of the reality, only, it is necessary to check if any simplification does not change the results too much.

The presented result of each solved implicit case is reached in several iterative steps. From the convergence course of the solution, it is possible to estimate that some cases with oscillating convergence could be unstable in real operation, i.e. with irregular spraying, pulsating flow, higher noise level, etc.

The main aim of the solution is not the quality of the spraying at the outlet, but a reduction of the dynamic effects of the outgoing flow, therefore, next simulations were not realized. In the given cases, the water level is relative low, so at relative large inlet and outlet it is not to expect uniform spraying along the whole outlet periphery. The initial design uses the vaulted shape of the valve body, for higher flows it is used a flat design, where the cross section is changed by the body width, not by body diameter.

b. Spraying quality

Influence of the outlet edge on the spray quality was tested on the case 4, only, see Figure 1 above. The best solution is given by the thin and sharp outlet edge, but the long cylindrical extension damps the spraying effect. Some improvement of it is given by the chamfered outer edge, where the flow adheres to the chamfering and dissipates in the radial direction. The effect is known from the so-called wall nozzles – for suitably designed dimensions, the outflow can be adhered of over 180°, see [12-13], etc.

c. Flow coefficient

The so-called flow coefficient of individual cases is evaluated as a check value and as an information item about the flow quality. It is defined as the rate of real (here simulated) flow and that of idealized flow. As an information item, the coefficient for the so-called quarter circle nozzle (well rounded) reaches a value of 0.95-0.97, for the simply drilled orifice in the wall a value of 0.64, for smaller vortex valves solved in [7] a value of 0.28, only. Here solved smaller and vaulted cases 1+3 have the coefficient of 0.50-0.58, larger and flat cases 2+4 that of 0.37-0.45. Higher influence on the flow quality has the vaulted shape, the influence of larger dimensions is not so important here.

d. Basic directives of the valve design

In a smaller valve body, the peripheral velocity is higher, so the spraying is better, too. But the inlet diameter must pass between the outlines of the valve body and the valve axial outlet to prevent short-circuit flow between the inlet and the outlet without rotation.

Both inlet and outlet diameters define approximately the possible flow range. If the outlet is smaller than the inlet, the flow tends to be straight and strong without spraying. If the outlet is equal or larger than the inlet, the water spraying will be good.

e. Error estimation

Hydrostatic pressure in the inlet of the models is simply defined as constant, but really there is some pressure difference of 0.35 kPa, approx., between the lowest and the highest point of the inlet cross section – in simulations [7] neglected. The difference between the simulated and the really measured flow is not more than 5 % approx. [7], due to slightly waved both inlet and outlet real water levels.

f. Operational characteristics

Inlet data for a valve designer are not only the main dimensions, but the so-called operational (resistance) characteristic $\Delta p = f(m)$, too, well known as a quadratic function $\Delta p = f(m, m^2)$.

In those cases 1-2-3-4 are solved high flows, only, just before the overflow over the security overfall. The lower branch of the characteristic is not solved here. The results are compared with the former results [7], see Figure 7.

For the valve body diameter of 780 mm (cases 2+4), it is possible to estimate that the transition between the lower and the upper branch of the characteristic is of 10 kPa approx. Therefore, simulations were made for the range of 10-16 kPa, also for valves of medium sizes (cases 1+3), where the transition area could be estimated under 10 kPa. Four solved points of each characteristic are used as the minimum for quadratic functions.

As found out in [7], the higher branch of the characteristic begins when the valve body is full of water. For here solved higher flows, i.e. larger valve bodies at the same maximum water level it could mean the operation in transitional area between both branches. So the spraying quality could not be good, but the main task of the observed valves is to suppress the dynamic effect of the flow, not the spraying quality.

All new characteristics, as a result of many simulated cases, are presented in the common graph in Figure 7, checked with the characteristics of small valves [7] (their inlet/outlet = 80/80-200 mm). Due to a large extent of flows, the scale of the x-axis is logarithmic. The graph confirms the preliminary estimations that the used lower point of the characteristic of 10 kPa is far from the transitional area for cases 1+3 (middle flows, middle valve sizes), but for cases 2+4 (large dimensions, large flows), the lower parts of their characteristics are bent, which signs the vicinity of the transitional area.

Note: For cases of small flows, solved in [7], the transitional area is of 3-4 kPa, only.

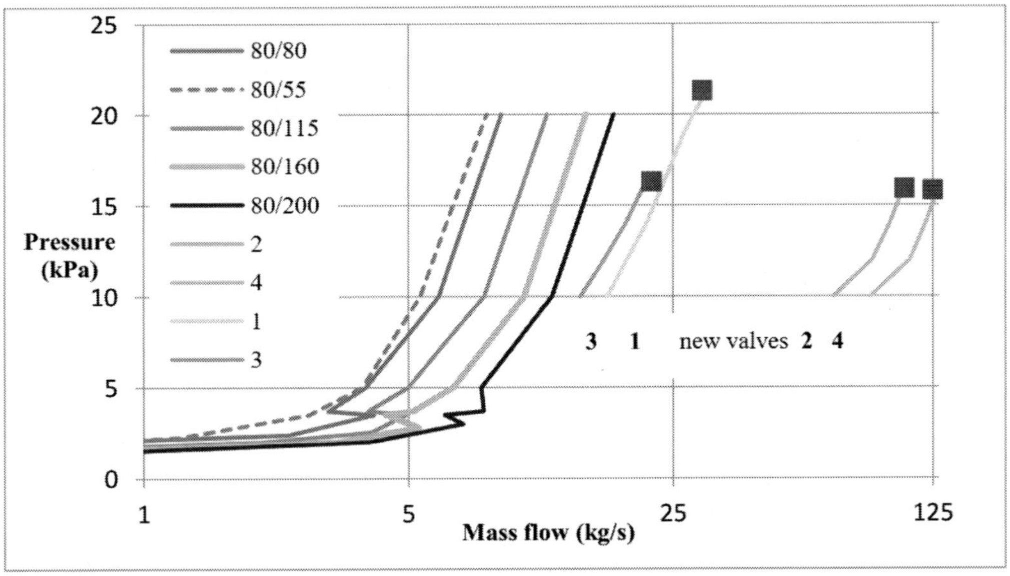

Fig. 7. Operational (resistance) characteristics of solved valves 1-2-3-4, together with valves [7].

Our acknowledgment is given to VUTS Liberec – Center for Development in Machinery Research for the support in the framework of the grant NPU-LO1213, granted by the Czech Ministry of Education, Youth and Sport.

References

1. Vortex valve, patent US 3219048 from 23.11.1965
2. D. Jarman, D. Butler, G. Tabor, R. Andoh, ICE proceedings – Engineering and computational mechanics, **168**, issue EM1

3. R. Andoh, M. Faram, K. Osey, Proc. of the 6th int. water sensitive urban design conference and hydropolis, Perth Australia (2009)
4. H. Steinriede, M. Merschdorf, Wirbeldrossel – Produkt information, UFT 2016
5. H. Brombach, H. Steinriede, Vertical vortex valve – product information, UFT 2014
6. V. Tesař, Superquadratic fluidic restrictors and their applications, Advances in fluid mechanics IX, WIT Transactions on Engineering Sciences, **74** (2012)
7. K. Adámek, J. Kolář, P. Peukert, Vortex valve, EFM Mikulov (2017)
8. Catalogue of semi-finished parts www.zyksal.cz
9. Catalogue of valves producer www.bmto.cz
10. DIN 28 011 – Vaulted foreheads, stainless steel
11. DIN 17 458 – Tubes seamless, stainless steel
12. H. Coanda, US Patent No. 2,052,869. Device for Deflecting a Stream of Elastic Fluid Projected into an Elastic Fluid (1936)
13. K. Adámek et. al, Wall flows and their applications, in: Int. Journal of Mechanical Engineering and Applications, special issue Numerical Flow simulations in textile technology, 3(2), 28-32 (2015)

Natural Convection Heat Transfer Around a Horizontal Circular Cylinder for the Case of Constant Surface Temperature of the Cylinder Near a Vertical Wall

Marcel Novomestsky[1,*], *Roman* Banovčan[1], *Andrej* Kapjor[1], and *Helena* Smatanová[1]

[1]University of Žilina, Department of Power Engineering, 01026 Žilina, Slovakia

Abstract. Natural convection heat transfer from a single horizontal cylinder near an isothermal wall is experimentally investigated in the environmental chamber. The cylinder is heated with constant temperature, the rising plum from the heated cylinder interacts with the side wall and significantly affects the surface heat transfer distribution. Thus, the distance between horizontal cylinder and side wall can influence the heat transfer. The heat transfer from heated cylinder with different distances were compared. Results are obtained for Ra = 10^4. It was found that natural convection heat transfer from horizontal cylinder depends on the distance between wall. Also, a heat transfer correlation has been proposed for a single horizontal cylinder near a wall and is compared with earlier work. The results indicate that the vertical wall affect the average Nusselt number at the cylinder. Therefore, the vertical wall has a positive effect on the Nusselt number at the cylinder.

1 Introduction

The natural convection flow in horizontal annuli has received increased attention in last decade because of their importance in engineering. Natural convection heat transfer is an inexpensive, reliable and noise-free method of heat transfer. For instance, horizontal cylinder has many engineering applications such as heat exchangers, radiators, thermal storage systems, and cooling of electronic components.

Past decade has witnessed extensive work on convective heat transfer from cylindrical surface. Morgan [1], Churchill and Chu [2] have determined empirical correlation equations which focus mainly on Nusselt number. Kuehn and Goldstein [3] investigated natural convection around an isothermal circular cylinder for Rayleigh numbers of < 10^7 and have solved the full Navier –Stokes and energy equations. Kuehn and Goldstien have shown that the heat transfer is at its maximum at the bottom of the cylinder and decreases toward the top of the cylinder. The decrease in heat transfer is attributed to an increase of the thermal boundary layer thickness.

* Corresponding author: marcel.novomestsky@fstroj.uniza.sk

Later, Fujii *et al.* [4] presented experimental results on natural convection heat transfer from horizontal wire. They used two computational domains, one is a domain near the cylinder and the other is one outside this domain. For order to check the validity of the boundary conditions they used an open and a solid boundary condition.

In particular, natural convection from horizontal cylinder has received considerable attention in the literature because of the fundamental nature of the geometry. The heat transfer from a cylinder near the side wall is quite different with respect to the one from a single cylinder due to the interaction of the side wall. This interaction is owing to the buoyant plume that is generated by cylinder.

The current research investigates the distribution of the heat transfer from horizontal heated cylinder near the side wall under natural convection conditions in air. This experimental study considers the influence of distance between wall and cylinder. The range of cylinder and wall spacing was taking from no inflation distance to distance closed to the wall.

In this article, we consider natural convection, alternatively called free convection, where any fluid motion occurs by natural means such as buoyancy. The buoyancy force is caused by the density difference between the heated fluid adjacent to the surface and the fluid surrounding it and is proportional to this density difference and the volume occupied by the warmer fluid.

2 Experimental apparatus

The experimental apparatus consists essentially of a heated circular cylinder mounted horizontally on holding construction and placed in the environmental chamber.

2.1 Experimental cylinder and installation of thermocouples

The experimental device is circular cylinder (1) Figure1 made of stainless steel with length of 1 m. The air flows due to the buoyancy effect. The cylinder is heated electrically by using an electrical heater as shown in Figure 1. It consists of nickel-chrome wire (3) electrically isolated by ceramic beads and wound uniformly along the cylinder as a coil in order to give uniform heat flux. For measuring the cylinder surface temperature were used NiCr-Ni (type K) thermocouples (2) mounted inside of the cylinder close to the surface in the middle of the cylinder. The outlet air temperature was measured by one resistance thermometer placed in the middle of the environmental chamber [6].

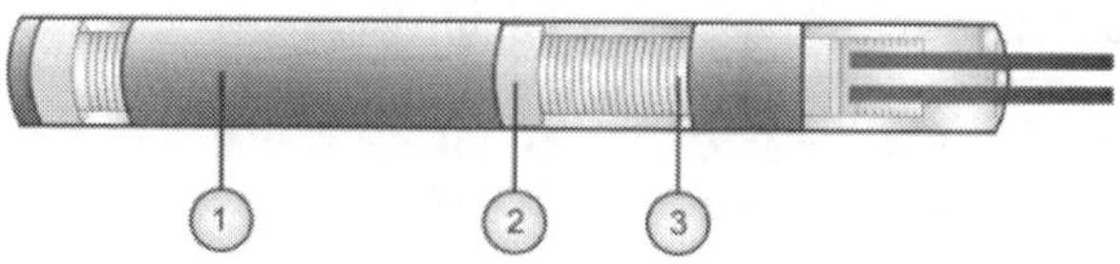

Fig. 1. Longitudinal section of experimental cylinder.

2.2 Environmental chamber

Experimental investigation was obtained in the environmental chamber Figure 2, where the surround temperature T_∞ was set to constant 20 °C.

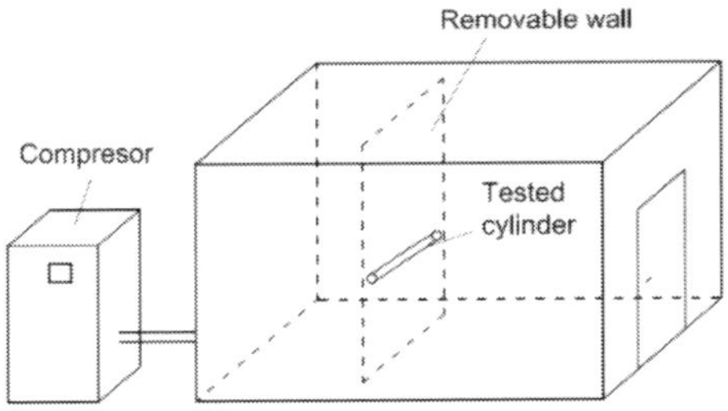

Fig. 2. An environmental chamber for performance tests.

The environmental chamber has almost constant conditions Table 1. The relative humidity was during the test between 52 – 67 % which has only small influence on the air density.

Table 1. Environmental chamber conditions.

T_∞ [°C]	P [kPa]	rH [%]
20±0.05	96.8	52 – 67

Thermal output measurement is based on the norm STN EN 442: 2000. According to this norm, the test area must be equipped with:

- closed chamber without ventilation with space for installation of the test specimen and a water-cooled surface in order to achieve a certain temperature conditions independent on the external environment,
- device for cooling the water which run in the wall of the chamber,
- primary heating circuit constructed to supply test specimen,
- measuring instruments and control equipment meeting the requirements of norm [5].

2.3 Formulation

It was assumed that the flow is steady, and no slip condition occurs. Thermo-physical properties of the air are assumed to be constant except for the density variation, which is approximated by the Boussinesq model. The cylinder surface is maintained at a constant temperature T_s which is higher than the outer temperature T_∞. The side wall temperature T_w is equal to the outer temperature T_∞. The momentum equation, for laminar natural convection flow over cylinder may be written under the Boussinesq approximation as follows [7]:

$$(\rho - \rho_\infty)\, g \approx - \rho_\infty \beta (T - T_\infty)\, g \tag{1}$$

where ρ_∞ is the (constant) density of the flow, T_∞ is the operating temperature, and β is the thermal expansion coefficient. Equation 1 is obtained by using the Boussinesq approximation $\rho = \rho_\infty (1 - \beta\Delta T)$ to eliminate ρ from the buoyancy term. This approximation is accurate as long as changes in actual density are small, specifically, the Boussinesq approximation is valid when $\beta (T - T_\infty) < 1$. The Boussinesq module should not be used if the temperature differences in domain are large.

2.4 Experimental procedure

A voltage regulator (matrix), digital ammeter and digital voltmeter were used to control and measure the input power to the working cylinder (Figure 3). The apparatus was allowed to run till pseudo-steady state conditions were achieved. The reading of thermometer and resistance thermometer were recorded every 2 min by digital electronic device until the reading became constant, and then, after half hour, the final reading was recorded [8].

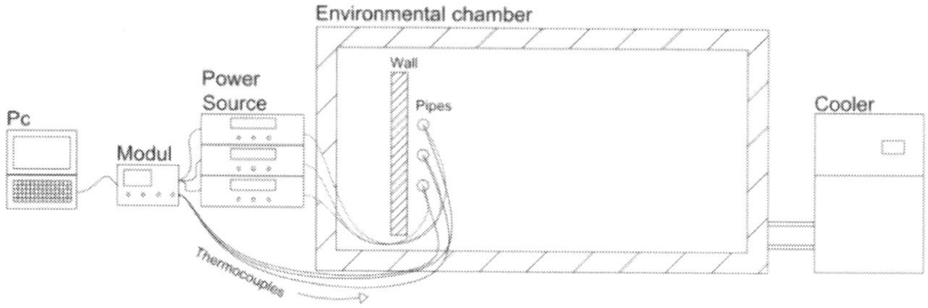

Fig. 3. Schematic of the data acquisition system.

2.5 Data analysis method

The confinement ratio C is the first control parameter of the experiments ($C = X/r$), where X is distance between the wall and the cylinder as shown Figure 4, and r is a cylinder radius). Nine different confinement ratios C will be investigated in this study: $C = 1.5$; 2; 2.5; 3; 4; 5; 8 and 10. The second control parameter (ΔT) is the difference between the cylinder surface temperature (T_s) and the air bulk (T_∞). The air bulk temperature is measured in the middle of the environmental chamber. This second parameter dictates the natural convection flow regime and is generally expressed in the form of the Rayleigh number define as [9]:

$$Ra = \beta g D^3 \Delta T Pr / \upsilon^2 \tag{2}$$

where β is the volumetric thermal expansion coefficient of the fluid, g is the gravitational constant, Pr is the Prandtl number and υ is the kinematic viscosity of air. All parameters are evaluated at the fluid temperature define as:

$$T_f = (Ts + T_\infty) / 2 \tag{3}$$

In the work reported here, the maximum Rayleigh number was of the order of 10^4, meaning that the flow is nominally still laminar.

The convection heat transferred from the heated cylindrical surface can be calculated as:

$$Q_{conv.} = Q_{IP} - Q_{rad.} \tag{4}$$

The total input power supplied to the heated cylinder can be calculated as

$$Q_{IP} = V \times I \tag{5}$$

The radiation heat transfer from the tube can be calculated from:

$$Q_{rad.} = A_s \varepsilon \sigma \,(T_s^4 - T_\infty^4) \tag{6}$$

where A_s is the surface area of the tube $= \pi \times d \times L$; ε is emissivity of given material; and σ is Stefan-Boltzmann constant $= 5.66 \times 10^{-8}$ W/m^2 K^4.

After calculating the radiative heat transfer from the cylinder, it was noticed that it is very small compared with the total heat transfer, and consequently, it can be ignored.

The convection heat transfer can be obtained from:

$$Q_{conv.} = Q_{conv.} / A_s \tag{7}$$

where $A_s = \pi \times d \times L$.

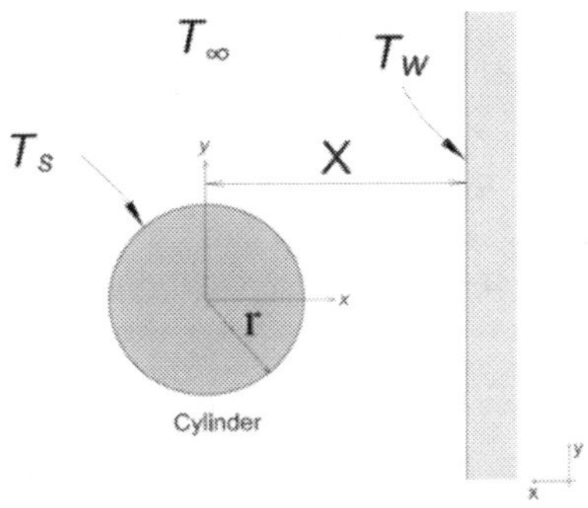

Fig. 4. Physical model and coordinates.

3 Experimental results

A total of 54 test runs were conducted to cover 9 different distances for heated circular cylinder with length of 1 m (L / D = 5). The range of surface temperature was varied from 40 °C to 90 °C.

3.1 Surface temperature

Generally, many variables such as heat flux and flow restriction situation may affect the variation of the surface temperature of the pipe. In this case, the surface temperature of the cylinder is constant for all distances. When the hot cylinder is moving closer to the wall, the temperature of the cylinder gradually decrease. This mean to maintain the constant temperature of the cylinder, the input power is increasing.

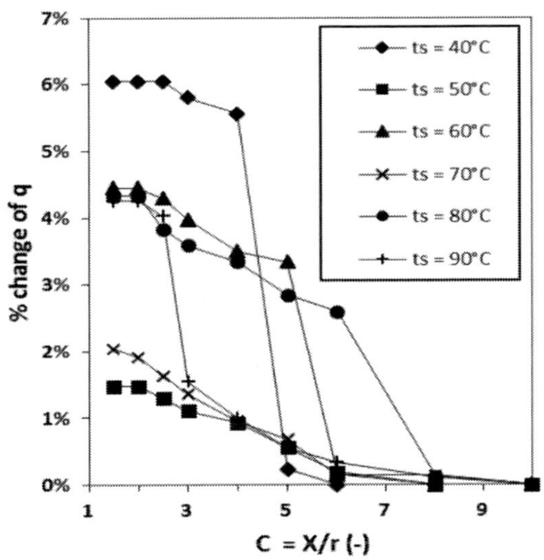

Fig. 5. Percentage change of input power with distance from the wall.

For analysis of the natural convection heat transfer around a horizontal cylinder with the constant surface temperature the temperature parameter was analysed for 10^4 Rayleigh number. The surface temperature of the cylinder is taken with one thermocouple placed in the middle of the cylinder close to the cylinder surface.

The results how the input power is change when the cylinder is moved are shown in the Figure 5. The heat flux exhibits a larger change when the distance between the cylinder and the wall is less than 60 mm for all tests.

3.2 Effect of wall confinement

The experimental data obtained from this investigation analyses the effects of the confining wall on the rate of heat rejection from surface of the heated cylinder. For this study, the maximum average Nusselt number occurs at a confinement ratio $C = 1.5$.

3.3 Average heat transfer correlation

The general correlation obtained by dimensional analysis from Incropera and DeWitt [10] for heat transfer by natural convection is:

$$Nu_D = f_1\,(Gr, Pr)^n \tag{8}$$

In the case of heat transfer from cylinder to vertical wall, one expects that there is effect of distance between the wall and the cylinder. For similarity with a horizontal cylinder, the characteristic linear dimension in the Nu and Gr may be taken as the cylinder diameter (D) and the effect of the distance between wall and cylinder may be taken as $C = X/r$. Then, Eq. (8) becomes:

$$Nu_D = f_2\,(Gr, Pr)^{n\cdot} C^m \tag{9}$$

Therefore, the following correlation was obtained from the present study for smallest and biggest surface temperature and a general correlation was obtained for all distances shown in Figures 6 – 7.

$$Nu_D = 0.57\,(Gr, Pr)^{0.25}\; C^{-0.035} \tag{10}$$

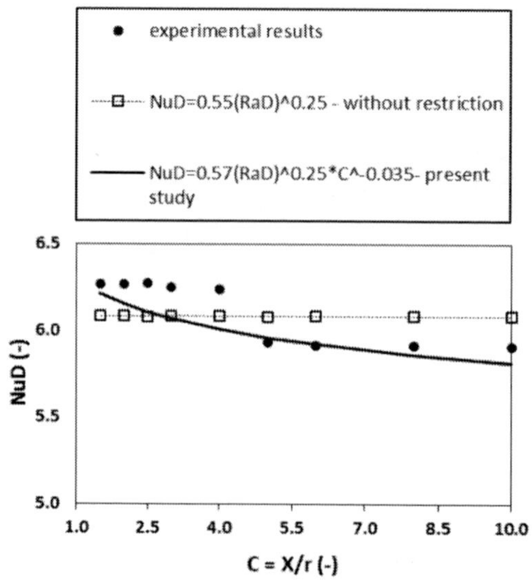

Fig. 6. Correlation between average Nusselt number versus confinement ratio for smallest Ts = 40 °C.

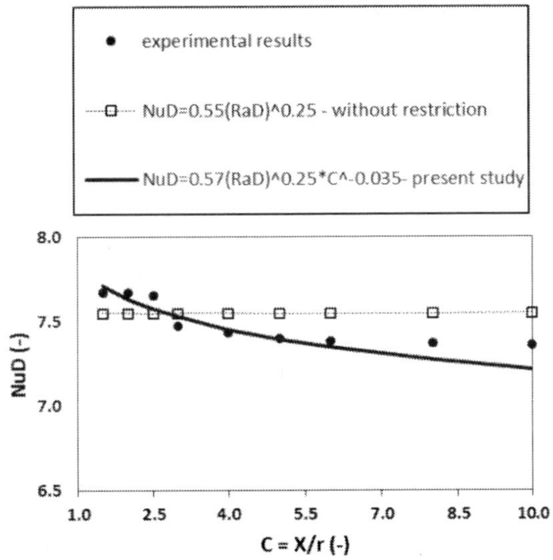

Fig. 7. Correlation between average Nusselt number versus confinement ratio for highest Ts = 90 °C.

As a comparison between the present study and normal case, which included a horizontal cylinder in the free air without any restrictions, the correlation obtained for the normal case from Incropera [10] is,

$$Nu_D = 0.55\,(Gr,\,Pr)^{0.25} \tag{11}$$

Comparing Eq. 10 with the Eq. 11 for this study shown in the Figures 6-7, is proven that the distance between the wall and the cylinder has an effect on the heat transfer results.

The difference of present correlation with Incropera work is about 5 % which is not counting with wall confinement.

4 Conclusions

An experimental study of a laminar natural convection heat transfer from a single horizontal cylinder near an isothermal wall was conducted. Laminar convection heat transfer causes flow because the density differences. The flow moved up around the hot cylinder and continued upward when the cylinder was not interacted with the wall. The flow moved across the wall when the upward flow is interacted with colder wall and began mainly when the distance between the wall and the cylinder of 20 mm diameter was less than 60 mm. The highest heat flux change occurs at confinement ratio 1.5. The improvement of correlation in this study is up to 5 %.

Nomenclature

A_s = Cylinder surface area, (m^2)
C = Confinement ratio, C = X/r (-)
D = Diameter of circular cylinder, (m)
g = Gravitational acceleration, (m/s^2)
Gr = Grashoff number, (-)
I = Heater current, (A)
k = Thermal conductivity, (W/m.k)
Nu = Average Nusselt number, (-)

Pr = Prandtl number, (-)
q = Heat flux, (W/m^2)
$Q_{conv.}$ = Convection heat transfer, (W)
Q_{IP} = Total input power, (W)
Q_{rad} = Radiant heat transfer, (W)
r = Radius of circular cylinder, (m)
Ra = Rayleigh number, (-)
T = Temperature, (°C)
T_s = Temperature of the surface, (°C)
T_W = Temperature of the wall, (°C)
T_∞ = Temperature of the cold surface, (°C)
ΔT = Temperature difference, (-)
k = Thermal conductivity, (W/m.k)
V = Heater voltage, (V)
X = Distance from centre of square cylinder to wall,
x = Cartesian coordinate in horizontal direction, (m),
y = Cartesian coordinate in vertical direction, (m)

Greek Symbols

α = Thermal diffusivity, (m^2/s)
β = Volumetric coefficient of thermal expansion, (1/K)
μ = dynamic viscosity of the fluid, (kg/m.s)
ε = Emissivity, (-)
σ = Stefan-Boltzmann constant, (W/m^2 K^4)
ρ = Density of the fluid, (kg/m^3)
υ = Kinematic viscosity of the fluid, (m^2/s)

This work was supported by Department of Power Engineering under APVV-15-0778 Limits of Radiative and Convective Cooling through the Phase Changes of Working Fluid in Loop Thermosyphon.

References

1. V.T. Morgan, Advances in Heat Transfer **11** (1975)
2. S.W. Churchill, H.H.S. Chu, Intern. Journal of Heat and Mass Transfer **18** (1975)
3. T.H. Kuehn, R.J. Goldstein, Intern. Journal of Heat and Mass Transfer **23** (1980)
4. T. Fujii, M. Fujii, T. Honda, Proc. of the 7th intern. heat transfer conference **6** (1982)
5. STN EN 442-2:2000, (1996)
6. A. Kapjor, Habilitation thesis (2012)
7. Ansys fluent, User's Guide **15** (2009)
8. R. Lenhard, J. Jandačka, M. Malcho, Vetranie a klimatizácia (2008)
9. M. Novomestský, H. Smatanova, A. Kapjor, AIP Proc. **1745** (2016)
10. F.P. Incropera, et al., Fundamentals of heat and mass transfer **7** (2011)
11. M. Čarnogurská, M. Příhoda, T. Brestovič, J. Molínek, R. Pyszko, Journal of Mechanical Science and Technology **26,** 1 (2012)

Analysis and Simulation of Forced Ventilation in Special Condition

Miroslav Rimár[1], *Andrii* Kulikov[1,*], *Marcel* Fedak[1], and *Milan* Abraham[1]

[1]Technical university of Kosice, Faculty of Manufacturing Technologies with a seat in Prešov, Bayerova 1, 080 01 Prešov, Slovak Republic,

Abstract. The research subject of the current work is heating of the closed area by the forced air ventilation system with heat recovery ventilator and floor radiators. The simulation model was made in ANSYS Fluent 14.0. In the simulation were taken to account the secondary thermal gains from the computers, monitors and humans. The results of the simulation approved that in the modern thermal passive houses heat balance calculations should observe secondary thermal gains from the installed equipment. Also were investigated the air circulation in the closed area and the influence of the different barriers which were installed in the laboratory.

1 Introduction

Nowadays air ventilation plays the great role not only for the human health but also for heat balance calculation of the building. But too hot or cold outdoor air makes interior parameters uncomfortable for some time periods. It is very difficult to build modern passive house without installing air recuperation systems or Heat recovery ventilators (HRV) [1-8].

In the heat balance calculation, it is necessary to know all heat loses and thermal gains. In a modern passive building the heat loses are small, approximately 2-3 kW per 100 m^2. According to this heat balance calculation should be expand by the thermal gains positions which were not present before, for example heat from the working electrical equipment or human body [9-11].

People lose heat in next proportion:

- 35 % through heat conduction and convection;
- 35 % through thermal radiation;
- 24 % through evaporation;
- 6 % with breathing.

Nevertheless for the private houses it is still very low figures but in the offices where 2-3 peoples are working on the computers is enough to change the heat [12-14].

* Corresponding author: andrii.kulikov@tuke.sk

In the current article we will simulate the ventilation of the air in the laboratory using of HRV and floor radiators in winter conditions with the impact of the small thermal gains such as computers and human body.

2 Materials and methods

Room under the study was a laboratory of the renewable energy sources of the Technical University in Kosice. The technical drawing of the laboratory is on the Figure 1.

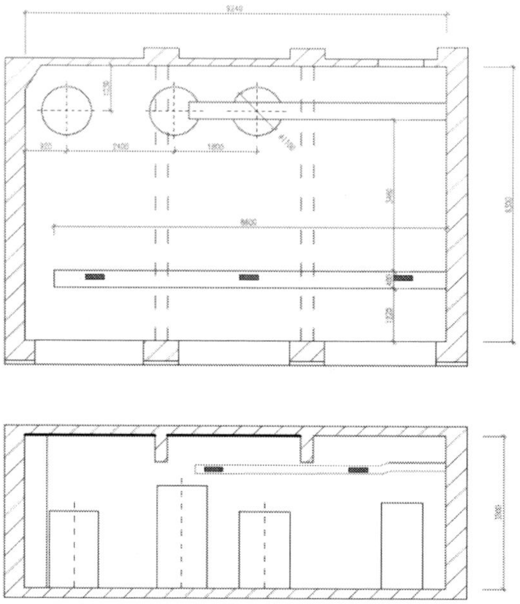

Fig. 1. Laboratory under the study.

In the laboratory there are situated tree thermal storage tanks for the hot and cold water for heating and cooling purposes. The ventilation of the laboratory is going through ventilation canals in the upper part of the room. Fresh air comes to the laboratory via canal 1 and goes out through the canal 2. As you can see from the picture first air canal has two nozzles while the second air canal has three situated approximately at the middle of each the windows. The HRV unit (Sabiana ENY 3) is situated in the room next door. The configuration of ENY 3 allows preheating fresh air in winter period by hot water or integrated electric heater when heat potential of exhaust air is not enough or intensity of ventilation is high.

The regulation equipment is totally controlled by the DESIGO by Siemens. View on the DESIGO program is on the Figure 2.

The input parameters were the same comparing with the article of prof. M. Rimar "Study of gaseous flows in closed area with forced ventilation" (SGFCAFV).

According to EN 12831 the temperature in the laboratory must be 20 °C.

Thermal characteristics of the windows, walls, floor and ceiling were chosen according to their technical documentation.

Sickness of internal walls is 200 mm and temperature is 22 °C.

External wall has 500 mm sickness and temperature in the middle of the wall is 18 °C.

Sickness of windows is 100 mm and external surface temperature is 12 °C.

Celling and floor have 300 mm sickness and 22 °C temperature.

 https://doi.org/10.1051/matecconf/201816802003

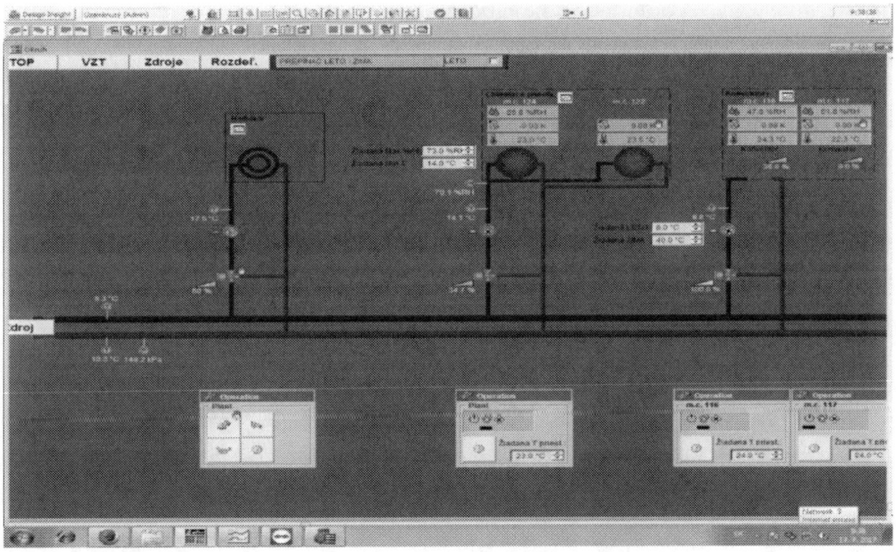

Fig. 2. DESIGO program.

The main source of heat in the laboratory is two floor radiators which are installed under the windows. The surface temperature of radiators is 42 – 45 °C in winter period. It is impossible to complete turn off the heating of the laboratory as through this flor radiators is going the whole heat to the building. Also heat achieved through the thermal storage tanks. The tanks are isolated by special fibrous material of 10 cm sickness. The surface temperature during the day of the first tank (in the corner) is around 24.5 °C, second – 23.5°C, third – 22 °C. The temperature of hot water inside the tanks is approximately 37.5°C for the first tank, 32 °C – second, 25 °C –- third. Also at the picture you can see six tables with computers and monitors. Computer boxes which are situated under the table are the working station so they have higher energy consumption and as a result higher thermal gains – 100 W/m^2. This parameter was chosen to simplify the simulation model. The thermal gains from the monitors are 50 W/m^2. Two cube models in front of the tables are the simplified models of the sitting human bodies. The surface temperature of the human model was 36 °C.

3D model was made in ANSYS design modeler to provide stable calculation and to eliminate inaccuracy in the geometry. ANSYS mesh solver was chosen for the mesh making purpose to reach the better quality. Proximity and curvature advanced size function with fine relevance center, active assembly initial size seed, high smoothing and slow transition also was used for these purpose. According to the settings model had 2,958,471 elements what is by 45 % higher comparing the model without tables, computers and people.

3 Simulation

Simulation of the heating and ventilation was made in ANSYS Fluent 14.0. As a setting for FLUENT was used energy equation model, standard K-epsilon model and radiation p1 model. Inlet type was chosen as velocity-inlet with velocity magnitude 0.2 m/s and 19 °C. For outlet was pressure outlet type. All the parameters were chosen according to the technical documentation of the HRV and the norm of the thermal comfort. The research of the heat recovery parameters of the ENY 3 is not a purpose of the current work. The results of the simulation are on the picture. At the picture illustrated the streamlines of fresh inlet air ant temperature contours of the laboratory (Figure 3).

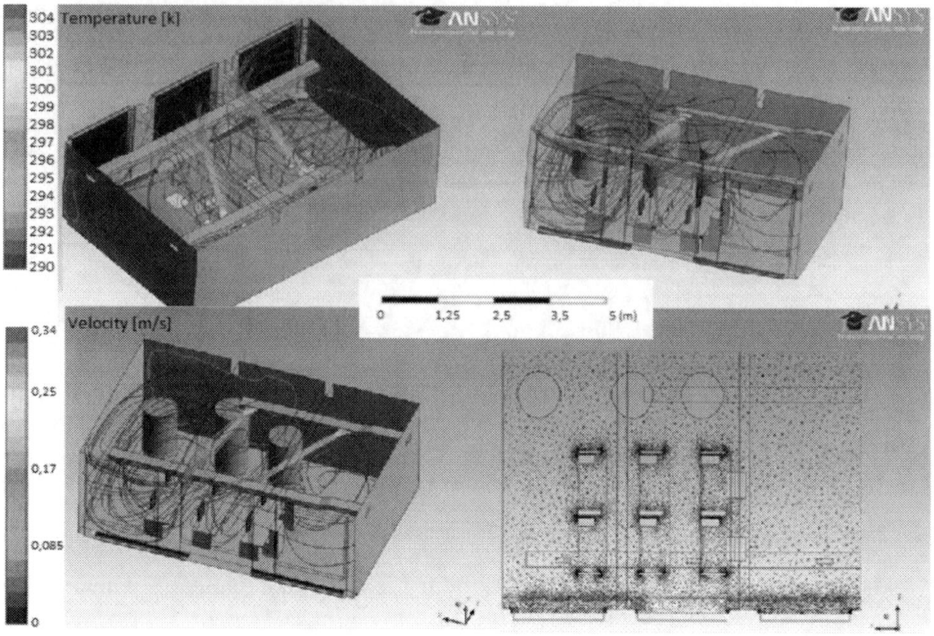

Fig. 3. Results of the first type simulations.

These simulations confirm the results of the previews research SGFCAFV. The main reasons of the turbulent stream at the half of the laboratory are the shape of the nozzles and the ventilation canal also as a velocity of the air. The maximum velocity of the fresh air is 0.36 m/s what is under the limits of EN 12831. At the down part of the picture illustrated temperature contours of walls, windows and floor radiators.

Comparing with the model in the previews article SGFCAFV current model has some differences in the down half of the laboratory. This is caused not only by presence of the barriers like tables and humans but also and extra heat sources like computer and monitors. The temperature difference in the vertical axis of the room is 0.7 °C/m what is appropriate according to the EN 12831.

Relatively warm fresh air from the HRV comes to the upper part of the laboratory. It encounters to the cold windows and fall down as its temperature reduce. In the down part of the windows warm air near radiators mixing with falling cool air. Than air flow divides by two streams. The first one (warmest) goes up and after mixing with warm extract air exhaust through outlet canal. The second one (coldest) circulate in the down part of the laboratory where it cooling the computers and monitors and after it heats goes up through outlet canals.

According to the simulation the internal temperature in the laboratory is 23 – 24 °C depending on the high. The internal temperature according to the thermometers is 24 °C. The difference between the values is caused by the fact that when was set the parameters of wall heat transfer were not consider with thin layers of the wall like paint. The surface temperatures of the computers and monitors are 45 °C and 41 °C respectively.

Also was made simulation of the air ventilation in the laboratory with turned off the heating from radiators (Figure 4).

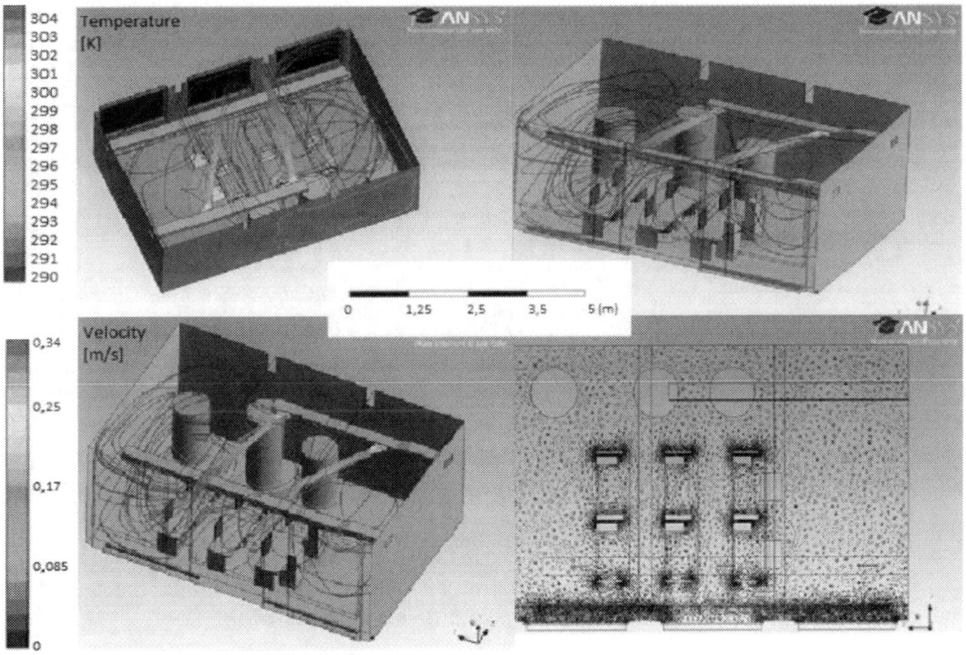

Fig. 4. Results of the first type simulations.

According to this simulation the internal temperature in the laboratory is 21.5 – 23.4 °C depending on the high. Such slight difference between the values in the first and second simulations is caused by the thermal gain from the next rooms. The temperature difference in the vertical axis of the room is 1.1 °C/m what is very close to the limit according to the EN 12831. Reduction of the internal temperature in the laboratory result slightly changes of the air ventilation. The main trends of air circulation are not changed but the path of the coldest air in the down part of the laboratory increased approximately by 15 – 20 %. Also the surface temperatures of the computers and monitors are decreased by 10 – 15 % to 42 °C and 37 °C respectively.

Comparison of the simulation results of the laboratory model with and without tables, computers, monitors and humans is on the figure (Figure 5,6).

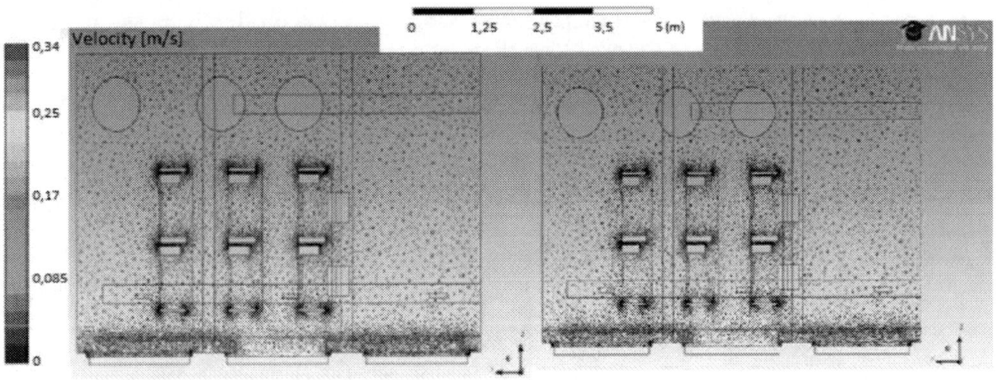

a (floor convectors turned off) b (floor convectors turned on)

Fig. 5. Air flow in the laboratory with barriers.

Air circulation in the down part of the laboratory changed significantly in the model with tables, computers, monitors and humans. Figure 5 presents turbulent flow in the down part of the room. This flow is caused by the barriers which are situated on the air flow path.

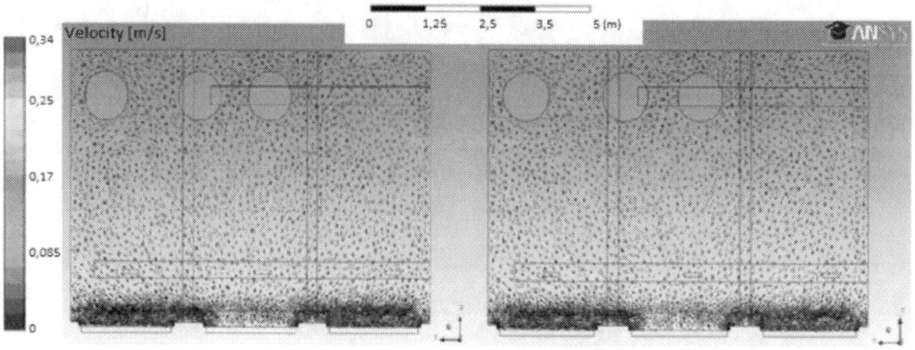

a (floor convectors turned off) b (floor convectors turned on)

Fig. 6. Air flow in the laboratory without barriers.

Figure 6 presents air flow in the laboratory without barriers. After the turbulent zone near the windows air flow stabilize and has laminar character of the flow.

4 Conclusions

The main idea of the current work was to make simulation and analysis of the influence of the secondary thermal gains in the room on its total heat balance system. Secondary thermal gains in the room under the study were computers, monitors and humans. Also was made an analysis of the air ventilation system with heat recovery ventilator. The conclusions of the study are as follows:

- In the modern buildings which belong to the energy passive standard is necessary to take into account with the secondary thermal gains like computers, monitors or other equipment that produces a waste heat;
- Were confirmed the results of the previews research that heating of the room by the only heat recovery ventilation results higher (more than 2 °C) differences of the temperature in the vertical axis of the room;
- It is necessary to use modern simulation programs to calculate internal climate of the room as one of the major factor which affect the effectiveness of the work in the closed areas with forced ventilation.

This work was supported by the Slovak Research and Development Agency under the contract No. APVV-16-0192

References

1. S. Driss, et al., *Energy and Buildings* (2016)
2. P. Kalliomaki, P. Saarinen, W. Tang J., H. Koskela, *Building and Environment* (2016)
3. A. Panda, et. al., *Applied Mechanics and Materials* (2014)
4. J. Zajac, I. Čorný, *Monitoring of processing fluids* (2004)
5. D. Micallef, *Energy and Buildings* (2016)

6. C. Zhang, P.K Heiselberg, *Building Simulation* (2017)
7. T.H. Pedersen, K.U. Nielsen, et al., *Building and Environment* (2017)
8. K. Hu, Q. Chen, *Building and Environment* (2015)
9. M. Napp, T. Kalamees, *Energy and Buildings* (2015)
10. X. Wu, et al., *Energy and Buildings* (2016)
11. T. Mikeska, J. Fan, *Energy and Buildings* (2015)
12. M. Flimel, D. Dupláková, *Applied Mechanics and Materials* (2014)
13. X. Ye, *Building and Environment* (2017)
14. J. Jandacka, et al., *10th International Conference on EFM* (2015)

Numerical simulation of induction heating thick-walled tubes

Richard Lenhard[1,*], *Milan* Malcho[1], and *Katarína* Kaduchová[1]
[1]University of Žilina, Department of Power Engineering, Univerzitná 1, 010 26 Žilina, Slovakia

Abstract. In the paper is shown the connection of two toolboxes in an Ansys Workbench solution for induction heating. In Ansys Workbench, Maxwell electromagnetism programs and Fluent have been linked. In Maxwell, a simulation of electromagnetic induction was performed, where data on the magnetic field distribution in the heated material was obtained and then transformed into the Fluent program in which the induction heating simulation was performed.

1 Induction heating

Induction heating is one of the heating variants, the principles of which are used mainly in industrial applications, melting, shaping, hardening, annealing of materials, but also in the home in the form of induction cookers and cooking plates.
Induction heating uses the principle of the action of the electromagnetic field described by Maxwell's equations on ferromagnetic material. The electrically conductive object is inserted into the alternating electromagnetic field of the induction coil during the induction heating through which the alternating current passes. As a result of electromagnetic induction, swirled currents are induced in the heated object, which have the opposite orientation as the current in the induction coil. Heating occurs due to resistive and hysteresis losses, the proportion of electrical and magnetic components depending on the electrical and magnetic properties of the heated material. The heat is generated directly inside the material, with the heated object and the inductor not in any mechanical contact.

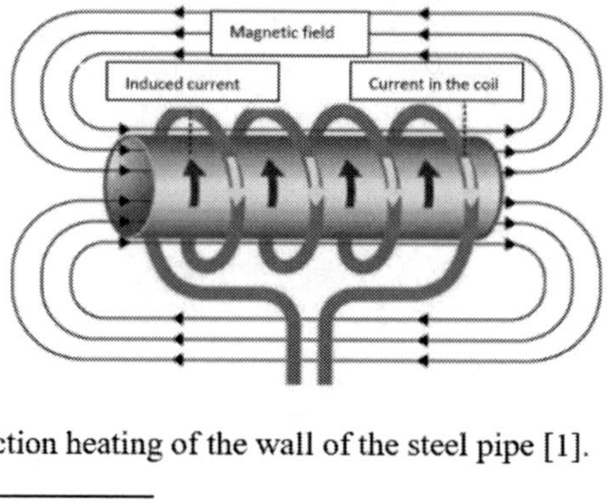

Fig. 1. Principle of the induction heating of the wall of the steel pipe [1].

* Corresponding author: richard.lenhard@fstroj.uniza.sk

The principle of induction heating of a ferromagnetic steel pipe is also documented in Figure 1.

The source of electromagnetic radiation is each conductor through which the alternating current flows. In the case of induction heaters, the conductor is usually adjusted to a suitable shape with respect to the heated article in order to ensure sufficient efficiency of energy transfer from the emitter to the heated object and the required heating parameters (depth and range of the heated layer, temperature,...) [1].

Induction heating is therefore principally based on three basic physical phenomena:

- electro-magnetic induction,
- skin effects,
- heat transfer.

As a result of the skin effect, the current is ejected into the surface areas of the driver. The induced stream thus flows only with a small surface portion of the total cross-sectional area of the heated article. The current density from the surface of the conductor decreases exponentially. The distance from the surface of the wire along which the current density exponentially decreases to 1/e = 0.368 the current density around the conductor, is referred to as the depth of penetration. In this layer, it is concentrated approximately 63 % induced current and 87 % heat generated. The depth of penetration depends on the electrical and magnetic properties of the material and also on the frequency of the current in the induction coil according equation: [2, 5]

$$\delta = \sqrt{\frac{\rho_{el}}{\pi \mu_0 \mu_r f}}, \tag{1}$$

where, f – is the frequency of the current (Hz), ρ_{el} – specific electrical resistance (Ω.m), μ_r – relative permeability of the environment (-), μ_0 – is the vacuum permeability (H/m).

From equation (1) it shows that the depth of penetration, inter alia, depends on the frequency. Due to the fact that electrical resistance and especially permeability changes significantly with temperature, the penetration depth also depends on the temperature. At low temperatures, the conductive materials are strongly ferromagnetic, the penetration depth is small, resulting in a rapid heating of the surface of the body. However, after the Curie temperature is exceeded, the relative permeability drops to a value μ_r =1 and at the same time, the value of the specific electrical resistance increases, increasing the depth of penetration. The penetration depth also increases with increasing magnetic field strength.

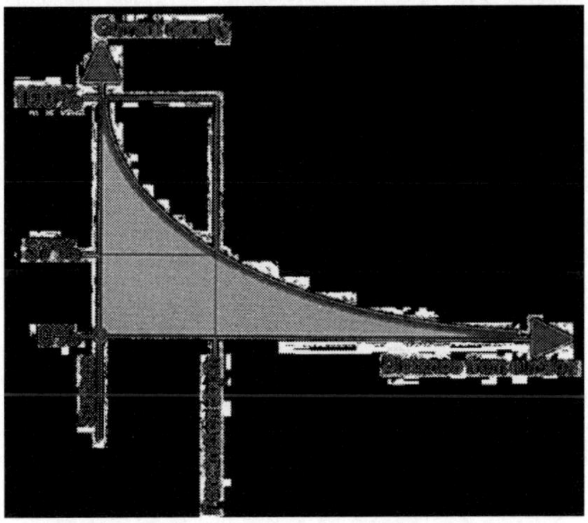

Fig. 2. Dependence of the electrical current density on the distance from the surface [2].

By induction heating, materials of often very different electromagnetic properties are heated. For example, copper, aluminium, steel. Obtaining a temperature dependence of relative permeability on ferromagnetic materials is very complicated. In practice, it is necessary to base the measurements on the parameters [3].

In our case, the induction heating of the roughness tube of the diameter was simulated 918 mm (Figure 3). Where the required temperature after induction heating is on the tube surface 1 030 °C and on its inner part 1 000 °C with a tolerance ±20 °C.

2 Simulation of induction heating

One of the methods of optimizing induction heating is also the simulation of electro-thermal events in various applications. For this purpose, ANSYS Maxwell simulation program is used [4].

Since the tube wall heating is an axially symmetric problem, was in Maxwell 2D from a 3D predefined model (Figure 3) a simplified 2D model was created (Figure 4). This is an induction heating model with defined electrical parameters of the roughness tube with the geometric parameters of Figure 3.

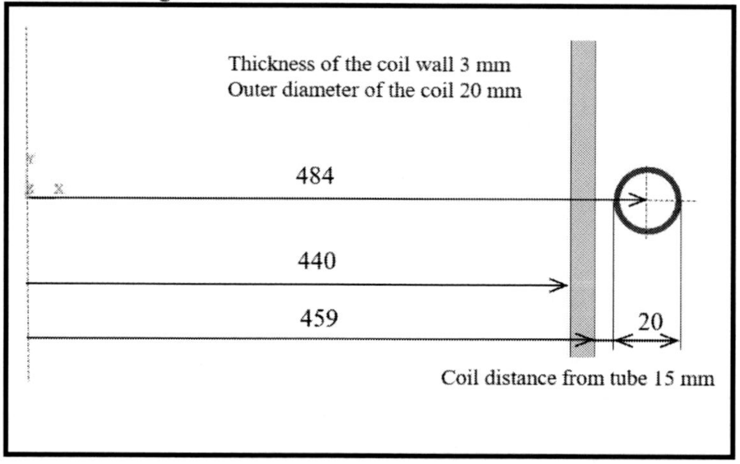

Fig. 3. Geometric parameters of inductor and tube.

In Maxwell 2D, the material requirements for each domain were defined as follows:

- for the surrounding area the selected material was Air,
- for a domain, presenting the tube is selected as Steel_1010,
- for a domain, presenting the coil, was selected the Cooper.

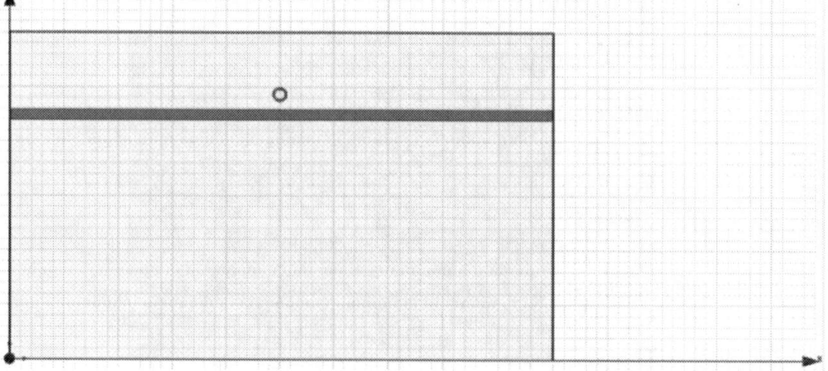

Fig. 4. Created 2D model in Maxwell 2D.

2.1 Model selection and boundary conditions

In Maxwell 2D, a Cartesian model, X, Y with Eddy Current was selected. The boundary condition of the Balloon for the boundaries of the surrounding space has been assigned and the matrix parameter assigned to the part of the model that represents the coil and will count the impedance. For the excitation, the magnitude of the current in the 9 kA coil was set.

The analysis setting allowed up to 15 calculations with 1 % error rate. In the solution, the frequency is set to 400 Hz and 0.0001 for nonlinear residues. Network creation has been left automated it was enough for these simulations.

2.2 Result from numerical simulation in Maxwell

The results thus obtained (Figure 5) from the numerical simulation were connected according to Figure 6 with Fluent Code Solution.

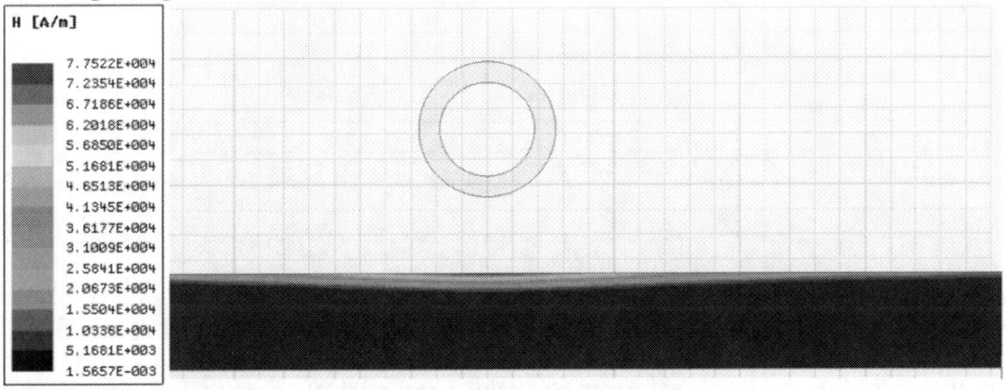

Fig. 5. Result of the magnetic field intensity distribution in the tube wall.

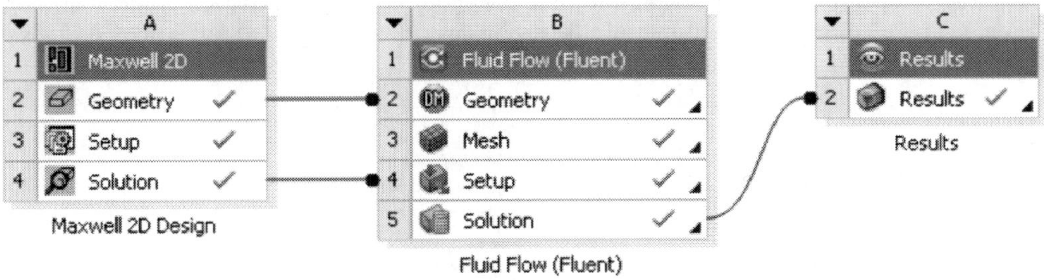

Fig. 6. Connect Maxwell 2D and Fluent toolboxs.

3 Settings of the numerical simulation in a program Fluent

The created model in Maxwell 2D has been transformed into a model designer, where the model has been supplemented with the geometry that is needed for another solution, which is the subsequent cooling of the material after an induction heating. The position and space for cooling air, the position of the inductor and the cooling water position. Of course, the cooling water space did not need to be modeled because the surface to which the cooling water actually enters is along the cooling part and will be replaced by the dependence of the heat transfer coefficient on the wall temperature. After the model was modified, a computational mesh was created which had 313 199 2D cells (Figure 7).

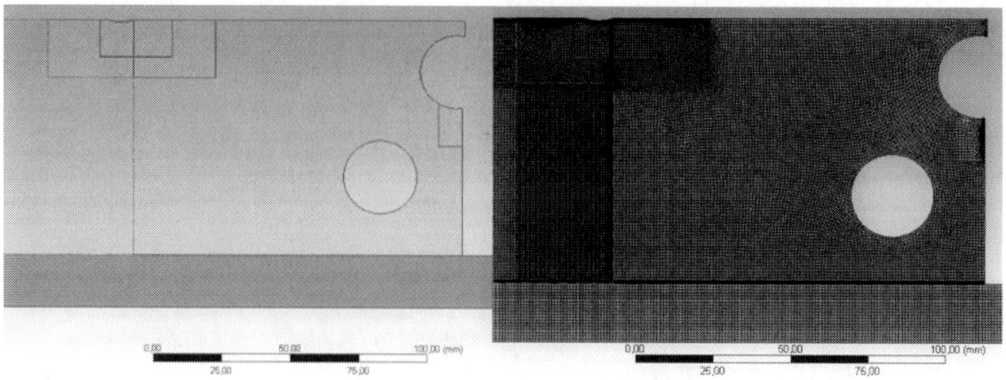

Fig. 7. 2D model after modification and created computational mesh.

After transferring the computing mesh in to Fluent was set up boundary conditions for the individual domains for induction heating Table 1.

Table 1. Boundary conditions for each domain.

Boundary conditions	Input parameter	Temperature	Radiation
Inlet_air	0 m/s	20 °C	
Wall_water	not entered	20 °C	1
Innet_wall	-	-	0.1
Walls	-	-	0.9

After the boundary conditions were set, it was necessary to assign the obtained electromagnetic field distribution from Maxwell to a particular domain using the function EM mapping – Volumetric Energy Source, which, after selecting a particular zone, has been assigned the desired electromagnetic field. The model thus prepared was launched as a time-dependent model first with energy and radiation only, which served to obtain the resulting temperature field distribution after the induction heating which could then begin to cool. The resulting field temperature distributions after the induction heating can be seen in Figure 8 and the temperature distribution in the tube wall (Figure 9).

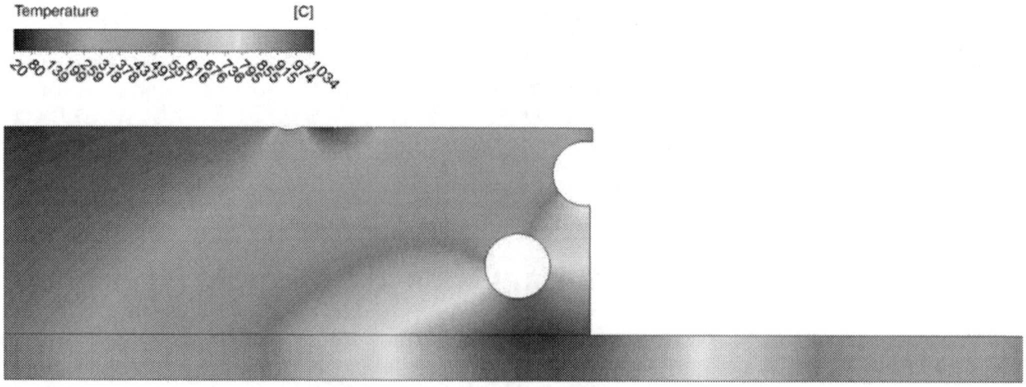

Fig. 8. Results of temperature field distribution after induction heating.

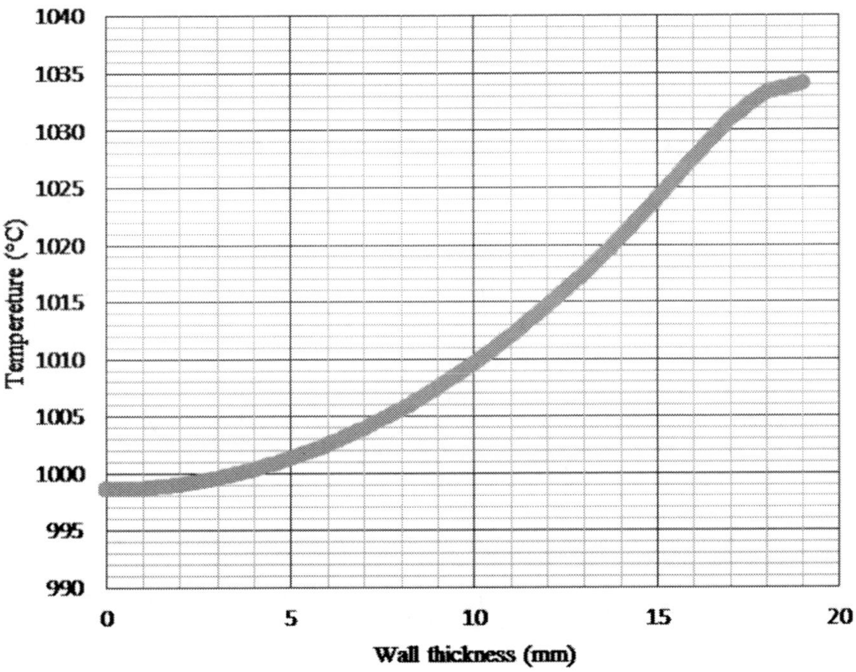

Fig. 9. Temperature distribution in the tube wall.

4 Conclusion

Numerical simulation of the temperature field change in the tube wall after induction heating, mild deformation and radiation showed that this physical, the technological process can be numerically modeled. This will further influence the process by, for example, cooling the surfaces of the oven after heating it by means of induction. It achieves further ways of reducing wall temperature after induction heating, which will ensure that the mechanical parameters of the pipe material are maintained even after the necessary technological induction heats.

The research is supported by project "Limits of Radiative and Convective Cooling through the Phase Changes of Working Fluid in Loop Thermosyphon", APVV-15-0778.

The research is supported by project: The storage of hydrocarbon gases in hydrate structures as an alternative method of energy accumulation, 063ŽU-4/2018.

References

1. V. Rudnev, *Handbook of induction heating.* New York: Marcel Dekker, (c2003)
2. T. Holm, P. Olsson, E. Troell, *Steel and its Heat Treatment* - a handbook, Mölndal: Swerea IVF, (2012)
3. http://www.allforpower.cz/clanek/aplikacni-omezeni-zaropevne-oceli-p92/
4. user's guide – Maxwell 3D, ANSOFT CORPORATION 225 West Station Square Dr. Suite 200 Pittsburgh, PA 15219-1119
5. Z. Hradílek, I. Lázničková, V. Král, *Elektrotepelná technika,* ČVUT Praha, (2011)

CFD prediction of flutter of turbine blades and comparison with an experimental test case

Vaclav Slama[1,*], *Bartolomej* Rudas[1], *Jiri* Ira[2], *Ales* Macalka[2], *Petr* Eret[3], and *Volodymyr* Tsymbalyuk[3]

[1]Doosan Skoda Power s.r.o., Tylova 1/57, Pilsen 30128, Czech Republic
[2]NUM solution s.r.o., Perucká 2482/7, Prague 12000, Czech Republic
[3]University of West Bohemia, Faculty of Mechanical Engineering, Department of Power System Engineering, Univerzitní 8, Pilsen 30614, Czech Republic

Abstract. Last stage blades are a key element of steam turbines and in many ways determine the turbine configuration alongside with the overall turbine performance. The total efficiency of the low pressure turbine section can be increased by extending the last stage blades. The design process of such long blades involves a flutter analysis using CFD tools which have to be validated by measurements in test facilities under various operating conditions. Experimental data obtained from a subsonic wind tunnel with an oscillating turbine blade cascade, which is available at the Department of Power System Engineering at the University of West Bohemia, was compared with simulations in ANSYS CFX currently used in the Doosan Škoda Power. The paper provides a brief summary of experimental rig description, CFD tool setup and the results for the case of a travelling wave mode with the pure torsion motion of amplitude of 0.5°, Ma = 0.2, reduced frequency of 0.38 and angle of attack +5°.

1 Test rig description

An in-draft wind tunnel which is shown in Figure 1 is utilized for experimental investigations of subsonic flutter in a blade cascade. Air enters to the system through a system of filters and turbulence reducing screens and then accelerates in a convergent nozzle before moving into a test section with a turbine blade cascade in a linear configuration which is shown in Figure 2. For the flutter investigation, this blade cascade consists of eight blades and the four central blades (3-6) are flexibly mounted and each has two degrees of freedom presenting bending and torsion motions. Because aerodynamic forces and moments are weak and can be supressed by blade inertial loading, the blades are made of carbon fibre. The construction of the test rig enables that the blade cascade can be adjusted for different angles of attack. In this paper, the case with the angle of attack +5° is presented.

Upstream of the cascade, there is a Pitot probe which is used for the steady flow inlet velocity measurement as well as the inlet static and the total pressure measurement.

* corresponding author: vaclav.slama@doosan.com

Inlet flow static temperature was obtained using a thermometer sensor also situated in the upstream part. Downstream of the cascade, there is a traversing mechanism with a Pitot-static probe in order to measure the pressure profiles downstream of the blade cascade. Before leaving the wind tunnel, the airflow discharges in the drum chamber and runs through the outlet duct towards the blower vacuum inlet.

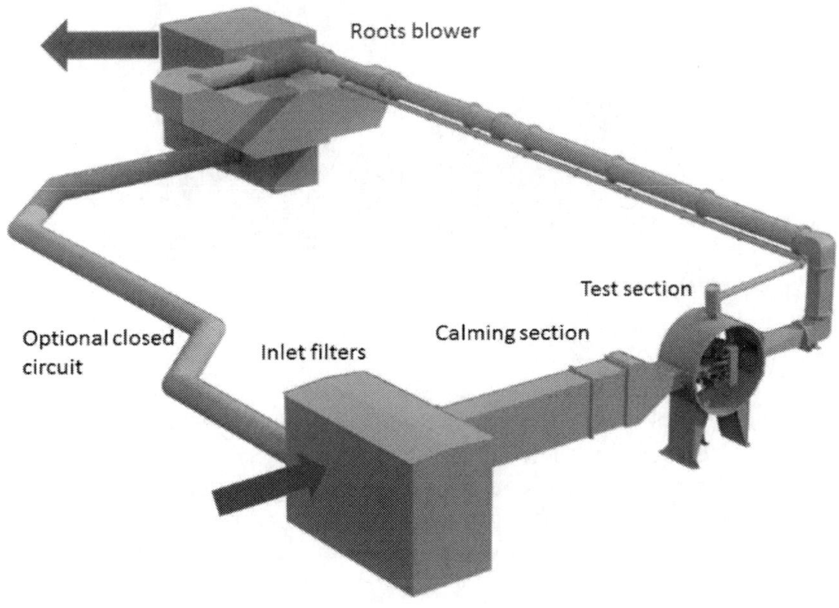

Fig. 1. The wind tunnel model.

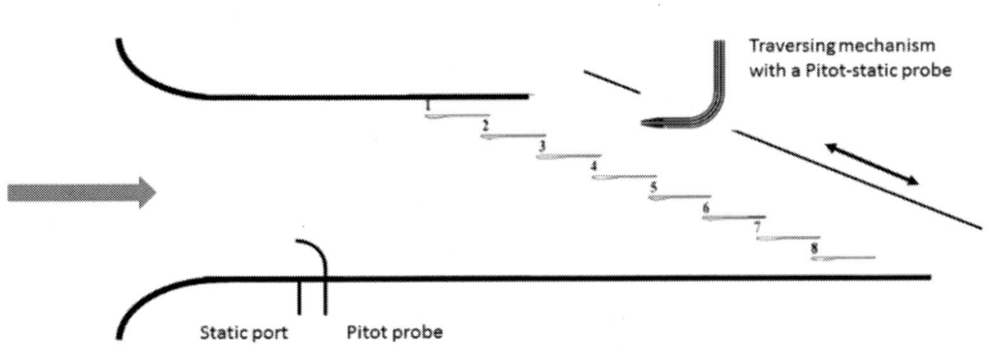

Fig. 2. Schematic of the test section, top view with the upstream probes, the linear blade cascade and the traversing plane.

The movements of central four blades (3-6) are controlled and simultaneously measured by four electromagnetic shakers, which are alternately attached at the top and bottom of a supporting frame as shown in Figure 3. From the measured signals the aerodynamic work per cycle done by the fluid on each moving blade can be calculated. Because flow has induced blade bending deformation during the measurement of the torsion motion, experimental data were corrected thus numerical simulations were performed on rigid blades [1].

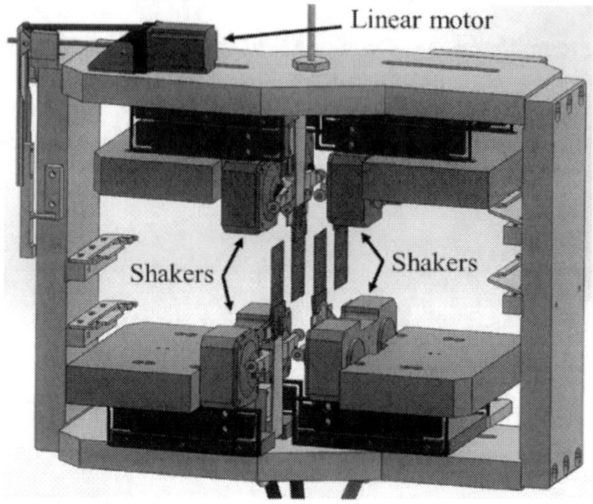

Fig. 3. Supporting frame with four electromagnetic shakers.

2 Computational part

A package of ANSYS 18.2 tools was used for performing CFD analyses. The numerical model was created for the angle of attack equal to +5°. The static pressure, the total pressure and the static temperature in the measurement points upstream of the cascade were monitored during entire calculations and then their final values were compared with the experimental data. The position of this upstream measurement can be seen in Figure 2. The traverse plane downstream of the cascade, also depicted in Figure 2, was another important reference point. Static and total pressures at 50 % of the blade height of the channel were monitored and their profiles were compared with the experimental data as well.

2.1 Domains for numerical analysis and used mesh

Before carrying out CFD calculations, it was necessary to simplify the 3D CAD model of the wind tunnel test section. Adjacent tunnel parts out of the main flow were removed and the model was separated into two sections (Figure 4). After several preliminary computations, the CFD model was further simplified and also improved (Figure. 5). Details about model simplifications and improvements can be found in [2].

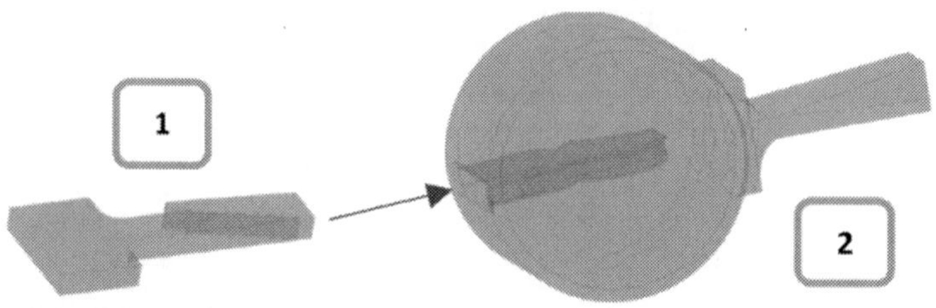

Fig. 4. Domain used during preliminary computations.

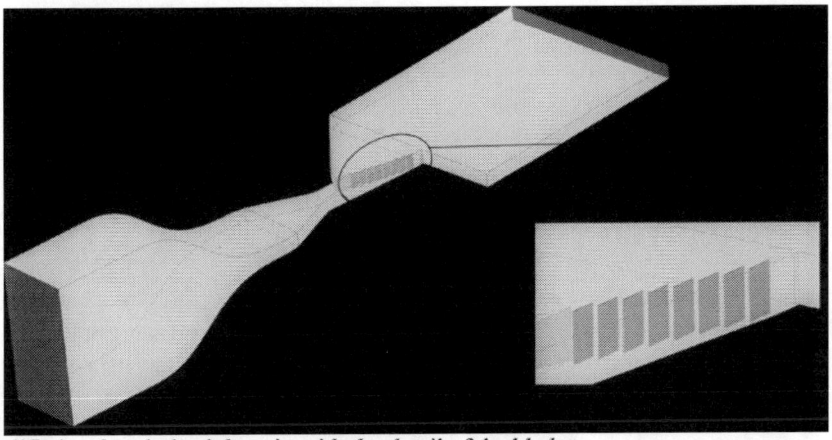

Fig. 5. Simplified and optimized domain with the detail of the blades.

A hexahedral mesh with an appropriate boundary layer, where the first element height is equal to 1.10^{-5} m, on the blades and the adjacent walls was created for the whole computational domain in ANSYS ICEM CFD 18.2. The mesh which was used for steady-state simulations can be seen in Figure 6. There is also shown that, for unsteady simulations, the domain was reduced.

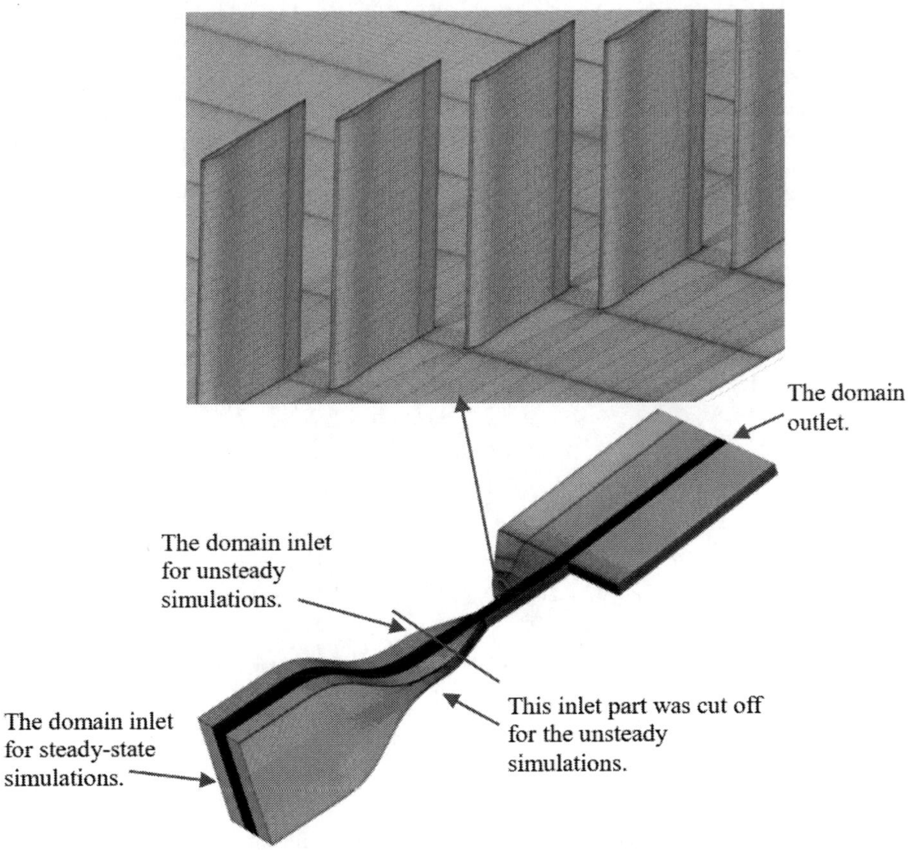

Fig. 6. Computational mesh for steady-state (16 million elements) and unsteady simulations (3.5 million elements) with the detail of blades.

2.2 Solver setup

ANSYS CFX 18.2 was used as the main computational tool. For steady-state simulations, the flow field was modelled as steady, viscous and fully turbulent. The medium was defined as air. The ideal gas law was used to describe its thermodynamic properties. The two-equation eddy-viscosity SST k-ω turbulent model with an automatic wall function was used for turbulence modelling. Second order accurate approximations for all terms in the governing equations were applied.

For unsteady simulations, the basic settings were the same as for the steady-state simulations. Moreover, a time period was specified as the reciprocal value of the tuned frequency (1/82.2 Hz). The total number of time steps per period was chosen to be equal to 100 which corresponds to the time step $\Delta t = 1.217 \cdot 10^{-4}$ s. The total number of periods per run was set to 6. The blade motion was defined according to the measurement as a pure torsion oscillation of amplitude of 0.5° around the middle of the blade chord. The calculations were performed for full range of inter blade phase angles (IBPA = 0 ÷ 360°) with the step of 30°.

2.3 Boundary conditions

For steady-state simulations, the total pressure and the total temperature were defined at the domain inlet (see Figure 6). Using an iterative process, these inlet parameters were determined in such a way that the total pressure in the traverse plane downstream of the blade cascade was approximately equal to the measured values and the inlet static temperature was equal to the upstream measured value. The mass flow rate value was adjusted in a reasonable range so that the static pressure distribution in the traverse plain matches the experimental data. For unsteady simulations, the boundary conditions at the inlet of reduced domain (see Figure 6) and at the outlet were taken from the steady-state simulation results. A total pressure and a total temperature profile as well as velocity vectors were defined at the inlet and the mass flow rate was defined at the outlet.

2.4 Steady-state CFD model validation

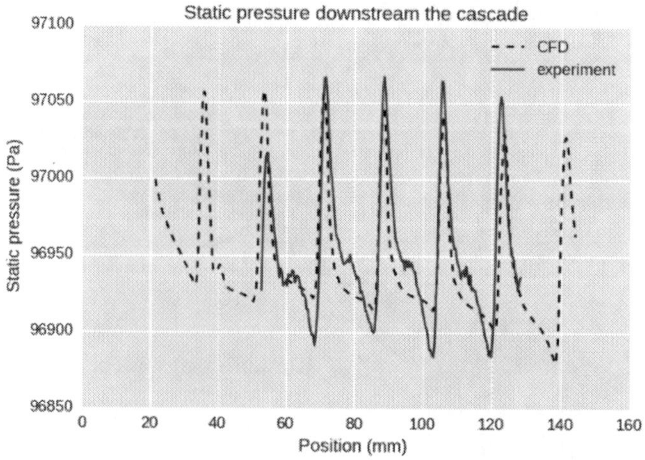

Fig. 7. Experimental and numerical results of static pressure in the traverse plane downstream of the cascade – see Figure 2) for defined conditions: Ma = 0.2, angle of attack +5°.

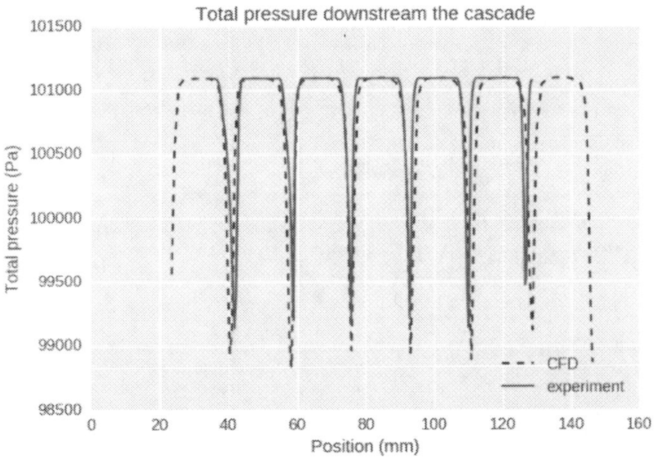

Fig. 8. Experimental and numerical results of total pressure in the traverse plane downstream of the cascade – see Figure 2) for defined conditions: Ma = 0.2, angle of attack +5°.

A comparison between the experimental and numerical results of the total and static pressures in the traverse plane, which is situated downstream of the cascade, is depicted in Figures 7 and 8. Both measurement and CDF simulation were carried out at 50 % blade height. As expected, the numerical results correspond reasonably well with the experimental data. Several possible explanations of the difference between the results are presented in [2].

3 Unsteady CFD and measurement comparison

As a common indicator of the potential danger of flutter, the aerodynamic work in Eq.1 is used. This work is defined as the work done by the fluid on the blade per vibration cycle:

$$AW_{cycle} = \int_{t_0}^{t_0+T} \int_{A} -p\mathbf{v}\mathbf{n}\,\mathrm{d}A\mathrm{d}t \tag{1}$$

where T is the period of one vibration cycle, t_0 is the time at the start of the vibration cycle, p is fluid pressure, $\mathbf{v}$ is the velocity of the blade caused by imposed vibrational displacement, A is the surface of the blade and $\mathbf{n}$ is the surface unit normal vector. When the aerodynamic work per cycle done on the blade is positive, the blade is aerodynamically unstable (i.e. flutter). The aerodynamic work calculated during the unsteady numerical simulations is in excellent agreement with the experimental values as shown in Figures. 9-12.

https://doi.org/10.1051/matecconf/201816802005

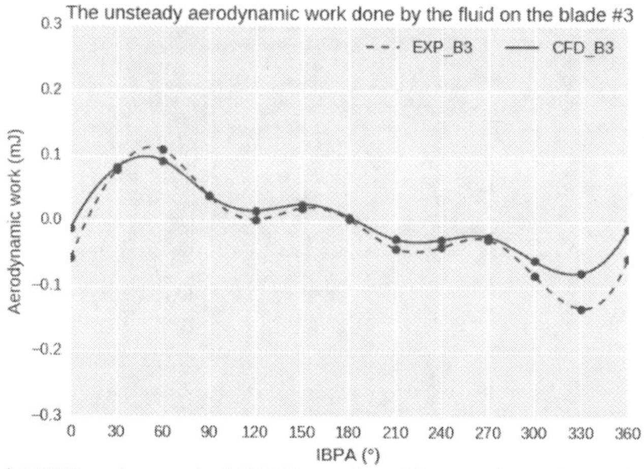

Fig. 9. Experimental (EXP) and numerical (CFD) results of the aerodynamic work done on blade #3.

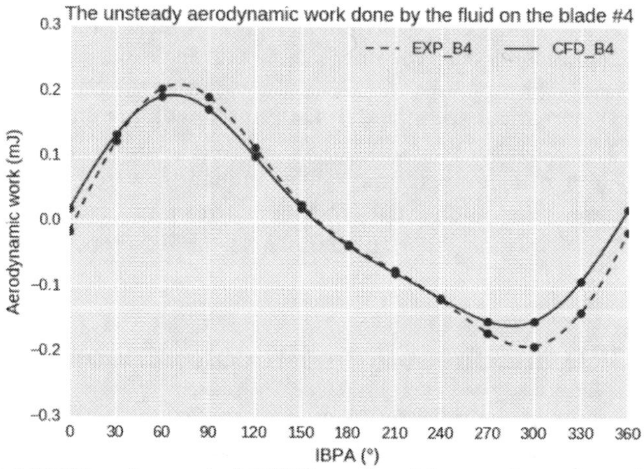

Fig. 10. Experimental (EXP) and numerical (CFD) results of the aerodynamic work done on blade #4.

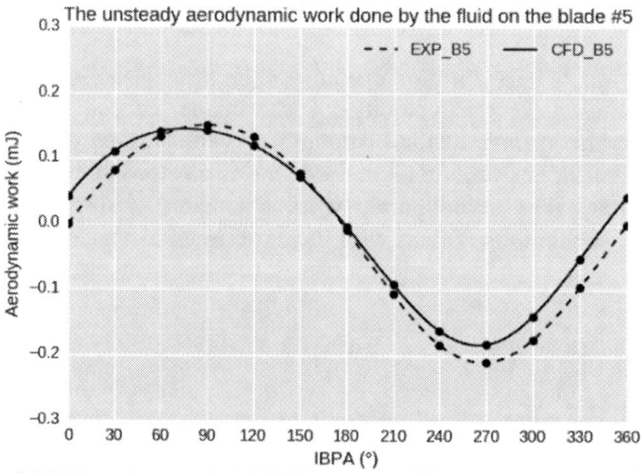

Fig. 11. Experimental (EXP) and numerical (CFD) results of the aerodynamic work done on blade #5.

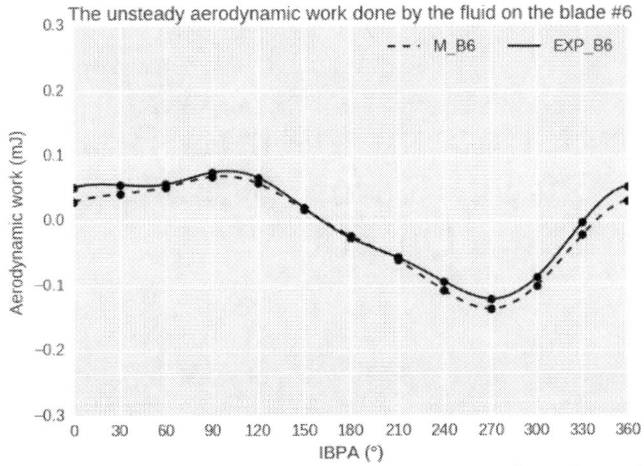

Fig. 12. Experimental (EXP) and numerical (CFD) results of the aerodynamic work done on blade #6.

4 Conclusions

Aerodynamic work, which is an important parameter indicating the potential danger of blade cascade flutter, was evaluated from experimental measurements and CFD simulations for the case of a travelling wave mode with the pure torsion motion of amplitude of 0.5°, Ma = 0.2, reduced frequency of 0.38 and angle of attack +5°. The experiments were carried out using an in-draft wind tunnel where a turbine blade cascade in a linear configuration with four flexibly mounted blades is installed. ANSYS CFX 18.2 was used to carry out the unsteady numerical analyses. In this test case, both CFD analysis and measurements showed excellent agreement. Further validation of flutter prediction and parametric study are necessary to verify dominant trends in flutter behaviour.

This research was founded by the European project Flexturbine, H2020 research and innovation programme under grant agreement No. 653941.

References

1. P. Eret, J. Ira, A. Macálka, B. Rudas, V. Sláma, V. Tsymbalyuk, ASME Turbo Expo, *The Validation of Flutter Prediction in a Linear Cascade on Non-Rigid Turbine Blades*, (to be published in 2018)
2. P. Eret, J. Ira, A. Macálka, V. Sláma, V. Tsymbalyuk, Experimental Fluid Mechanics, *Experimental and CFD analysis of steady subsonic airflow in a stationary cascade of turbine blades*, 230-234 (2017)
3. T. Tanuma, Woodhead Publishing, *Advances in Steam Turbines for Modern Power Plants, Development of last-stage long blades for steam turbines*, **13**, 279-305 (2017)
4. K. Hourigan, F. Liu, I. McBean, M. Thompson, J. Fluids Eng., *Prediction of Flutter of Turbine Blades in a Transonic Annular Cascade*, 127-133 (2005)

The Sonic Surface in the Inter-Blade Channel of the Last Stage Rotor Wheel in the Steam Turbine of Large Output

Martin Luxa[1,*]

[1]Institute of Thermomechanics of the Czech Academy of Sciences, Dolejškova 5, 18200 Prague, The Czech Republic

Abstract. The paper deals with sonic surface in a modern turbine wheel consisting of non-prismatic ultra long blades. The whole inter-blade channel is choked. Different positions and shapes of the sonic line in particular cross-sections along the span are observed. The sensitivity of sonic line formation to small changes of effective shape of the inter-blade channel in the root section and the influence of inlet angle, stagger angle and pitch/chord ratio in the tip section are discussed. The problematic of sonic line development in the case of supersonic inlet flow filed is also described. The presented work is based on results of theoretical, experimental and numerical approaches.

1 Introduction

The modern rotor blade wheels in the last stages of low-pressure cylinders, that are embodied in the steam turbines of large energy output, usually consist of ultra long blades. The thermodynamic conditions in the last stage of the low-pressure cylinder, sizeable diameter of the cylinder, finally the common nominal rotational frequency ($50.s^{-1}$) of the wheel, result in transonic flow field in the inter-blade channel and outlet part of the blades assembly. Under specific conditions, the transonic (or even supersonic) flow field appears at the blade wheel inlet in the tip region. This description is well illustrated in the Figure 1, showing an example of numerical simulation of the flow field in the last stage consisting of ultra long turbine titanium blades 1 375 mm.

The knowledge of sonic surface shape (or sonic line - regarding the particular section in the 3D inter-blade channel) is important. It usually represents the boundary condition for the supersonic flow field solution and is the boundary line of the region in the transonic flow, wherein the solution of hyperbolic partial differential equations is supposed. The shape of sonic line and thus the entire development of supersonic flow is sharply dependent on the geometrical parameters of the cascade, in particular on the pitch to chord ratio, on the angle of flow turning etc. In the Prandtl- Meyer problem (simple supersonic expansion), the sonic line is always orthogonal to the stream lines upstream. It means that

* Corresponding author: luxa@it.cas.cz

the stream line curvature is zero in the sonic point. This assumption is often used in such a way, that the sonic line is identical with the geometric throat of the considered channel.
This condition is fulfilled rarely in real flows. The solution of the shape of sonic line in a channel is proposed e.g. by Safarik [2], [3], Shapiro [4] or Thompson [5]. Interesting overview of various analytical solutions of sonic line shape in symmetric and asymmetric channels provides the work of Dvorak [6]. The solution of the shape of sonic line in the narrowest part of an inter-blade channel is based on knowledge of the channel curvature in the region of channel throat. The concept of a steady flow of ideal inviscid compressible fluid is applied, which is described by linearised potential equation. These analytical solutions of sonic line usually approximate the curvature of channel surfaces defining the aerodynamic throat by circular curves.

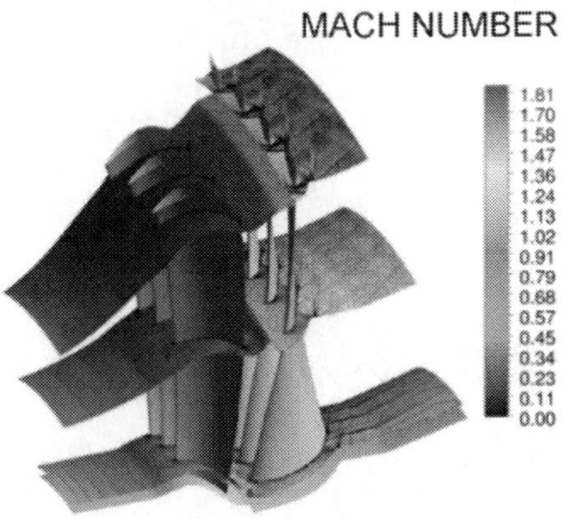

Fig. 1. Result of numerical simulation of flow in the last stage of the low pressure cylinder in the steam turbine Škoda consisting of ultra long titanium rotor blades *1 375 mm* – visualisation of the Mach number distribution [1].

The data regarding the shape and position of sonic line are often adopted from numerical simulation or from experiment. Very useful experimental method for the sonic line detection is interferometric method (in 2D flow fields modelling). Advantageous seems application of so called infinite interferometric fringe method [4], where the fringes in interferometric chart are directly related to regions of constant refractive index. The mutual relation of refractive index and density of the fluid is given by Lorenz -Lorentz law. This non-linear law can be replaced by simpler linear Gladstone - Dale law if the flowing fluid is dry air. Thus, we can match the appropriate density with each fringe in the interferometric picture and this procedure makes it naturally also possible to determine the position and shape of sonic fringe in the flow field. It is necessary to bear the isoentropic condition in mind, i.e. the evaluation of interferometric fringes is usually related to the total pressure at inlet of the investigated inter-blade channel. Distortion of resulting interferometric picture is especially important in boundary layers and shear layers.

2 Sonic lines in selected sections of an inter-blade channel

The sonic line joins the surfaces of adjacent blades under nominal conditions of the flow in all sections along the inter-blade channel in rotor row of the last stage. The exception can represent the tip region of ultra long rotor blades with supersonic flow at the inlet. The sonic line closes around the leading edge in this case (started supersonic flow), near the stagnation point. The situation between the subsonic and started supersonic inlet flow (unstarted supersonic flow) represents much more complex issue [6] with unsteady behaviour and manifestation of flow field instability. The next paragraphs are focused mainly on flow fields in such turbine cascade geometries, where the inlet flow is subsonic, and thus the sonic line connects the pressure and suction side of the adjacent profiles.

2.1 Root section

A considerable sensitivity of sonic line shape on the inter-blade channel shaping is typicall for geometries representing root sections of rotor rows of the last stage. The inlet Mach number in nominal regimes is usually relatively very low. Its value commonly lies under $M_1 = 0.4$. The convergent part of the inter-blade channel is formed so that downstream the inlet part, relatively very long part of inter-blade channel with significantly small narrowing is situated. This convergent part ends in the region of geometrical throat and the following divergent part of the inter-blade channel is usually very short. The supersonic expansions take place mainly in the exit part of the profile cascade. Considering small changes of inlet/throat cross section ratio along the central line of practically convergent inter-blade channel (Figure 2), the small changes of profile shape in region of aerodynamic throat together with different parameters of boundary layers have considerable impact not only on the shape of the sonic line but also on the supersonic expansion downstream.

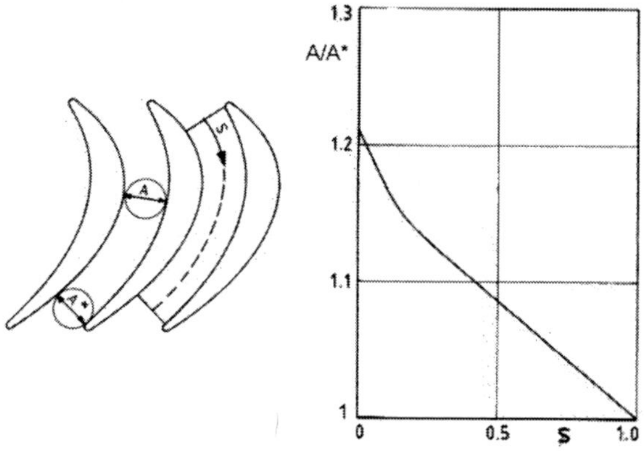

Fig. 2. Example of root section designed in VKI with depicted dependence of inter-blade channel contraction on the position along the channel central line [7].

An example of experimental modelling of flow fields in two variants of root section supplemented by data from numerical simulation is shown in the Figures 3 and 4. Both variants were intended for the same diameter of rotor wheel and the contours of profiles were similar; however, the number of blades in the wheels was different. The small difference in the number of blades generates very small changes in the pitch/chord ratio parameter (the second decimal place). The interferograms of both variants are in the Figure 3 with highlighted sonic line. The experimental data regarding the measurement of the variant with smaller pitch/chord ratio ($s/c = 0.2603$) corresponds well with numerical simulations in the Figure 4. However, the sonic line in the numerical simulation data of flow with larger pitch/chord ratio ($s/c = 0.2732$) is bended in a different way in comparison with experimental data. The reasons of this difference can reside in an application of chosen calculation method - the flow field was considered fully turbulent. The turbulent model *kω-SST* with non-reflective boundary condition in commercial software was applied. The more detailed numerical modelling of the boundary layer on the suction side of the profile with application of transition models of turbulence might bring more accurate results. This research was performed for nominal inlet flow angle where the value of incidence angle was positive ($\iota = +14°$). This positive incidence is linked to the application of the profile on straight fir-tree dovetail. The positive incidence leads to small decreasing of static pressure along the suction side of the profile in a restricted region downstream the stagnation point. This expansion is downstream followed by large region of practically constant static pressure. This results not only in complicated processes in the boundary layer on the suction side, but also in important changes of effective shape of the inter-blade channel.

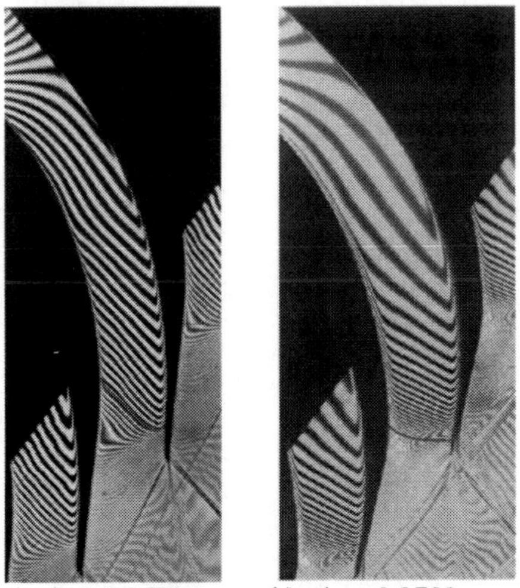

a) $s/c = 0.2603$ b) $s/c = 0.2732$

Fig. 3. Impact of small changes of pitch/chord ratio on the shape of sonic line in the root section, interferometric measurements, exit Mach number $M_{2is} \sim 1.8$ [8].

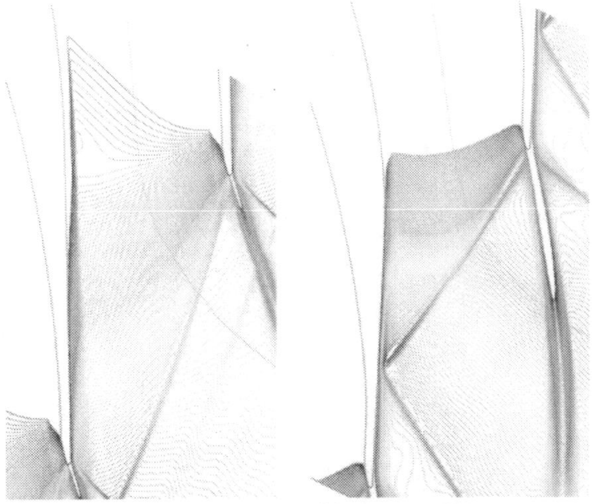

a) $s/c = 0.2603$ b) $s/c = 0.2732$

Fig. 4. Impact of small changes of pitch/chord ratio on the shape of sonic line in the root section, numerical simulation, exit Mach number $M_{2is} \sim 1.6$, isolines of isoentropic Mach number [9].

The next example shows other variant of this root section. The pitch/chord ratio is the same as in the variant *b* ($s/c = 0.2732$), the shape of new profile is slightly more arched in the suction side. The arrangement of new profiles in the cascade is different, i.e. stagger angle is set in such a way, that the incidence is zero in the contrast to the original variant.

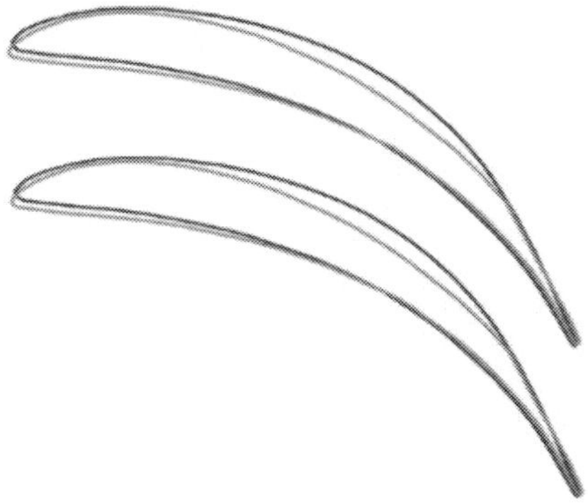

Fig. 5. Comparison of original (green line) and improved (red line) variant of the root profile.

Both profiles are compared in the Figure 5, the cascades in the figure are mutually turned. The interferometric picture together with result of numerical simulation is compared in the Figure 6. The sonic line is highlighted by red line. The calculation was carried out with commercial software with application of turbulent model *kω-SST* with non-reflective boundary condition (the same procedure as in the foregoing example). One can observe the agreement of the position of sonic point on the suction side in both research approaches. This good agreement between numerical study and experimental data is certainly caused by more favourable static pressure distribution along the suction side in the subsonic part of the inter-blade channel - the region of static pressure stagnation is eliminated. Substantial is also the neutral incidence angle for boundary layer development on the suction side.

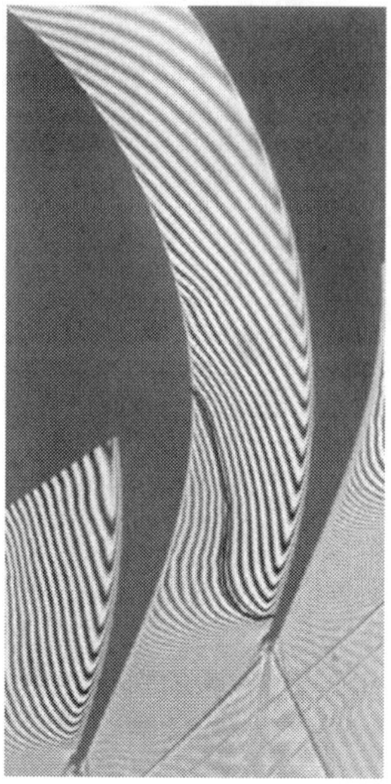

Fig. 6. The shape and position of sonic line in improved version of root section (*s/c = 0.2732*). Interferometric measurement, exit Mach number $M_{2is} \sim 1.7$, incidence angle $\iota = 0°$ [10].

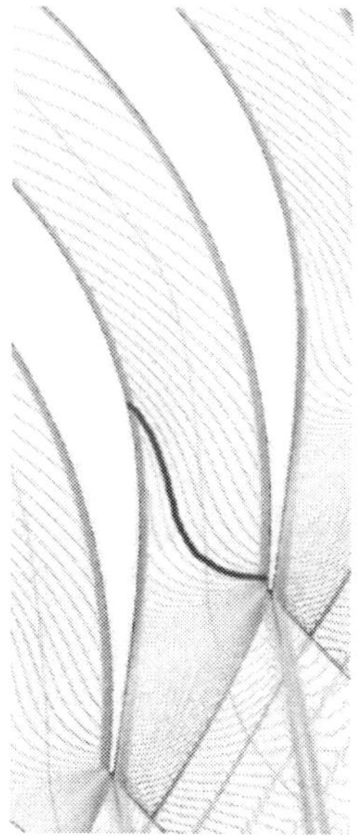

Fig. 7. The shape and position of sonic line in improved version of root section (*s/c = 0.2732*). CFD calculation, exit Mach number $M_{2is} \sim 1.7$, incidence angle $\iota = 0°$ [11].

These examples of sonic line in the root section well illustrate the extraordinary sensitivity of the sonic lane shaping in the inter-blade channels not only on profile geometry but also on small changes of effective channel contours, which are related mainly to boundary layer development. These root channels are generally characterized as relatively narrow, relatively long and mildly convergent (or convergent - divergent) channels. The experience with application of numerical methods shows the necessity to use proper models of turbulence (including transition models). The CFD simulations, which consider fully turbulent flow field, can lead to inaccurate results in terms of the sonic line and pressure distribution along the profile.

2.2 Middle section

Along the inter-blade channel in the radial direction, the pitch/chord ratio increases and contrarily, the profile chord usually somewhat decreases. These two facts together with changes of stagger angle (the long rotor blade are non-prismatic) lead to increasing of inter-blade channel width and concurrently to considerable reduction of its length. The aerodynamic throat moves upstream to the middle area of the profile cascade (see Figure 8). The maximum of the curvature of the camber line moves in the leading edge direction. The inlet Mach number increases along the blade height. The real values of the inlet Mach number ranges still in subsonic region. The example of common shape of sonic line in the mid-section is shown in the interferometric picture in the Figure 9. A typical distribution of isentropic Mach number near the profile in profile cascade representing the mid-section is depicted in the Figure 10.

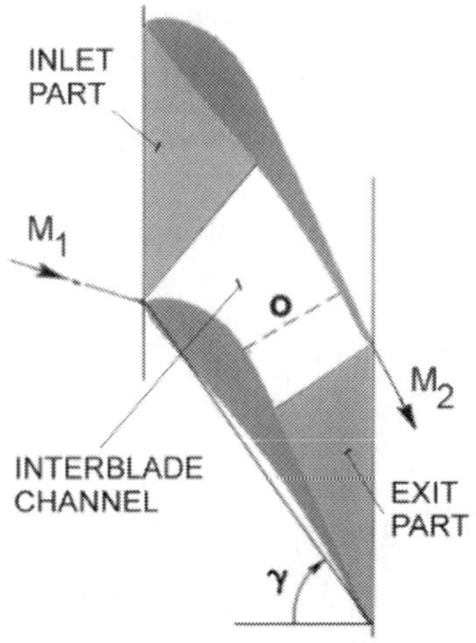

Fig. 8. Geometric configuration of typical middle section in the rotor wheel consisting of long blades.

Fig. 9. The sonic line in the inter-blade channel of the middle section. Interferometric measurement, exit Mach number $M_{2is} \sim 1.35$, incidence angle $\iota = 0°$.

Approximately in the middle of blade height the special damping device tie-boss is often situated. The influence of this body on the sonic line in the adjoining sections along the inter-blade channel is described in [12].

2.2.1 Recompression

The shaping of the profiles intended for application in the middle of the long rotor blade is often a reason for progressive development of recompression phenomenon in the supersonic part of the flow field next to the sonic line.

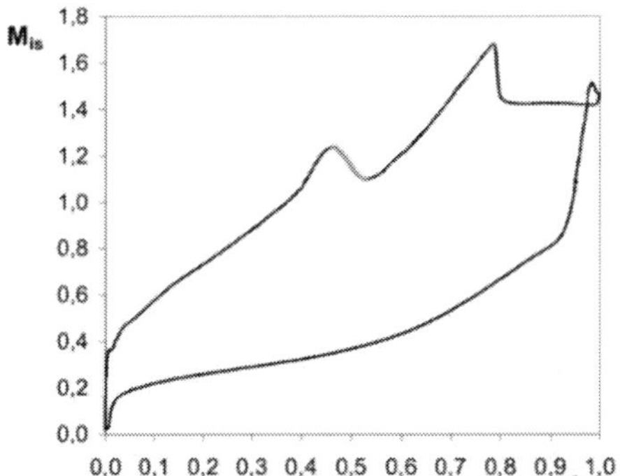

Fig. 10. The isentropic Mach number distribution along the blade of the middle section. Numerical simulation, exit Mach number $M_{2is} \sim 1.323$, incidence angle $\iota = 0°$, the recompression highlighted in green [13].

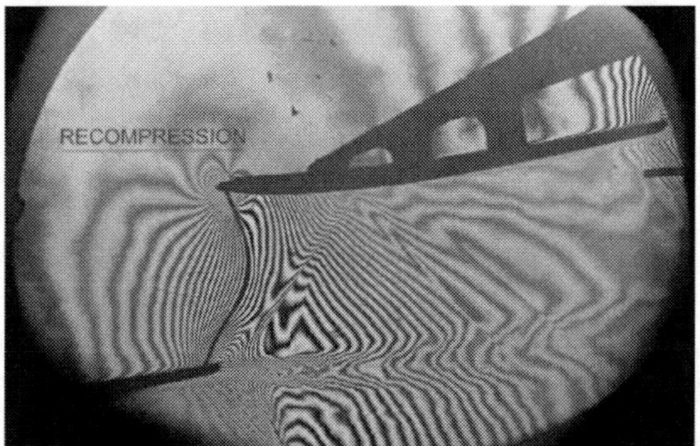

Fig. 11. The sonic line and recompression field in a tip section of the rotor wheel consisting of long blades. Interferometric measurement, exit isentropic Mach number $M_{2is} \sim 1.28$, inlet Mach number $M_1 \sim 0.790$ [14].

The recompression area often appears in supersonic flow field in different sections along the ultra long blade, even in the tip sections - (if the tip profile has usually bended

central line and the inlet Mach number is subsonic). The supersonic compression under transonic expansion (so called recompression) is a characteristic consequence of application of line segment in the shape of the suction side. This segment is attached up the upstream previous convex curve and is placed usually near the geometrical throat of the profile cascade. Although the sudden change of the shape curvature lies in the supersonic part of the flow field downstream the sonic line, the suitable geometrical configuration can guarantee the smooth increasing of static pressure, which is followed by ongoing supersonic expansion. The dissipative shock wave does not appear. On the other hand, the local increasing of the static pressure causes moderate decreasing of pressure force on the profile. The recompression appears also in other sections of wheels with long blades (example of the tip section with recompression area is shown in the Figure 11).
The sudden change of the suction side curvature just downstream the sonic line leads to the origin of reflected compression waves. The expansion waves, which originate on the convex part of the suction side, reflect from the sonic line as the compression waves.

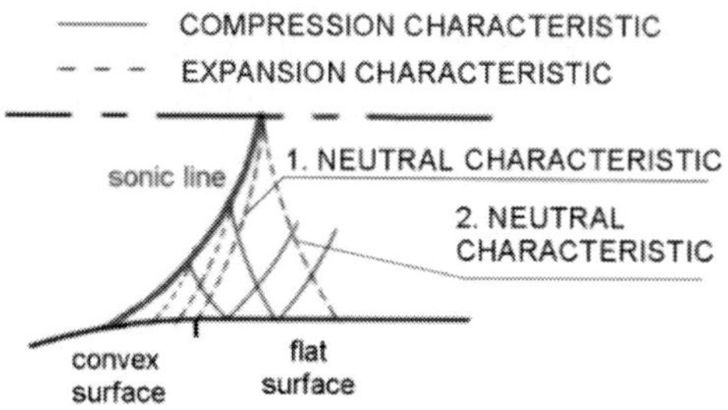

Fig. 12. Principle of the compression field formation in the supersonic flow downstream the sonic line.

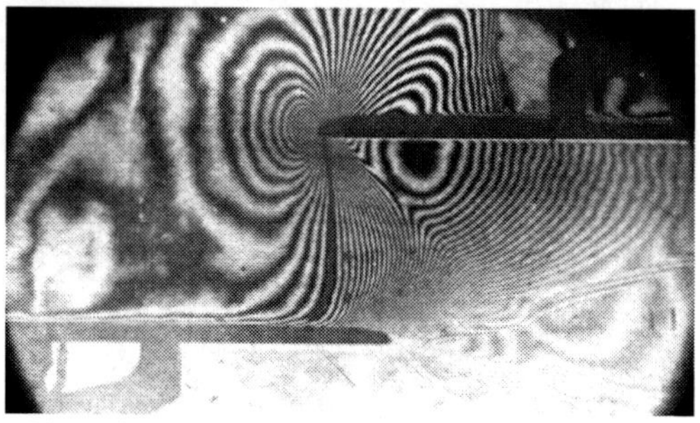

Fig. 13. The sonic line in the tip section of the rotor wheel consisting of blade *1080mm* long. Interferometric measurement, exit isentropic Mach number $M_{2is} = 1.83$, inlet Mach number $M_1 = 0.906$ [15].

These reflected compression characteristics are reflected again from the straight surface of the profile and keep the compression character. This way the region of supersonic compression arises between the point of discontinuous curvature of the suction side and the point of impact of the second neutral characteristic (see Figure 12). It is valid in general that the extent of recompression area depends on the shape of the sonic line.

2.3 Tip section

The pitch/chord ratio (*s/c*) reaches larger values in the tip sections, it often exceeds the value $s/c = 1$, what means that the inter-blade channel vanishes. The stagger angle γ reaches high values and often even exceeds the value $\gamma = 80°$. The inlet velocity in the tip region significantly grows. Not only is the velocity approaching the sonic value, it often exceeds it. If the inlet Mach number M_1 is lower than the value $M_1 = 1$, the sonic line links the adjacent profiles and is shifted in the inlet part of the inter-blade channel. It is rising next to the leading edge on the suction side and ends on the pressure side just upstream the trailing edge (see examples on the Figures 11 and 13). The shape of sonic line in the case of cascades consisting of flat profiles (similar to the profile in the Figure 13) depends on stagger angle, pitch/chord ratio and inlet flow angle. The selected results of experimental research are shown in the Figure 14.

The sets of profile cascades intended for even longer rotor blades usually work with supersonic inlet flow field. The process of the further sonic line formation is well known and has been described by Dvorak [6].

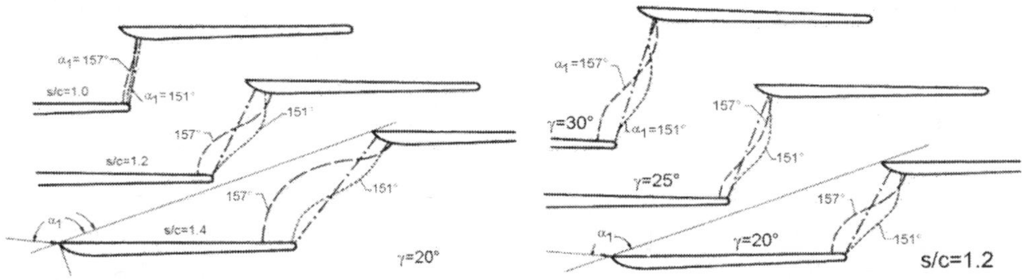

Fig. 14. The dependence of the sonic line shape in the tip section cascade on different values of inlet flow angle α_1 and different geometric parameters of the cascade (stagger angle γ, pitch/chord ratio s/c), subsonic inlet Mach number [7].

The topologic description of sonic line in the case of supersonic inlet is different for unstarted and started inlet supersonic flow.

The unstarted inlet supersonic flow field is characterized usually by relatively distant position of the inlet shock wave from the leading edge. The inner branch of the shock wave is ordinarily approximately normal to the pressure side of the adjacent profile in the beginning of unstarted supersonic flow regime. The sonic line is closed in a typical form around the leading edge. An example is shown in the interferometric picture in the Figure 15.

Fig. 15. The sonic line in a tip section with supersonic inlet. Unstarted supersonic inlet flow, exit isentropic Mach number $M_{2is} = 1.815$, inlet Mach number $M_1 = 1.089$ [16].

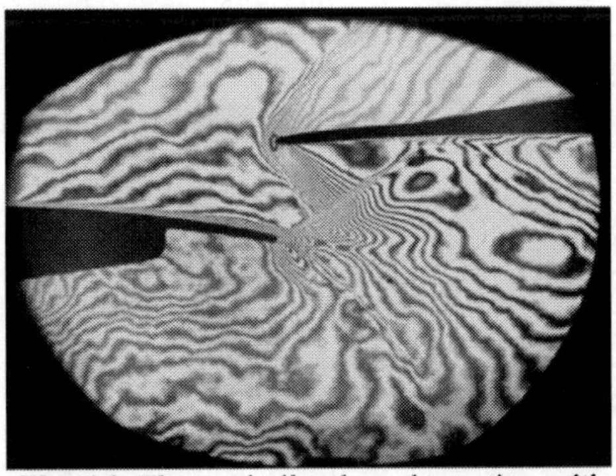

Fig. 16. The sonic line in a tip section with supersonic inlet. Started supersonic inlet flow, exit isentropic Mach number $M_{2is} = 1.800$, inlet Mach number $M_1 = 1.665$ [16].

The transition from unstarted to started flow is unsteady process in which even the bistability effects appear [6]. The unsteady behaviour of the flow field naturally has an impact on the sonic line. When the inlet Mach number is sufficiently high and the backpressure downstream the cascade is fitting low, the started supersonic flow field takes place in the cascade. Then the sonic line surrounds a small subsonic area near the leading edge (see Figure 16). This subsonic area bordered with sonic line is stable structure and does not disappear even if the supersonic inlet Mach number noticeably increases. The reason lies in the thick leading edge, which is typical for the tip section of rotor long blades intended for last stages of steam turbines.

3 Conclusions

The sonic surface is important aerodynamic threshold in the complex flow field, which occurs in the inter-blade channel of modern last stage rotor wheels blading consisting of long blades. The position and shape of sonic line depends not only on the cascade geometry and inlet and outlet boundary condition (inlet Mach number, inlet flow angle, backpressure) but often also on the effective geometric shape of the cascade profiles. The notion of sonic line as a boundary of constant static pressure is important for proper design of supersonic parts of the blade profiles with line segment in the transonic part of the suction side side of the profile.

The sonic line in the cascade with subsonic inlet Mach number spans across the inter-blade channel. If there is no overlapping of profiles and inter-blade channel is "missing" in the cascade geometry, the sonic line links the regions of leading and trailing edges of adjacent profiles.

The rapid unsteady transition from unstarted to started inlet supersonic flow field results in small stable subsonic area near the edge surrounded by sonic line (in cascades with thick profiles leading edges).

Acknowledgements of the author belong to the Technology Agency of the Czech Republic, which supported this research under grant No. TH02020057. An institutional support RVO 61388998 is also gratefully acknowledged.

References

1. M. Luxa, D. Simurda, J. Fort, et al., *Aerodynamic Investigation of the the Tip-Section for Titanium Blade 54''*. Proceedings of the 11th European Conference on Turbomachinery, Fluid Dynamics and Thermodynamics (Madrid, 2015)
2. P. Safarik, *Application of the Method of Characteristics in Calculation of the Flow in Blade Cascades* (in Czech). Semináře o problémech parních turbín **3**, ČVTS (Prague, 1976)
3. P. Safarik, *Flow in Supersonic Exit Part of the Blade Cascade* (in Czech). Research report IT CAS No.: **Z-409/73** (Prague, 1973)
4. A.H. Shapiro, *The Dynamics and Thermodynamics of Compressible Fluid Flow* (1953)
5. P.A. Thompson, *Transonic Flow in Curved Channels*. Transactions of ASME, J. of Basic Engineering **12** (1967)
6. R. Dvorak, *Transonic Flow* (in Czech), (Prague, 1987)
7. C. Sieverding, *Transonic Turbine Bladings*. Privat communications
8. D. Simurda, M. Luxa, P. Safarik, et al., *Aerodynamic Data for Two Variants of Root Turbine Blade Sections for a 54'' Turbine Rotor Blade*. Proceedings of the ASME Turbo Expo, Vol. **2C**, No: V02CT38A007 (Düsseldorf, 2014)
9. M. Luxa, D. Šimurda, *Optical Measurement on TR-P-3 Cascade* (in Czech). Research report IT CAS No.: **Z -1492/13** (Prague, 2013)
10. J. Hala, M. Luxa, D. Simurda, et al., *Optimization of Root Section for Ultra-Long Turbine Rotor Blade*. J. of Thermal Science, Vol. 27, No. 2 (2018)
11. M. Luxa, J. Hala, D. Simurda, *Optical Measurement on TR-P-4 Cascade*. Research report IT CAS No.: **Z-1558/16** (Prague, 2016)
12. T. Radnic, J. Furst, M. Luxa, et al., *Aerodynamic Effects of Tie-Boss in Extremely Long Turbine Blades*. J Eng Gas Turb Power. **GTP-17-1218** (2018)

13. M. Luxa, D. Simurda, P. Safarik, et al., *High-Speed Aerodynamics Investigation of the Midsection of 48'' Rotor Blade for the Last Stage of Steam Turbine.* Proceedings of the 10th European Conference on Turbomachinery - Fluid Dynamics and Thermodynamics, pp. 360-369, (Lappeenranta, 2013)
14. R. Dvorak, *The Aerodyanmic Properties of Profile Cascades (TR-B-1,2,3) Representing Tip Sections of Turbine Rotor Blading* (in Czech). Research report IT CAS No.: **Z-423/73** (Prague, 1973)
15. R. Dvorak, P. Safarik, *Aerodynamic Research of the Tip Section of Last Stage Rotor Blades for Steam Turbines od Large Output.* In: ASME Paper No.: **C185/79** (ASME, 1979)
16. M. Luxa, D. Simurda, *Optical Measurement on TR-U-6 Cascade.* Research report IT CAS No.: **Z-1535/15** (Prague, 2015)

Effect of inlet turbulence on compressible flow through the tip-section turbine blade cascade with supersonic inlet

Petr Straka[1,*], *Jaromír* Příhoda[2], *David* Fenderl[3], and *Bartoloměj* Rudas[3]

[1]VZLU Czech Aerospace Centre, Plc, Prague, Czech Republic
[2]Czech Academy of Sciences, Institute of Thermomechanics, Prague, Czech Republic
[3]Doosan Škoda Power Co., Ltd., Pilsen, Czech Republic

Abstract. The contribution deals with the numerical simulation of 2D compressible flow though the tip-section turbine blade cascade with the supersonic inlet boundary conditions. The simulation was carried out by the in-house numerical code using the explicit algebraic Reynolds stress model completed by the bypass transition model with the algebraic equation for the intermittency coefficient. The γ-*Re* model implemented in the commercial code Fluent was used for the comparison. Predictions carried out for the nominal conditions were focused on the effect of inlet free-stream turbulence on the flow structure in the blade cascade under supersonic inlet conditions. Numerical results were compared with experimental data.

1 Introduction

Flows in turbine blade cascades are influenced by many parameters which should be considered in numerical simulations. Such parameters are particularly the effect of free-stream turbulence and possibly wall roughness that influence noticeably the transition to turbulence and the turbulent heat transfer. Moreover, the transonic and supersonic flows though the turbine blade cascade are considerably influenced by the interaction of shock waves with the boundary layer on the blade surface that can cause the transition to turbulence in separated flow. The character of the interaction depends not only on the shock wave intensity but also in large measure on the type of the boundary layer. At the interaction of the shock wave with the laminar boundary layer, the flow separation usually comes out followed by the transition in separated flow, while the turbulent boundary layer remains attached. Therefore, simulations were focused mainly on the effect of the inlet free-stream turbulence on the flow structure in the blade cascade.

The contribution deals with the numerical simulation of compressible flow through the linear turbine blade cascade with blade profiles corresponding to the tip-section of the low-pressure part of the large steam turbine. Numerical simulations were carried out for the nominal regime partly by the in-house numerical code based on the EARSM turbulence

* Corresponding author: straka@vzlu.cz

model with the algebraic bypass transition model proposed by Straka and Příhoda [1] and partly by the γ-*Re* transition model connected with the SST turbulence model according to Langtry and Menter [2] implemented in the commercial code ANSYS Fluent.

2 Mathematical model

The mathematical model of compressible flow is based on conditionally-averaged Navier-Stokes equations completed by constitutional relations for the ideal gas, the explicit algebraic turbulence model (EARSM) by Hellsten [3], the algebraic transition model proposed by Straka and Příhoda [1] and by the model of turbulent heat transfer based on the generalized gradient hypothesis according to Launder [4]. The EARSM model is modified into the form corresponding to models with the turbulent viscosity. The turbulent stress is given by the relation

$$\tau_{ij} = \mu_t \left(2S_{ij} - \frac{2}{3}\delta_{ij}\frac{\partial \bar{U}_k}{\partial x_k} \right) - \frac{2}{3}\delta_{ij}\rho k - a_{ij}^{(ex)}\rho k \tag{1}$$

where the extra-anisotropy tensor $a_{ij}^{(ex)}$ is dependent on strain-rate and vorticity tensors and their invariants. Turbulent scales are determined by the k-ω/k-ε turbulence model of Menter [5]. Turbulent time scale $\tau \sim 1/\omega$ is replaced near the wall by the Kolmogorov viscous time scale. The production term in the turbulent energy equation is modified for the reduction of the over-production of the turbulent energy in the stagnation region. Production and destruction terms in the k-equation are multiplied by the intermittency coefficient γ. Similarly, the effective viscosity is given by $\mu_{ef} = \mu + \gamma\mu_t$ in the transition region. The EARSM model including the prescription of boundary conditions is described by Hellsten [3]. Some modifications of the model are given by Straka and Příhoda [6].

The generalized gradient hypothesis is used for modelling of the turbulent heat transfer in the form

$$q_i = -C_t \frac{\tau_{ij}}{\omega}\frac{\partial \bar{T}}{\partial x_j} \qquad C_t = 0.3 \tag{2}$$

The transition model is based on the concept of different values of the intermittency coefficient in the boundary layer γ_i and in the free stream γ_e. The intermittency coefficient in the boundary layer γ_i is expressed by the relation

$$\gamma_i = 1 - \exp\left[-\hat{n}\sigma \left(Re_x - Re_{xt} \right)^2 \right] \tag{3}$$

The transition onset is given by the empirical correlation for the momentum Reynolds number

$$Re_{\theta t} = Re_{\theta to}\left[1 + \mathrm{F}(Tu)\frac{1-\exp(-40\lambda)}{1+0.4\exp(-40\lambda)} \right] \tag{4}$$

where the transition onset on the flat plate without the pressure gradient is given by the relations

$$Re_{\theta to} = 975.8 - 497.2Tu + \frac{11.4}{Tu} \qquad \text{for} \quad Tu \le 1\% \tag{5}$$

$$Re_{\theta to} = 96.7 + \frac{340}{Tu} + \frac{53.3}{Tu^2} \qquad \text{for} \quad Tu > 1\% \tag{6}$$

depending on the free-stream turbulence level *Tu* (%) and the pressure-gradient parameter λ_t. The length of the transition region is expressed using the parameter

$$N = \hat{n}\sigma\, Re_{\theta t}^3 \tag{7}$$

combining the spot generation rate $\hat{n}$ and the spot propagation rate σ introduced by Narasimha [7]. The parameter N for the attached flow is given by the empirical relation $N = \mathrm{f}(Tu, \lambda_t)$ proposed by Solomon et al. [8]. The onset of transition in separated flow is given by the correlation according to Mayle [9] in the form

$$Re_{xt} = 300\,Re_{\theta s}^{0.7} + Re_{xs} \tag{8}$$

where $Re_{\theta s}$ is the momentum Reynolds number at the separation and Re_{xs} is the Reynolds number related to the distance of the separation from the leading edge. The transition length of the short separation bubble is given according to Walker [10] by the relation

$$\hat{n}\sigma = \frac{Re_{xt}^{-1.34}}{40} \tag{9}$$

and so the same approach can be applied as in the attached flow.

The application of local variables is necessary for prediction of transitional flows in complex geometries. Therefore, the momentum Reynolds number is replaced by the maximum of the vorticity Reynolds number according to Langtry and Menter [3]. The vorticity Reynolds number is given by the relation

$$Re_{\Omega} = y^2 |\Omega| / \nu \tag{10}$$

where y is the distance from the wall and Ω is the absolute value of the vorticity tensor. The link between both Reynolds numbers is expressed by the relation $Re_\theta = Re_{\Omega max}/C$ where the parameter C depends on the pressure gradient expressed by the parameter

$$L = Re_{\Omega}^2 \frac{\nu}{U_e^2} \frac{\mathrm{d}U_e}{\mathrm{d}x} \tag{11}$$

The variation of the parameter C with the pressure gradient parameter L was estimated by means of similar solutions of Falkner-Skan velocity profiles.

The system of governing equations was implemented into the in-house numerical code based on the finite volume method of the cell-centered type. The governing equations are discretized using a multi-block quadrilateral structured grid with a block overlapping implementation. The algebraic transition model is described in detail by Fürst et al. [11].

The correlation-based γ-Re_θ model of the bypass transition proposed by Langtry and Menter [3] with the two-equation SST turbulence model implemented into the commercial code ANSYS Fluent was used for comparison. The transition model is based on transport equations for the intermittency coefficient and the transitional momentum Reynolds number. The intermittency coefficient γ is given by the transport equation

$$\frac{\partial(\rho\gamma)}{\partial t} + \frac{\partial(\rho U_j \gamma)}{\partial x_j} = P_\gamma - E_\gamma + \frac{\partial}{\partial x_j}\left[\left(\mu + \frac{\mu_t}{\sigma_\gamma}\right)\frac{\partial \gamma}{\partial x_j}\right] \tag{12}$$

where

$$P_\gamma = F_{length} C_{a1} \rho S (\gamma F_{onset})^{1/2} (1 - C_{e1}\gamma) \tag{13}$$

is the production term with empirical constants F_{onset} for the transition onset and F_{length} for the transition length. The application of local variables is allowed by the relation between the Reynolds number related to the momentum thickness and the maximum of the vorticity Reynolds number used in the form $Re_\theta = Re_{vmax}/2.193$ valid for the Blasius boundary layer. The empirical correlation for the transition onset F_{onset} depends on the critical Reynolds number $Re_{\theta c}$ giving the position where the turbulent energy begins to grow up, see Langtry [12]. This Reynolds number $Re_{\theta c}$ is expressed by means of the local Reynolds number $\tilde{Re}_{\theta t}$ for the transition onset determined from the transport equation

$$\frac{\partial\left(\rho\tilde{R}e_{\theta t}\right)}{\partial t}+\frac{\partial\left(\rho U_j\tilde{R}e_{\theta t}\right)}{\partial x_j}=P_{\theta t}+\frac{\partial}{\partial x_j}\left[\sigma_{\theta t}\left(\mu+\mu_t\right)\frac{\partial\tilde{R}e_{\theta t}}{\partial x_j}\right] \tag{14}$$

The onset and length of the transition region, i.e. parameters $Re_{\theta c}$ and L_{length} are expressed by means of the local Reynolds number $\tilde{R}e_{\theta t}$. The model is switched over to a simple algebraic transition model in the case of the transition in separated flow. The onset and the length of the transition region are controlled by threshold functions based on empirical correlations. The transition model is based on local variables only and can be used for the simulation of complex flows using unstructured computational grids. The γ-Re_θ model with all necessary empirical correlations was published by Langtry and Menter [3].

3 Results

Numerical simulations of two-dimensional compressible flow through the linear tip-section turbine blade cascade TR-U-6 were carried out for practically nominal conditions, i.e. for the inlet Mach number M_1 = 1.68 and the isentropic outlet Mach number M_{2isSC} = 2.05 corresponding to the pressure in the settling chamber. The relevant isentropic outlet Mach number in the traversing plane is M_{2is} = 1.97 and the isentropic outlet Reynolds number Re_{2is} = 1.98×10^6. Numerical results were compared with results of optical and pressure measurements, see Luxa et al. [13, 14].

The scheme of the computational domain for simulations by the EARSM turbulence model with the algebraic transition model is shown in Figure 1. The detail of the computational mesh of the "chimera" type near the leading and the trailing edge is given in Figure 2. The mesh consists of the O-type block around the blade profile and the H-type block which covers one pitch of blade cascade. The computational grid is refined near the wall with the first grid node in the distance from the wall about y^+ = 1.

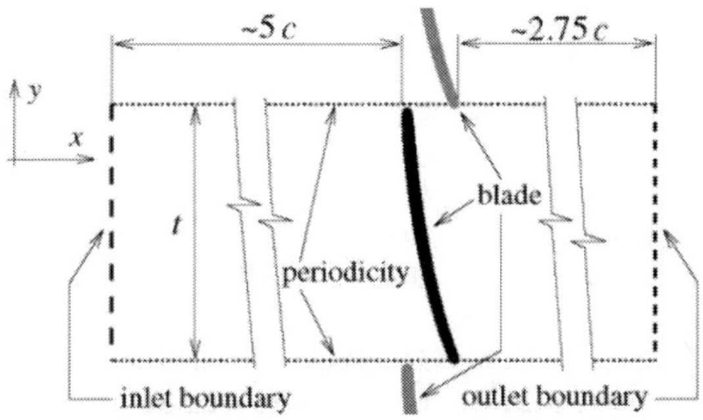

Fig. 1. Scheme of the computational domain.

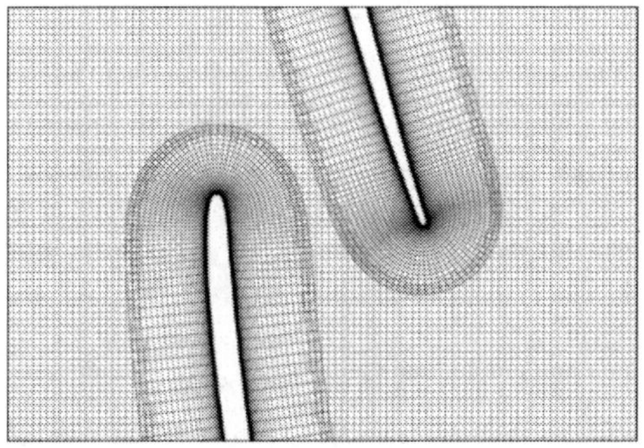
Fig. 2. Detail of computational mesh.

Because of supersonic inlet conditions the flow through the blade cascade is influenced by parasitic shock waves arising by the reflection from boundaries of the domain. The application of the quadrilateral block-structured computational grids with their overlapping leads moreover to the reflection of shock waves on the block boundaries. Therefore the domain was extended in x-direction upstream and downstream to suppress these reflections. The constant values of the total pressure p_{01}, total temperature T_{01} and incidence angle α_1 were prescribed as inlet conditions. The static pressure was extrapolated from the computational domain. The static pressure given by the outlet isentropic Mach number was prescribed as the outlet condition. Periodicity conditions were applied on corresponding free side boundaries of the computation domain. The inlet turbulence parameters, i.e. turbulence level Tu and the ratio of turbulent and molecular viscosity μ_t/μ are chosen due to the turbulence decay according to the length of the inlet part. The relevant values of the turbulence level in the distance about one spacing upstream of the plane of the leading edges were Tu = 1, 1.5 and 2.5 % where the value Tu = 1.5 % corresponds to experimental data.

The computational domain used in calculations by means of the commercial code ANSYS Fluent is shown in Figure 3. The domain was extended before and behind the blade cascade to 3*c*. The mesh was formed by quadrilateral elements refined near walls and by triangular elements in the extended parts of the domain to suppress the effect of shock waves reflections. The detail of the grid near the blade is shown in Figure 4. The same inlet conditions were used as for the numerical simulation by means of the in-house code.

Due to supersonic inlet the prescription of inlet boundary conditions is rather complicated. The flow angle is prescribed at the inlet boundary in numerical simulations. The relation between values of the inlet Mach number and the inlet angle is given by the unique incidence rule (see e.g. Luxa et al. [15]). The relevant inlet conditions were established in the distance 50 mm upstream of the leading edge plane.

The data reduction method proposed by Amecke and Šafařík [16] was used for the evaluation of experimental data and results of CFD simulations as well. Due to the stagger angle of blades and to the inlet boundary conditions the concave side of the blade is the suction side and the convex side is the pressure side.

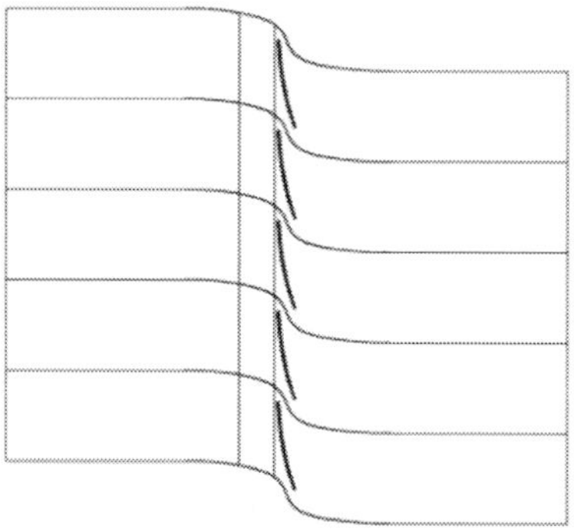

Fig. 3. Scheme of the computational domain (Fluent).

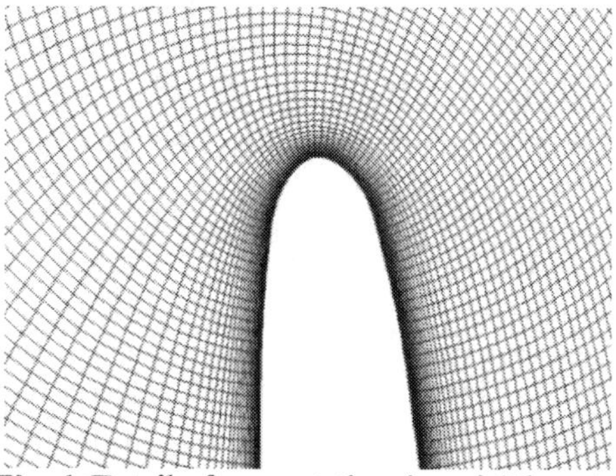

Fig. 4. Detail of computational mesh(Fluent).

Simulations carried out for the nominal regime were focused mainly on the effect of the inlet free-stream turbulence on the flow field in the blade cascade especially on the interaction of the impinging shock wave with the boundary layer on the suction side. Simulations are to-follow up to predictions carried out by Straka et al. [17] for different flow conditions where the effect of the inlet angle and the free-stream turbulence was studied.

The flow field in the tip-section blade cascade is influenced partly by the interaction of internal branch of exit shock wave with boundary layer on the suction side of neighbouring blade and partly by the interaction of internal branch of inlet shock wave upstream the blade leading edge with boundary layer on pressure side of previous profile. At higher inlet Mach number this inlet shock wave is to that extent curved that it pasts the suction side of the previous blade and propagates downstream, see Luxa et al. [18].

The interferometric and schlieren pictures obtained for nominal conditions, i.e. for the inlet Mach number M_1 = 1.68 and the isentropic outlet Mach number for M_{2is} = 1.97 are shown in Figure 5 and 6. Due to a relative high inlet Mach number M_1 only the interaction of internal branch of exit shock wave with boundary layer on the suction side of neighbouring blade occurs. This interaction with the laminar boundary layer leads to flow separation.

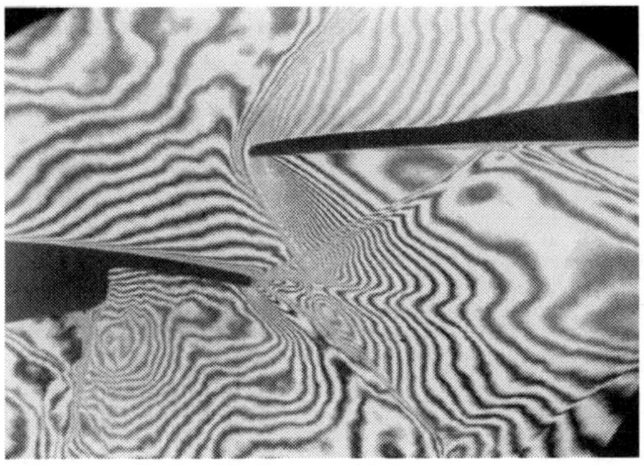

Fig. 5. Interferometric picture for M_{2is} = 1.97.

Fig. 6. Schlieren picture for M_{2is} = 1.97.

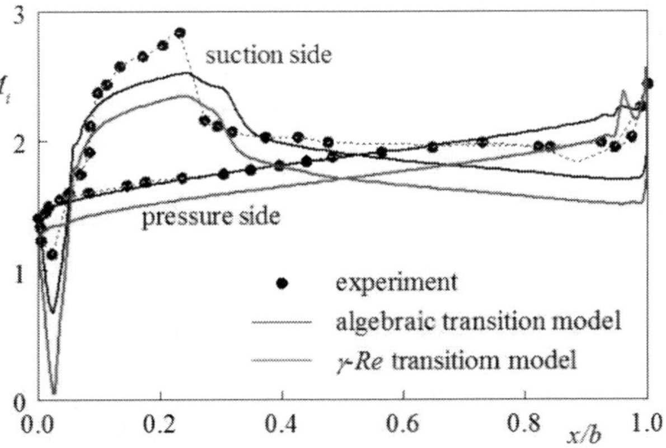

Fig. 7. Distribution of the isentropic Mach number.

The distribution of the isentropic Mach number along the blade for $Tu \approx 1.5$ % is compared with experimental data in Figure 7. The distance b represents the axial chord length. The flow field in the interblade channel is supersonic with a small subsonic region around the leading edge. The EARSM model with the algebraic transition model quite well corresponds with experi-mental data except the part upstream of the interaction. The SST turbulence model with γ-Re transition model gives generally rather lower values M_i with a similar form as the algebraic model. It should be caused by the used isentropic outlet Mach number M_{2is} = 1.94 instead of the nominal value. Only small changes in the distribution of the isentropic Mach number occur with changes of turbulence on the suction side in the sudden decrease of the velocity. The Mach number isolines at M_{2is} = 1.97 obtained by the algebraic transition model are shown in Figure 8.

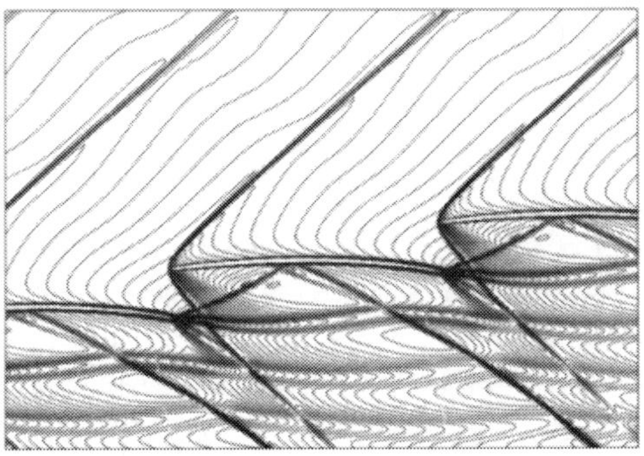

Fig. 8. Mach number isolines in the blade cascade.

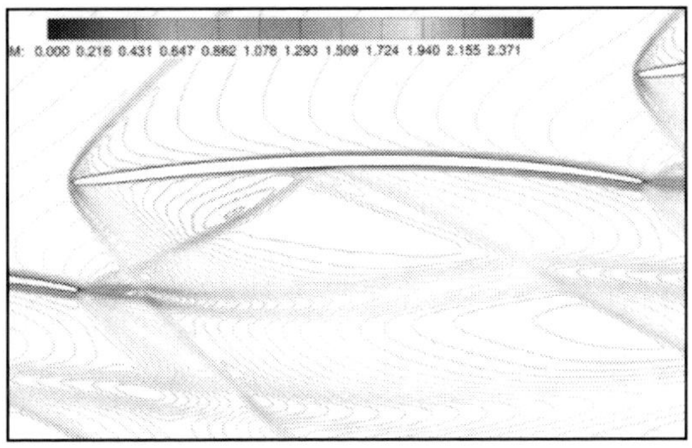

Fig. 9. Detail of Mach number isolines.

The detail of Mach number isolines showing the interaction of the inner branch of the exit shock wave with the boundary layer on the suction side of the blade is presented in Figure 9. The field of Mach isolines well corresponds with the interferometric picture. The impact of the shock wave and the interaction with the boundary layer induces flow separation connected with the transition to turbulence. The similar flow field structure was obtained as well by means of the γ-Re transition model as can be seen in Figure 10. The flow structure in the blade cascade apparent in the field of the Mach isolines can be well documented by means of the distribution of the skin friction coefficient.

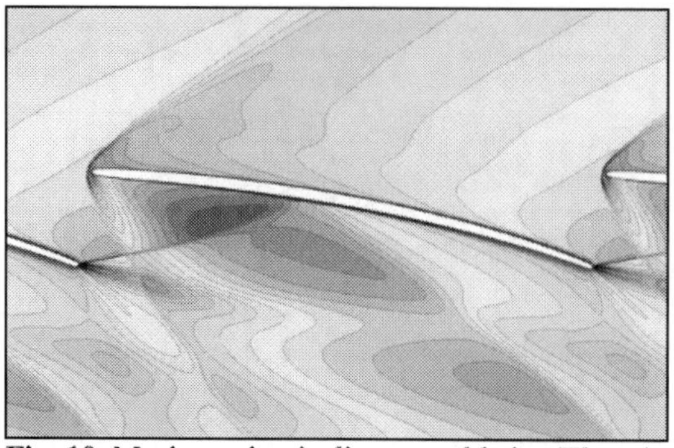

Fig. 10. Mach number isolines near blades (Fluent).

The skin friction coefficient obtained by the algebraic transition model for various free-stream turbulence levels is shown in Figure 11. The skin friction coefficient is defined by the relation $C_f = \tau_w/p_{d1}$ where p_{d1} is the inlet dynamic pressure. The extent of the separation zone estimated from experimental data is marked by the rectangle and is approximately from $x/b \approx 0.28$ to 0.40. The transition onset on the pressure side moves upstream with the increasing free-stream turbulence. A quite different situation is on the suction side where the transition is caused by the shock-wave interaction with the laminar boundary layer. This interaction has as a consequence the flow separation and the transition in separated flow. The extent of the separation zone becomes shorter with the increasing turbulence level.

The skin friction distribution obtained by both transition models for the free-stream turbulence about $Tu \approx 1.5$ % are compared in Figure 12. Results for the suction side are very similar including the modelling of the separation zone. Nevertheless, the γ-Re model demonstrates some oscillations of the skin friction near the trailing edge. On the pressure side, the algebraic transition model predicts the transition onset near the leading edge at $x_t/b \approx 0.03$ while γ-Re model gives the transition onset about at $x_t/b \approx 0.18$ with a rather long transition region.

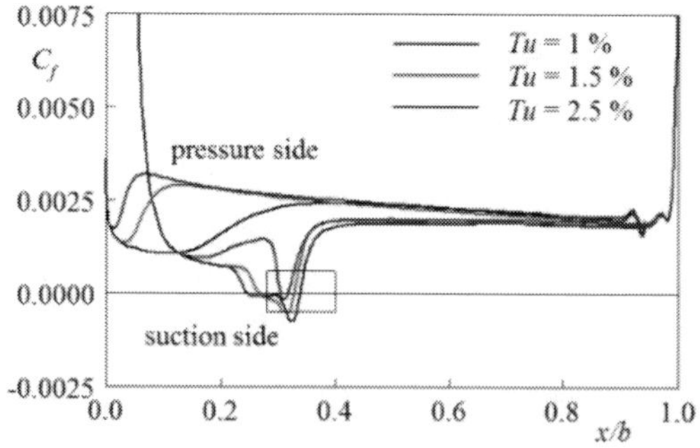

Fig. 11. Skin friction distribution distribution (algebraic model).

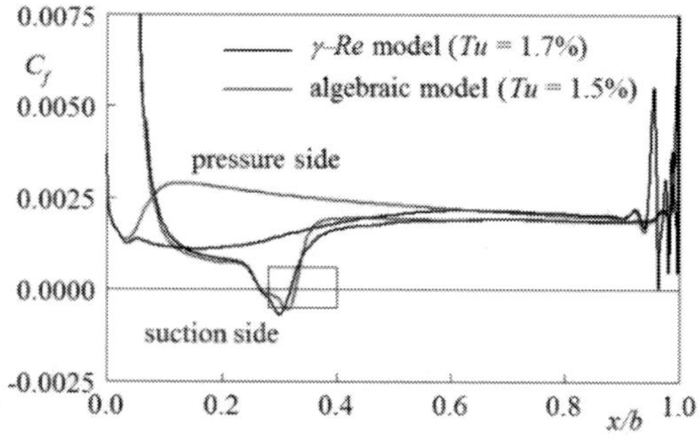

Fig. 12. Comparison of the skin friction for both transition models.

The field of the turbulent energy around the blade obtained by the EARSM turbulence model with the algebraic transition model is shown for various turbulence levels in Figures 13 a) and b). The transition in the separated flow for $Tu = 1.5$ % results in the greater growth of the turbulent energy on the suction side towards to the trailing edge while the transition zone for $Tu = 2.5$ % includes a very small separation and so the growth of the turbulent energy is rather smaller. On the contrary, the transition onset on the pressure side moves upstream with the increasing turbulence and so the length of turbulent boundary layer becomes longer which leads a higher turbulent energy.

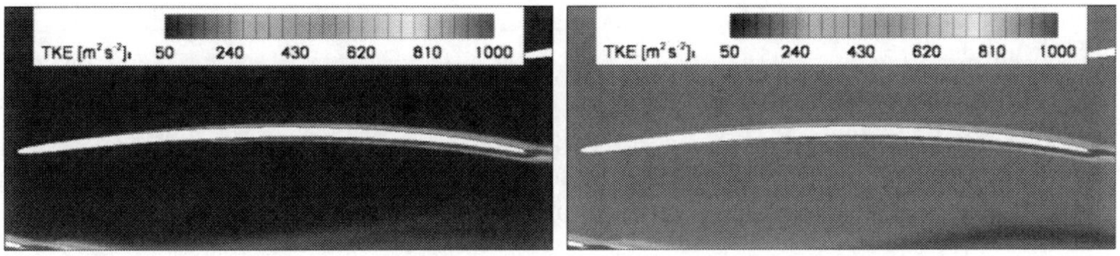

a) $Tu = 1.5$ % b) $Tu = 2.5$ %

Fig. 13. Field of the turbulent energy around the blade (algebraic transition model).

4 Conclusion

The explicit algebraic model of Reynolds stresses with the algebraic transition model based on local variables was used for the numerical simulation of compressible flow through the tip-section turbine blade cascade TR-U-6 with supersonic inlet conditions. The calculations by the γ-*Re* transition model with the SST turbulence model implemented in the commercial software ANSYS Fluent was carried for the comparison. Simulations were focused especially on the effect of the inlet free-stream turbulence on the interaction of the exit shock wave with the boundary layer on the blade suction side. Obtained results indicate the significant effect of the transition modelling on the adequate description of the compressible flow through the blade cascade. The application of the advanced turbulence models with the adequate transition model for attached and separated flow is necessary.

The work was supported by the Technology Agency of the Czech Republic under the grant TA02020057 and by the Long-term Framework Advancement Plan provided by the Ministry of Industry and Trade of the Czech Republic. Institutional support RVO 61388998 is also gratefully acknowledged.

References

1. P. Straka, J. Příhoda, Proc. Conf. Experimental Fluid Mechanics, Liberec, 636-641 (2010)
2. R.B. Langtry, F.R. Menter, AIAA Jour., **47**, 2894-2906 (2009)
3. A. Hellsten, *New two-equation turbulence model for aerodynamics applications*, PhD. thesis, Helsinki University of Technology (2004)
4. B.E. Launder, Jour. of Heat Transfer, **110**, 1112-1128 (1988)
5. F.R. Menter, AIAA Jour., **32**, 1598-1605 (1994)
6. P. Straka, J. Příhoda, Proc. Conf. Topical Problems of Fluid Mechanics, Praha, 131-134 (2014)
7. R. Narasimha, Jour. Aerospace Science, **24**, 711-712 (1957)
8. W.J. Solomon, G.J. Walker, J.P. Gostelow, Jour. Turbomachinery, **118**, 744-751 (1996)
9. R.E. Mayle, Jour. Turbomachinery, **113**, 509-537 (1991)
10. G.J. Walker, AIAA Jour., **27**, 595-607 (1989)
11. J. Fürst, J. Příhoda, P. Straka, Computing, **95**, S163-S182 (2013)
12. R.B. Langtry, *A correlation-based transition model using local variables for unstructured parallelized CFD codes,* PhD. thesis, Univ. Stuttgart (2006)

13. M. Luxa, D. Šimurda, J. Fořt, J Fürst, P. Šafařík, J. Synáč, B. Rudas, Proc. 11th European Conf. on Turbo-machinery, Fluid Dynamics and Thermodynamics (ETC11), Madrid (2015)
14. M. Luxa, J. Příhoda, D. Šimurda, P. Straka, J. Synáč, Jour. Thermal Science, **25**, 138-144 (2016)
15. M. Luxa, B. Rudas, J. Synáč, P. Šafařík, D. Šimurda, Proc. Conf. Application of Experimental and Numerical Methods in Fluid Mechanics and Energy 2014, 149-152, Žilina (2014)
16. J. Amecke, P. Šafařík, *Date reduction of wake flow measurements with injection of an other gas*, DLR-Forschungsbericht 95-32, Göttingen (1995)
17. P. Straka, J. Příhoda, M. Bobšík, Proc. Conf. Topical Problems of Fluid Mechanics, Praha, 207-2014 (2016)
18. M. Luxa, D. Šimurda, J. Příhoda, J. Váchová, VII European Congress on Computational Methods in Applied Sciences and Engineering (ECCOMAS), Crete, 10 p. (2016)

Numerical modelling of transonic centripetal flow in radial turbine blade cascade

Petr Straka[1,*]

[1]VZLU – Czech Aerospace Research Centre, Beranových 130, 19905 Prague 9, Czech Republic

Abstract. Numerical modelling of transonic centripetal turbulent flow in radial blade cascade is described in this paper. Method of the confusor buffer zone is applied to overcome some numerical obstacles related with specifical properties of the outlet confusor. Kinetic energy loss coeficient of the radial blade cascade is compared with its linear representation and with experimental data.

1 Introduction

Two dimensional flow field in a radial blade cascade is (in case of centripetal flow) limited with specific properties of the outlet confusor. Maximal achievable Mach number on radius of the trailing edges $M_{TE,max}$ is limited by the outlet confusor geometry (specifically by the outlet radius r_{out}) and by the flow angle on the radius of the trailing edges α_{TE} (which is given by the blade cascade geometry). Relation between $M_{TE,max}$ and r_{out} is significant namely for high values of the outlet Mach number. In numerical simulation the outlet confusor is that part of the computational domain, which is bounded with the trailing edges radius r_{TE} and the outlet radius r_{out}. In case of supersonic outlet achieving of higher and higher outlet Mach number is possible only by modification of the outlet confusor geometry – by increasing the outlet radius r_{out}. By other words it means placing the outlet boundary of the computational domain closer and closer to the trailing edges, which may lead to some numerical obstacles related with unphysical shock waves reflections from the outlet boundary.

This paper deals with numerical simulation of transonic centripetal flow through radial blade cascade in range of the outlet Mach number on the trailing edge radius $1.025 \leq M_{TE} \leq 1.88$ and for the chord based Reynolds number $Re = 8.5\times10^5$. Geometry of the radial blade cascade is derived from the linear blade cascade via conformal mapping as shown in Figure 1. The original linear blade cascade of short axial length was developed for transonic regimes in range of the outlet isentropic Mach number $1.6 \leq M_{out,is} \leq 2.1$ (see [1]).

As mentioned above, achieving high supersonic values of the outlet Mach number in case of centripetal flow is possible only by increasing the outlet radius r_{out} of the outlet confusor. Beside way of placing the outlet boundary closer and closer to the trailing edges, there is another method called *confusor buffer zone* proposed in [2]. Method of the confusor buffer zone respects relation between $M_{TE,max}$ and the outlet confusor geometry and allows

* Corresponding author: straka@vzlu.cz

to overcome numerical obstacles related with unphysical shock waves reflections. Method of the confusor buffer zone used in this work is a slight modification of that which is proposed in [2].

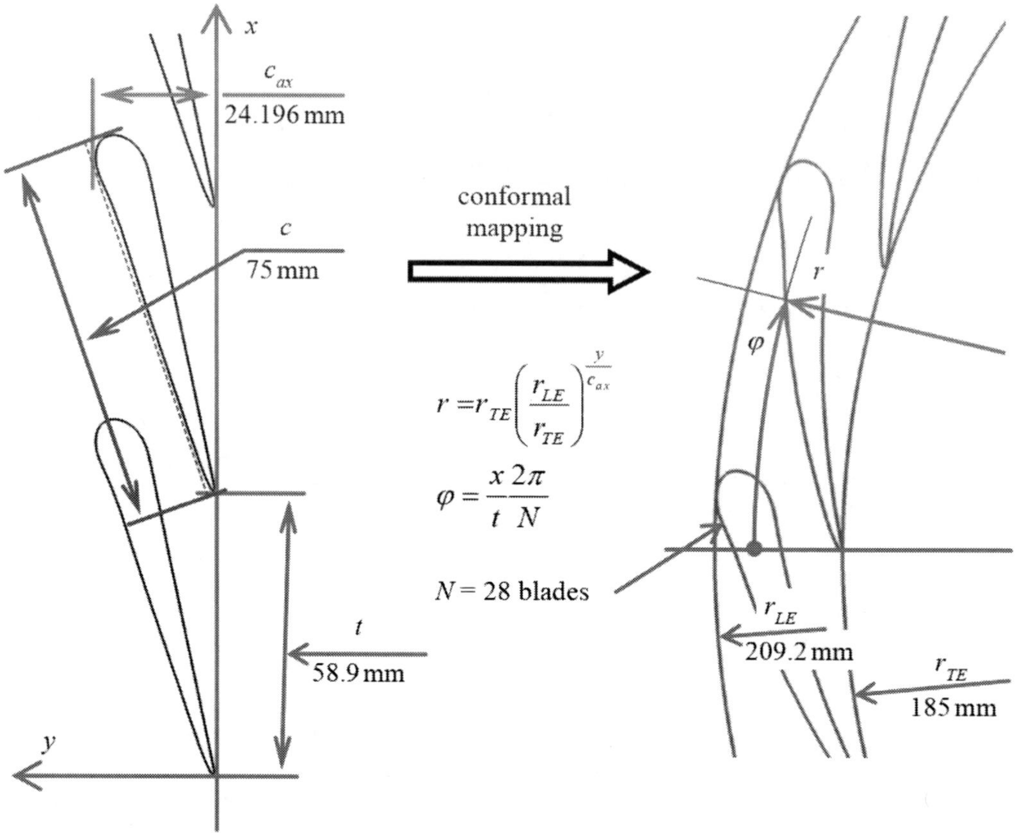

Fig. 1. Scheme of the conformal mapping of the linear (left) to the radial (right) blade cascade.

2 Outlet confusor

As shown in Figure 2, the outlet confusor is bounded by inlet radius r_{TE} and outlet radius r_{out}. The confusor inlet radius r_{TE} corresponds with the radius over the trailing edges, the outlet radius corresponds with the outlet boundary of the computational domain in numerical simulation.

The flow through the radial confusor was solved in [3, 4] theoretically using model of the potential sink-vortex flow. This model is described by system (1) - (2) which is derived from conservation laws of mass, momentum (vortex conservation condition) and energy:

$$\frac{\partial u_t}{\partial r} = -\frac{u_t}{r}, \tag{1}$$

$$\frac{\partial u_r}{\partial r} = -\frac{u_r}{r}\frac{1+M_t^2}{1-M_r^2}. \tag{2}$$

In system (1) - (2) u_t and u_r are tangential and radial velocity vector components, M_t and M_r are tangential and radial Mach number components. From (1) it is evidence that in case

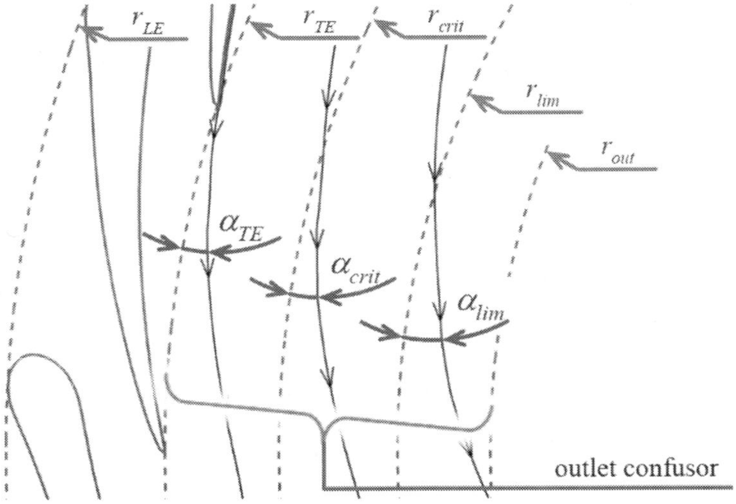

Fig. 2. Scheme of the outlet confusor.

of centripetal flow the tangential velocity component increase towards to the outlet radius r_{out}. Solution of (2) has two branches: for $M_r < 1$ and for $M_r > 1$. We will deal with first branch ($M_r < 1$) only in this work. From (2) it is clear that in case of centripetal flow and $M_r < 1$ also the radial velocity component must increase towards the outlet radius, where limit velocity is given by condition $M_r \to 1$. More details can be found in [3, 4].

An important parameter for solution of flow in radial confusor is the flow angle α_{crit} at the critical radius (see F 2). Parameter α_{crit} is characteristic of the cascade exit geometry and is related to the confusor inlet parameters by relation

$$\tan\alpha_{crit} = \left[1 + \frac{\kappa - 1}{2}(1 - M_{*TE}^2)\right]^{\frac{1}{\kappa - 1}} \tan\alpha_{TE} \,. \tag{3}$$

Note, that relation between the Mach number M and the normalized velocity M_* is given as

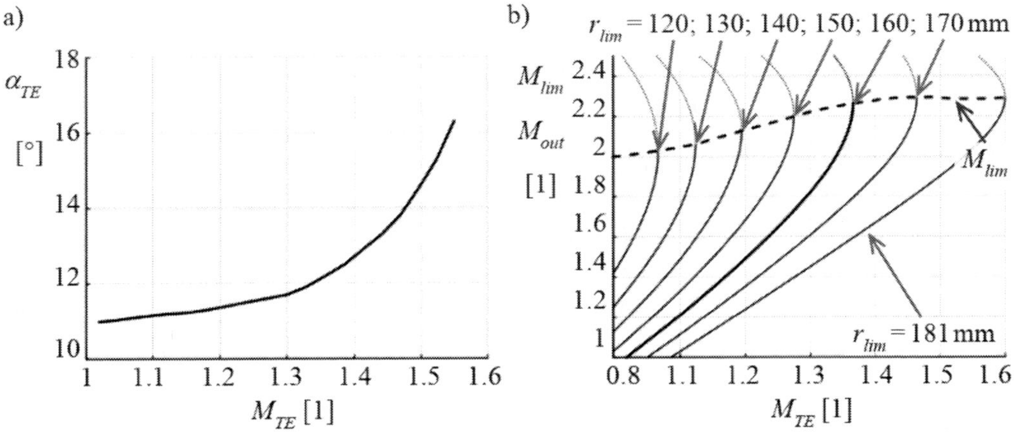

Fig. 3. a) dependency of the flow angle on the Mach number on the trailing edge radius; b) relation between Mach number on the trailing edge radius and the outlet Mach number and the limit Mach number in dependence on the limit radius.

$$M_*^2 = \left(\frac{\kappa+1}{2}M^2\right)\Big/\left(1+\frac{\kappa-1}{2}M^2\right). \tag{4}$$

Figure 3a shows dependency of the outlet flow angle on the outlet Mach number for the linear blade cascade. With regard to properties of the conformal mapping, this data can be used in relation (3) for evaluation of parameter α_{crit}.

As mentioned above the condition $M_r \rightarrow 1$ defines the limit velocity which can be solved from following system:

$$M_{*lim}^2 = 1 - \frac{2}{\kappa-1}\left[\left(\frac{\tan\alpha_{crit}}{\tan\alpha_{lim}}\right)^{\kappa-1} - 1\right], \tag{5}$$

$$\tan\alpha_{lim} = \sqrt{\frac{(\kappa+1)-(\kappa-1)M_{*lim}^2}{(\kappa+1)(M_{*lim}^2-1)}}. \tag{6}$$

Finally the model of the potential sink-vortex flow defines relation for the limit confusor radius r_{lim}:

$$r_{lim} = r_{TE}\Big/\sqrt{\frac{M_{*lim}^2\, g(M_{*lim})[\tan^2\alpha_{crit} + g(M_{*TE})]}{M_{*TE}^2\, g(M_{*TE})[\tan^2\alpha_{crit} + g(M_{*lim})]}}, \tag{7}$$

where $g(M_*)$ is given as

$$g(M_*) = \left[1 + \frac{\kappa-1}{2}(1-M_*^2)\right]^{\frac{2}{\kappa-1}}. \tag{8}$$

Relation between the limit Mach number M_{lim}, the limit confusor radius r_{lim} and the Mach number M_{TE} for presented radial blade cascade is shown in Figure 3b.

For given outlet radius of the computational domain r_{out} the Mach number M_{TE} at the

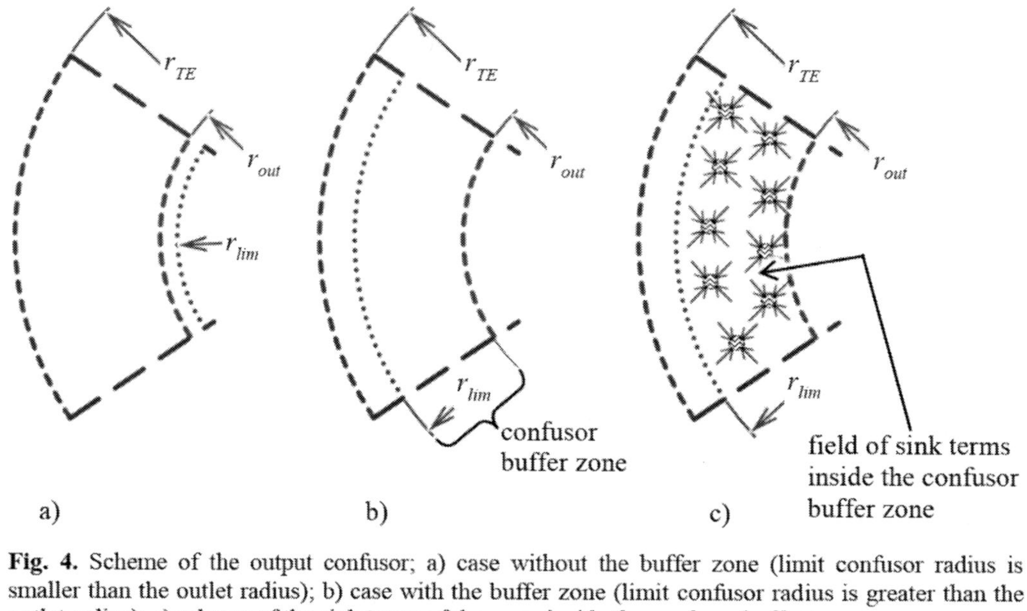

Fig. 4. Scheme of the output confusor; a) case without the buffer zone (limit confusor radius is smaller than the outlet radius); b) case with the buffer zone (limit confusor radius is greater than the outlet radius); c) scheme of the sink terms of the mass inside the confusor buffer zone

radius r_{TE} can be increased by decreasing of the pressure at the outlet boundary up to limit state given by (5) and (6). The limit state occurs at the outlet radius r_{out}, it means that $r_{out} = r_{lim}$. Any further decreasing of the pressure at the outlet boundary (outlet radius) has no effect on flow in radial confusor and consequently in radial blade cascade. Next increasing of the Mach number M_{TE} at radius r_{TE} is possible only through increasing of the limit radius r_{lim}. It can be done in two ways. One way is to keep the limit state at the outlet boundary ($r_{out} = r_{lim}$) and increase the outlet radius. This means modifying the computational domain and the computational mesh for each regime. Another way to increase the limit radius is to construct a confusor buffer zone that is bounded by the limit radius and the outlet radius (see middle part of Figure 4). The role of confusor buffer zone is to overcome the limit condition (singularity) given by relation (2) and to avoid above mentioned numerical obstacles related with high supersonic outlet Mach numbers. Construction of such buffer zone using the sink term of the mass will be discussed in following part. The advantage of this approach is that there is no need to modify the computational domain and the computational mesh, because the outlet radius can be fixed for all regimes.

3 Mathematical models

Flow in radial blade cascade is modeled as compressible viscous turbulent flow of the perfect gas. Model is described by system of Favre averaged Navier-Stokes equation:

$$\frac{\partial \rho}{\partial t} + \nabla(\rho \mathbf{U}) = q \,, \tag{9}$$

$$\frac{\partial(\rho U)}{\partial t} + \nabla(\rho U \otimes U) + \nabla p = \nabla \tau + qU \,, \tag{10}$$

$$\frac{\partial(\rho E)}{\partial t} + \nabla[(\rho E + p)U] = \nabla(\tau U) + \nabla(\lambda \nabla T) \,, \tag{11}$$

where ρ is the density, **U** is the velocity vector, p is the pressure, τ is the stress tensor, E is the specific total energy, λ is the thermal conductivity and T is the temperature. The system is closed by the equation of state for the perfect gas $p / \rho = r\,T$ and constant specific heat capacity c_p and by two equation nonlinear EARSM k–ω turbulence model [5].

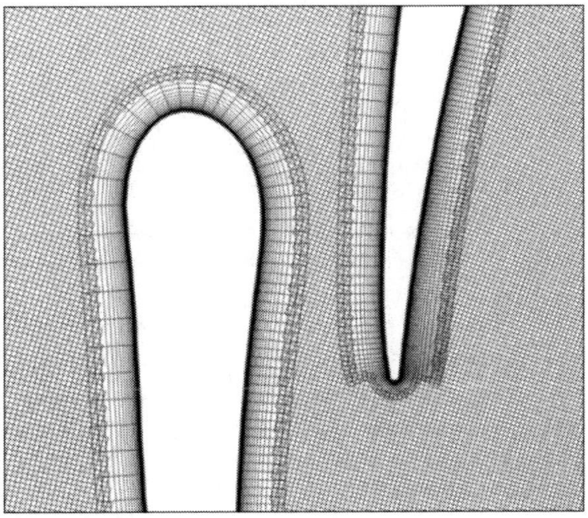

Fig. 5. Detail of the overset "O-H" computational mesh.

The sink terms of the mass q in equation (9) which is activated inside the confusor buffer zone (see fig. 4) is given as follows:

$$q = \begin{cases} 0\,, & \text{K for } r \geq r_{\lim}\,, \\ -\sigma_S \sqrt{2\rho\Delta p}\,, & \text{K for } r < r_{\lim}\,, \end{cases} \tag{12}$$

where r_{lim} is given by relation (7). The pressure difference in (15) is $\Delta p = \max\,(p - p_{out}, 0)$ where p is local pressure and p_{out} is the outlet pressure. The parameter σ_S (m^{-1}) controls the mass sink rate. The role of the mass sink term q is to exhaust the area $r < r_{lim}$ enough to overcome the fact, that the outlet radius r_{out} is smaller than the limit radius r_{lim}. Contrary to [2] in present work there is used the sink term $q\boldsymbol{U}$ also in equation (10). It means that not only mass is removed from the confusor buffer zone but also the momentum. It prevents unwanted spinning of the flow field inside the confusor buffer zone.

The mathematical model is implemented into the in-house numerical code based on the finite-volume discretization method [6] using the overlapped computational mesh [7] (see Figure 5).

4 Computation setup

Solution of flow past presented radial blade cascade was done for the total inlet pressure 114.1 kPa, the total inlet temperature 300 K, the inlet turbulence intensity 2 % and radial flow direction. Outlet Mach number M_{TE} at radius r_{TE} is in range $1.025 \leq M_{TE} \leq 1.88$ and the Reynolds number related to outlet velocity and the blade chord length is $Re < 8.5\times10^5$. Because of the lower Reynolds number, the algebraic bypass and separation induced boundary layer transition model [8] is used beside the two-equation turbulence model mentioned in previous section.

Figure 5 shows detail of the overlapped computational mesh which consists from "O"-type block over the blade profile and from "H"-type block which cowers periodical section of one pitch. The "O"-type block is refined near the blade surface to ensure that the viscous sublayer is covered at least by five cells. Total number of cells of the computational mesh is 76180.

Dimension of the outlet boundary is $r_{out} = 135$ mm whereas the trailing edges radius is $r_{TE} = 185$ mm. Radius of the inlet boundary is $r_{in} = 260$ mm and the leading edges radius is $r_{LE} = 209.2$ mm.

5 Results

Development of the transonic flow field with increasing the value of M_{TE} is shown in Figure 6 in form of numerical reconstruction of the shadow-graph pictures ($\nabla^2\rho$). The limit radius r_{lim} is for $M_{TE} = 1.025$ identical with radius of the outlet boundary, thus $r_{lim} - r_{out}$. Increasing of M_{TE} is possible only by increasing of the radius r_{lim}. This is shown in Figure 6. For Mach number $M_{TE} = 1.880$ the limit radius lies 2.5 mm behind the trailing edges. Note that the limit radius r_{lim} is the inlet boundary to the confusor buffer zone and that the conditions on the limit radius asymptotically approach state $M_r \rightarrow 1$ (see equation (2)). Next, let's note that (according to [3, 4]) the two-dimensional centripetal flow can only exist outside the limit radius. It means that for $r < r_{lim}$, thus inside the confusor buffer zone, there exists a flow different from the two-dimensional centripetal. That's because the flow field inside the confusor buffer zone is driven by the sink term (12). However, evaluation of aerodynamic and energy properties of the radial blade cascade is possible only outside the limit radius, thus outside the confusor buffer zone.

In Figure 7 there is depicted the kinetic energy loss coefficient dependence on the outlet Mach number. In the radial blade cascade the kinetic energy loss coefficient was evaluated in distance 2.5 mm behind the trailing edges. Figure 7 compares CFD prediction of the kinetic energy loss coefficient in radial and linear blade cascades in distance 2.5 mm behind the trailing edges and CFD prediction of the kinetic energy loss coefficient in linear blade cascade with the experimental data in distance 20 mm behind the trailing edges. The distance 2.5 mm behind the trailing edges in case of the radial blade cascade was chosen because flow inside the outlet confusor accelerates and value of the Mach number in a larger distance behind the trailing edges isgnificantly differs from M_{TE} value. Next reason, why it was chosen so little distance behind the trailing edge, is the presence of the confusor buffer zone. The kinetic energy loss coefficient was evaluated using the methodology

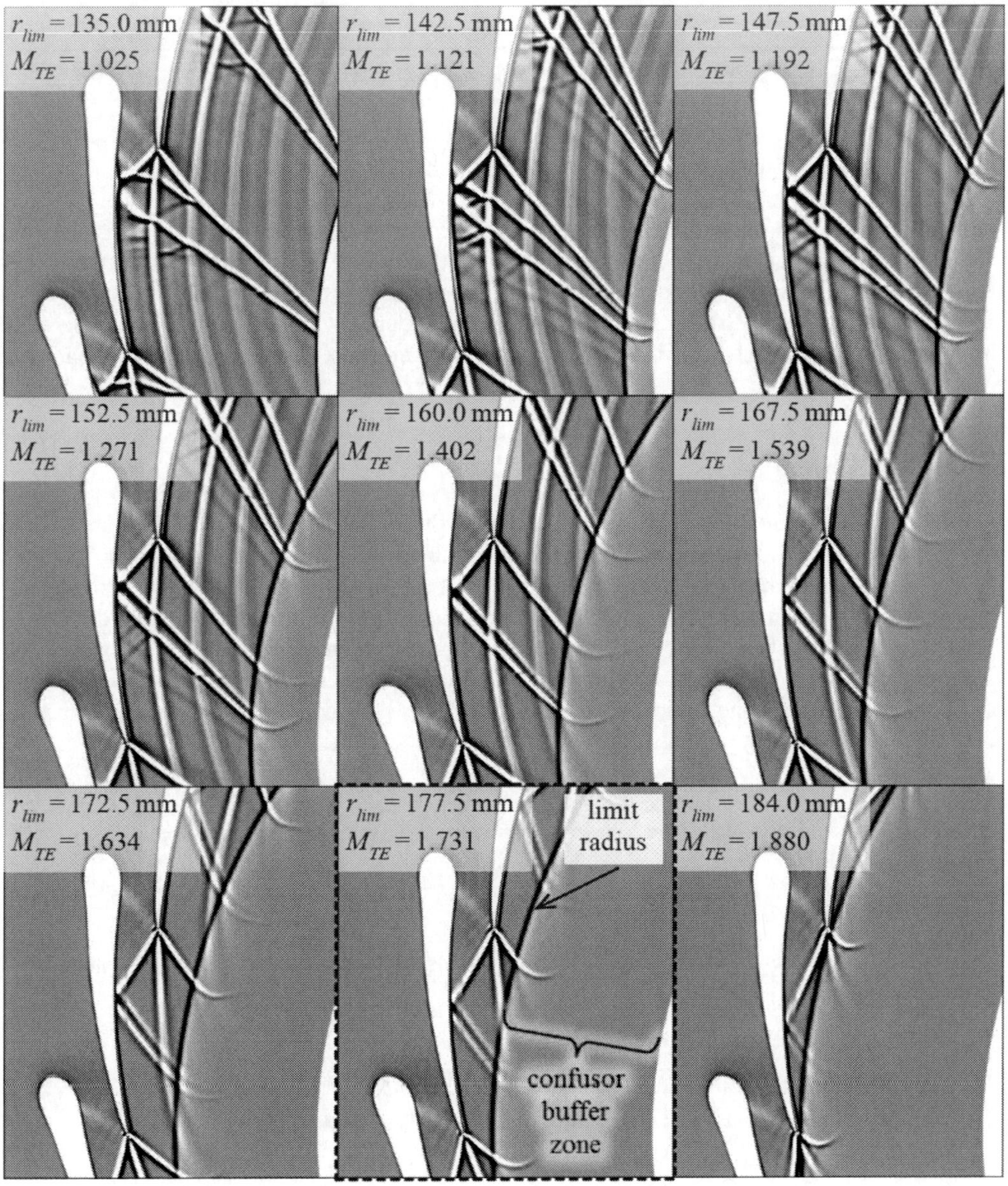

Fig. 6. Numerical shadowgraph reconstruction - development of the transonic flow field depending on the outlet confusor limit radius r_{lim}.

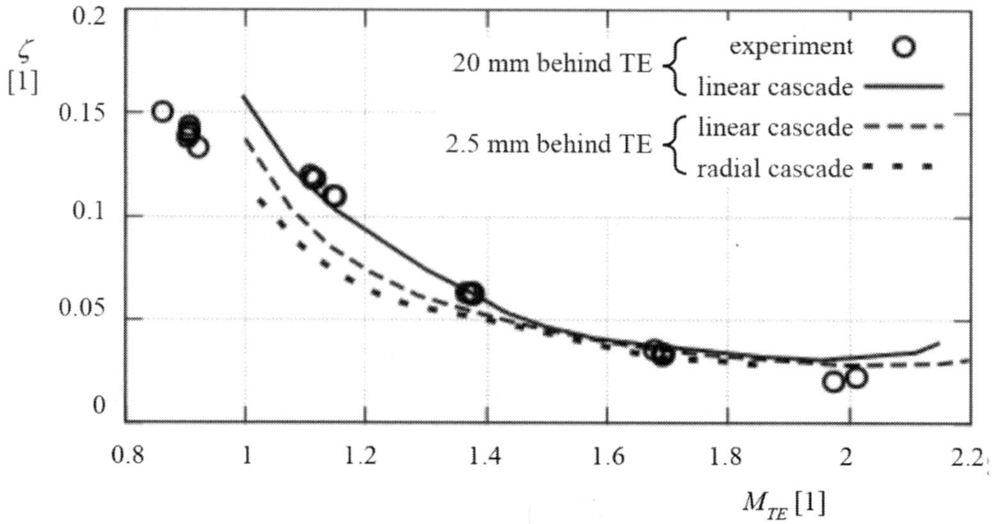

Fig. 7. Dependence of the kinetic energy losses coefficient ζ on the Mach number M_{TE}.

according to [9] but in the arrangement for the polar coordinate system. From Figure 7 it is clear that the kinetic energy loss coefficient of the radial blade cascade is little bit less compared to the linear blade cascade. This is given by the stabilization effect of the accelerated centripetal flow inside the inter-blade channel which slows the growing of the boundary layers thickness. It leads to little bit thinner wake compared to the linear blade cascade.

6 Conclusions

Numerical simulations of transonic flow through the radial blade cascade with short radial length were carried out for the outlet Mach numbers M_{TE} in the range from 1.025 up to 1.88. The aerodynamics and energy parameters of the presented radial blade cascade were compared with parameters of the linear blade cascade and with the experimental data.

The confusor buffer zone was proposed to avoid some numerical obstacles related with solution of regimes with high supersonic outlet Mach numbers M_{TE}. The main advantage of the confusor buffer zone approach is that there is no need to modify the computational domain and the computational mesh.

This work was supported by the Long-Term Development Concept of Research Organization DKRVO-IPOWER provided by the Ministry of Industry and Trade of the Czech Republic.

References

1. M. Němec, P. Straka, T. Jelínek, M. Babák, ASME Paper No. GT2010-22490, 1221-1230 (2010)
2. P. Straka, AIP Conf. Proceedings **1889**, 020041 (2017)
3. M. Luxa, R. Dvorak, D. Simurda, J. Vimmr, Jour. Thermal Science, **19**, 42-46 (2010)
4. M. Luxa, PhD Thesis, Czech Technical University, Prague (2005) (in Czech)
5. A. Hellsten, PhD Thesis, Helsinki University of Technology, Helsinki (2004)
6. P. Straka, VZLÚ Report R-4321, Prague (2008) (in Czech)

7. P. Straka, Proc. Conf. Finite Volumes for Complex Application VI, Prague, 751-759 (2011)
8. P. Straka, J. Příhoda, Proc. Conf. EFM 2010, Liberec, 636-641 (2010)
9. M. Němec, T. Jelínek, M. Mackovič, VZLÚ Report R-5163, Prague (2011) (in Czech)

Flow modelling of particulate matter by using baffles placed in the flue tract of wood stove

Nikola Kantová[1,*], *Alexander* Čaja[1], *Michal* Holubčík[1], and *Jozef* Jandačka[1]

[1]University of Žilina, Department of Power Engineering, Univerzitna 1, 010 26 Žilina, Slovakia

Abstract. It is very important to separate particulate matter before their leaving into the air because of their harmful impact on human health. It exists a lot of devices, which can separate these particles. But their implementation and operation in small heat source can be considerably difficult on finances, and often have complex maintenance. Therefore, it is important to look for solutions of reduction the producing of particulate matter with easier maintenance and also less financial difficulty. This article deals with flow modelling of particulate matter by using baffles placed in the flue tract of wood stove. Influence of baffles is observed by using CFD simulations.

1 Introduction

Flue gases from biomass combustion contains relatively high amount of particles smaller than 10 μm (PM10). Epidemiology studies have shown a clear correlation between the particle concentration in the air and severe health effects on human being [1]. Particles less than 2.5 μm (PM2.5) can penetrate deep into human lungs, deposit in the alveolus, and even may be transported to systemic apparatus or tissue leading [2].

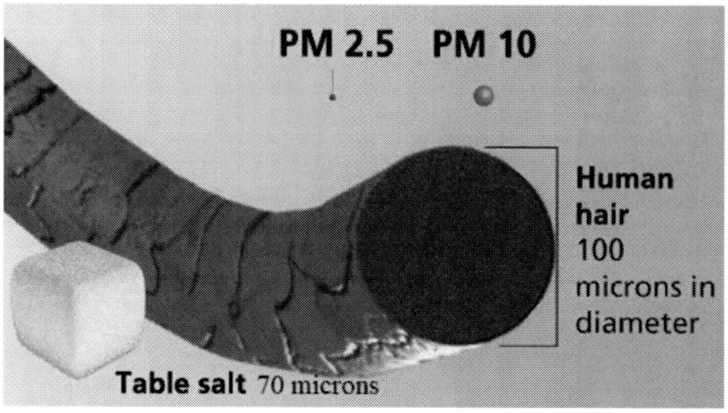

Fig. 1. Particulate matter size.

* Corresponding author: nikola.kantova@fstroj.uniza.sk

It is very important to separate particulate matter before their leaving into the air. Reduction of particulate matters can be achieved primarily by optimizing the air flow and flue gases flow in the combustion chamber, secondary by use of separator [3]. There are five major approaches for removing particulate matter based on different mechanisms. The first is settling and momentum separators by gravity, inertia and centrifugal effects such as cyclones, baffle chambers and settling chambers. The second is filtration separators by diffusion, interception, such as fabric filters, paper filters and baghouse filter. The third are wet collectors mainly scrubbers by impingement, diffusion, thermal gradients and electrostatic attraction. The fourth are acoustic separators by agglomeration resulting from sound waves and the fifth is electrostatic precipitation by electrostatic forces [4, 5]. These implementation and operation of separators in small heat source can be considerably difficult on finances, and often have complex maintenance. There are looking for solutions of reduction the producing of particulate matter with easier maintenance and also less financial difficulty [6].

For separation of particulate matter in wood stove is designed a model of flue tract with baffles. Their effect is observed by using CFD simulations. CFD is an abbreviation for computational fluid dynamics, which is a branch of fluid mechanics that uses numerical analysis and algorithms to solve and analyse problems that involve fluid flows [7, 9].

2 Mathematical simulation

Analysis was carried out in Ansys program by using baffles for separation. There were used three baffles placed in the flue tract of wood stove. Optimized flue tract was realized on 3D model with network of 194787 elements, which is shown in the Figure 2. All elements had aspect ratio lower than 10. Flue tract was thought as a cube with length of edge 400 mm. The chimney had diameter of 160 mm. Baffle was placed perpendicular to the flue track with length of 350 mm. Particles entered into a model through a hole with dimensions 135 x 400 mm.

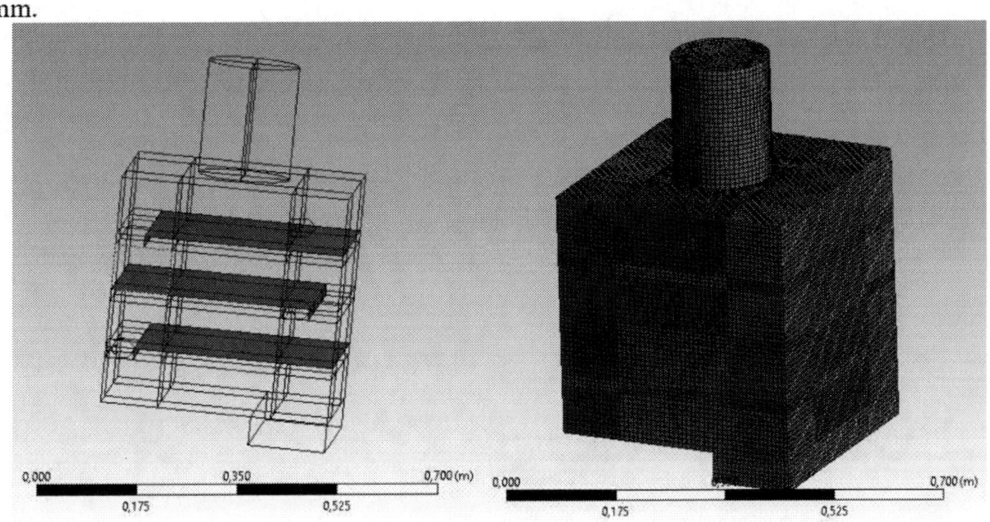

Fig. 2. 3D and mathematical model of flue tract with baffles.

The object was simulated by the k-ε model with standard wall function. The flowing of particles was realized using by the trajectory of a discrete phase particle by integrating the force balance on the particle, which is written in a Lagrangian reference frame [8].

It was important to realize size distribution of particles and then use Rosin-Rammler distribution function because of determine mean parameter and spread parameter in Ansys program. Size distribution, which you can see in the Figure 3, was experimental determined by using of vibratory sieve shaker machine. Used sieves had the size of mesh 500 μm, 250 μm, 150 μm, 100 μm, 50 μm and 20 μm. Sample of particulate matter was obtained from electrostatic precipitator after combustion of wood chips from spruce wood in the boiler with moving grate.

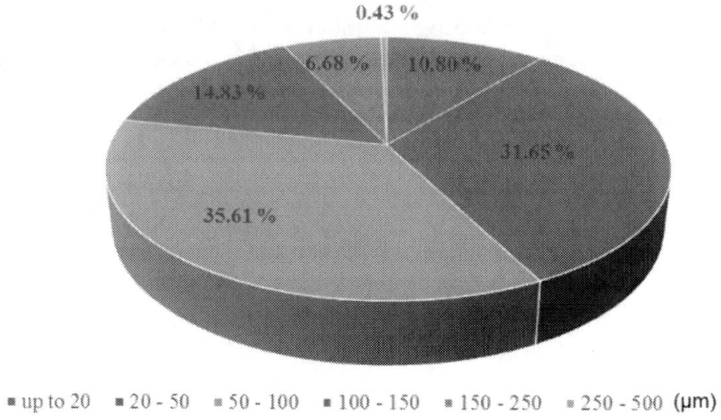

Fig. 3. Size distribution.

Particles were considered spherical and also the influence of turbulence on particle motion. Flowing was time steady, without combustion and the influence of gravity was considered. Used materials were air with a density of 0.6 kg.m^{-3} and particles with a density of 600 kg.m^{-3}. Based on the stoichiometric calculations was mass flow set on 0.0168 kg. s^{-1}.

3 Results

In Ansys program were created velocity profile and particle diameter profile, which you can see in the Figure 4.

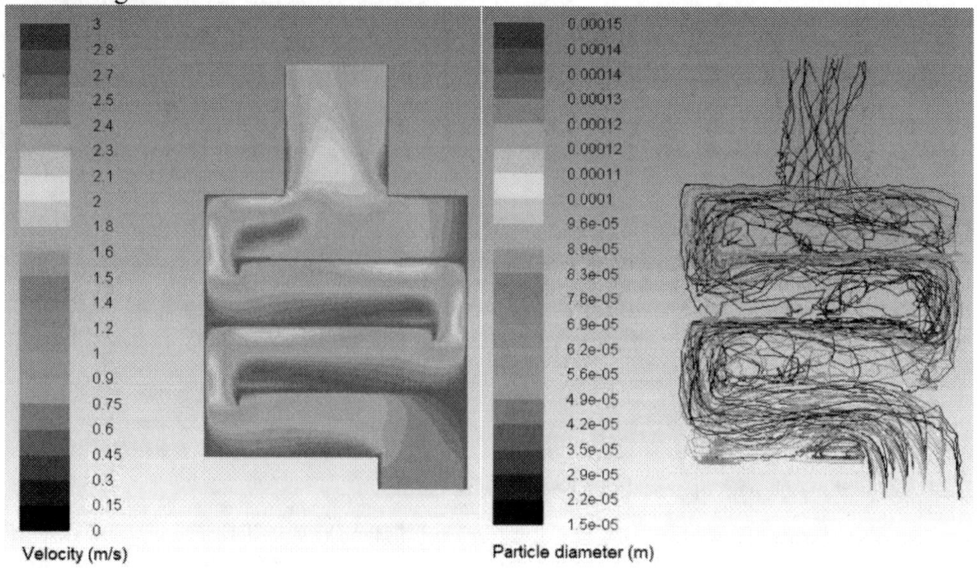

Fig. 4. Velocity and particle diameter profile.

It is important to do not realize difficult shape of baffles, where could be big narrowing. It would cause large pressure loss and increasing in flow velocity. Maximum flow velocity was 2.5 $m.s^{-1}$ in this flue tract. Particles achieved bigger velocities in a places of boundary walls after a transition of the baffle and also on the bottom parts of baffles. The largest particles did not pass through first baffle. Based on the number of particles, 44.83 % particles escaped and 55.17 % particles captured. Based on the mass flow, 81.6 % escaped and 18.4 % particles captured. It is caused that the smallest particles escaped in the flue gas stream and larger particles were captured.

4 Conclusion

Decrease of particulate matter is necessary topic for air protection. Elimination of particulate matter can be achieved by various ways. But many separators are difficult on finances, and often have complex maintenance. Therefore, it is important to look for more useful solutions. This paper deals with flow modelling of particulate matter by using baffles placed in the flue tract of wood stove. Their effect was observed and analysed by using CFD simulations. Used baffles were effective for larger particles, which were captured. Bigger amount of smaller particles escaped in the flue gas stream.

This work has been supported by the project KEGA 033ŽU-4/2018 "Heat sources and pollution of the environment".

References

1. K.G. Nigusie, Master's thesis, online
2. Who, Health effects of particulate matter, Policy implications for countries in eastern Europe, Caucasus and central Asia, online, (2013)
3. P. Ďurčanský, M. Holubčík, M. Malcho, Meeting of Departments of Fluid Mechanics and Thermomechanics (2006)
4. L. Guan, G. Harvel, S. Park, J.S. Chang, 11th International Conference on Electrostatic Precipitation (2008)
5. J. Moravec, Technical Gazette, 24, 143-147
6. K. Sulovcová, J. Jandačka, R. Nosek, Power control and optimization, (2013)
7. A. Ryfa, R. Buczynski, M. Chabinski, A. Szlek, R.A. Bialecki, Thermal Engineering **73** (2014)
8. Fluent user' guide, online
9. S. Gavlas, P. Ďurčanský, R. Lenhard, J. Jandačka, *EPJ Web of Conferences* **92**, (2015)
10. P. Nemec, M. Smitka, M. Malcho, Scientific world journal, **2014** (2014)
11. P. Nemec, M. Malcho, J. Jandačka, *AIP Conference Proceedings*, **1558**, (2013)

Flow over evaporator in electrotechnical box

Tomáš Puchor[1,*], *Richard* Lenhard[1]

[1]University of Žilina, Department of Power Engineering, Univerzitná 1, 010 26 Žilina, Slovakia

Abstract. The article focus on heat flow generated by heat points over evaporator in eletrotechnical box. The evaporator consists of horizontal pipes supplemented with vertical ribs to enlarge the heat exchanger surface. This ribs were changed as porous zone for simulation process. The evaporator is located in a hermetically sealed electrotechnical box. Observes the spread of heat in the porous materials and air over evaporator and space inside the electrical box.

1 Introduction

Electronic components inside all devices are producing heat. Some of them are producing more heat than an others and can cause failure. This heat need to be carried out to make device work properly. This work focus for heat distribution over evaporator inside electro technical box by visualisation with CFD simulation using ANSYS workbench. Model consist from model of box and inside it is evaporator with ribs. This ribs are a computational load and are replaced by a porous model. Optimized model with porous zone is better for simulation and less time consuming process. This simulation can be used to optimize a more complicated model in the future and provides a preview of the problem of solving the exchanger with ribs by using a porous model.

2 Model of evaporator

On Figure 1 is shown model of evaporator with ribs and model with porous zone. The design of the cooling device was based on a comparison of the current work on the refrigeration equipment for electronic components and by space inside electrotechnical box. This evaporator consists of horizontal tubes supplemented by vertical ribs to increase the heat exchange surface [1]. Evaporator is placed inside box at back wall with 30 mm offset. At Table 1 are dimension parameter of evaporator with ribs. Because simulation process with ribs is very time consuming, the model with porous part was chosen.

* Corresponding author: tomas.puchor@fstroj.uniza.sk

Table 1. Dimensions of exchanger.

Fins		Pipes	
Quantity [pcs]	265	Quantity [pcs]	33
Dimensions [mm]	50 x 1 600	Diameter [mm]	15
Thickness [mm]	0.25	Length [mm]	660
Distance [mm]	2.25	Material	copper
Material	aluminium		

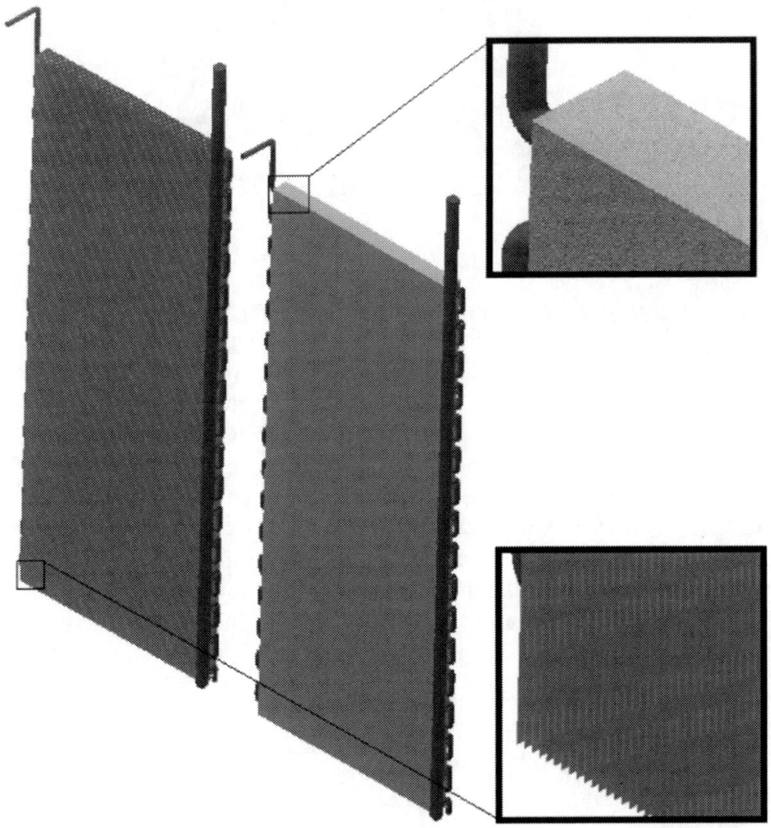

Fig. 1. Model of evaporator with porous material (right) and with ribs (left) and detail.

3 Meshing

Meshing setup was simple. Set up for generating mesh using ANSYS CFD meshing was to get the basic result with standard quality. Size function was set to curvature with fine relevance center. Initial size seed was set for active assembly. Smoothing, transition and span angle center was set as low, slow and fine. Curvature normal angle was set from default to 15° to get better mesh. All size setup was left to default. Part of generated mesh can be seen on Figure 2.

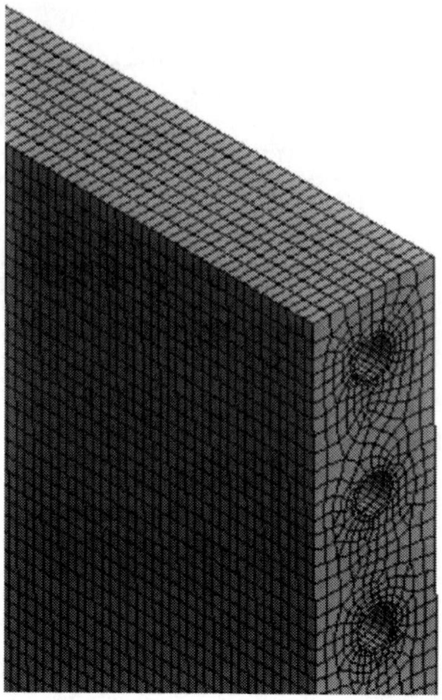

Fig. 2. Meshed model.

4 Pressure drop over porous evaporator

To correctly determine boundary conditions for successful simulation and result, the porous zone was needed to calculate. A porous zone is used in cases where a flow has to be defined, for example, through filters and heat exchangers. Instead of a 3D detailed model, the component is set on a simplified spatial model or porosity on the surface. As material with porosity was chosen air. Porosity is a ratio between solid volume and air volume to the total examined volume. It was calculated based on ribs and air space. The resulting porosity was 98 %. Consequently, it was necessary to determine the coefficients of porosity [2]. These coefficients were calculated on the basis of the pressure loss in the ribbed exchanger for both cases of air distribution over evaporator with different velocity as seen in Table 2.

Table 2. Pressure loss.

Velocity [m/s]	Case	
	1	2
	Pressure [Pa]	
2.5	113.96	12.98
3.5	160.13	22.30
4.5	206.63	33.39
5.5	253.45	46.34
6.5	300.59	61.06

The space between the two ribs was used for the calculation. However, for the accuracy of the calculation, pressures from three intercostal spaces were used [2]. The procedure for deriving the loss coefficients of the porous zone is given by.

$$\Delta p = -\left(\frac{\mu}{\alpha}\cdot v + C_2 \cdot \frac{1}{2}\cdot \rho \cdot v^2\right)t \tag{1}$$

Where μ is kinematic viscosity of air, ρ is air density, t is thickness of porous media, α is the permeability and C_2 is the inertial resistance factor.

From the data obtained from the airflow simulation in the intersection space, a course and then a trend equation are created [3].

$$y = -7.2416\, x^2 + 94.55\, x \tag{2}$$

Coefficients in the trend equation (Figure 3) are used in the following formulas and the loss constants for the porous medium are calculated [4, 5].

$$-7.2416 = C_2 \,.\, ½ \,.\, \rho \,.\, t \tag{3}$$

$$64.689 = \mu/\alpha \,.\, t \tag{4}$$

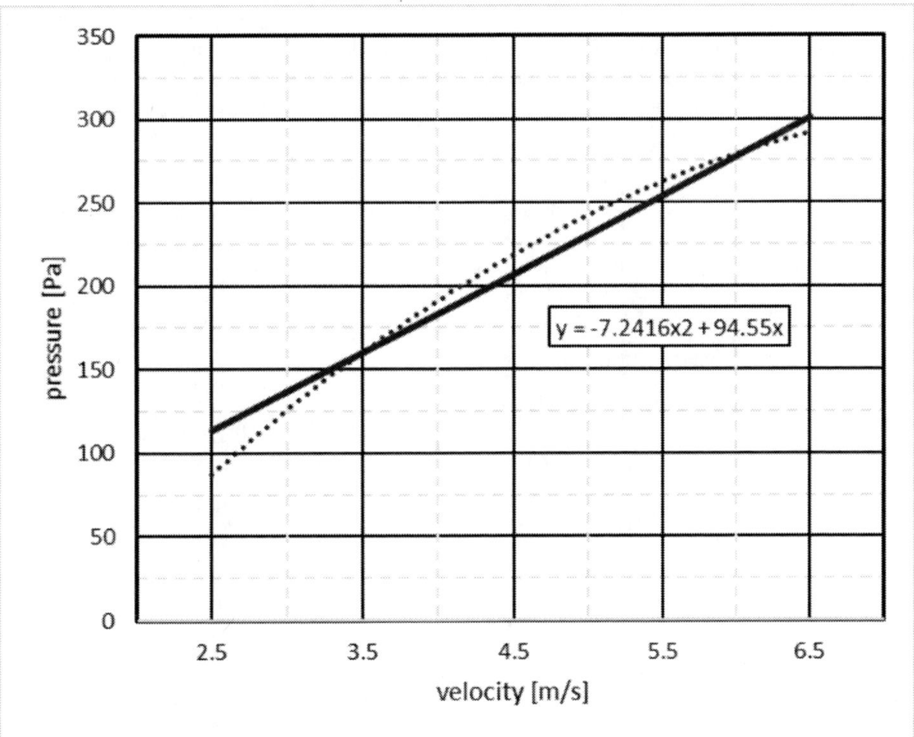

Fig. 3. The process of pressure loss through a porous environment.

The loss coefficients was then used for setting up the porous parameters in cell zone conditions for material air as seen in Table 3.

Table 3. Resulting loss coefficients.

Relative Velocity Resistance Formulation		
Viscous Resistance (Inverse absolute Permeability)		
Diretion-1 [1/m²]	1.056779e+08	constant
Diretion-2 [1/m²]	1.056779e+08	
Diretion-3 [1/m²]	1.056779e+08	
Inertial Resistance		
Diretion-1 [1/m]	236	constant
Diretion-2 [1/m]	236	
Diretion-3 [1/m]	236	

Viscous Resistance is the effect of surface friction between a particle and a liquid when the particle moves through the liquid. In the context of fluid flow, viscous resistance is energy lost due to the fact that the fluid has to shear to get out of the way of some obstacle, and it doesn't like shearing because it has Viscosity.

Conversely, inertial resistance is energy loss due to that the fluid has to be given kinetic energy to get out of the way of the obstacle, and although in principle this could be recovered and recycled, it commonly ends up being lost through turbulence in the body of fluid. The inertial resistance is proportional to the fluid density times the flow rate. The inertial resistance is significant with gas because flow rates can be high [4].

The significance of inertial resistance is that it tends to be proportional to v^2 (whereas viscous resistance is proportional to v). Thus it dominates at high flow rates.

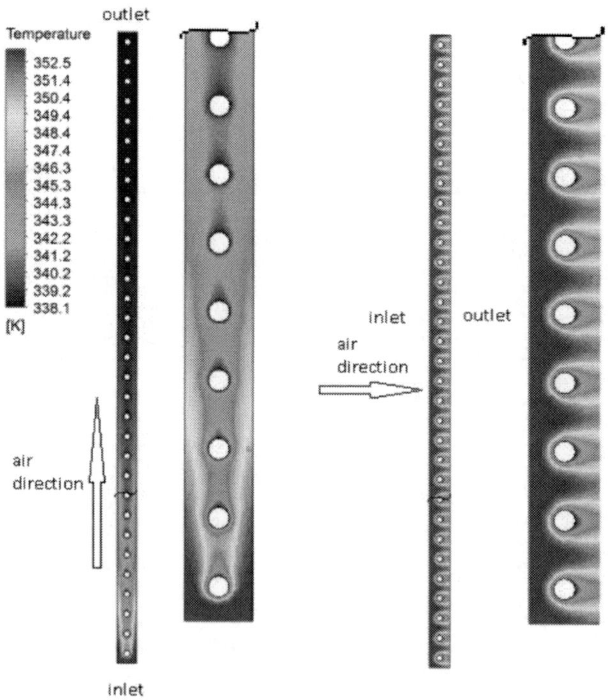

Fig. 4. Temperature results.

5 Basic setup for boundary conditions

The boundary conditions settings were as follows. As the entry area for air was used bottom surface and named as inlet and air velocity entering the intercostal space (in this case porous material) was set at 2.5 m/s [5]. Turbulence was set as Intensity with a value of 5 % and a hydraulic diameter of 0.917 m and the temperature was set to 353.15 K. The exchanger air outlet area was set for upper surface was named as the outlet and set as a pressure-outlet with the same values for the intensity and the hydraulic diameter as the inlet setting. The walls of the Exchanger tubes were set at a temperature of 338.15 K. In second case of exchanger model for inlet and outlet was used from front to back flow of air. All other setup was left as previous.

6 Results

Air flow over evaporator heat exchanger was simulated for both cases. Case 1 respond with best result for getting lowest outlet temperature, as shown in Table 4. In Figure 4 is shown flow over exchanger for each case.

Table 4. Temperature results.

Temperature		
case	inlet	outlet
1	353.15 K	638.15 K
2	353.15 K	348.67 K

7 Conclusion

The correct functionality of most electronic components is dependent on the operating temperature range. In practice, it means that the components and also the devices must be cooled.

The aim of the work was to determine the best distribution of air over heat exchanger using porous zone model in ANSYS using CFD. The porous zone in the simulation of the flow is preferably used wherever the actual geometry could not be used. For correct adjustment of loss coefficients, it is necessary to perform a test and measure the pressure loss created by the physical part or another model. The Ansys Fluent manual can be used to derive coefficients for porous zone.

The pressure loss simulations were performed for both cases to calculate the coefficients of porosity. Best airflow cooling results have been obtained when the air flows from the bottom of the heat exchanger upwards.

The research is supported by project "Limits of Radiative and Convective Cooling through the Phase Changes of Working Fluid in Loop Thermosyphon", APVV-15-0778.

The research is supported by project: The storage of hydrocarbon gases in hydrate structures as an alternative method of energy accumulation, 063ŽU 4/2018.

References

1. R. Lenhard, M. Malcho, J. Jandačka, Heat transfer engineering, **40**, 3 (2019)
2. R. Lenhard, J. Jandačka, *AIP Conference Proceedings*, **1558,** 2138 (2013)
3. P. Hrabovský, Š. Papučík, K. Kaduchová, *AIP Conference Proceedings*, **1745**, 020014 (2016)
4. P. Ďurčanský, Š. Papučík, M. Malcho, *AIP Conference Proceedings*, **1768,** 020028 (2016)
5. J. Raszka, Mechanical Series, **LXI**, 1996 (2015)

Investigation of hydraulic fitting losses

Marian Bojko[1,*], and *Milada* Kozubková[1]

[1]Department of Hydromechanics and Hydraulic Equipment, Faculty of Mechanical Engineering, VŠB-Technical University of Ostrava, 17. listopadu 15, 70833 Ostrava – Poruba, Czech Republic

Abstract. Fluid flow in hydraulic systems, fittings, and piping causes hydraulic losses due to the change of flow direction and friction in the fluid. The main consequence of the resulting losses is the increase in the overall pressure gradient of the circuit. The paper is focused on the investigation of valve losses and the determination of resistance coefficients, the definition of which depends on the valve type. Methods of determination are of two types, i.e. experimental methods and mathematical methods. In the case of experimental methods, the procedures prescribed by the standards and regulations must be respected. Mathematical methods are related to defining an appropriate mathematical model and numerical approach. Both methods are applied to the ball valve and model verification is performed at the same time. The output of the thesis is the methodology for determination of resistance coefficients of valves with other diameters and other flowing media based on verified mathematical modeling.

1 Introduction

When fluid flows through the hydraulic circuint, there are various hydraulic losses due to the viscosity of the fluid and the change of direction of flow. As a result, it is necessary to select a pump with appropriate performance parameters. This consequence must not be neglected in the design and calculation of the circuit, so that the sum of all the pressure losses in the circuit will be greater than the pressure supplied by the pump. The secondary adverse effect is the transformation of the hydraulic loss energy into heat and hence the heating of the circuit elements and the fluid, which is associated with other demands on the choice of appropriate cooler. The resulting hydraulic resistances are defined by various types of loss coefficients that are specific to the individual circuit elements [1, 2].

- **local loss coefficient** ζ is determined when fluid is flowing in areas where the flow direction of the medium (knees, arcs, branches, compensating loops, etc.) changes,
- **coefficient of friction losses** λ is typical coefficient for the fluid flow in pipelines,
- **flow coefficient** μ is typical in the areas where the flow is throttled (check valves, pressure reducing valves, ball valves, etc.).

* Corresponding author: marian.bojko@vsb.cz

In nowday of the modern computational technology, it is possible to identify hydraulic resistances in addition to the methodology of experimental approach by a methodology based on mathematical approach and using various software (ANSYS Fluent, Matlab, Maple, MathCAD, etc.). It is important to verify this methodology as each solution can be burdened a certain computational error due to the mathematical model used. When modeling the flow in hydraulic systems, it is necessary to know the characteristics and resistance coefficients of the individual elements and, if necessary, to adapt them to a particular variant of the mathematical model. The admission of mathematical modeling into the technical sector has brought a number of advantages, for example, to find out all the information about the behavior of the whole mechanism, to perform simulation of flows in elements where physical experiments appear to be very difficult, to simplify the design the entire hydraulic system [3, 4].

2 Design, measurement of characteristics, specific valve parameters

Hydraulic circuits include valves of various designs and designs. Their distribution is according to the type and construction of the control element, which may be a ball, a bow, a plate, a saddle and the like. Then the valves according to the design can perform various functions in the hydraulic circuits, especially control and flow control. The most common types include ball valves, bbutterfly valves, gate valves, globe valves and others, see Figure 1.

Fig. 1. Ball valves, Butterfly valves, gate valves, globe valves [5].

The evaluation of the characteristic of the valves and the respective loss factors (loss coefficient ζ, flow coefficient μ, effective surface S) and other parameters can be realized by experimental measurements on the hydraulic circuit according to the requirements of the standard. The results of experimental measurements of different types of valves are presented in publications [6, 7] including the relevant characteristics. The basic characteristic of the valve is the dependence of the pressure drop Δp on the flow rate Q. This characteristic consists of sub-characteristics, which are determined for selected strokes of the poppet control valve or for selected angles of rotation of the ball in the case of ball valves. Typical characteristics ($\Delta p=f(Q)$, $\zeta=f(z)$) for the poppet valve obtained from the measurements are shown in Figure 2, Figure 3.

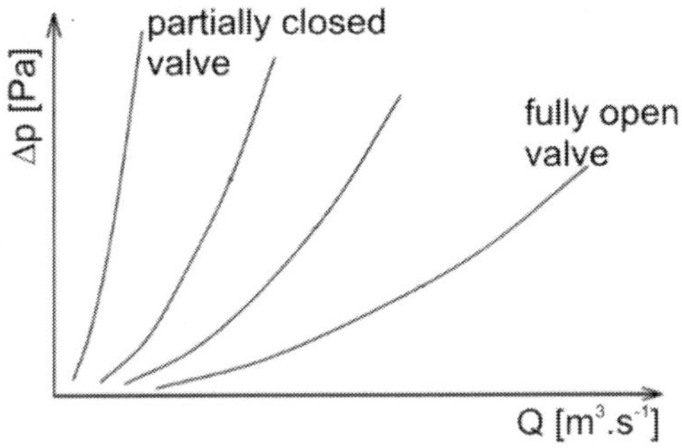

Fig. 2. Valve characteristic.

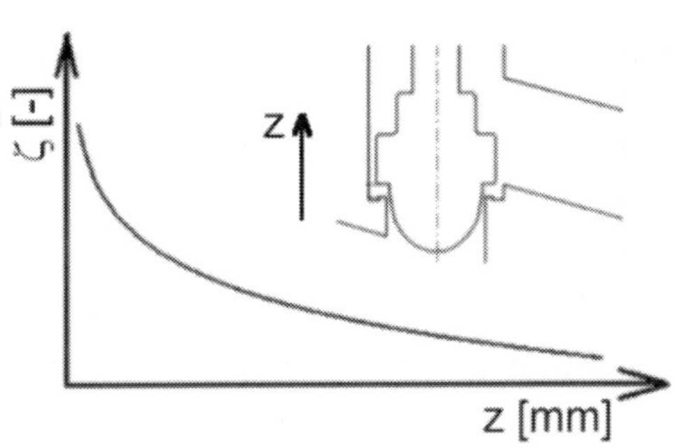

Fig. 3. Loss coefficient's dependence on the valve stroke.

From graphical plotting of the measured values, it is possible to determine the behavior of the valve at different flow rates and to select a suitable valve for the given application based on this information. When creating of the circuit and the following measurement, the principles given by the standard [8] have to be met so that the loss factors are universally used. Scheme of the measuring system according to ČSN EN 60534-2-3 is shown in Figure 4.

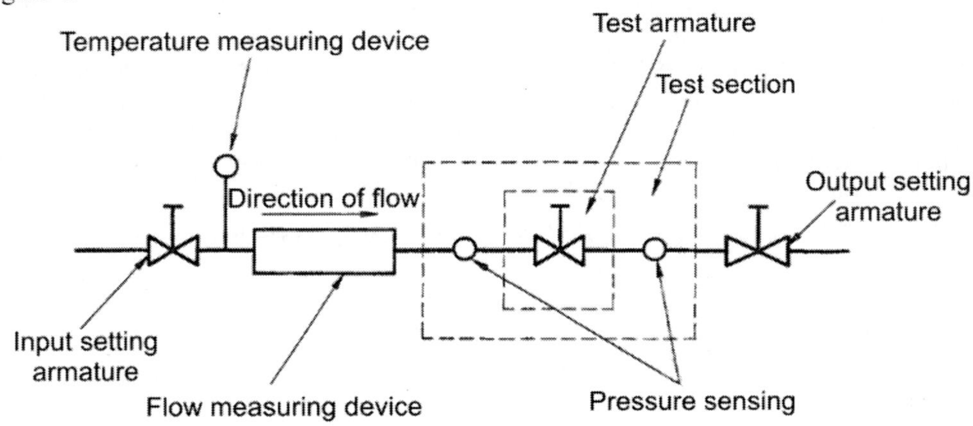

Fig. 4. Diagram of flow test system.

A second approach to investigating the characteristics of valves of different constructions and types is mathematical flow modeling or CFD analysis. CFD analysis can be performed on the 3D models of individual valves and the application of suitable turbulent models to obtain the relevant characteristics. Then, it can be verified by experimental measurements. Studies on mathematical modeling are presented in papers [9, 10, 11, 12].

2.1 Basic valve parameters

The basic parameters of the valve are nominal diameter DN, nominal pressure PN, maximum working temperature of the valve and medium, loss ζ and discharge coefficient μ, effective area S and others, including the cavitation coefficient.
The loss coefficient is derived from the Bernoulli equation and $\Delta p = f(Q)$ characteristic.

$$\Delta p = \rho \zeta \frac{u^2}{2} = \rho \zeta \frac{Q^2}{2S^2} \quad (1)$$

For the measured sub-characteristic, the loss coefficient is determined by the least squares statistical method. In the case of a flow through a narrowed area (valve, diaphragm, nozzle, etc.) the inverse relation to equation (1) is used more frequently

$$Q = \mu S \sqrt{\frac{2}{\rho}} \cdot \sqrt{\Delta p} \tag{2}$$

Comparing both equations $\frac{\Delta p}{Q^2} = \frac{\rho}{2}\frac{\zeta}{S^2}$ a $\frac{\sqrt{\Delta p}}{Q} = \sqrt{\frac{\rho}{2}} \cdot \frac{1}{\mu S} \Rightarrow \frac{\Delta p}{Q^2} = \frac{\rho}{2}\frac{1}{\mu^2 S^2}$ the following **dependence between the loss and discharge coefficient** can be derived

$$\mu = \frac{1}{\sqrt{\zeta}} \tag{3}$$

For hydraulic element with variable flow area (using Matlab-Simulink-SimHydraulics software) instead of the loss or discharge coefficient, the effective area depending on the position and, respectively opening is used. Variable area of the orifice is not identical to the geometric area because it is modified by contraction and swirling when the fluid flows around the edges of the orifice. If the valve is completely open, then the orifice area is maximum area and is equal to the flow area of the connected tubes S_0. Discharge coefficient can then be determined from the loss coefficient ζ obtained from the $\Delta p=f(Q)$ sub-characteristic for the full opening of the valve and will be constant even for the sub-characteristics when opening the valve (eg 20°, 40° and 60°). For other opening therefore, the area of the orifice, which is determined from the experimentally measured characteristic (discharge coefficients), must be changed. The following effective area for partial closing of the valve eg 20°, 40° and 60° is determined from the other measured sub-characteristics $\Delta p=f(Q)$, but the discharge coefficient μ will be constant ($\mu_0 = \mu$). Thus, for a partial closing of the valve, for example by 20°, the effective area is determined:

$$S_{20} = \frac{\mu_{20}}{\mu} \cdot S_0 \tag{4}$$

Further effective areas of the orifice are determined analogously.
Depending on the standard or use in different countries with different physical unit systems, other variants of the discharge coefficient labeled as A_v, K_v a C_v are used.

3 Definition of mathematical model of flow through the ball valve

The defined mathematical model characterizes the flow rate of water in the general spatial geometry presenting the inner space of the ball valve at different closure positions. The general model specification is as follows:

- 3D flow of one phase (water).
- Flow of incompressible liquid, ρ=const.
- Steady flow, $\frac{\partial}{\partial t} = 0$.
- Isothermal flow, T=const.
- Turbulent flow.

In the case of turbulent flow, and due to swirling when the fluid flows around the edges of the ball valve orifice, it appears to be an optimal the two-equation k-ε RNG mathematical model, which consists of the following basic system of balance equations in vector notation [3]:

continuity equation

$$\frac{\partial \rho}{\partial t}+\nabla\cdot\left(\rho\vec{u}\right)=S_z \tag{5}$$

Navier-Stokes equation

$$\frac{\partial\left(\rho\vec{u}\right)}{\partial t}+\nabla\cdot\left(\rho\vec{u}\vec{u}\right)=-\nabla\overline{p}+\rho\vec{a}+\nabla\cdot\left(\overline{\overline{\tau_t}}\right)+S_m \tag{6}$$

Equation for turbulent kinetic energy k

$$\frac{\partial}{\partial t}(\rho k)+\nabla\cdot\left(\vec{u}\,k\right)=\nabla\cdot\left(\frac{\nu_t}{\sigma_k}\nabla k\right)+\rho P-\rho\varepsilon \tag{7}$$

Equation for tubulent dissipation rate

$$\frac{\partial}{\partial t}(\rho\varepsilon)+\nabla\cdot\left(\vec{u}\,\varepsilon\right)=\nabla\cdot\left(\frac{\nu_t}{\sigma_\varepsilon}\nabla\varepsilon\right)+C_{1s}\rho P-C_{2s}\rho\frac{\varepsilon^2}{k} \tag{8}$$

and Boussineq's analogy is used to calculate Reynolds' strains.

The RNG k-ε model as a modification of the Standard k-e turbulent model using the renormalization group method RNG [3] was chosen and was solved by final volume numerical method. At the input of the computation area, the VELOCITY INLET conditions were defined and at the outlet the PRESSURE OUTLET were used. The flowing substance was defined as water with corresponding physical properties (density, kinematic viscosity).

4 Results of the experimental part and mathematical modeling

In the experimental part were measured $\Delta p=f(Q)$ of the ball valve characteristic for each valve position (0°, 20°, 40°, 60°). Ball valves are hydraulic elements in the circuit that are intended to close of the flow rate through piping with help of rotating the ball with the hole (see Figure 1). Depending on its opening or closing, the flow rate is controlled, but the valves can not be used as throttle or regulating valves. This valve produces a minimum pressure drop when fully opened. The range of use of valve depends on the type and temperature of the flowing medium and material. Measurement of the static characteristics of the ball valve was performed in a laboratory on a water line of plastic components. The ball valve in the circuit has established sampling points before and after the element according to EN 60534-2-3, which are connected to an inverted U-tube for measuring differential pressure. The flow rate through the circuit is measured by the clone (cross-sectional gauge) through the loss height on the U-tube of the clone and the given calibration curve of the clone.

Fig. 5. Ball valve model, valve postions.

4.1 Evaluation of experimental measurements

Based on measured and calculated values, it is possible to plot *Δp=f(Q)* characteristics of the ball valve for different angles of closure see Figure 6. From the distribution, there is a visible difference between the individual characteristics for the individual (0°, 20°, 40°, 60°).

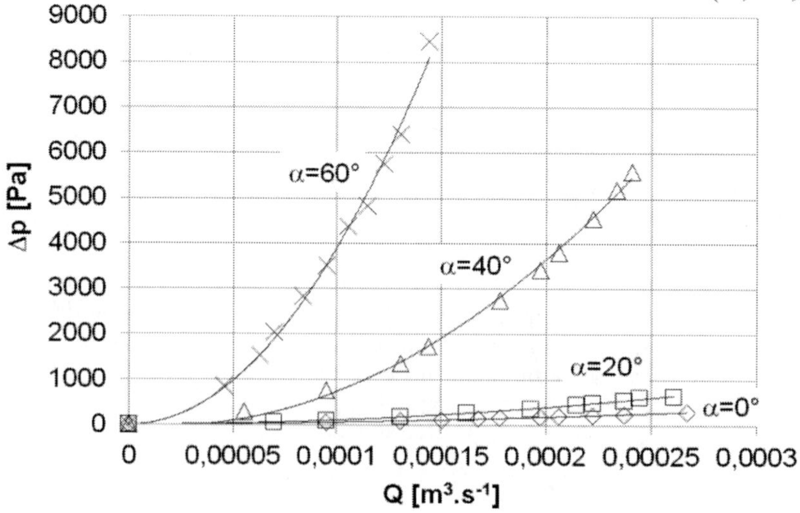

Fig. 6. Experimentally measured characteristic *Δp=f(Q)* ball valve for individual closures (0°, 20°, 40°, 60°).

Then from the individual *Δp=f(Q)* characteristic were calculated by the least squares method, the loss coefficients *ζ*, the flow coefficients *μ* and the effective area *S*, which will then be compared with the CFD analysis of the flow.

4.2 Evaluation of mathematical modeling

4.2.1 Model ball valve, computational grid

For the CFD analysis of the ball valve flow, a model corresponding to the internal volume of flowing water was created for the subsequent creation of the computational grid. A 3D design model of the ball valve from Inventor (Figure 5) was used to create the final model. This model was subsequently opened in ANSYS DesignModeler in the appropriate format for CFD analysis requirements. In this way four models (Figure 7) corresponding to the fully open valve, the valve in the open position 20°, 40° and 60° were created [13]. Then, a corresponding computational grid was generated in these models. Software ANSYS Meshing was used to create the computational grid. In the region of the valve itself and in particular in the area of closure, a finer grid was created than in the areas of valve connection to the pipeline.

Fig. 7. Resulting models in position 0°,20°, 40° a 60°, computational grid for the model in position 60° [13].

Then the basic current variables such as pressure and velocity in individual cuts were evaluated in detail to map the flow characteristics of each ball valve position. The longitudinal section pressure field for the valve closing position of 20° is shown on Figure 8.

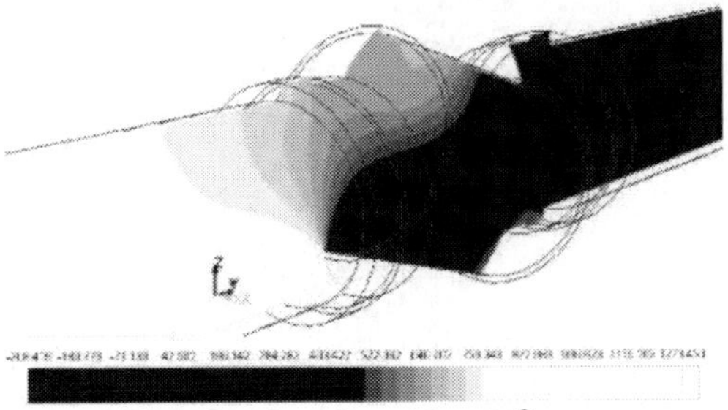

Fig. 8. pressure field (-300 ÷ 1300)Pa for closed position 20°.

In the next phase the $\Delta p=f(Q)$ characteristics from the mathematical modeling of the flow for the different valve closing positions are evaluated. Then loss factors ζ, flow coefficients μ and specific area S are calculated, which will then be compared with experimental measurement.

5 Comparison of experimental measurement and mathematical modeling

Experimental measurements and mathematical modeling were performed under the same boundary conditions. The resulting experimentally measured *Δp=f(Q)* ball valve characteristic for the different closing positions is shown in Figure 5. Subsequent comparison of the experimental measurement with the results of the mathematical modeling is focused on the evaluation of the loss coefficient ζ the flow coefficient μ and the reference area S using the experimental measurement and CFD analysis. The resulting values for the individual closing positions are shown in Table 1.

Table 1. Evaluation ζ, μ, S based on experimental measurement and mathematical modeling.

	Experimental measurements			
	Closure angle			
	0°	20°	40°	60°
Loss coefficient ζ [1]	0.579	1.271	11.928	50.545
Flow coefficient μ [1]	1.314	0.887	0.290	0.141
Efective a S [mm^2]	254.469	171.699	56.055	27.231
	Mathematical modeling			
Loss coefficient ζ [1]	0.672	1.450	12.813	52.124
Flow coefficient μ [1]	1.220	0.830	0.279	0.139
Efective a S [mm^2]	0.672	1.450	12.813	52.124

Further, the loss coefficient α and the effective area S for the individual closure positions are plotted by graphical dependencies $\zeta=f(\alpha)$ a $S=f(\alpha)$, Figure 9.

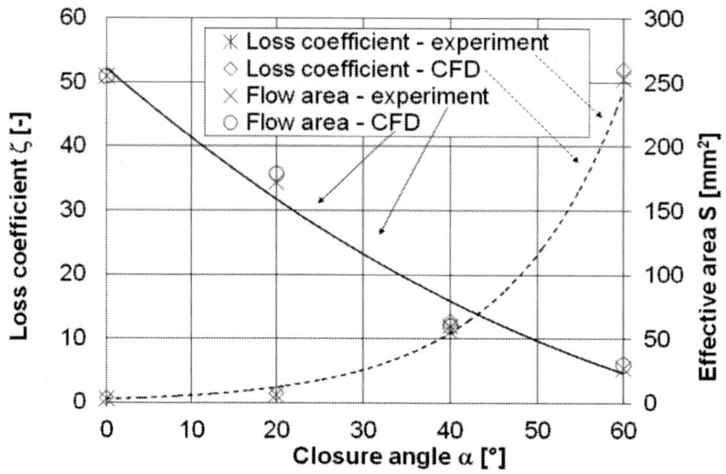

Fig. 9. Dependency of the loss factor ζ.

The difference between the experimental and mathematically determined values of the resistance factors was up to 15 %. The maximum value was reached for maximum opening because the mathematical model did not include the transition between the pipe and the valve. This transition is determined by the solid welding of plastic elements and therefore its geometry is undefined. In the case of partial closure of the valve, the effect of this local loss is negligible compared to local loss due to flou around the edges of the valve hole.

6 Conclusion

The aim of the paper was to create an experimental device and a mathematical model to determine the characteristics of the ball valve. In the introductory part the theory of characteristics and resistance coefficients of different types of valves is treated in detail. Measurement of characteristics for the selected ball valve and mathematical modeling under the same boundary conditions was performed. The optimal turbulent mathematical model was chosen by the testing, which took into account the problematic swirl flow at partial opening of the valve. The results of mathematical modeling were compared with experimental measurements in terms of loss coefficient, discharge coefficient and effective area. Analogously, it is possible to investigate the characteristics of valves of different designs in connection with the evaluation of specific resistance coefficients. Mathematical modeling of resistive coefficients is important in cases of large-diameter fittings and for specific media where experimental testing is problematic.

Work was supported by specific research SP2018/157 (Modelování a experimentální ověření dynamických jevů v tekutinových a vakuových systémech) and by the European Regional Development Fund in the Research Centre of Advanced Mechatronic Systems project, project number CZ.02.1.01/0.0/0.0/16_019/0000867 within the Operational Programme Research, Development and Education.

References

1. F.P. Incopera, D.P. Dewitt, T.L. Bergman, A.S. Lavine, *Fundamentals of heat and mass transfer. 6th* ed., 997 (2006)
2. S. Drábková, a kol. *Mechanika tekutin*. VŠB-TU Ostrava, 260 (2007)
3. M. Kozubková, *Modelování proudĕn FLUENT, CFX*, VŠB-TU Ostrava, 142 (2008)
4. M. Kozubková, T. Blejchař, M. Bojko, *Modelování přenosu tepla, hmoty a hybnosti*, VŠB-TU Ostrava, 174 (2011)
5. http://www.armaturygroup.cz/en/valves
6. B. Ulanicki, L. Picinali, T. Janus, Measurements and analysis of cavitation in a pressure reducing valve during operation – a case study. *Procedia Engineering*, 119, 270-279 (2015)
7. J. Jablonská, M. Kozubková, Flow characteristics of control valve for different strokes. EPJ Web of Conferences, 114, 5 (2016)
8. ČSN EN 60534-2-3 Regulační armatury pro průmyslové procesy – Část 2-3: Průtok – Zkušební postupy. Prosinec (1999)
9. A.S. Tabrizi, M. Asadi, G. XIE, G. Lorenzini, C. Biserni, Computational Fluid-Dynamics-Based Analysis of a Ball Valve Performance in the Presence of Cavitation. *Journal of Engineering Thermophysics*. 23, 1, 27-38 (2014)

10. S.C. Zhang, Y.L. Zhang, Z.M. Fang, Numerical simulation and analysis of ball valve three-dimensional flow based on CFD. *26th IAHR Symposium on Hydraulic Machinery and Systems*. 1-5 (2012)
11. A. Azad, D. Baranwal, R. Arya, N. Diwakar, Flow Analysis of Butterfly Valve Using CFD. *International Journal Of Modern Engineering Research*. 4, 6, 50-56 (2014)
12. V.J. Sonawane, T.J. Rane, A.D. Monde, R.V. Vajarinkar, P.C. Gawade, Design and Analysis of Globe Valve as Control Valve Using CFD Software. *IOSR Journal of Mechanical and Civil Engineering (IOSR-JMCE)*. 63-71 (2014)
13. J. Rak, *Matematické modelování proudění v kulovém ventilu*. Diplomová práce, (2014)

Vortex Tube: A Comparison of Experimental and CFD Analysis Featuring Different RANS Models

Radomír Chýlek[1,*], *Ladislav* Šnajdárek[1], and *Jiří* Pospíšil[1]

[1]Brno University of Technology, Faculty of Mechanical Engineering, Dept. of Power Eng., Technická 2896/2 Brno, Czech Republic

Abstract. The Ranque–Hilsch vortex tube represents a device for both cooling and heating applications. It uses compressed gas as drive medium. The temperature separation is affected by fluid flow behaviour inside the tube. It has not been sufficiently examined in detail yet and has the potential for further investigation. The aim of this paper is to compare results of numerical simulations of the vortex tube with obtained experimental data. The numerical study was using computational fluid dynamics (CFD), namely computational code STAR-CCM+. For the numerical study, a three-dimensional geometry model, and various turbulence physics models were used. For the validation of carried out calculations, an experimental device of the vortex tube of identical geometrical and operating conditions was created and tested. The numerical simulation results have been obtained for five different turbulence models, namely Standard k-ε, Realizable k-ε, Standard k-ω, SST k-ω and Reynolds stress model (RSM), were compared with experimental results. The most important evaluation factor was the temperature field in the vortex tube. All named models of turbulence were able to predict the general flow behaviour in the vortex tube with satisfactory precision. Standard k-ε turbulence model predicted temperature distribution in the best accordance with the obtained experimental data.

1 Introduction

Cooling and heating performance of a Ranque–Hilsch vortex tube is a subject of research nearly a century after it was accidentally invented by Ranque. This simple and compact device operates without any moving mechanical parts and simultaneously separates a pressurized gas into cold and hot fractions. A simplified model of the counter-flow vortex tube with displayed streamlines is shown in Figure 1. Experimental, analytical and numerical studies have been developed for better understanding of the problem. These studies were focused on gas thermal separation and energy transformations within the vortex tube, with a target to achieve better cooling and heating effects. In contrast with the simplicity of the device itself, the observed energy and flow separation processes are very challenging to describe it analytically or numerically.

* Corresponding author: radomir.chylek@vut.cz

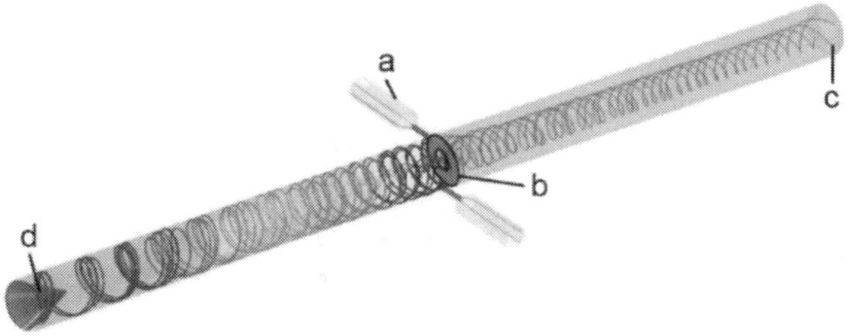

Fig. 1. Schematic model of the vortex tube; a-tangential inlet nozzle, b-cold exit orifice, c-cold outlet, d-hot control valve and hot outlet.

There is no general agreement on the theory of such a complex process. However, radial pressure gradient, generated by gas entering the tube after leaving nozzles, is considered to be the main reason for the occurring temperature and flow separation effects. A rapidly spinning vortex of the air is created, when air expands after passing the nozzles, at the outer diameter of a hot section of the vortex tube. A fraction of the air, named as a hot fraction, leaves the hot section at the hot end, after passing through a control valve. Remaining second fraction of the air is due to the control valve forced to move in an opposite direction to the hot fraction. It forms an inner vortex which is heading to the opposite side than the outer vortex of the hot fraction. Gas of the inner flow transfers heat energy to the outer flow when passing the hot section. Then the cooled inner flow passes through the cold exit orifice, enters a cold section and leaves the vortex tube colder as a cold fraction. Difference between temperature of the air at inlet T_{IN} and at the cold end T_C is given as:

$$\Delta T_C = |T_{IN} - T_C| \tag{1}$$

Hot end temperature T_H difference is defined similarly, as:

$$\Delta T_H = |T_{IN} - T_H| \tag{2}$$

All numerical studies of the vortex tube are bound by a selection of a suitable turbulent model. The selected numerical model must describe a continuum of the gas moving through the vortex tube as accurately as possible in terms of flow description, energy separation and temperature and pressure field. In the present state of development of computing power, a Computational Fluid Dynamics (CFD) numerical analysis can be used for fully three-dimensional simulations of the vortex tube including complex turbulent and physics models.

A large variety of both RANS and LES turbulent models was used for two-dimensional simulations, or partially and fully three-dimensional simulations, to produce results of differing accuracy when compared to those obtained by an experiment. However, from the previous research it is not clear which turbulent model has achieved the best agreement with experimental results in terms of temperature fields. Models used most frequently were two-equation turbulence models k-ε Standard and k-ε Realizable, Reynolds stress model (RSM) and Large eddy simulation model (LES) with the Smagorinsky–Lilly variant. Numerical studies mostly focused on the optimization of the vortex tube parameters and enhancing its performance using some of these turbulence models. A list of previously used turbulent models for selected numerical analysis of the vortex tube is shown in Table 1.

Table 1 indicates that nowadays authors are turning to more demanding three-dimensional models, either fully three-dimensional or with axis-symmetrical conditions. Different authors tried to apply Large Eddy Simulation (LES) and RSM models, which are computationally more expensive than two-equation RANS methods. The most commonly used turbulent model is Standard k-ε model due to the accuracy of some reported results and lower computational costs.

Table 1. Summary of selected numerical studies on vortex tubes.

Year	Name of the author	Geometry	Used turbulence model	Software
1999	Fröhlingsdorf [1]	2D	Keyes k-ε	CFX
2005	Aljuwayhel [2]	2D	Standard k-ε, RNG k-ε	Fluent
2006	Eiamsa-ard [3]	2D	Algebraic RSM (ASM)	TEFESS
2006	Skye [4]	2D	Standard k-ε, RNG k-ε	Fluent
2008	Behera [5]	3D	RNG k-ε	STAR-CD
2009	Farouk [6]	2D	LES	CFD-ACE+
2010	Shamsoddini [7]	3D	RNG k-ε	Fluent
2013	Ouadha [8]	3D	RSM	Fluent
2014	Bovand [9]	3D	RNG k-ε	Fluent
2014	Alekhin [10]	3D	SAT-SST	CFX
2014	Liu [11]	3D	Realizable k-ε	Fluent
2016	Bianco [12]	3D	RSM-LRR, LES	CFX
2017	Thakare [13]	2D	Standard k-ε	Fluent

The following numerical studies are directly devoted to the problem of selecting a suitable turbulent model for the counter-flow vortex tubes. A two-dimensional axis-symmetrical model was used for various turbulence models and discretization schemes by Thakare [14]. It was found that the most suitable turbulence model was k-ε Standard. It was observed that the deviation in the prediction of a total temperature of air in the cold region, as well as other parameters of the flow, were in an acceptable range using k-ε Standard model.

Eimsa-ard et al. [15] used Algebraic Stress Model (ASM) and Standard k-ε turbulence model for simulating the thermal separation. ASM had better agreement with experimental results.

Liu and Liu [11] compared numerical results of five turbulence models (SST, Standard k-ε, RNG k-ε, Realizable k-ε and RSM) with obtained experimental data. The distinction between experimental results and the numerical model was smallest using k-ε Realizable and RSM models. Similar studies with similar results were conducted by Baghdad [16] and Dutta [17], but with different criteria for choosing the appropriate turbulent model.

Farouk et al. [18] carried out Large Eddy Simulation (LES) and observed that energy separation by LES was in better agreement with experiment than results obtained using k-ε. However, LES under predicted the cold temperature drop.

Secchiaroli et al. [19] handled LES Smagorinsky sub-grid scale closure in the simulations of the vortex tube. They also employed RSM and RNG k-ε models to compare temperature and velocity plots. It was found that RSM agrees with LES better than RNG k-ε in case of temperature and velocity profiles.

In terms of a double-circuit vortex tube, Alekhin et al. [10] and Bianco et al. [12] investigated and compared Standard k-ε, RNG k-ε, RSM, SAS-SST and LES turbulent models for a more complex double-circuit vortex tube. It was found that RSM and LES turbulence models are the most suitable for this more complicated device.

However, several important factors can influence the choice of turbulence model for vortex tube numerical analysis. Mainly the selection of a fully three-dimensional model can play a key role. Choosing specific geometrical parameters of the tube and its inlet and outlet operating parameters should also be evaluated.

The main aim of this work is to choose the most accurate turbulence model for the fully three-dimensional model of the vortex tube in terms of temperature distribution and flow separation within the tube. STAR-CCM+ computational code was used for the numerical study, and the selection of turbulence was directed to RANS-based turbulence models, specifically to two-equation RANS models and Reynolds Stress Transport model. To compare turbulent models, an experimental vortex tube of corresponding geometrical and operating conditions was created and analysed.

2 Experimental study

For comparison and validation of numerical simulations, the experimental device was set up. Its essential geometrical parameters are listed in Table 2. The same parameters were used in the numerical simulations. The material of the vortex tube casing was Polymethyl methacrylate-PMMA. Six tangential nozzles were extruded to the nozzle module of Polyoxymethylene (POM).

Table 2. Geometric parameters of the experimental vortex tube.

Parameter	Value	Unit
Hot section length	360	[mm]
Cold section length	200	[mm]
Number or inlet nozzles	6	[-]
Vortex tube diameter	24	[mm]
Cold exit orifice diameter	8.5	[mm]
Hot control valve diameter	23.4	[mm]
Nozzle diameter	2	[mm]

Pressurized air was supplied by the compressor, accumulated in the air tank and dried in the air dryer. Before the air enters the nozzles of the vortex tube, its pressure is regulated by the reducing valve. Temperature (thermocouple type T, accuracy ±0.25 °C), volume flow (accuracy ±5 %), absolute pressure (accuracy ±0.35 %) and relative pressure (accuracy ±0.6 %) of the inlet air were measured. The volume flow of the cold fraction is measured at the cold end of the vortex tube by the anemometer (accuracy ±1 %). The schematic diagram of the experimental setup is presented in Figure 2. The experimental setup of the vortex tube is shown in Figure 3.

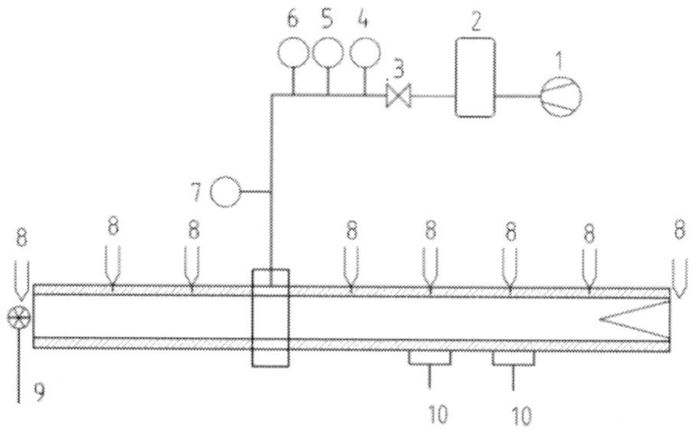

Fig. 2. Schematic diagram of the experimental setup. 1-compressor, 2-air tank and air dryer, 3-reducing valve, 4-relative pressure sensor, 5-thermocouple, 6-flowmeter, 7-absolute pressure sensor, 8-type T thermocouples, 9-anemometer, 10-heat flux sensor.

Fig. 3. The experimental setup of the vortex tube. 7-absolute pressure sensor, 8-type T thermocouples, 9-anemometer, 10-heat flux sensors.

The temperature of the cold and hot fraction is measured by type T thermocouples (accuracy ±0.1 %) at both ends of the vortex tube. The thermocouples were also embedded into drilled holes of the tube casing at regular distances along the length of the tube and placed 1 mm from the inner surface of the vortex tube. Tube casing is also equipped with heat flux sensors (accuracy ±3 %) at the hot section to monitor and determine the steady state of the measurement.

After the compressor pressurized the air tank to 8 bar, reducing valve was released until the desired relative pressure of 4 bar was reached for the inlet nozzles. The experimental setup is tested for any leakage in the system. Software LabVIEW and National Instruments instrumentation were used to monitor and regulate the system. The temperature of the air at the inlet is 20 °C. Cone control valve is centered and the distance between two created concentric circles is set to 0.3 mm. The experimental observation requires 30 – 45 min. until all temperature and heat flux sensors reported steady-state values.

3 Numerical study

3.1 Geometry

A fully three-dimensional model was used for the numerical simulations. Geometric parameters for the numerical model correspond with parameters of the experimental device in Table 2. Relaxation section was added beyond the hot end orifice to determine the temperature of the hot fraction. The geometry of the vortex tube is shown in Figure 4.

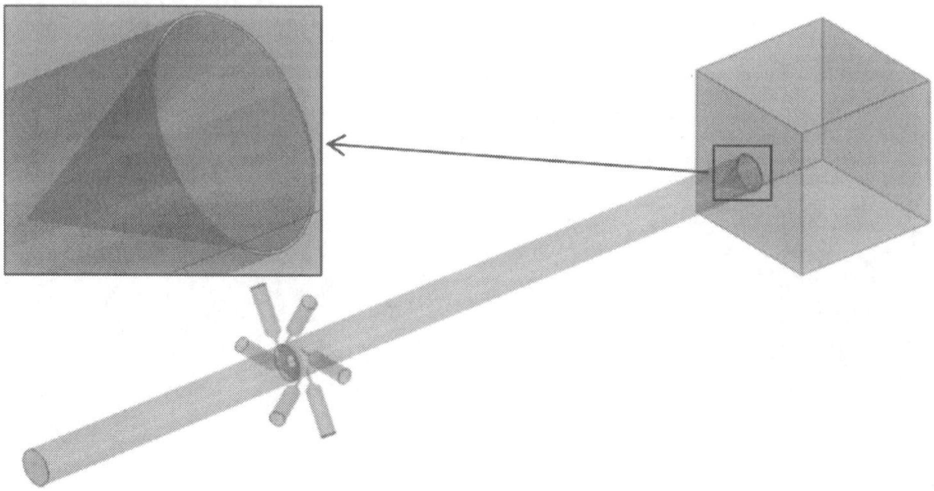

Fig. 4. The geometry of the vortex tube with the hot orifice in the detail.

3.2 Mesh

Automatic mesher integrated into STAR-CCM+ was used to create a partially structured mesh. The polyhedral mesher and the prism layer mesher modules were used. The resulting mesh contains about 8.2 million cells. The number of cells was slightly different according to the used turbulent module and its requirements for the boundary layer. A grid independence study of solution was conducted. For fewer than 1 million cells, the fully 3D simulation was unstable for some turbulence models. For 8 million to 13 million of grid cells, temperature values for hot and cold fraction were within a one–percent range for the individual turbulent model. The volumetric control refinements were used in the sections of higher pressure, higher temperature gradients, and with transonic velocity. Refinements were primarily used in the section of the inlet nozzles and near the hot end orifice. High velocity of air is reached near the nozzle inlets (Ma > 0.8) and hot end refinements, see Figure 5.

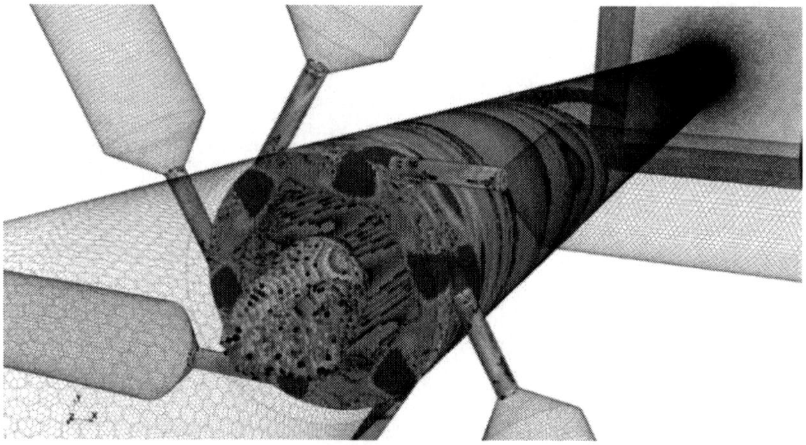

Fig. 5. High velocity of air at the nozzle inlets (Ma > 0.8) and hot end refinement.

3.3 Governing equations

The turbulent, compressible flow of real gas (Redlich-Kwong equation) was considered for the carried out numerical analysis. Governing equations of continuity, momentum, and energy and mass are:

$$\frac{\partial \rho}{\partial t}+\nabla\cdot(\rho \boldsymbol{v})=0 \tag{3}$$

$$\frac{\partial(\rho \boldsymbol{v})}{\partial t}+\rho(\boldsymbol{v}\cdot\nabla)\boldsymbol{v}=-\nabla p+\nabla\cdot\tau \tag{4}$$

$$\frac{\partial}{\partial t}(\rho H)+\nabla(\rho \boldsymbol{v} H)-\nabla\left(\frac{\lambda+\lambda_t}{c_p}\nabla h\right)=\frac{\partial p}{\partial t} \tag{5}$$

$$p=\frac{RT}{V-b_1}-\frac{a_1}{T^{0,5}V(V+b_1)} \tag{6}$$

Where ρ is air density, H is total enthalpy, h is static enthalpy, p is static pressure, τ is stress tensor, v is velocity, λ is coefficient of molecular thermal conductivity, c_p is heat capacity, V is volume, R is gas constant and constants a_1 and b_1 are defined as:

$$a_1=0,42748\frac{R^2T_{cr}^{2,5}}{p_{cr}} \tag{7}$$

$$b_1=0,08664\frac{RT_{cr}}{p_{cr}} \tag{8}$$

Where T_{CR} [K] is critical temperature, p_{CR} is cirtical pressure [Pa] and Turbulent thermal conductivity λ_t is calculated from turbulent Prandtl number Pr_t as:

$$\frac{\lambda_t}{c_p}=\frac{\mu_t}{\mathrm{Pr}_t} \tag{9}$$

Where the coefficient of turbulent dynamic viscosity μ_t is given by means of turbulence models (k-ε, k-ω, RSM Linear Pressure Strain). Turbulent Prandtl number Pr_t was set to 0.8.

Used RANS models offer the most economic approach for computing complex turbulent flows. RANS models simplify the problem of turbulence to the solution of two additional transport equations and introduce a turbulent viscosity to compute Reynolds stresses. One of the transported variable is the turbulent kinetic energy k (it determines the energy of the turbulence), the second is for certain cases Turbulent disipation ε (k-ε Standard and k-ε Realizable) or the Specific turbulent disipation ω (k-ω Standard or k-ω SST) and it determines the scale of turbulence. More complex RANS models that solves an individual equation for each of the six independent Reynolds Stresses directly are Reynolds Stress Models (RSM).

3.4 Boundary conditions

Numerical simulation was initialized in a steady-state regime with the coupled solver. After it sufficiently advanced, it was shifted to an implicit unsteady regime. Reference pressure was set to 101325 Pa. The initial static temperature was 20 °C. Boundary conditions are shown in Table 3.

Table 3. Boundary conditions for numerical models.

Boundary	Quantity	Value	Unit
Inlet nozzles	Relative pressure	4	[bar]
	Temperature	20	[°C]
	Turbulence intensity	5 %	[-]
Hot outlet	Relative pressure	0	[mm]
Cold outlet	Relative pressure	0	[-]
Walls	Shear stress	No slip	[-]
	Thermal specification	Adiabatic	[-]

3.5 Numerical solution

All solutions were performed in STAR-CCM+ software. The coupled solver was used for all carried out calculations with turbulence model based on the hybrid LSQ-Gauss gradient method. Second-order upwind discretization scheme was selected. First-order temporal scheme (Euler Implicit) was chosen for transient calculations.

Simulations were executed on the cluster on sixteen processor cores arranged in a computational cluster. One transient simulation took from 10–30 days to simulate 0.05–0.2 second of the physical time. Calculation times were dependent on the turbulent model, convergence status and rate of stabilization of temperature field in the vortex tube. Time-step was set to values from 1.0E-6 to 1.0E-5 with the number of inner iterations from 5 to 10. Residual values were kept below 1.0E-4 with this procedure. Temperatures at the hot and cold end were monitored in order to provide monitor-based judging criteria of convergence, together with the residual criterion.

4 Results and discussion

This section is dedicated to the validation of used numerical models. Results from individual numerical models were compared with obtained experimental data. The comparison was focused on the temperature distribution. Firstly, the total temperature distribution in the mid–plane of the tube is shown in Figure 6–10 for every turbulence model. Minimal and maximal range values of total temperature were varied based on maximum and minimum temperature achieved by the individual turbulence model. It was observed, that the general flow behaviour creates hot and cold sections in every case, not depending on the used turbulence model. Energy separation mechanism was therefore not affected by used turbulent model. However, the maximum and minimum temperature values were dependent on the turbulent model.

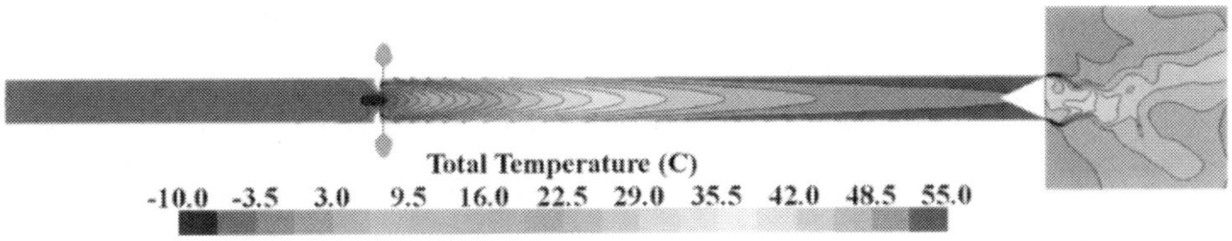

Fig. 6. Total temperature distribution for k-ε Standard turbulence model.

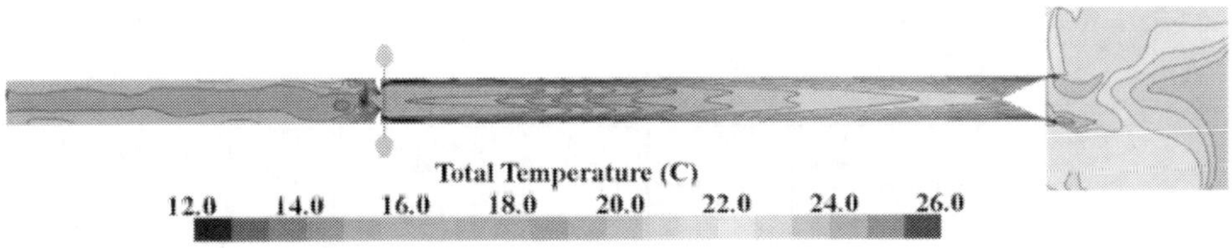

Fig. 7. Total temperature distribution for k-ε Realizable turbulence model.

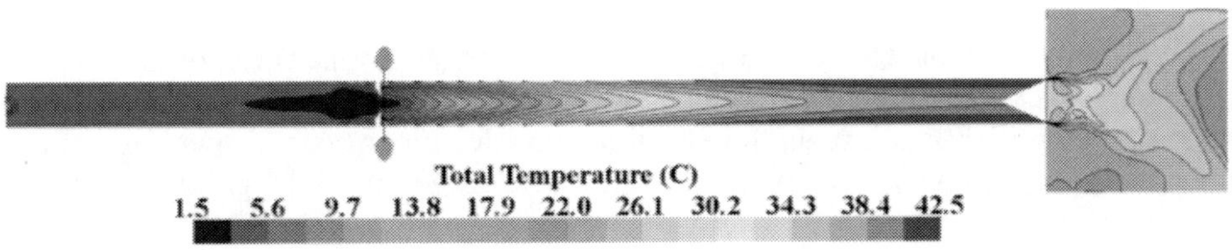

Fig. 8. Total temperature distribution for k-ω SST turbulence model.

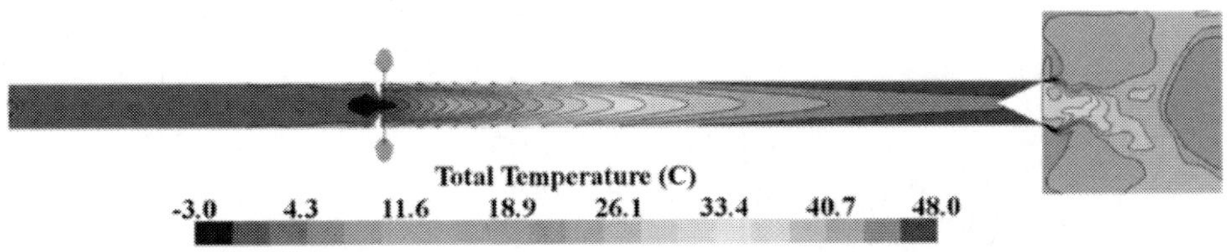

Fig. 9. Total temperature distribution for k-ω Standard (Wilcox) turbulence model.

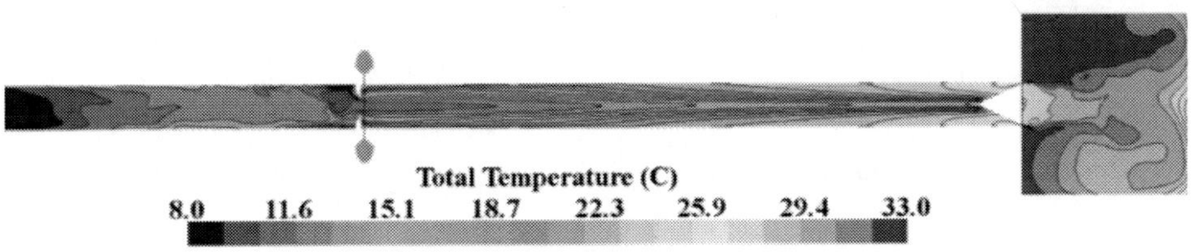

Fig. 10. Total temperature distribution for RSM turbulence model.

In order to compare the values of total temperature of the hot and cold fraction, % cold and hot deviation is declared as:

$$\%\left(\Delta T_C\right)=\left|\frac{\left(\Delta T_C\right)_{exp}-\left(\Delta T_C\right)_{cfd}}{\left(\Delta T_C\right)_{exp}}\right|\times 100 \tag{10}$$

$$\%\left(\Delta T_H\right)=\left|\frac{\left(\Delta T_H\right)_{exp}-\left(\Delta T_H\right)_{cfd}}{\left(\Delta T_H\right)_{exp}}\right|\times 100 \tag{11}$$

These % deviations between experimental and CFD total temperature values of hot and cold fraction are shown in Table 4.

Table 4. % deviations of the cold and hot end temperatures.

	T_C	T_H	% (ΔT_C)	% (ΔT_H)
Experiment	-1.8	57.2	[-]	[-]
k-ε Standard	-3.0	54.9	5.50	6.18
k-ε Realizable	15.4	25.3	78.90	85.75
k-ω SST	4.6	41.0	29.36	43.54
k-ω Standard	1.8	47.0	16.51	27.42
RSM	8.1	32.3	45.41	66.94

Temperatures along the tube were also compared in Figure 11. For the experiment, temperatures were measured by thermocouples located in drilled holes on the surface of the plastic tube in points 1–6, and with two thermocouples measuring temperature of the cold and hot fraction (T_C and T_H). Numerical values were gathered from a points at a distance of 1 mm from the wall.

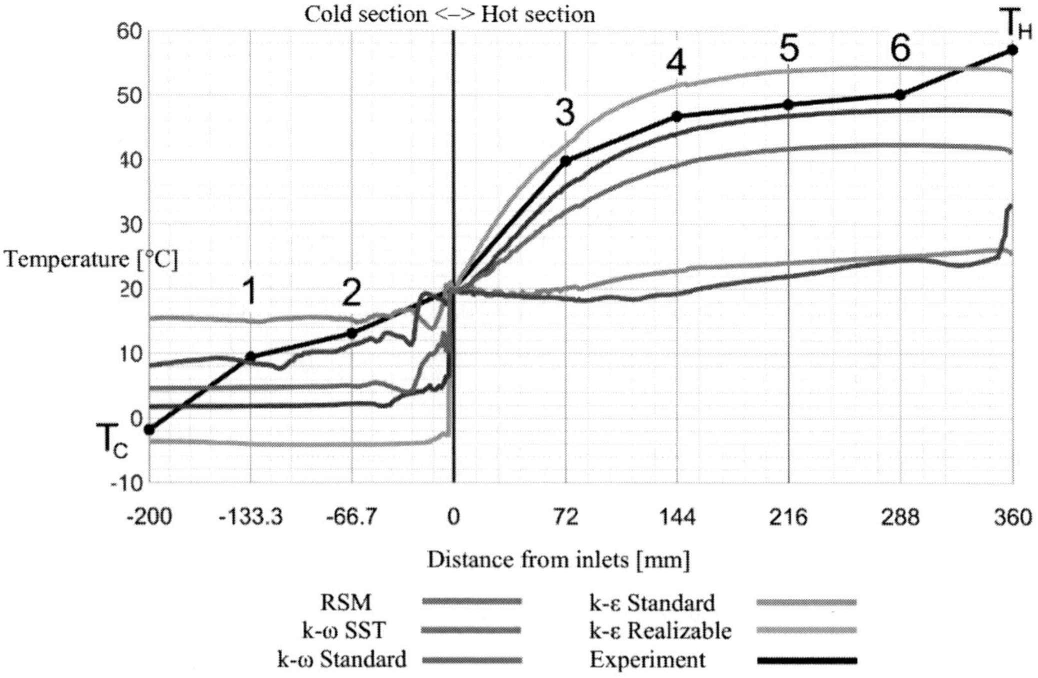

Fig. 11. Distribution of total temperatures.

As can be observed from Table 4 and Figure 11, the smallest deviation was achieved with turbulence model k-ε Standard. Both for the cold and hot fraction % deviations. The greater discrepancy in points 1 and 2 in the Figure 11 was caused by heat conduction of the plastic tube used for the experimental study. Nothing suggests that the cold section causes the cold fraction to cool down more. For this reason, it can be assumed that the temperature of the air T_C of the cold fraction is characteristic for the entire cold section.

5 Conclusion

For the numerical study, a three-dimensional turbulent real gas model was used. Numerical simulations were carried out in computational code STAR-CCM+. Simulation were using various turbulence models, namely: Standard k-ε, Realizable k-ε, Standard k-ω, SST k-ω and RSM. Presented numerical models were validated and compared with the obtained experimental data. The following major conclusions were summarized:

The general temperature distribution, energy separation effects, and flow behaviour worked with every used turbulence model to some extent.

However, magnitudes of ΔT_C and ΔT_H were different depending on the used turbulence model in the numerical simulation.

Significant deviations in cold and hot fraction temperature differences of Realizable k-ε turbulence model were observed.

More demanding RSM model should produce more complex and accurate velocity fields thanks to the anisotropy and six more partial differential equations for each component of Reynolds stress tensor solved. However, it underpredicted magnitudes of ΔT_C and ΔT_H as compared with RANS models, except k-ε Realizable model.

Models k-ε Standard and k-ω Standard predicted temperatures of both fractions in the better accordance with the obtained experimental data than other models covered in this paper.

The smallest deviations were achieved by k-ε Standard model with 5.5 % for the cold fraction and 6.2 % for the hot fraction in a the comparison with experimental values.

This research was financially supported by the Brno University of Technology, Faculty of Mechanical Engineering in frame of the Specific Research Fund, grant no.: FSI-S-17-4531

References

1. W. Fröhlingsdorf, H. Unger, Int J Heat Mass Transf, **42**, (1999)
2. N. Aljuwayhel, G. Nellis, S. Klein, Int J Refrig, **28**, (2005)
3. S. Eiamsa-ard, P. Promvonge, Int J Heat Mass Transf, **50**, (2007)
4. H. Skye, G. Nellis, S. Klein, Int J Refrig, **29**, (2006)
5. U. Behera, P. Paul, K. Dinesh, S. Jacob, Int J Heat Mass Transf, **51**, (2008)
6. T. Farouk, B. Farouk, A. Gutsol, Int J Heat Mass Transf, **52**, (2009)
7. R. Shamsoddini, A. Nezhad, Int J Refrig, **33,** (2010)
8. A. Ouadha, M. Baghdad, Y. Addad, Int J Refrig, **36**, (2013)
9. M. Bovand, M. Valipour, S. Eiamsa-ard, A. Tamayol, Int Commun Heat Mass Transf, **50,** (2014)
10. V. Alekhin, V. Bianco, A. Khait, A. Noskov, Int J Therm Sci, **89,** (2015)
11. X. Liu, Z. Liu, Appl Therm Eng, **67**, (2014)
12. V. Bianco, A. Khait, A. Noskov, V. Alekhin, Appl Therm Eng, **106,** (2016)
13. H. Thakare, A. Parekh, Energy, **133,** (2017)
14. H. Thakare, A. Parekh, Int J Heat Mass Transf, **78,** (2014)
15. S. Eiamsa-ard, P. Promvonge, Int Commun Heat Mass Transf, **35,** (2008)

16. M. Baghdad, A. Ouadha, O. Imine, Y. Addad, Int J Therm Sci, **50,** (2011)
17. T. Dutta, K. Sinhamahapatra, S. Bandyopdhyay, Int J Refrig, **33,** (2010)
18. T. Farouk, B. Farouk, Int J Heat Mass Transf, **50**, (2007)
19. A. Secchiaroli, R. Ricci, S. Montelpare, V. D 'alessandro, Int J Heat Mass Transf, **52**, (2009)

Pressure and velocity conditions in home health care accessory for babies

Blanka Skočilasová[1,*], *Jan* Skočilas[2], and *Josef* Soukup[1]

[1]University J. E. Purkyně in Ústí nad Labem, Faculty of Mechanical Engineering, Institute of Machinery and Energetics, Na Okraji 1001, 400 96 Ústí nad Labem, Czech Republic
[2]Czech Technical University in Prague, Faculty of Mechanical Engineering, Department of Process Engineering, Technická 4, 166 07 Praha 6, Czech Republic

Abstract. The home health care accessory for babies is selected for investigation. The objectives are to find out the maximal under-pressure and air velocity, which is possible to reach by accessory Nosko® using standard home vacuum cleaner recommended by manufacturer of the accessory. CFD software is used for prediction of the flow characteristics of the air in the body of the accessory. Also simple experiment with the trace is performed in the transparent body. The flow characteristic of the Nosko® is obtained by systematic simulation and can be used for further engineering calculation and detailed investigation. Maximum flow rate of the air through the accessory under maximal under pressure generated by home vacuum cleaner was 2.6 litres. s^{-1} which is comparable with results of investigation of the human breathing in the literature source.

1 Introduction

The progress in home health care is obvious in the every following human generation, which is reflected by utilization of the new materials and technologies. Above all the evolution of the new preparations, gadgets and supplements or accessories is the most prominent in the field of baby care. One of these accessories is the tool for removing of the phlegm from baby sinus nasal for example. We have chosen it for our investigation. The Nosko® is the product of the company Ranyak Gyyrgy, Budapest, Hungary. It is plastic draft tube extender, which is inserted into the baby nose. The second end is connected to the home vacuum cleaner. However, it seems that it is drastic or torturous tool, the manufacturer is boasting about the high efficiency of the phlegm removal and prevention to the phlegmasia of upper respiratory tract or otitis media (inflammatory diseases of the middle ear). The product passed by clinical tests and moreover, according to long time users references, the satisfaction of using this as the prevention of diseases (also for adults) prevails. We stood in front of the decision (in the role of the parents as well as engineers) to use standard methods or new surroundings recommended accessory Nosko®. Despite of all positive references, we firstly investigated and tested the accessory before first application to baby. The results of the investigations are presented in this contribution.

* Corresponding author: blanka.skocilasova@ujep.cz

In the field of the fluid flow in nozzles, there are many basic and specialized literature [1-3]. This issue is deeply investigated and continuously broaden out till present. The many theoretical and experimental studies exist, focused on the pressure and velocity profiles in or around the nozzle for different purposes, flow regimes or method of the nozzle operation (discharge or suction). Today, CFD models are standard tools of the scientists, which save the time and cost of experiment, however the result of the simulation is necessary at least partially to compare with experiment or basic theory. Many author use CFD software for investigation of the fluid flow through nozzle [4], e.g. distribution of the air in the building or room for air-condition [5-6], cavitation in the fuel injection nozzles [7] or fluid flow together with heat transfer of the impinging stream on the cooled or heated wall [8]. From these articles the general rules and recommendation raised, which deals with set up of the mathematical models from proper geometry generation and simplification, CGI analysis [9], density and quality of the mesh including suitability of the different type of the mesh elements, up to recommendation for different models of turbulence and relationship between primitive variables in Navier-Stokes equations and continuity equations.

Nosko® is utilized together with dome vacuum cleaner. The European standard [10] exists for these kind of the cleaners commonly available in the market, which above all specifies the limits of operation parameters for dry or wet cleaner. Maximal reachable uder pressure is limited to 30 kPa for dry cleaning. This standard also states the maximal volumetric flow-rate of the air in the cleaner, which is from 20 up to 40 litres.s^{-1} for standard vacuum cleaners.

From above mentioned information the doubt is arising. Is it ever reasonable to interconnect the Nosko® with home vacuum cleaner on the one side with baby sinus nasal on the other side? Is there any threat of baby health impediment? Several authors describe standard pressure relation in the sinus nasal of adults under normal breath condition. The flow rate of the air via the nose during maximum-effort vital capacity maneuvers reached 4.32 litres.s^{-1} on inspiration and 7.09 litres.s^{-1} on expiration period [11]. The most often these studies are published in the special medical journals, however we have doubt about the clinical tests existence, which deals with investigation of maximal under pressure of the air in the baby sinus nasal limited by baby health impediment. Therefore this article deals with investigation of the maximal available operation conditions reachable by Nosko® accessory together with home vacuum cleaner. Pertinent examination of these results from baby health point of view is conveyed to medical experts with respect.

Above all the article objective is to find out the maximal under-pressure and air velocity, which is possible to reach by Nosko® using standard home vacuum cleaner. The simulation results of air flow in accessory by CFD tool were partially compared with simple experiment considering the observation of the flow contours – streamlines of the tracer. The flow characteristic on the accessory was also obtained from simulations.

2 Materials and methods

For the investigation the standard home health care accessory for babies, Nosko® was chosen for this study, which is commonly available in the market. Nosko® is the product of the company Ranyak Gyyrgy, Budapest, Hungary. It is transparent plastic draft tube extender, which is inserted into the baby nose. The second end is connected to the home vacuum cleaner. The general look of this tool is presented in the Figure 1. The narrowed part is inserted into the baby nose. The sleeve which is perpendicular to the axis of the accessory is connected by silicon tube and reducer with the hose of the home vacuum cleaner. The inlet jet cavity is prolonged and its diameter is extended by inserted straight tube with constant thickness almost to the bottom end of the outer shell of the accessory body. The air enters to the Nosko® by narrow nozzle. The stream channel is gradually enlarged to the diameter of inserted tube. The air axially continues to the end bottom, where it is completely turning round to the counter flow direction. Then the air flows by annular section between outer shell and inserted tube and finally it leaves the body in perpendicular direction to the axis by sleeve. The phlegm is sucked in with air but it doesn't leave the Nosko® and remains in the body, because the described geometry also serves as gas-liquid separator. The body is easy demountable to three pieces for water cleaning.

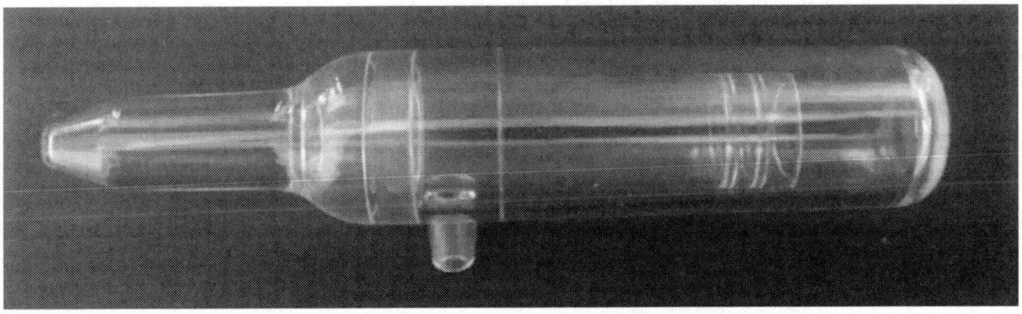

Fig. 1. Side view of the Nosko®.

The geometry for CFD simulation was generated according to the measured dimensions of the body in CAD/CAM modeller. The wall were completely remove because only the negative of the inner volume was used for simulations, see Figure 2.

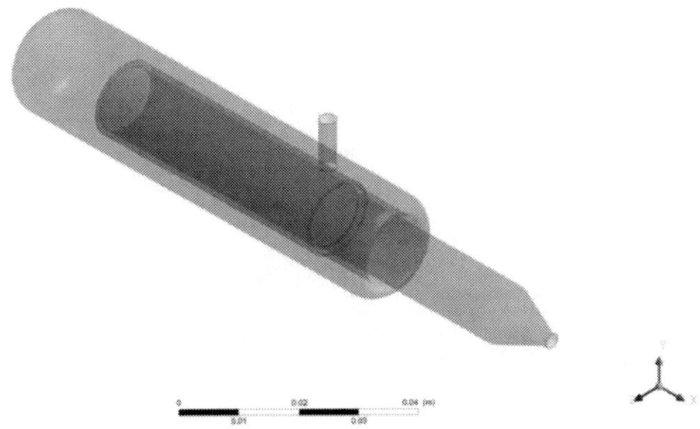

Fig. 2. Final geometry of model for CFD simulation.

For CFD simulation and mesh generation the packages of the software ANSYS® were used. The simulations were performed in Fluent ver.16.0 software. The CGI analysis [9] was also performed before systematic investigation of the flow characteristics. The final mesh has approximately 500.000 element which was acceptable for further simulation. Due to rounded geometry the tetrahedral elements prevails. Quality of the mesh was inspected (especially orthogonal quality and skewness) and the mesh was optimized replacing the worst elements and by using improving mesh quality tool in Fluent software. The gravity forces and compressibility of the air were not assumed in the calculations. The physical properties of the air was considered under standard pressure and temperature conditions. The boundary conditions were set up to pressure inlet and outlet respecting the maximal under pressure available for home vacuum cleaners [10]. After the maximum flowrate was obtained, the boundary conditions were changed to mass inlet and pressure outlet to investigate whole range of the flowrates and flow characteristic of the Nosko®. The k-ε SST model was used for turbulence regime of air flow. The solver was set upped to double precision, and second order upwind for momentum and pressure. SIMPLE scheme was used for relation between primitives' variables (velocity and pressure). At least 5 000 iterations were performed at each simulation for reaching the stable solution, however the oscillation of the residuals occurred. Therefore the average value of pressure from last 500 iteration was calculated for assessing the pressure drop.

Also simple experiments were carried out to observe the air flow streams in the body of the Nosko®. The basic home aerosol/fog generator Ultrasonic Nebulizer LD-250U, Nantong Co. Ltd., China, was used for production of the fog which was sucked into the body. The flow of the tracer was recorded by camera SONY HDR-CX115, Sony Co., China against the black background. The recorder file was investigated frame by frame and the best results of the tracer streams are presented. The velocity of the incoming stream into the Nosko® connected to home vacuum cleaner Miele Compact C1 PowerLine was also measured using the AHLBORN Almemo 2590 anemometer with hot wire probe AHLBORN ALMEMO FVAD15S240, but with unsuccessful result, because the stream velocity at the inlet exceed the measuring range of the probe which was 40 $m.s^{-1}$.

3 Results and discussion

The results of the simulations as well as the result of the simple experiment are presented here. The main graphical outputs of the simulation are the pressure and velocity distribution presented in the Figure 3 and 4. The investigated range of the pressure drop was from 20 Pa up to 56 kPa and corresponding flow rate from 0.03 litres. s^{-1} up to 2.6 litres. s^{-1}. All results presented here in the graphical form represent the flow condition 2.6 litres. s^{-1} and pressure drop 56 kPa (absolute pressure 47 kPa).

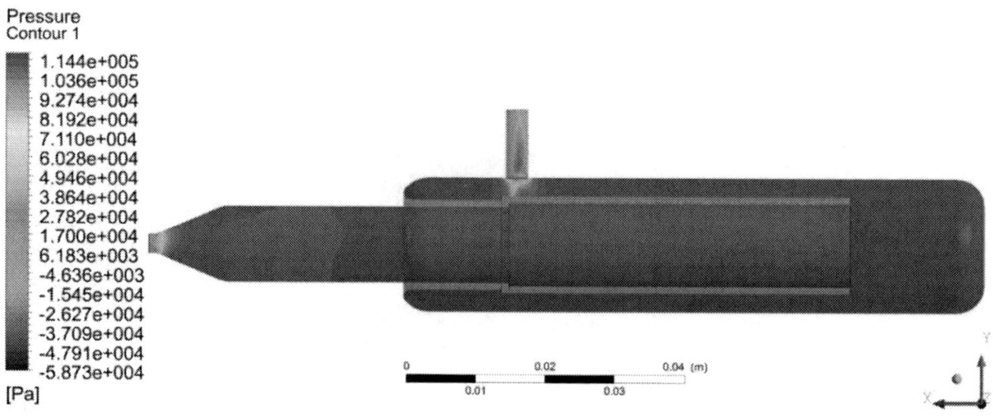

Fig. 3. Pressure contours for 2.6 $litres.s^{-1}$.

The maximal local pressure drops which are the main part of the global pressure drop of the investigated body are located at the inlet and outlet to the system. There are extreme reduction and expansion of the cross-section area, see Figure 3. It seems that the bending of the main stream by turning round at the end of the outer shell together with distribution of the stream from circular to annular cross section and bending and collecting of the whole stream into the outlet have no significant effect on the overall pressure drop respecting the uniform colour of the pressure in the main body.

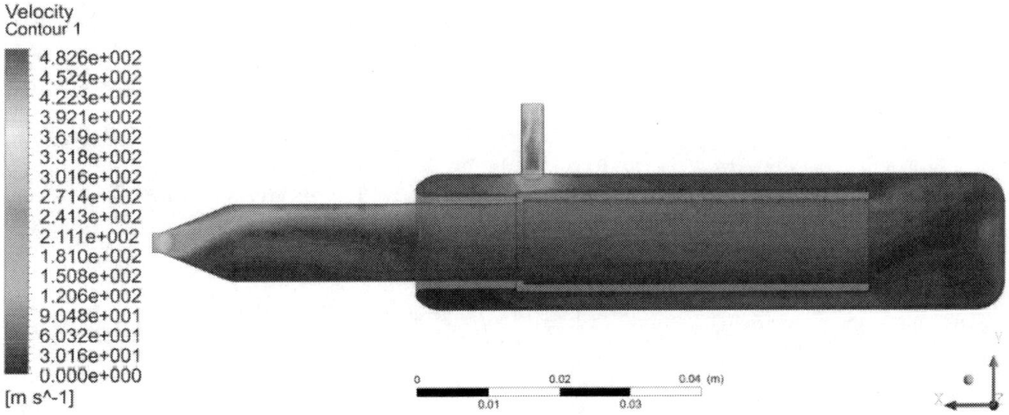

Fig. 4. Velocity contours for 2.6 litres.s^{-1}.

In the Figure 4 and 5 the contour of the velocity profile and streamlines are presented respectively. According to asymmetry of the body geometry in the XY plane, the stream of the air is oscillate and hasn't got the straight direction though the inner tube. After the inlet it is immediately bended to the vicinity of the wall. The movement of the air in the annular part of the body after the outlet tube is almost immobile. By the effect of the geometry the output stream is rotating in the outlet tube. Maximal velocity at the inlet is 427 m.s^{-1} and at the outlet is 424 m.s^{-1} for the maximal operation condition of the Nosko® derived from vacuum cleaner limitation.

The results of the simple experiment investigation the streamlines of the trace in the body is presented in the Figure 6. This streamlines partially agree with those obtained by simulation. The bending of the input stream can be observed in the both results. The white region in the left of the Figure 6 represent generated fog which is sucked into the body by arranged under pressure. In the right side of the Figure 6 the backflow in the central tube which is generated by reflection of the part of the main stream is slightly visible (left top near the inner wall). This backflow can be observed also in the Figure 4 of velocity contours as well as in the Figure 5 of stream lines.

In the Figure 7, the kinetic energy of turbulence contour is presented. The highest value of the kinetic energy of turbulence occurs next to the inlet region, where mostly by expansion of the stream the central vortex in the velocity filed is created. This central vortex push out the streamlines of the flux near to the wall. This vortex passes through central inner tube and it breaks down by viscous forces into the small vortices and it continuously dissipates into the heat. High intensity of the turbulence also occurs at the outlet of the body, which is expected due to flow contraction and complicated fluid flow from annular cross section into the small outlet tube.

The described and investigated complicated flow condition of the air affected by the body design of manufacturer serves primarily for phlegm separation from the air. This body design ensures that the phlegm remains in the body and eliminates the suction of the phlegm into the tube of the vacuum cleaner.

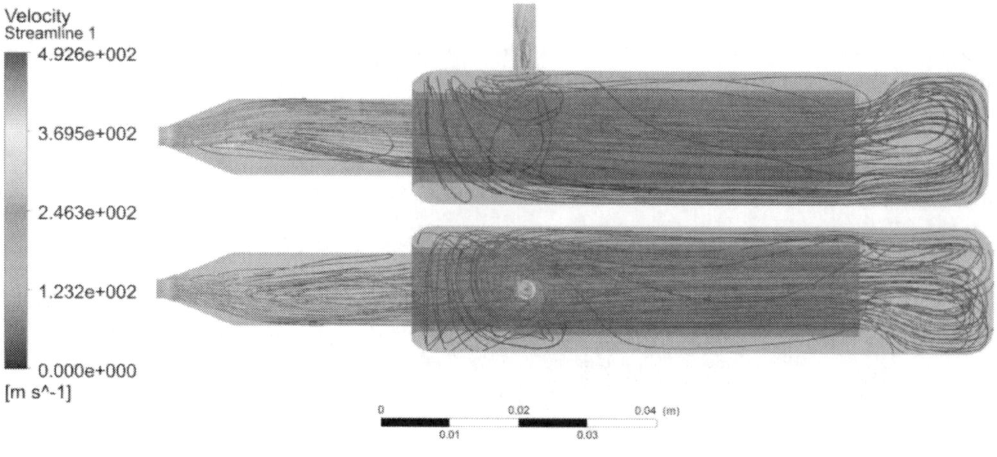

Fig. 5. Streamlines for 2.6 litres.s^{-1}.

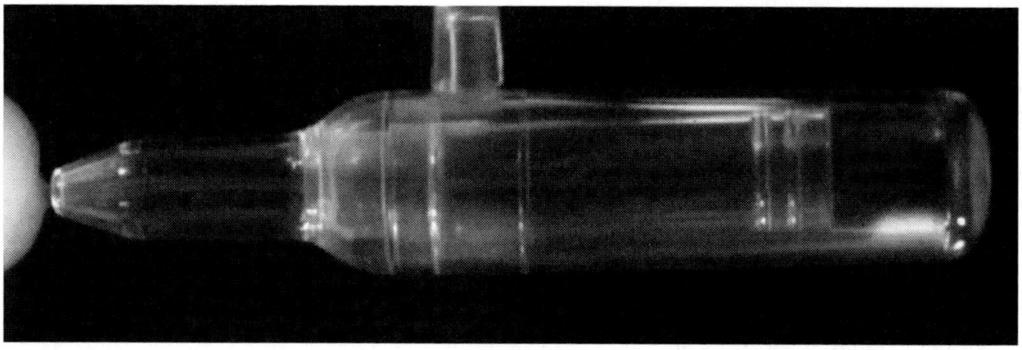

Fig. 6. Record of the tracer streamline obtained by experiment.

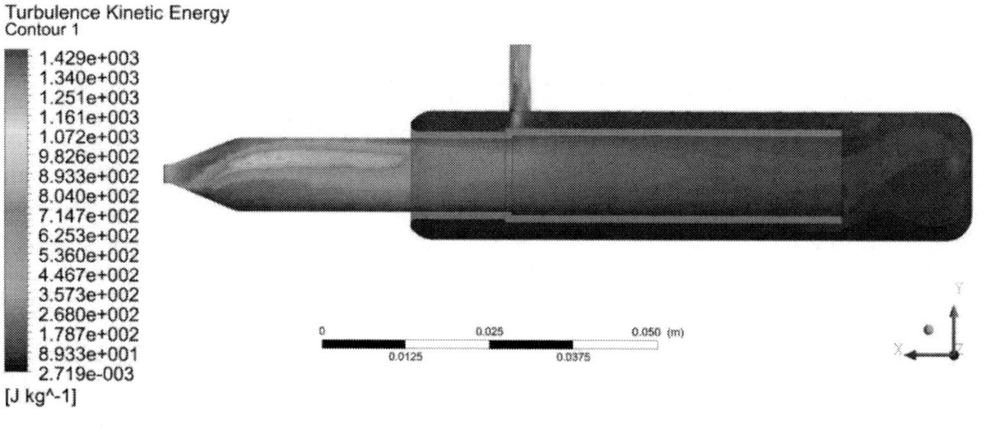

Fig. 7. Kinetic energy of turbulence flow contours for 2.6 litres.s^{-1}.

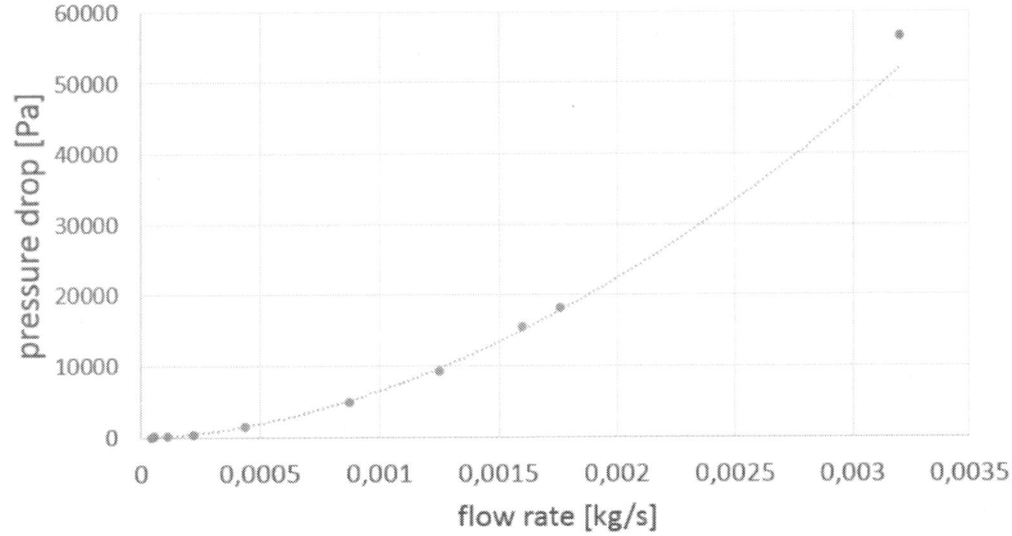

Fig. 8. Flow characteristic of the Nosko®, points represent particular simulation results and line is presentation of the equation (1).

The CFD model of the Nosko® was utilized for assessment of flow characteristic, see Figure 8. Several flowrates were investigated to obtain related value of pressure drop. By regression analysis the final curve of the flow characteristic or local pressure drop calculation was obtained, and it mathematic formula is presented by equation (1)

$$\Delta p = a \cdot m^b \tag{1}$$

where $a = 2.9 \times 10^9$ Pa.(kg. s^{-1})$^{-b}$ with 95 % confidence bounds (2.6 to 3.2)$\times 10^9$ and exponent $b = 1.89$ with 98 % confidence bounds (1.87 to 1.91). Root Mean Square Error is RMSE = 122.9 and coefficient of determination $R^2 = 1$.

Maximal flow rate for the pressure drop 30 kPa is 2.6 litres.s^{-1} and mean velocities in the inlet is 427 m.s^{-1} and outlet is 353 m.s^{-1}. However, these results look extreme and can be considered as harmful condition they have to be interpreted carefully. These values are extremely influenced by phlegm presence, pressure drop of the sinus nasal of individual person as well as the operation parameters set up of the vacuum cleaner. These values can be reached by condition before connection to the human nose and the hydraulic resistance of the sinus nasal can significantly reduce the value of the flow rate. With knowledge of working curve (pneumatic characteristic) of the vacuum cleaner together with hydraulic resistance of the sinus nasal and obtained hydraulic characteristic of the Nosko® by this article, the working point of the connected system can be achieved. It is evident that real values of working point will be then considerably lower. On the other hand, the value of the air flow rate is not so far from the values presented by article [11] for adult human breathing.

4 Conclusion

The article deals with investigation of the pressure and velocity conditions in selected home health care accessory for babies. The objective was to find out the maximal under-pressure and air velocity, which is possible to reach by accessory Nosko® using standard home vacuum cleaner recommended by manufacturer of the accessory. CFD software was used for prediction of the flow characteristics of the air in the body of the accessory. Also simple experiment with the trace was performed with the Nosko® to observe the streamline inside the transparent body. Standard techniques of CFD tool were used. The obtained results were compared and the agreement between simulation and experiment is satisfactory. The assessed flow behaviour of the air predicted by CFD tool was utilized for explanation of the main function of the accessory. Also the flow characteristic of the Nosko® was obtained by systematic simulation and can be applied for further engineering calculation and investigation of the system human nose - Nosko® - home vacuum cleaner. Maximum flow rate of the air through the accessory under maximal under pressure generated by home vacuum cleaner was 2.6 litres.s^{-1} which is comparable with results of investigation of the human breathing presented by [11].

This work was supported by SGA (Student Grant Agency) Jan Evangelista Purkyně University in Ústí nad Labem, Czech Republic (UJEP-SGS-2017-48-003-3).

References

1. R.P. Benedict, *Fundamentals of pipe flow* (John Wiley & Sons, New York, 1980)
2. R.W. Miller, *Flow Measurement Engineering Handbook* (Osborne-McGraw-Hill, New York, 1983)
3. U.G. Pirumov, G.S. Roslykov, *Gas flow in Nozzles* (Springer-Verlag Berlin Heidelberg, New York, 1986)
4. Y. Yu. M. Shademan, R. M. Barron, R. Balachandar, Eng. App. Comp. Fluid Mechanics, **6**(3), 412-425 (2012)
5. H. Koskela, Energy and Buildings, **30**, 1011-102 (2004)
6. R. Lenhard, T. Puchor, *EPJ Web of Conferences*, **143** (2017)
7. C. Badock, R. Wirth, A. Fath, A. Leipertz, Intl. J Heat and Fluid Flow **20**, 538-544 (1999)
8. K. Petera, M. Dostal, M. Věříšová, T. Jirout, Chem. Biochem. Eng. Q., **31**(3), 343-352 (2017)
9. I.B. Celik, U. Ghia, P.J. Roache, Ch. J. Freitas, J. Fluids Eng **130**(7), (2008)
10. IEC 312 / DS/EN 60312 Standard, *Vacuum Cleaners for Use in the Household – Methods for Measuring the Utility Properties*
11. J. Pertuze, A. Watson, N.B. Pride, J. Appl Physiol, **70**(3), 1369-76 (1991)

Basic CFD model of heat pipe

Peter Hrabovský[1,*], *Zuzana* Kolková[1], *Marián* Mokrý[1] and *Patrik* Nemec[2]

[1]University of Zilina – Reasearch Centre, Univerzitná 8215/1, 010 26 Zilina, Slovakia
[2]University of Zilina – Department of Power Energy, Univerzitná 8215/1, 010 26 Zilina, Slovakia

Abstract. The article deals with the creation of the basis of the CFD model of the heat pipe, which is designed to simulate and show the predicted course of the work-fluid-activity within the heat pipe. This simulation is part of the study of the heat pipe device. This device, based on CFD simulation, can be constructed to describe the phenomena running inside the heat pipe based on real-world experimental measurements. A comparison of the CFD and the experimental model can explain the problems of areas and phenomena in the heat pipe. Based on this comparison, it is possible to optimize single device errors. The optimization transferred to the CFD model will show us how to solve problems and how to improve the experimental models of heat pipes.

1 Introduction

Heat pipes are now mainly used in refrigeration engineering, to remove heat from electrical components. They can be found mainly in computing, which is mounted on the processor (CPU, GPU, chipset ...) from which dissipates heat through active or passive cooler. The heat pipes are mainly used in the industry where it is necessary to comply with dust-proof, air tightness, water tightness. From such a closed space can be removed heat by heat pipes [1].

Heat pipes are used in solar system for heating and heat transfer with heat exchangers. Heat pipes are devices that transfer heat efficiently. They can be used to transfer heat from the source at a higher energy potential for example built-in fireplaces in the home [2]. Of these, it is possible to capture the radiant heat component, which can be used for hot water or heating other rooms.

2 Basis theory of experimental Heat Pipe

We used a heat pipe to transfer heat from the heat source to the point of consumption of heat with high efficiency and low temperature difference. It is a sealed device in which the net substance (water) exist in the liquid and the gas phase at a pressure below atmospheric pressure (vacuum). Cleanliness of the working fluid is an important parameter for the efficiency of the heat pipe, in terms of heat transfer [3, 4]. The actual heat pipe consists of evaporation, condensation, and the adiabatic section. In the evaporator the working fluid is

* Corresponding author: peter.hrabovsky@rc.uniza.sk

evaporated at a temperature below normal conditions. In the adiabatic section is not exchanged heat transfer through the ambient area, due to the isolation of a few working fluid to the wall of the heat pipe condensation film of condensed working fluid. In the condenser the vapour of working fluid condenses on the walls of the heat pipe, which transmit latent (condensing) heat through the walls of the heat pipe to the heat consumption point [5, 6, 7]. Figure 1 presents heat pipe model and heat pipe schema.

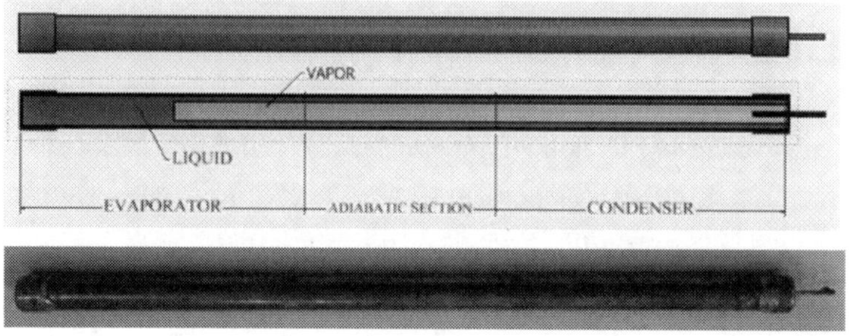

Fig. 1. Heat pipe.

3 Description of Experimental Device Set

A case situated in the upper part, is the part of a cooling device. It is connect to the cooling device, the cooling medium is water. The water circulates in a closed circuit between the cooling device and the case. In the cooling circuit included the safe cases, in which there are devices for measuring the temperature at entry and exit of water. In the circuit is included flow meter too that shows speed of flow of the refrigerant in a closed circuit. A case situated at the bottom part is the heating of the device. It is connected to the heating device, the working medium is water. In a closed heater circuit included thermocouples for temperature measurement inlet and outlet water from the case, flow meter for flow measurement. The circuit contents a valves that get disconnect the heater case when it is need for cooling with cold water, while avoiding the cooling of heating up the water in the heater circuit. Measured samples are mounted in the cases by caulked joints with seals, which are not being relaxed, possible leakage of cooling or heating of the working fluid. Figure 2 presents model of the gravity heat pipe and schema of measuring device.

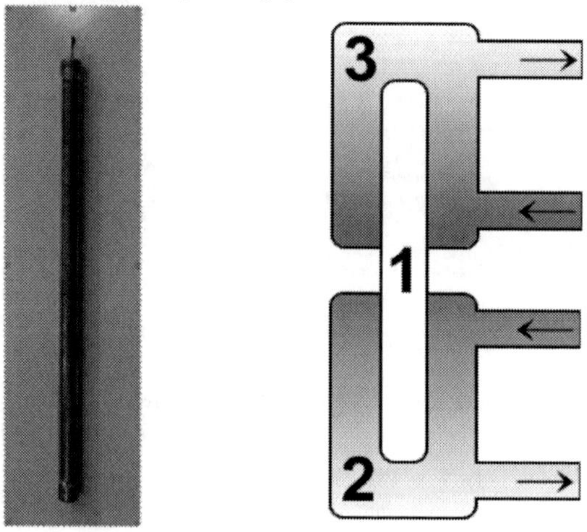

Fig. 2. Gravity heat pipe and scheme of experimental device 1 – heat pipe, 2 – heater, 3 – cooler.

4 Basic parameters of CFD Simulation by heat pipe

The experiment was carried out in a mathematical simulation of CFD ANSYS.

The resulting mesh in the Figure 3 is a simplified model of polyhedral and has parameters 239 302 cells, 1 552 487 faces, 1 306 224 nodes.

Individual volumes created appear cold water, hot water, copper rod, heat pipes filled with water-vapor – its thermal conductivity was changed in steps – 500, 1000, 2500, 5000, 10000, 15000, 20000, 25000, 30000, 40000, 45000, 50000.

Boundary conditions are the same as physical experiments (entered from physical experiment) – energy on, laminar flow – revealed by the Re f(v,d), time-independent model.

Input parameters of cooling water - temperature 10 °C, the mass flow rate 0.0287 kg.s^{-1}. The outlet of cooling water was determined with parameter outflow. Input parameters of heater – temperature 80 °C, the mass flow rate 0.0781 kg.s^{-1}. The outlet of the heating water was determined with parameter outflow.

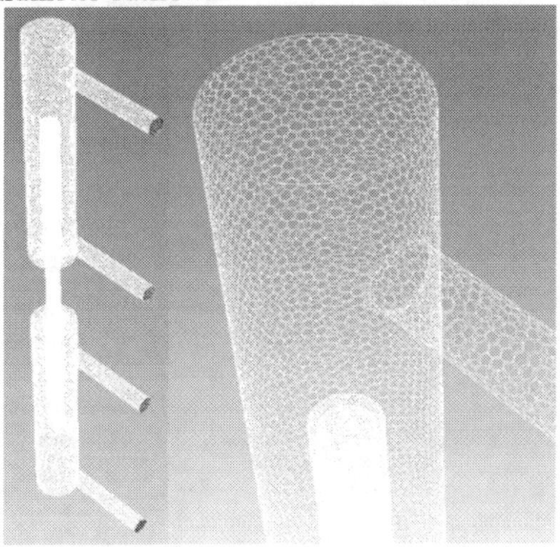

Fig. 3. Created mesh of CFD model of heat pipe.

Table 1. Result of simulation of heat pipe.

Thermal coductivity	In cold water	Out cold water	In cold water	Out cold water	Heat capacity	Power
[W.m^{-1}.K^{-1}]	[K]	[K]	[°C]	[°C]	[J.kg^{-1}.K^{-1}]	[W]
500	283.15	283.42	10	10.27	4197.62	32.92
2500	283.15	284.30	10	11.15	4197.15	138.52
5000	283.15	284.96	10	11.81	4196.80	217.71
10000	283.15	285.75	10	12.60	4196.39	313.40
15000	283.15	286.22	10	13.07	4196.14	369.55
20000	283.15	286.53	10	13.38	4195.98	406.78
25000	283.15	286.74	10	13.59	4195.87	433.04
30000	283.15	286.90	10	13.75	4195.79	451.71
40000	283.15	287.07	10	13.92	4195.70	472.79
45000	283.15	287.10	10	13.95	4195.68	476.53
50000	283.15	287.14	10	13.99	4195.67	480.26

The results of simulation of heat pipe are shown in the picture Figure 4. We can see, that the result of simulation are similarly as the result of physical model. When we use the water-vapor as the working fluid of heat pipe, the result of simulation are approaching to the result of physical model. We had changed one property of water-vapor, which is thermal conductivity.

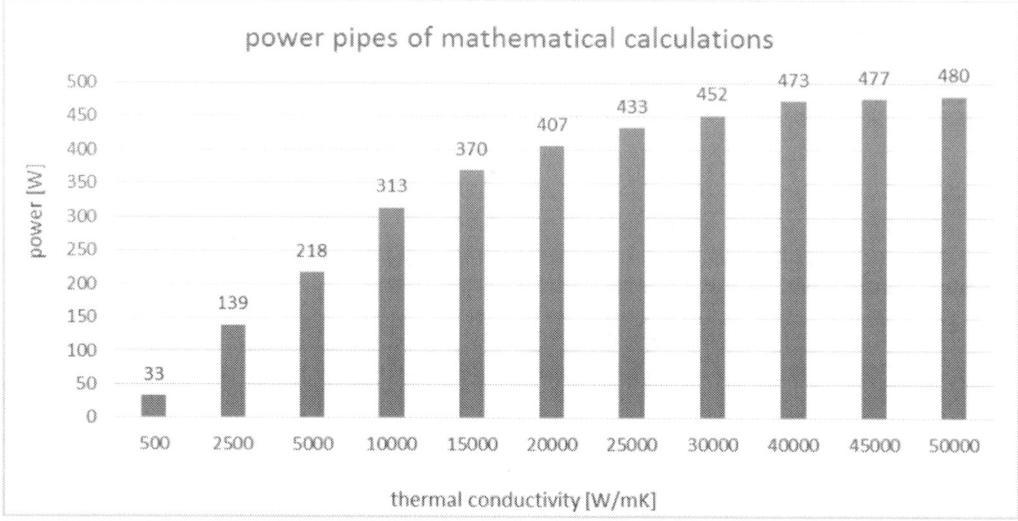

Fig. 4. Depend the power of thermal conductivity for CFD simulation.

In Figure 5 we can see progress of water-vapour, which has changeable property of thermal conductivity. Efficiency of heat transfer is directly depend on the value of thermal conductivity.

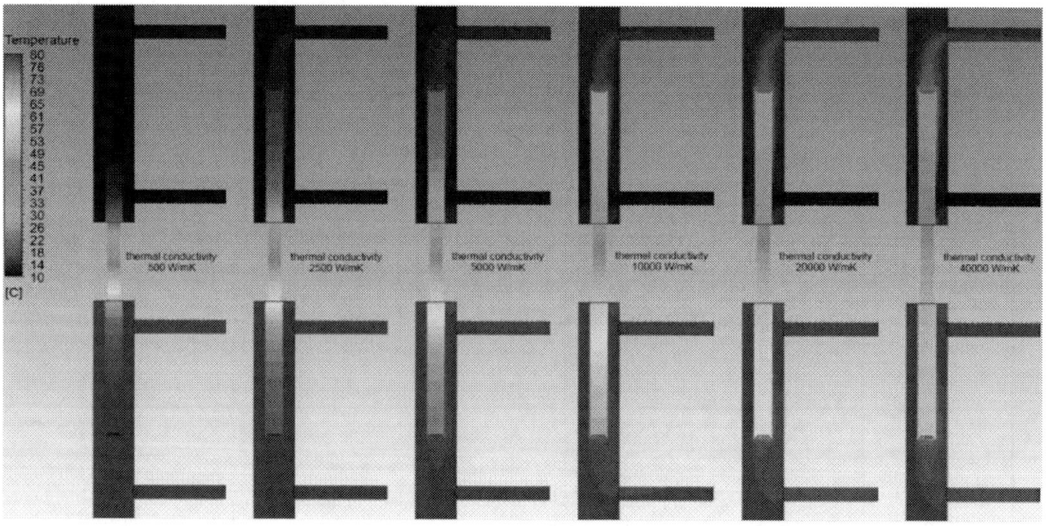

Fig. 5. Heat transfer at various values of thermal conductivity for CFD simulation of heat pipe.

In Table 1 are value of thermal conductivity, temperature of cooling water (in, out), heat capacity at temperature outer cooling water and power that the new material transferred in form of heat.

5 Conclusion

This experiment was conducted for the heat transfer heat transfer basic research. The basic factor for this research is mathematical simulations in the Ansys program in Fluent. For the need to define the heat transfer process through the heat pipe in this program, it is necessary to examine the basic phenomena occurring in the heat pipe. Based on this analysis, it is possible to design a material with similar properties to the heat pipe. The newly created heat pipe replacement material will show us the basics of heat transfer through a heat pipe under different conditions. At this simulation, we determine a possibility of substitution of heat pipe with another material with changed of some parameter. We have approximated to real conditions of real heat pipe. This simulation is valid only for specific heat pipe, which is overwritten upper in section 3. This simulation is steady time. The new transient time simulation is in progress now.

Using simulation programs (such as the ANSYS CFD module) helps us to get an idea of the course of action activities of the action members and experimental phenomena, which can be optimized on the basis of the results obtained, thus reducing the problem of the experimental measurement.

This article forms the basis for creating a series of physical experiments based on CFDs and mathematical models to identify the unrecognized in the area of the use of heat pipes in their activities and in the ongoing phenomena.

The research is supported by the European Structural Fund - project ITMS 26220220048 (50%).

The research is supported by project "Limits of Radiative and Convective Cooling through the Phase Changes of Working Fluid in Loop Thermosyphon", APVV-15-0778 (50%).

References

1. R. Nosek, S. Gavlas, R. Lenhard, M. Malcho, V. Sedlak, S. Teie, AIP conference proceedings, 1917, (2017)
2. P. Ďurčansky, Š. Papučík, J. Jandačka, M. Holubčík, R. Nosek, The scientific world journal 2014, (2014)
3. J. Jandačka, J. Mičieta, M. Holubčík, R. Nosek, Energy & fuels. **31**, 3 (2017)
4. R. Lenhard, K. Kaduchová, Š. Papučík, J. Jandačka, EPJ Web of Conferences 67, (2014)
5. P. Ďurčanský, M. Patsch, J. Jandačka, AIP conference proceedings, **1745** (2016)
6. P. Pilát, M. Patsch, Š. Papučík, M. Vantúch, AIP Conference Proceedings, **1608**, (2014)
7. R. Lenhard, M. Malcho, J. Jandačka, Heat transfer engineering, **40**, 3 (2019)

Simulation of Combustion Air Flow in the Gasification Biomass Boiler

Marek Patsch[1,*], and *Peter* Pilát[1]

[1]University of Žilina, Department of Power Engineering, Univerzitná 1, 010 26 Žilina, Slovakia

Abstract. The article deals with the optimization of biomass combustion in a small heat source by means of an optimal distribution of combustion air. The uneven distribution of combustion air has been observed in certification tests and in real operation of used heat source and it has an influence on uneven combustion of biomass in the gasification chamber, on increase emissions and combustion losses. At this stage of the research, optimization of the combustion air distribution is performed by CFD simulations, which will be later verified by PIV measuring of the velocity fields in gasification and combustion chambers of the experimental heat source. CFD simulations and subsequent PIV measurements on the experimental device are realized without real combustion, only the air flow in the empty gasification chamber and in the combustion chamber is investigated. This approach has been chosen to simplify calculations and experiments, and on the assumption that when the combustion air distribution is optimal in empty chambers, it will be optimal even during real combustion. The primary air flow in the gasification chamber is in real operation affected by the size and shape of the inserted biomass and its placement in chamber and this effect is accidental and difficult to verifiable.

1 Introduction

Gasification heat sources are a modern way of biomass energy utilization where combustion is with high efficiency and low emissions [6]. The combustion takes place in a three-stage process and it can be divided into three zones. In the first zone there is drying and gasification of the wood with access of the primary combustion air, volatile substances are released from the solid fuel. Gasification is, in principle, the thermal decomposition of organic and inorganic substances in the closed chamber of the boiler under a slight overpressure produced by the combustion air fan. This gasification process is carried out in the fuel container, above the refractory nozzle. In the second zone, released gases with preheated secondary air are mixed in the nozzle space to form a mixture of combustion gas. In the third zone there is a total burning of gases in the boiler's combustion chamber and the removal of flue gases through the pipe heat exchanger into the chimney. In this controlled combustion process a high efficiency is guaranteed - 80-90 %, the boiler's heat output is continuously adjustable in the range of 40-100 % of nominal power [1].

* Corresponding author: marek.patsch@fstroj.uniza.sk

CFD simulations and subsequent measurements on the experimental heat source were made on a model based on biomass gasification boiler Vigas 16 (Figure 1). Operation of this heat source in laboratory conditions showed uneven distribution of combustion air, resulting in residual unburned fuel remaining, increased emissions and reduced nominal heat power.

Important factors affecting emission and ash generation are the distribution of combustion air to primary and secondary, its distribution to the air inlets (the heat source has four primary and six secondary air inlets) and the velocity of combustion air [3-5]. Optimal flow of primary combustion air in the gasification chamber and secondary combustion air in the nozzle at the inlet to the combustion chamber has a major impact on the process of gasification and subsequent combustion of biomass. The amount of supplied combustion air is given by the fan speed, the portion of the total amount of air being taken away as secondary air, and the rest is the primary combustion air. The amount of secondary air is determined by the setting of the secondary air valve.

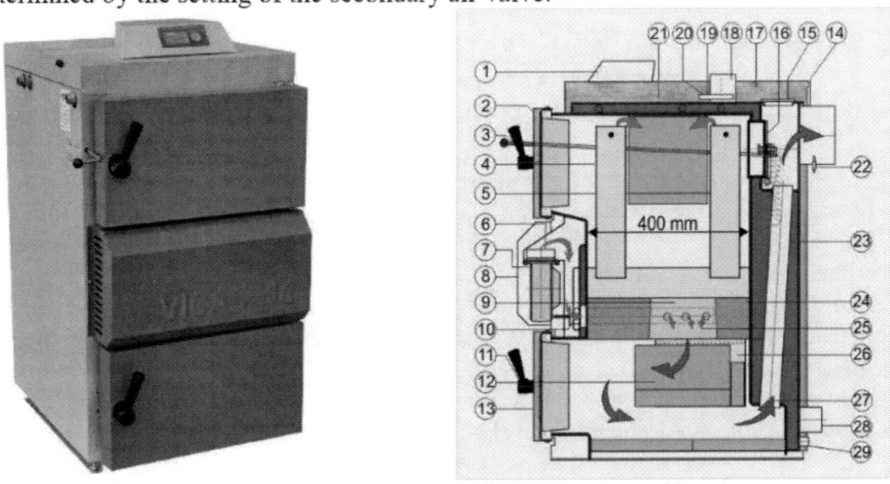

a) b)

Fig. 1. a) Gasification boiler Vigas 16 [2], **b)** section of the gasification boiler: 1. regulation, 2. upper door, 3. chimney flap rod, 4. gasification chamber, 5. primary air line, 6. damper actuator, 7. combustion air fan, 8. fan cover, 9. refractory concrete block, 10. secondary air valve, 11. door closure, 12. refractory hearth lining, 13. lower door, 14. chimney, 15. exchanger cover, 16. firing flap, 17. top back cover, 18. water outlet, 19. thermal fuse, 20. boiler thermometer, 21 front cover, 22. Flue gas thermometer, 23. pipe heat exchanger, 24. primary air chamber, 25. secondary air, 26. combustion chamber, 27. direction of the flue gas, 28. water return, 29. filling water throat [2].

2 Numerical design for CFD simulations

The 3D model (Figure 2) was constructed according to the design documentation, followed by the generation of a calculation mesh with approximately 1 400 000 elements. The calculation was performed in Fluent, the standard k-ε model and the standard wall functions were used. Simulated was only the combustion air flow without combustion, used were a constant air temperature of 288.16 K, a constant density of 1.225 kg.m^{-3} and a viscosity of 1.7894.10^{-5} Pa.s, considered with a turbulent flow.

In the evaluation, all computational volume was divided into twelve sections, and for each section the flow rate was evaluated at fifteen combustion air fan settings, corresponding to a volume flow rate of 10 to 150 m^{3}.h^{-1}.

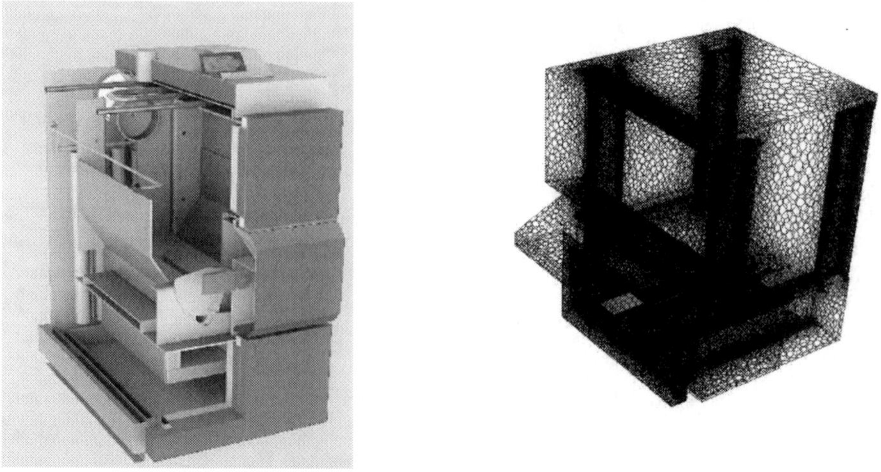

a) b)

Fig. 2. a) 3D model of gasification boiler Vigas 16, **b)** Calculation mesh for CFD analysis.

3 Evaluation of CFD simulations

Figure 3 shows an example of calculated flow in the gasification and combustion chamber in a vertical cross-section which is guided by the axis of the heat source. Simulations have shown in particular the uneven distribution of primary air flow in the gasification chamber at higher volumetric air flow through the combustion air fan. This is result of higher pressure and length losses from air fan to primary air inlets. All inlets have the same cross-sectional area, so the optimization step should be to change the cross-sectional area (distant air inlets have to have higher cross-section area to reduce pressure losses, or closer air inlets have to have smaller cross-section area to increase pressure losses). The distribution of secondary air appears to be uniform in this cross-section [7, 9].

The more uneven distribution of primary air is more apparent in vertical cross-sections which are guided closer to the wall of the gasification chamber. Figure 4 shows strong effect of air flow reflection from the bottom of the empty gasification chamber. This effect would not be so pronounced in the wood-filled chamber.

Figure 5 shows an example of a flow distribution and velocities in a gasification chamber in a horizontal cross section which is guided approximately in half the height of the gasification chamber. At higher volumetric flows through the fan, the highest velocity of the primary combustion air is achieved at the right front inlet. It results from the location of the combustion air fan which is located on right side of heat source axis, i.e. closest to that air inlet. Also, the air flow to the inlet is larger on the right side, closer to the fan, which causes less pressure and length losses on this side. As already mentioned all primary air inlets have the same cross-sectional area, so there are two ways to the optimization. The first option is to change the cross-sectional area of the distant or closer inlets, the second option is to change the position of combustion air fan into the vertical axis of the heat source [10, 11].

Figure 6 shows an example of a flow distribution in the gasification and combustion chamber in a vertical cross-section which is guided through the second pair of secondary air inlets (perpendicular to the longitudinal axis of the heat source), viewed from the rear side of the heat source. At higher volumes of combustion air through the fan there is an apparent uneven distribution of primary air in the gasification chamber, the flow being diverted from the chamber axis, which is caused by a larger flow rate of primary air through the right inlets closer to the fan. It is also possible to observe a slight diversion of the secondary air flow in the nozzle into the combustion chamber. The maximum nozzle speed is 6 m.s^{-1}.

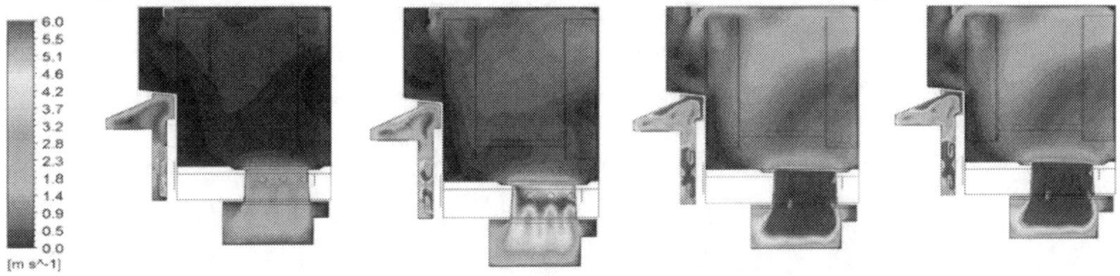

40 m^3.h^{-1} 80 m^3.h^{-1} 90 m^3.h^{-1} 150 m^3.h^{-1}

Fig. 3. Example of calculated velocity fields in gasification and combustion chamber in different volumetric air flows (vertical cross-section in device axis).

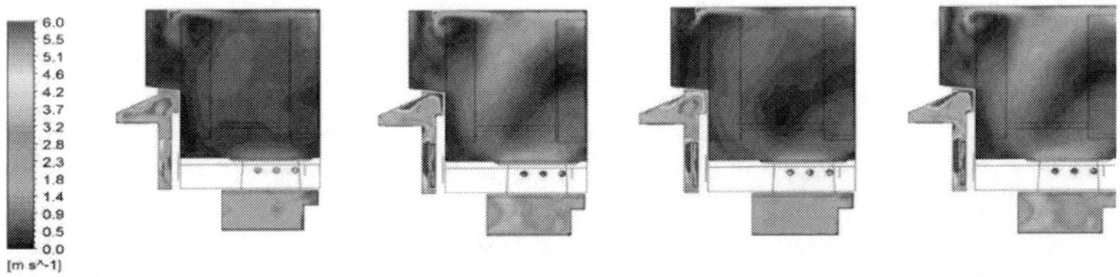

80 m^3.h^{-1} 100 m^3.h^{-1} 110 m^3.h^{-1} 150 m^3.h^{-1}

Fig. 4. Example of calculated velocity fields in gasification and combustion chamber in different volumetric air flows (vertical cross-section closer to wall of gasification chamber).

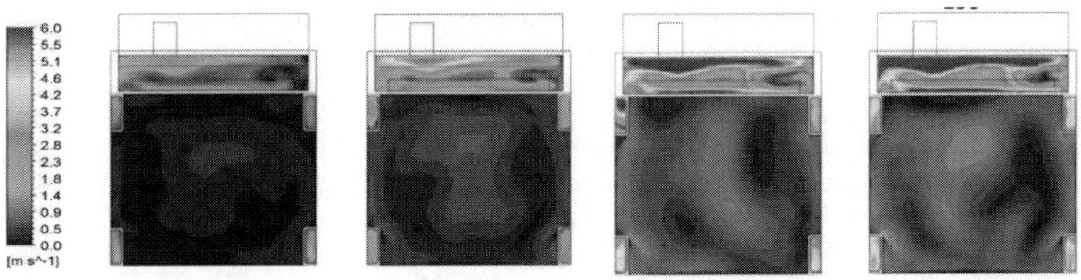

40 m^3.h^{-1} 80 m^3.h^{-1} 130 m^3.h^{-1} 150 m^3.h^{-1}

Fig. 5. Example of calculated velocity fields in gasification chamber in different volumetric air flows (horizontal cross-section in half the height of gasification chamber).

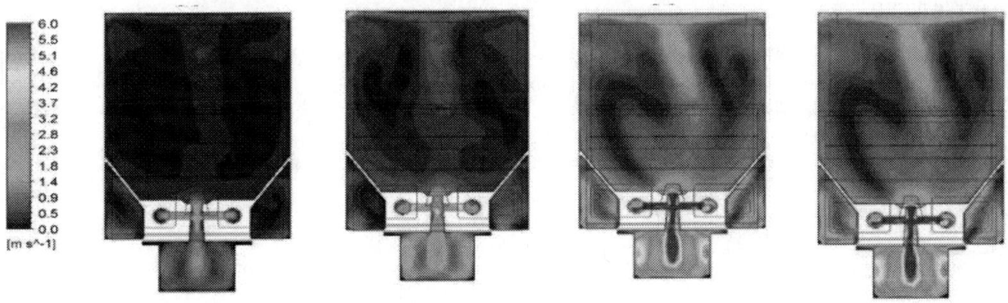

30 m^3.h^{-1} 60 m^3.h^{-1} 90 m^3.h^{-1} 150 m^3.h^{-1}

Fig. 6. Example of calculated velocity fields in gasification and combustion chamber in different volumetric air flows (vertical cross-section through the secondary combustion air inlets, viewed from the rear side).

The uneven distribution of primary air in the gasification chamber is also apparent from the graph on Figure 7, which shows the percentage distribution of air at the inlets of the right and left pairs of primary air inlets. Through the right side, which is closer to the fan, at higher volumetric air flows through the combustion air fan more than 5 % primary combustion air flows.

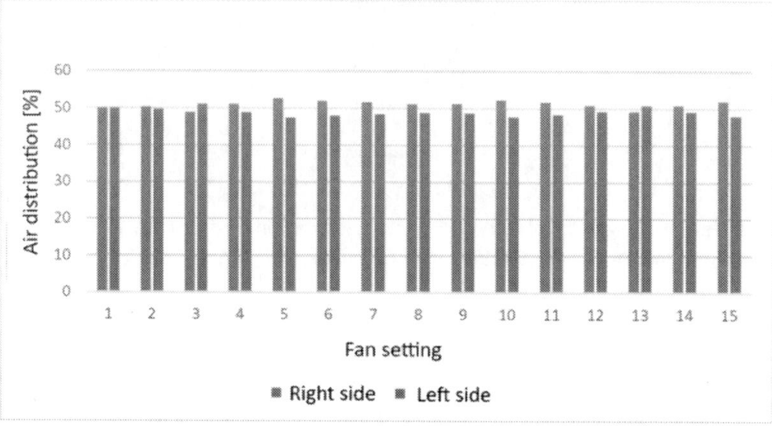

Fig. 7. Percentage distribution of primary air at right side and left side inlets (fan setting scale: 1 means 10 $m^3.h^{-1}$ to 15 means 150 $m^3.h^{-1}$).

4 Conclusion

Modelling and CFD simulations confirmed the uneven distribution of primary air in the gasification chamber and partly confirmed the uneven distribution of secondary air in the combustion chamber nozzle, which were also observed in the real operation of a small heat source. The uneven distribution of secondary air needs to be dealt with in more detail later and, in particular, to compress the computing mesh at a given location because it is a small space compared to the overall computing volume. The computational CFD model will be verified and optimized by PIV velocity fields measurements on the real gasification biomass boiler model. Flow optimization will be achieved by moving the combustion air fan and changing the cross sectional areas of the primary and secondary air inlets.

This article was realised in the frames of project VEGA 1/0864/16 "Analysis and optimization of input factors in the process of dendromass combustion in small heat sources".

This article was created with grant KEGA 042ŽU-4/2016 „Cooling on the basis of physical and chemical processes".

This publication is the result of the Project implementation: Competency Centre for Knowledge technologies applied in Innovation of Production Systems in Industry and Services, ITMS: 26220220155, supported by the Research & Development Operational Programme funded by the ERDF.

Európska únia

References

1. J. Jandačka, M. Malcho, M. Mikulík, *Technológie pre prípravu a energetické využitie biomasy* (2007)
2. Technical documentation of gasification biomass boiler Vigas 16 (2016)
3. T. Nussbaumer, M. Kiener, *21st European Biomass Conference* (2013)
4. M. Kiener, T. Nussbaumer, *20th European Biomass Conference* (2012)
5. T. Nussbaumer, M. Keiner, P. Horat, *Biomass and Bioenergy*, **76** (May 2015)
6. S. Van Loo, J. Koppejan, *The Handbook of Biomass Combustion and Co-firing* (Earthscan, London, 2008)
7. K. Sulovcová, R. Nosek, J. Jandačka, M. Holubčík, Emis. Cont. Sc. and Tech. (2018)
8. M. Holubčík, J. Jandačka, R. Nosek, J. Baranski, Emis. Cont. Sc. and Tech. (2018)
9. N. Kantová, J. Jandačka, M. Holubčík, A. Čaja, *36th Meeting of Departments of Fluid Mechanics and Thermodynamics* (2017)
10. P. Oršanský, B. Ftorek, P. Ďurčanský, *AEaNMiFMaE 2014* (2014)
11. Š. Papučík, R. Lenhard, et al., *AIP Conference Proceedings* **1608** (2014)

Effect of a blade shape on hydraulic and mechanical properties of a single - blade impeller

Matej Kurilla[1,*], *Branislav* Knížat[1], *Zoltán* Csuka[1], and *Maroš* Hyriak[2]

[1]Slovak University of Technology in Bratislava, Nám. slobody 17, 812 31 Bratislava, Slovakia
[2]PRAKTIKPUMP, Jesenského 63, 96001 Zvolen, Slovakia

Abstract. Single-blade centrifugal pumps with a screw-shaped blade are perspective hydraulic machines with possible usage in many areas. According to the non-symmetrical impeller shape, mechanical properties, resulting from a dynamic balancing, are not fulfilled automatically. The way of designing of an impeller which enables fulfilled main requests such as good hydraulic, mechanical properties and the ball passage diameter, is described in this academic paper.

1 Introduction

Centrifugal pumps with a single-blade impeller represent an important segment of the hydraulic machines in the wastewater pumping area. The advantages include a large passed solids size and good cavitation properties by remaining a relatively high efficiency in the wide flowrate range [4]. The construction is generally simple and massive, so single-blade centrifugal pumps are characterized by long lifetime and high reliability during the running process. Solids with diameter equal to the diameter of the suction pipe are able to pass through the impeller without any damage. That's why this type of pumps is used in food processing for pumping fruits, fish, etc. Many types of single-blade centrifugal pumps can be recognized, differing mainly in the blade shape. This paper is devoted to hydrodynamic pumps with the screw-shaped blade.

2 Flow in a single-blade impeller

Flow in a single-blade impeller has got its own specific features that differentiate this pump from other conventional centrifugal pumps with multi-blade impellers. In respect of a long wrap angle, the blade outlet angle accomplishes relatively low values. As a result, the Q-Y curves are considerably steep and also linear, resulting in the stability in the pump operation. Due to the single blade, the impeller is geometrically asymmetric with respect to the rotational axis. The same applies to a diffuser which is the volute casing.

Rotational, asymmetrical pressure and velocity fields interact with also non-symmetrical fluxional field in the stator. Accordingly, non-zero radial thrust acts on an impeller, even in

* Corresponding author: matej.kurilla@stuba.sk

the design point. The radial thrust increases along with a higher blade outlet angle and also with higher flowrates [5]. The result is that the radial force problem is much stronger than with conventional centrifugal pumps. Considerable researcher's effort is focused on the knowledge of energy transformation mechanism in single-blade pumps. The aim is decreasing of losses and therefore raising the efficiency. According to the fluxional fields asymmetry it is possible to observe the substantial fluctuation of performance parameters in the pump during one revolution. Performance parameters are torque, specific energy, flowrate, and shaft power.

A designing process uses standard design methods [2] which don't provide as accurate methodology as necessary. Consequently, these methods are combined with CFD computation methods [3]. We consider a shape of the dimensionless function of specific energy, in the design process - equation (1):

$$\Psi = \Psi_t - \Theta(\Phi_2) \tag{1}$$

The dimensionless flowrate Φ_2 and theoretical specific energy Ψ_t are given by the following equations [7]:

$$\Phi_2 = \frac{Q}{\eta_V A_2 U_2} \tag{2}$$

$$\Psi_t = \sigma_Y - \frac{\Phi_2}{\tan \beta_2} \tag{3}$$

In equations (1) to (3) is Ψ_t – theoretical specific energy, β_2 – blade outlet angle, η_V – volume efficiency, A_2 – outlet area a U_2 – tangential outlet velocity.

The curve Ψ-Φ_2, respectively *Y-Q* (specific energy - flowrate) is shaped by the σ_Y and $\Theta(\Phi_2)$ parameter, i.e., the slip factor [7] and the hydraulic losses function.

The slip factor of the flow in the impeller σ_Y, in the equation (3), is about 0,4 for single-blade impellers. It characterizes imperfect guidance of a liquid [6]. We can estimate hydraulic losses in the optimal point, based on supposed efficiency of the pump.

A subject of the proposal is meridional cut of the pump as well as a shape of the blade. Considerably important fact needs to be considered here, in contrast with conventional centrifugal pumps. Owing to the asymmetric blade shape, the computed blade designed to fulfil energetic parameters, causes dynamic unbalance of the whole rotational part of the pump. The impeller loses its mechanical properties as a rotor. Accordingly, the impeller has to be balanced by adding of additional material. It influences retroactively the shape of the blade channel and the whole hydraulic solution. We will refer to influence of the dynamic balancing of the impeller on hydraulic parameters of the pump in this paper.

2.1 Hydraulic parameters of the pump

Mechanical and hydraulic properties of the impeller were researched on a single-blade spiral pump with parameters in the Table 1. For these parameters a pump with two impeller alternatives has been designed, hereinafter referred as A and B. Both impellers have the same base what does concern geometrical parameters. They differ from each other in the level of balance when the impeller A is unbalanced and the impeller B is dynamically balanced. A diameter of the suction, as well as discharge pipe of the pump, was 50 mm. Single-blade pumps with such hydraulic parameters have an expected efficiency approximately in the interval between 50 % and 60 %. All of the results are given for clean water. Internal spaces of the pump from suction to discharge, including the impeller and the spiral casing were designed so a ball with the diameter of 50 mm passes through the pump.

Table 1. Hydraulic parameters of the pump.

Parameter	Value	Description
Q [m^3/s]	0.011	Flowrate
Y [J/kg]	105	Specific energy
n [s^{-1}]	2 900	Speed
n_b [-]	0.15	Specific speed

The spiral casing were made of welded metal plates.

2.2 Geometric parameters of impellers

Two impellers marked as A and B were made to test their mechanical and hydraulic properties. Both impellers have the same meridional cut and the suction side of blades is the same expandable line surface. They are manufactured as semi-open impellers. The impeller A is dynamically and statically unbalanced. Mechanical properties of this impeller are substantially adverse. However, the hydraulic properties are considerably good because the blade channel shape directly responses to the calculation. The basic blade shape is illustrated in the Figure 1. The red point C represents the location of the center of gravity of the impeller A. In the design, the basic course of the blade and its mass distribution is chosen, so the center of the gravity lies as close as possible to the rotational axis. This is achieved by the inclination of cut lines A-B, i.e. by the angle ϑ , exactly like the Figure 1 shows. The curves a and b represent then streamlines on the front respectively back plate. It isn't possible to totally balance the impeller only with the blade shaping. The impeller has to be dynamically balanced by adding of material.

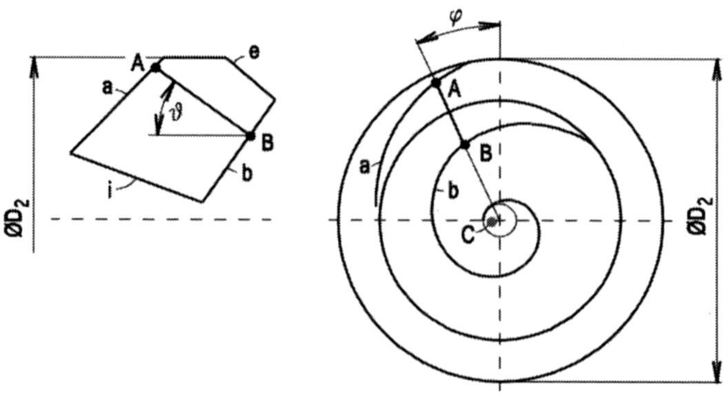

Fig. 1. The impeller blade as a ruled surface.

The geometrical parameters of both impellers are shown in the Table 2. The impeller's photo is in the Figure 2. The impeller A was made by welding of a steel plate (the blade) and the impeller B was drawn in a CAD programme and printed on a 3D printer from ABS plastic.

Table 2. Geometry parameters of impellers.

	Impeller A	Impeller B	Description
D_1 [mm]	50		Inlet diameter
D_2 [mm]	138		Outlet diameter
β_2 [°]	8.2		Outlet blade angle
φ_2 [°]	486		Wrap angle
Balanced	No	Yes	Dynamic balancing

The solids passage diameter is ensured up to 50 mm by both impellers. The impeller B is dynamically balanced so the center of gravity lies on the rotational axis and the main axis of inertia is identical with the rotational axis.

Fig. 2. Impeller A (left) and impeller B (right).

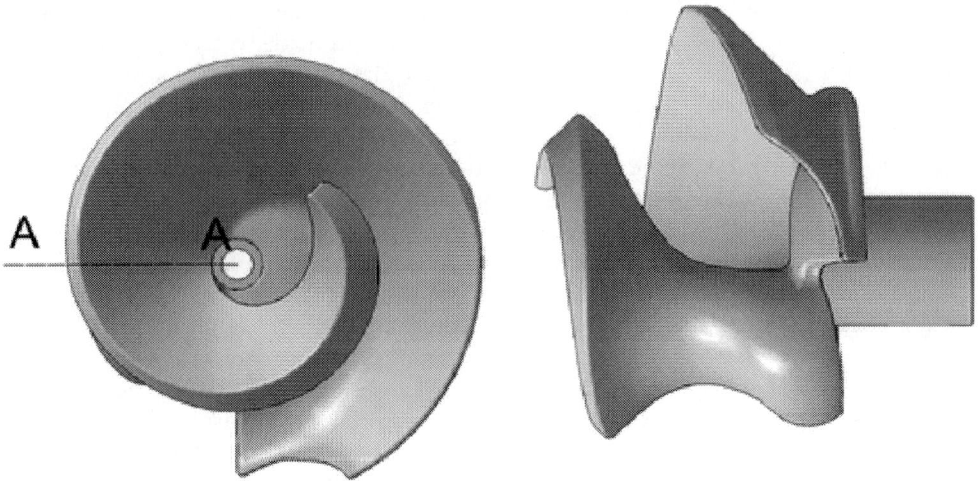

Fig. 3. Balancing of the impeller B.

The way of adding a mass on a 3D model of the impeller B in a CAD software is illustrated in the Figure 3. The balancing mass was added so that hydraulic properties and the solids passage diameter wouldn't get worse. The mass was added in the cut planes passing through the rotational axis – example is the cut A-A indicated in the Figure 3.

3 A comparison of properties of the impellers

Hydraulic properties of the impeller A and the impeller B were on the one hand analysed numerically using CFD methods and, on the other hand, experimentally.
CFD simulations were executed with the following solver setup, see also [1]: unstructured mesh, Menter SST turbulence model, steady-state calculation. The mesh was generated by the software ANSYS Meshing. General criteria for the mesh generation were applied in order to achieve a numerical accuracy. Total number of elements of the numerical model of the pump was 19.2 millions (impeller – 15.2 millions; spiral casing – 2.1 millions; suction pipe – 1.9 millions). In the mesh occurred 5.2 millions tetrahedral elements; 13.9 millions three prism elements; and 2.1 million pyramid elements. The calculations were executed in the steady state using frozen rotor method. The impeller was sequentially set up turning into three various positions. The results were subsequently averaged. This approach is accurate enough only close to the optimal point. Therefore, calculation curves are displayed on graphs only in the area close to the optimal point.

Fig. 4. A model of the tested pump.

Experiments were carried out on a test rig with a closed water circuit in a laboratory of hydraulic machinery. A body of the tested pump model was made by welding and it's in the Figure 4. An axial gap between the impeller and the stator was considered 0.3 mm during the measurement and CFD simulations. As a transport medium clean water was used during the experiments as well as during the CFD simulations.

Results of calculations and experiments are shown in the form of characteristic curves *Q-Y* and *Q-η* in the Figure 5. Blue curves represent CFD calculations, while red curves are measured experimentally. The Figure 5 shows that adding of the balancing mass didn't influence hydraulic parameters of the pump in the best efficiency point. BEP location Q_{opt}, Y_{opt} and the magnitude of efficiency η_{opt} are differing only negligibly within the measurement uncertainty. Measured curves (blue colour) are differing only at lower flowrates when the impeller B has higher specific energy. Measurement of the impeller B wasn't executed at lower flowrates than 6 l/s. The reason for this is that the radial as well as the axial force acting on impellers is considerable, as shown, for example, the study [5]. According to the fact that the balanced impeller was made from ABS plastic, it might acquire a mechanical destruction of the rotor at lower flowrates.

The fact that the added mass didn't affect hydraulic parameters is confirmed by CFD simulations (red curves). Performance curves Q-Y of both impellers are identical according to calculations. The problem is a certain difference between calculated and measured curves

Q-Y. This is caused by CFD computational method, which is called "frozen rotor". This method is not an optimal one for single-blade pumps.

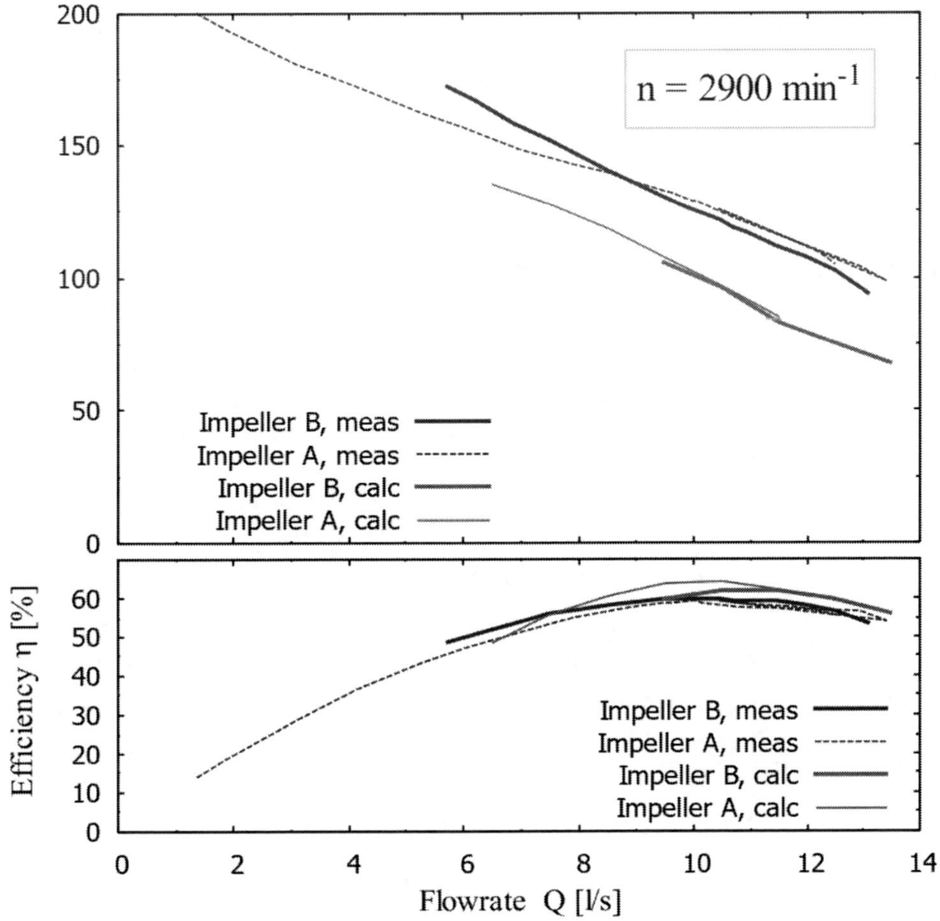

Fig. 5. Comparison of measured and calculated data.

Table 3. Summary of results – values in BEP.

	Impeller A		**Impeller B**	
	Meas	**Calc**	**Meas**	**Calc**
Q_{opt} [m³/s]	0.010	0.0105	0.0095	0.0105
Y_{opt} [J/kg]	128.6	96.5	130.3	95.8
η_{opt} [%]	58.9	64.2	59.5	61.7
Vibrations	YES	-	NO	-

Achieved results are shown in the Table 3. It should be added that the unbalanced impeller A caused vibration of the whole test rig. Despite that, the dynamic balanced impeller B had silent running without vibrations. However, the vibration intensity wasn't measured.

4 Conclusions

This paper describes the way of designing the single-blade impeller pump so that mechanical and hydraulic properties requests such as efficiency, BEP location, solids passage diameter and dynamic balancing were fulfilled. The results of the presented research show the following outcomes:

• The shape of blade surface is possible to be designed using basic hydraulic proposal methods, however, it is not possible to ensure dynamic balancing of the impeller. This issue needs to be done additionally, by adding the balancing mass with the help of CAD software. The effect of balancing masses on the flow needs to be monitored by CFD flow simulations.

• It is possible, in case of single-blade pumps, to add the balancing mass so that the total dynamic balancing is achieved and the hydraulic parameters remain unchanged.

• It is possible to achieve high passage diameter unless the balancing mass is added in a position where the flow channel is wide enough.

• CFD methods of flow simulations are giving a relatively good prediction of hydraulic parameters during balancing process of the impeller.

References

1. B. Knížat, Z. Csuka, M. Hyriak, AIP Conference Proceedings **1768**, 020034 (2016)
2. X. Cheng, R. Li, Procedia Engineering **31**, 914-921 (2012)
3. J.H. Kim, B.M. Cho, Y.S. Kim, Y.S. Choi, K.Y. Kim, J.H. Kim, Y. Cho, International Journal of Fluid Machinery and Systems, Vol. **9**, No. 4, (october-december 2016)
4. H. Quan, R. Li, Q. Su, W. Han, P. Wang, Advances in Mechanical Engineering, Volume **2013** Atricle ID 512523 (2013)
5. Y. Nishi, J. Fukutomi, R. Fujiwara, IOP Conf. Series: Earth and Environmental Science **15** (2012) 072039
6. Y. Nishi, R. Fujiwara, J. Fukutomi, Journal of Fluid Science and Technology **4** 3 (2009)
7. Y.L. Zhang, Z.C. Zhu, H.S. Dou, B.L. Cui, Y. L. J.J. Xiao Int. J. Turbo Jet-Engines (2015); **32**(1): 59-64

Reduction of the suction pressure of a liquid ring vacuum pump with a supersonic gas ejector

Róbert Olšiak[1,*], *Zoltán* Fuszko[1], and *Zoltán* Csuka[1]

[1]Slovak University of Technology, Faculty of M.E., Institute of energy machinery, Námestie slobody 17, 812 31 Bratislava, Slovakia

Abstract. A supersonic gas ejector in conjunction with a liquid ring vacuum pump is used for creating and maintaining vacuum in a chamber for technological purposes. In this paper the authors submit an overview about the problematics of suction pressure reduction with a supersonic gas ejector used as a pre-stage of a liquid ring vacuum pump. This system has also the function of a cavitation protection due to the higher pressure present at the suction throat of the vacuum pump. A part of this paper is devoted to the governing equations used at the definition of the flow through an ejector. The experimental studies are then carried out in or own laboratory for verification purposes.

1 Introduction

The vacuum pumps are technological devices used to create and maintain vacuum for technological purposes. In practice the vacuum means the state of diluted gas, which absolute pressure is lower than the atmospheric. Liquid ring vacuum pumps can be used at extraction of condensated vapor and gas mixtures. In this case a part of the pumped mixture is condensating in form of vapor inside the vacuum pump, while the condensate is drained with the working fluid. At the other vacuum pump types this phenomenon is undesirable due to the present lubricating grease. The relatively low efficiency and the high minimal suction pressure can be considered as a disadvantage. This minimal pressure is limited by the used working fluid inside the vacuum pump. Is problematic to determine this correlation, because the parameters of the liquid ring vacuum pump are bounded not only with the type of used fluid (density, viscosity) but with the change of these properties in dependence on the temperature and pressure also.

The barrier of the minimal suction pressure, at which the vacuum pump can be used, is depending on the fluid properties, construction type, number of stages etc. This pressure is according to the following authors stated as: Bannwarth 5 kPa, Faragallah 15 kPa, Szabó 4-14 kPa. In the practice is a common requirement to reach a suction pressure, which is much lower than the listed values. This is not possible due to the occurrence of cavitation inside the vacuum pump. This can be partially solved by lowering the pressure ratio, so using a

* Corresponding author: robert.olsiak@stuba.sk

multistage liquid ring vacuum pump, however even this solution is unacceptable from the viewpoint of long-term run. The suction pressure can be lowered using a gas ejector as a pre-stage of the vacuum pump. This arrangement can reduce the pressure up to 1/5 of the initial pressure.

2 The options of the suction pressure reduction

The ejector as a technological device is due to the simple construction and operational reliability are widely used at different applications. It is basically a device, in which the primary medium entrains the secondary driven medium, accelerates it to supersonic speed at reduced pressure. The pressure measured at the ejector inlet p_2 is decisive to achieve the design pressure in the considered pressure chamber. From the viewpoint of ejector design, there are some limiting parameters (Figure 1): pressure measured at the ejector outlet p_3, mass flow of the mixture m_3, which are corresponding to the occurrence of cavitation inside the vacuum pump. The values of the primary (driving) fluid p_1 and m_1 are used at the design of the primary supersonic nozzle. In case that the ejector is correctly designed, it will extend the working range of the liquid ring vacuum pump as seen on the figure below.

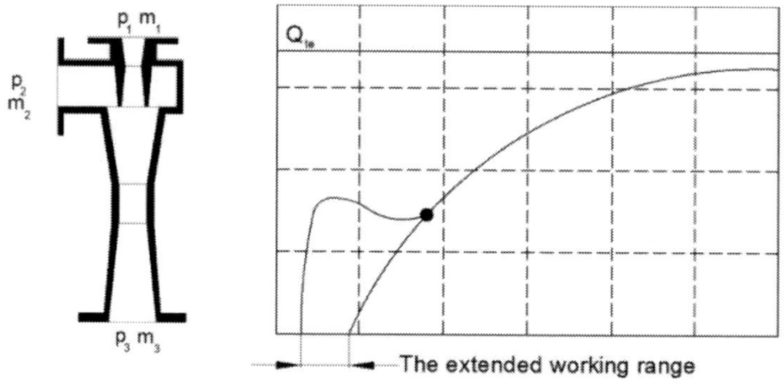

Fig. 1. The construction of a supersonic ejector (left) and the performance characteristics of a liquid ring vacuum pump with and without a supersonic ejector pre-stage.

From this simple analysis results that for the correct design of a supersonic ejector is necessary to know the following parameters:

- The required minimal suction pressure p_2 at the chamber throat.
- The required mixture mass flow m_2 at the chamber throat.
- The minimal suction pressure p_3 of the vacuum source.

It is obvious that the pressure p_2 and mass flow m_2 are the chosen parameters of the designed ejector, the suction pressure p_3 and mass flow m_3 can be specified experimentally, which is the most commonly used method, or with theoretical calculation according to the geometrical dimensions of the device, operational data and properties of the working fluid.

3 The governing equations

The mostly used ejector design method is based on a 1-dimensional analysis of compressible gas flow. This method was first time proposed by Keenan (1942) and Neumann (1950) and it uses the following assumptions:

- The kinetic energy on the inlet of both fluids is negligible.
- The two working fluids are considered as ideal gases.
- At the end of the mixing chamber are both of the fluid uniformly mixed, without occurring shocks inside the mixing chamber.

Based on these assumptions the equations describing the character of the flow from the inlets to the outlet can be given as follows [1]:

-the momentum equation

$$p_1A_1 + \rho_1Q_1v_1 + p_2A_2 + \rho_2Q_2v_2 = p_3A_3 + \rho_3Q_3v_3 \tag{1}$$

-the continuity equation

$$\rho_1Q_1 + \rho_2Q_2 = \rho_3Q_3 \tag{2}$$

-the energy equation for adiabatic flow

$$\dot{m}_1\left(h_1 + \frac{v_1^2}{2}\right) + \dot{m}_2\left(h_2 + \frac{v_2^2}{2}\right) = \dot{m}_3\left(h_3 + \frac{v_3^2}{2}\right) \tag{3}$$

The required compression ratio (CR) of an ejector determines the ratio of the static pressures on the ejector exit (identic to the pressure measured at the inlet throat of the vacuum pump) to the pressure measured at the secondary suction throat of the ejector. The entrainment ratio (ER) [2] expresses the relationship between the desired mass flow rates through the secondary and primary inlets of the ejector.

$$CR = \frac{p_3}{p_2}, \qquad ER = \frac{\dot{m}_2}{\dot{m}_1} = \frac{p_2}{p_1}\frac{A_2}{A_1}\left(\frac{T_1}{T_2}\right)^{\frac{1}{2}}\frac{f_b(\kappa_2, M_2)}{f_a(\kappa_1, M_1)} \tag{4}$$

where: $$f(\kappa, M) = \frac{\dot{m}}{pA}(RT_0)^{\frac{1}{2}} = M\left[\kappa\left(1 + \frac{\kappa - 1}{2}M^2\right)\right]^{\frac{1}{2}} \tag{5}$$

4 Design method

The motion of the working fluid is given by the pressure difference between the atmospheric air at the primary inlet and the minimal suction pressure given by the vacuum pump. Due to the small molecular weight of the used working fluid (atmospheric air), the value of the kinetic energy of the primary fluid depends on the outlet speed of the supersonic nozzle. For this reason is the nozzle considered as the most important part of an ejector. The mostly used nozzle geometry due to his simple design is the *conical*, with constant wall angle (12°-18°). Since the outlet velocity field has a radial component also, it is necessary to introduce a correction factor, which takes into account the ratio of the ideal momentum of the conical nozzle, with the axial component of the velocity.

Parabolic *de Laval* nozzles take into account the presence of Prandtl-Mayer expansion. The shape of the nozzle is obtained either by an empirical method, or by the method of characteristics (MoC) (Figure 2).

The Prandtl-Meyer formula (6) is used to determine the Mach number at a given point of the flow field followed by the Mach angle calculation (7).

$$\nu(M) = \sqrt{\frac{\kappa + 1}{\kappa - 1}}\arctan\sqrt{\frac{\kappa - 1}{\kappa + 1}\left(M^2 - 1\right)} - \arctan\sqrt{M^2 - 1} \tag{6}$$

$$\mu = \arcsin\frac{1}{M} \tag{7}$$

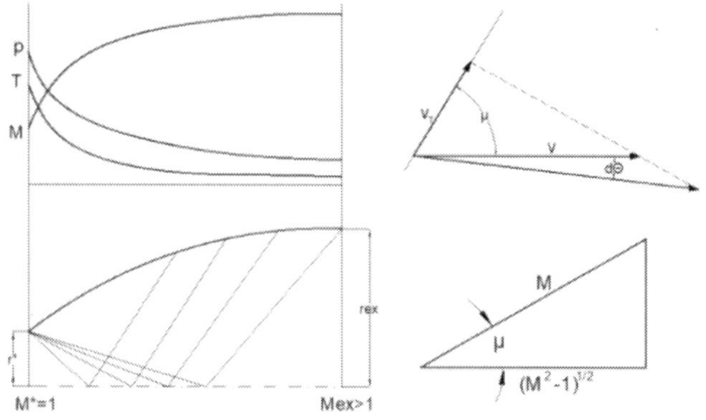

Fig. 2. The divergent portion of a parabolic de Laval nozzle with characteristic lines and distribution of the observed parameters (left), the velocity components with the angles used by the method of characteristics.

These equations alongside with the equations of quasi one-dimensional-isentropic flow of compressible fluids in ducts was an automated software developed [3]. The flowchart of the calculation can be seen on the following figure.

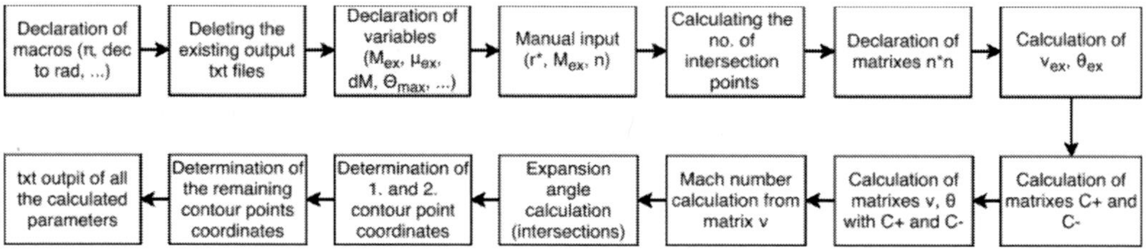

Fig. 3. Flowchart of the automated calculation process.

The high speed mixture of the primary air and secondary working fluid is entering the mixing chamber. The 1-D design model uses the constant-area mixing model, for which is the supersonic nozzle exit located directly on the cross section plane of the mixing chamber inlet (Figure 4).

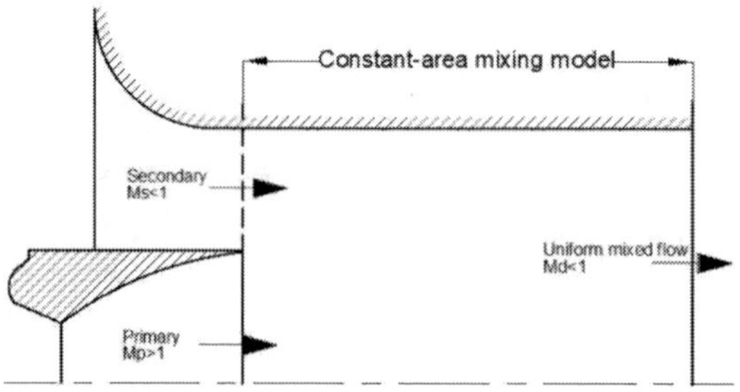

Fig. 4. Constant-area-mixing chamber.

With use of the ER alongside with equation (1) and the known flow rate at the diffuser outlet, the velocity at the exit of the mixing chamber can be obtained:

$$v_{mch,ex} = \eta_{mixing}\left(\frac{\dot{m}_1 v_1 + \dot{m}_2 v_2}{\dot{m}_1 + \dot{m}_2}\right) \tag{8}$$

5 The experimental setup

The experimental station consists of 2 main parts: a two-stage supersonic ejector and a liquid ring vacuum pump. The ejector stages are connected in series, so the outlet of the first stage is connected to the secondary suction of the second stage. The two-stage ejector is then added as a pre-stage to the vacuum pump as shown on the following figure.

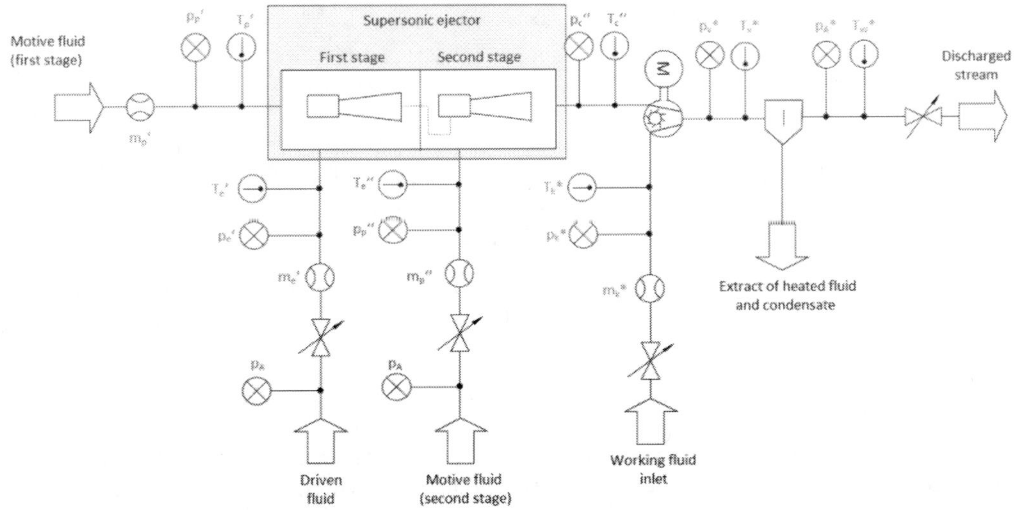

Fig. 5. The experimental setup (installed sensors and meters: m-flow rate, T-temperature, p-pressure).

The used liquid-ring vacuum pump type SZO is designed to extract air, chemically active and non-active vapors in metallurgical, chemical and food industry. The pump consists of a stationary housing in which is an eccentrically rotating impeller placed. The working liquid is fed through a separate circuit to the pump, which by the centrifugal force creates a cylindrical ring inside the casing which acts as a liquid piston. The gradually reducing volumes of the individual interblade chambers are compressing the pumped air, which reports to the discharge port of the pump. The working fluid is along with the pumped gas entering the separator, where the extraction of heated fluid and condensate occurs.

As the first step the digitization of the signals from the installed sensors was made [4]. Since the signal in the digital form has an electrical value, it was necessary to convert them to the values of the actual measured parameters. This step was done in the LABView environment with the known ranges of measuring sensors. The experiment has been made in 2 phases: at first was the vacuum pump itself measured, than the supersonic ejector was added to compare the characteristics of the both setups [5].

The experiment begins by running the pump of the external circuit to constantly deliver water to the pump. As the housing of the vacuum pump fills up to the desired level, the operating speed was set with the frequency converter. The shut-off valve located on the motive fluid inlet of both stages is fully opened. After the stabilization of the monitored values, the first point of the Q-p diagram was recorded. By gradually closing the valve on the secondary inlet port was the suction pressure reduced. At each stop was the actual point data recorded. The measured Q-p values were then imported to MS Excel, approximated, than the characteristics were plotted as shown on the following figure.

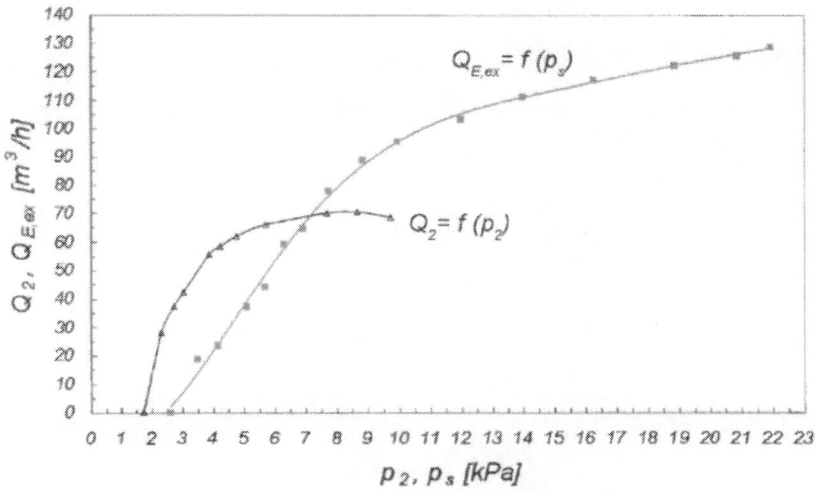

Fig. 6. Comparison of performance characteristics of the liquid ring vacuum pump with and without the supersonic ejector at 1400 RPM.

6 Evaluation of the obtained experimental results

At the analysis of the projected ejector performance characteristics should be remembered, that the performance is not given by the flow rate of air entering the suction throat of the vacuum pump. In case of an ejector used as a pre-stage, the flow rate of air measured at the secondary suction port (inlet no. 2 from Figure 1) can be considered as effective performance. This amount of air is exhausted from the vacuumed chamber. Also, the effective suction pressure is referred analogically to the secondary suction port. From the comparison of the performance characteristics is evident, that the added ejector had a positive impact on the performance of the whole vacuum system. At lower suction pressures (approx. under 10 kPa abs.) is a significant performance increase present. The added ejector also extended the working range of the vacuum pump, so the vacuum system can be operated in a wider range of suction pressures in cavitation-free regime.

References

1. Ch. Liao, *Gas ejector modeling* (dissertation thesis, Texas, 2008)
2. P. Vician, M. Palacka, M. Holubčík, *Effect of ext. temp. on performance of bin* dryer, AIP Conference Proceedings, **1889**, 020064 (2016)
3. M. Puškár, T. Brestovič, N. Jasminská, *Num. sim. and exp. analysis of acoustic wave influences on brake mean eff. pressure in thrust-ejector inlet pipe of combustion engine,* Int. Journal of Vehicle Design, **67**, 63-76 (2015)
4. J. Jablonská, M. Mahdal, M. Kozubková, *Spectral an. of pressure, noise and vibration velocity measurement in cavitation,* The Journal of Institute of Measurement Science of Slovak Academy of Sciences, **17**, 250-256 (2017)
5. S. Daneshmand, C. Aghanajafi, A. Bahrami, *Analytical and experimental methods of design for supersonic two-stage ejectors*, Int. Journal of Aerospace and Mech. Eng., **3**, 2 (2009)

Numerical simulation of heat exchanger operation

Peter Durcansky[1,*], and *Pavol* Orsansky[2]

[1]University of Žilina, Faculty of Mechanical Engineering, Department of Power Engineering, Univerzitna 1, 010 26 Žilina, Slovakia
[2]University of Žilina, Faculty of Mechanical Engineering, Department of Applied mathematics, Univerzitna 1, 010 26 Žilina, Slovakia

Abstract. Numerical simulations have opened wide field of possible applications and opportunities to be used not only in science but also in the industry. With new developed numerical methods we can not only describe existing systems, but we can also predict possible behaviour or possible state of inspected system. This article deals with numerical simulation of proposed heat exchanger for hot air engine and with simulation of its possible states and behaviour with certain conditions.

1 Heat source for hot air engine

Cogeneration unit with hot air engine should be used as an independent source of electrical and heat energy for the household to ensure an efficient energy supply with respect to high reliability. The source of heat is determined by the furnace for burning biomass, in form of a wood pieces, or pellets. Transport energy from the combustion of biomass to a working fluid in hot air engine is providing a heat exchanger connected to that source. In the design of the heat exchanger were taken into account all the conditions imposed on the heat exchanger and the desired properties. Most often the investigation of the heat transfer is the exchanger heat surface [1, 2]. From this are derived all other parameters, external dimensions, connection dimensions, such as flanges, pipe connecting dimensions, mounting flange dimensions consoles supporting structures. Draft of heat exchanger for hot-air engine had to take into account a wide range of specifications and operating conditions. We have established basic dimensions using the criterion equations. To determine the boundary conditions have been transferred several measurements that confirm and specify the chosen boundary conditions. The basic schema of our proposed micro cogeneration unit is in the Figure 1.

* Corresponding author: peter.durcansky@fstroj.uniza.sk

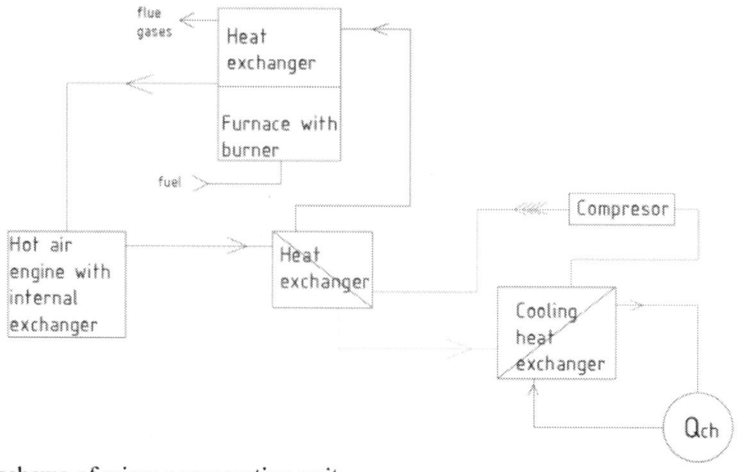

Fig. 1. Principle schema of micro cogeneration unit.

The boiler is fired by biomass. Hot gases are heating the compressed air that flows through the flue gas heat exchanger. In order to increase the efficiency of the system, the flue gas after energy transfer in the heat exchanger, do not leave into atmosphere, but is led through water cooled boiler, which increases the thermal energy transmitted to water. This boiler therefore serves as a heat exchanger and provides an opportunity to gain heat in case of increased demands on thermal energy, for example, in the cold winter months. The compressor is connected into a closed circuit, allowing the use of other working medium such as dry air at a higher pressure. The problem is the need to observe the temperature of the working fluid at the inlet to the compressor at a value prescribed by the manufacturer [3, 4]. Therefore there are included two heat exchangers. Heat exchanger for hot-air engine is used for recovery of thermal energy from the engine of the medium to the medium that flows from a compressor, heat exchanger serves as a heat receiving regenerator. Next, the cooling heat exchanger, is used for cooling the operating medium before entering the compressor to the required temperature.

2 Measuring system

Experimental measurements of the heat source are focused on verification of the boundary conditions in the selected theoretical calculations but also to determine the real characteristics of the proposed device. Also, attention was paid to setting the boiler parameters, especially the amount of air supplied to the boiler. The boiler was placed in the test room and plugged in, as shown in Figure 2. In the combustion chamber, thermocouples are placed, which allow to follow course of temperature in the plane before entering hot flue gas exchanger. At air inlet were placed anemometers, which recorded a rate of air flow to the boiler. Air flow is therefore measured by the indirect method. As heat source was installed into the furnace a gas burner.

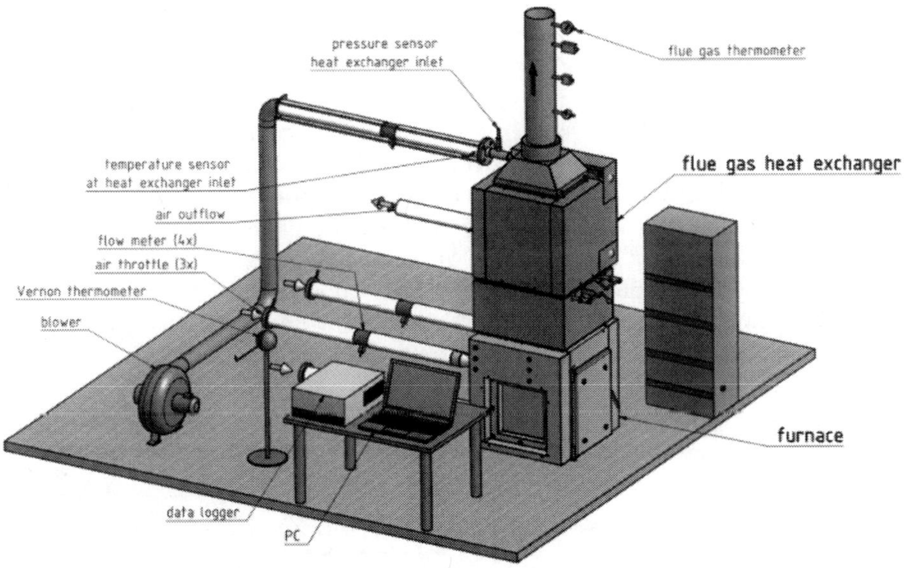

Fig. 2. Measuring system of heat exchanger.

Heat exchanger located in the upper part of the furnace was connected with blower, which feeds the air into the heat exchanger, thus ensuring the flow of air through the exchanger. Observed was the speed of air flow into the heat exchanger, the air temperature at the inlet and outlet of the heat exchanger and also the pressure at the exchanger. All data from the sensors were recorded with logger.

3 Numerical simulation

In our simulation we use MatLab – Simulink, which is signal orientated graphical programming environment for dynamic simulations. Blocks are connected via signals.

As we mentioned above, if we have to simulate thermo dynamical systems, it is good to have special tools that enable work in this specific scientific areas. For MatLab – Simulink we used the tool Thermolib. To represent all thermodynamic information in an easy to access way, Thermolib uses Simulink busses to represent media flows and thermodynamic states.In models that are based on the Thermolib, the signal flow describing the medium flow between the blocks is described by a so-called flow bus. The flow bus contains the primary information on the flowing media [5, 6]. Apart from the total molar flow and the chemical composition, it contains the thermodynamic properties temperature, pressure and vapor fraction.

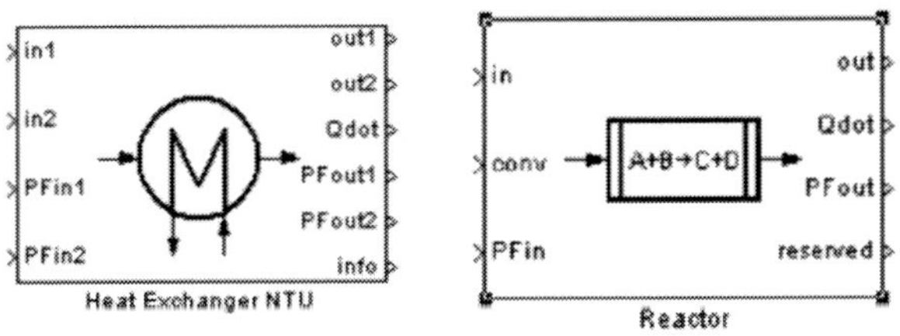

Fig. 3. Two basic parts of proposed Matlab simulation.

Moreover, it is possible to specify the chemical composition of the medium. Chemical media are the chemical compounds that are available in Thermolib simulations. Some of the default species available are: methane, water, hydrogen gas, oxygen gas, carbon dioxide, among others.

This means that it is sufficient to select individual blocks of the system from the library (e.g.: turbine, expander, cooler and warmers), put it into the model and set the parameters [7, 8]. Since the simulations runs in real time, the parameters can be changed during the entire simulation. Also the total time of the simulation can be set as finite or infinite, where we have the opportunity to stop the simulation change any parameter and again run the simulation. In each moment of the simulation we are able to know every to obtain each value of parameters of the flow bus, namely: total molar flow, temperature, pressure, enthalpy flow, entropy flow, Gibbs energy rate, heat capacity rate, vapor fraction of all compounds and molar fractions of all compounds [9].

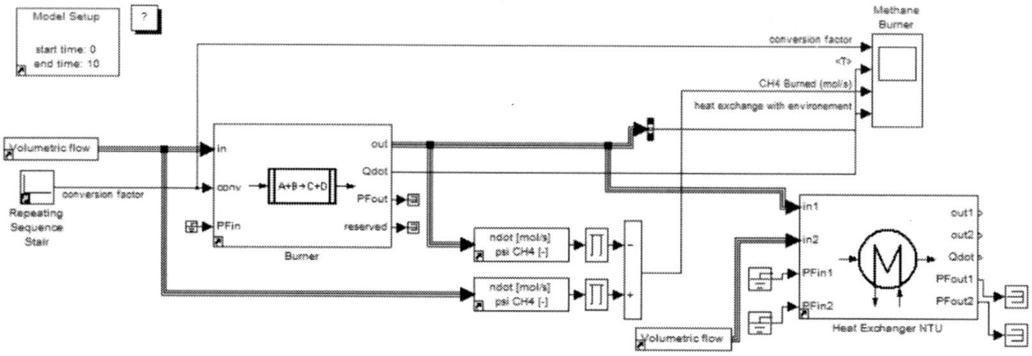

Fig. 4. Simplified numerical model in Matlab.

4 Conclusion

Simulation of thermodynamic processes in MatLab environment using block library Thermolib gives many various opportunities to create very useful models. There are countless possibilities of setting individual block components as well as the medium itself. The numerical model can be used for many applications. In our work we focused on exploring various options of involvement and use of this numerical method to describe an existing heat source. First we have made measurements with specified settings to obtain the real behavior of this dynamic system at various conditions. We have observed the system with changed burner power output. After this, the numerical model was adapted to the existing system. Tuned was for all the overall heat transfer rate and also the particular coefficients of heat transfer rate between the two mediums. The power output of the exchanger was then calculated for every setting, as in measurement steps. The results can be seen in the Figure 5. It is clearly to see, that the model is case sensitive and is sensitive at the inputs by small power range. The model produces also an error around 20 kW. But with higher power outputs, the simulation has very good accuracy with small errors.
As next step was added to the model the polynomial regression curve both for measurement and for simulation data, that allowed to predict power output of the furnace. The maximal possible power of the furnace is 35 kW. The numerical model with prediction showed divergence only of 3 kW at 38 kW predicted maximal power.

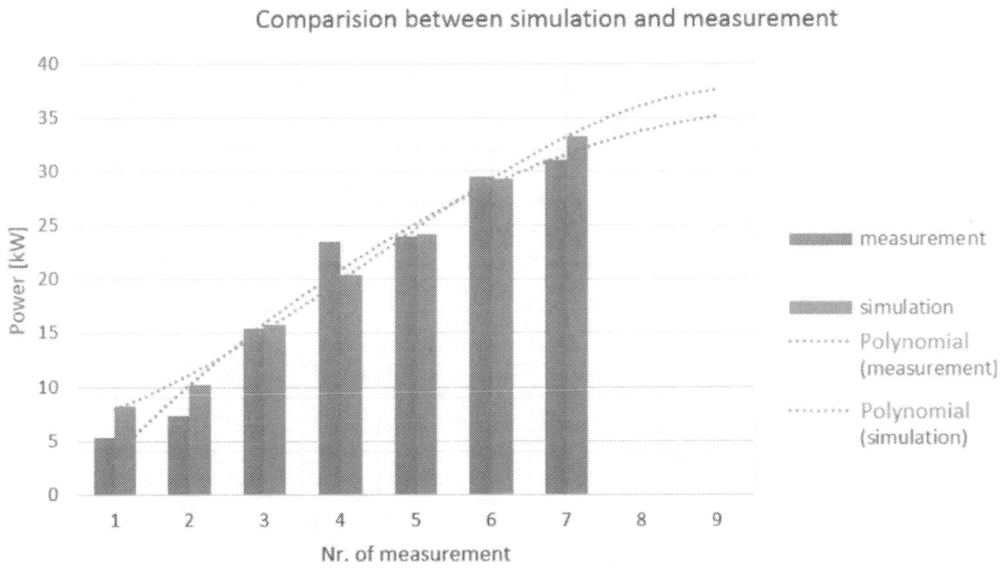

Fig. 5. Results of simplified numerical model.

The research is supported from the project "APVV-15-0778 - The limits of conventional radiation and cooling through the phase change of the working fluid in the thermo-syphon loop" and APVV-15-0790 „Optimization of biomass combustion process with low ash melting temperature". The project is co-financed by EU funds. Project title: "Research on new methods of heat conversion from RES to electricity using new progressive thermal cycles" ITMS 26220220117

References

1. R. Nosek, S. Gavlas, R. Lenhard, V. Sedlak, H.M. Arvesen, Communications, **16** (2014)
2. M. Malcho, J. Jandačka, R. Lenhard, S. Gavlas, T. Ochodek, Communications, **18** 115-121 (2016)
3. A. Kapjor, K. Kaduchová, R. Lenhard, H. Smatanová, *Prenos tepla z orientovaných teplovýmenných plôch pri prirodzenej konvekcii.* (2017)
4. T. Brestovič, M. Čarnogurská, R. Pyszko, M. Kubik, *XVIII AEaNMiFMaE*, 19-26 (2012)
5. B. Skočilasová, J. Skočilas, *AIP,* **1608**, 206-210 (2014)
6. Z. Trávníček, P. Dančová, J.H. Lam, V. Timchenko, J. Reizes,. EPJ Web of Conferences (2012)
7. J. Kňourek, R. Matas, O. Prokeš, D. Tenkrát, *EPJ Web of Conferences*, 67 (2013)
8. M. Brůna, A. Sládek, Technológ (2011)
9. Michalková M., Michalková L, Applied Mathematical Sciences, **10**, 57 (2016)

Analysis of thermodynamic parameters and their influence on the thermal comfort in the working environment

Zuzana Kolková[1,*], *Peter* Hrabovský[1], *Jozef* Matušov[1], and *Radovan* Nosek[2]

[1]University of Žilina, Research centre, Univerzitná 8215/1, 010 26 Žilina, Slovakia
[2]University of Žilina, Department of Power Energy, Univerzitná 8215/1, 010 26 Žilina, Slovakia

Abstract. Air quality inside the building is an important measure of comfort for people living and working in it. Achieving optimal air quality requires a regulated indoor air control system, often in combination with thermal regeneration. Typical short ventilation, irregular window and door ventilation are rarely adequate. From the point of view of the thermal comfort of the environment, it is necessary to predict the development and selection of suitable systems for the thermal comfort of the environment to predict how many thermal factors will affect the population, room temperature uniformity, asymmetry of the radiation temperature, and turbulence intensity of the flowing air. People are capable of detecting heat, temperature changes can feel not only globally, but also as local heat loss on the parts of the body. Thermal comfort is given by the air temperature and radiant heat, air velocity and relative humidity in the room. The main impact of these factors on the comfort level of individuals depends on the activity and type of clothing. The bad thermal comfort is detrimental to workers' health, increasing the risk of accidents and affects adverse effects on the psyche. The assessment of impact of the working environment to human form comparing actual and optimal parameter values of the working environment.

1 Introduction

The human body is a continuous source of heat. This metabolic heat production is divided into basal and muscle metabolism. Basic metabolism is the heat produced by biological processes. Muscle metabolism arises in human activity. Heat produced by the organism must be taken to the environment or body temperature changes. The temperature inside the human body is around 37 °C, while the skin temperature may range from 31 to 34 °C depending on the surrounding environment. Differences arise over time but also by parts of the human body [1].

Indoor environments in buildings create many components. Thermal humidity, smell, aerosol, microbial and acoustic components are essential. The basic influence on people has the thermal humidity parameters of the living room. In terms of impact on human health, air

* Corresponding author: zuzana.kolkova@rc.uniza.sk

quality seems to be the most important factor. Most of these environmental components are affected by ventilation [2].

Space temperature uniformity, asymmetry of the radiation temperature, and turbulence intensity of the flow air affect thermal comfort. People are capable of detecting heat and they can feel changes in temperature globally on the entire surface of the body but also locally on body parts. It has been shown that humans can acclimate to ambient temperature physiologically and psychologically. It is therefore important to address the issues of thermal comfort in the interiors (buildings) and to experimentally determine a suitable way of creating the optimal conditions of the environment. The goal is to ensure a sense of satisfaction from the environment and also to increase work performance and a sense of comfort [3].

2 Dantec system

ComfortSense is a multichannel system for measurement air velocity and air temperature, humidity and operative temperature. It calculates statistical values – Draught rate (DR), Predicted Mean Vote (PMV) and Predicted Percentage Dissatisfied (PPD). ComfortSense can be used for many applications including HVAC research and development, testing of ventilation equipment, building research, passenger comfort studies in the automotive and aviation industries. Thermal comfort measurements with ComfortSense comply with International Standards EN 13182, ISO 7726, 7730, ASHRAE standard 113 and ASHRAE standard 55. The ComfortSense system consists of a main frame with input channels for up to 16 probes. The omnidirectional probes measure both air velocity and temperature. The ruggedly designed probes and cables are perfectly suited for large test rooms. The draft probe is equipped with an omnidirectional thin-film sensor for measuring air velocity and a small fast-response thermistor for measuring air temperature (Figure 1).

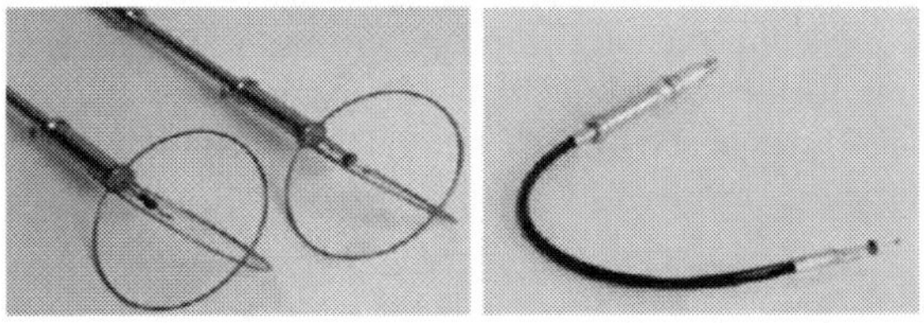

Fig. 1. The draft probe, robust velocity temperature probe and manikin version.

The manikin version of the draft probe is very compact, with a flexible cable connection to the probe tip, making it suitable for building into a manikin for passenger comfort applications. Velocity range is 0.05 to 5 m/s, indicates up to 10 m/s and temperature is -20 to 80 °C for both types. Robust velocity and temperature probe has velocity range 0.1 to 30m/s. Humidity range is 0 – 100 %. Dantec has also operative temperature probe. The sensor element simulates a standing person when it is vertical, a sitting person when tilted 30 °C from vertical and a reclining person when in the horizontal position. Temperature range is 0 – 45 °C (Figure 2) [4].

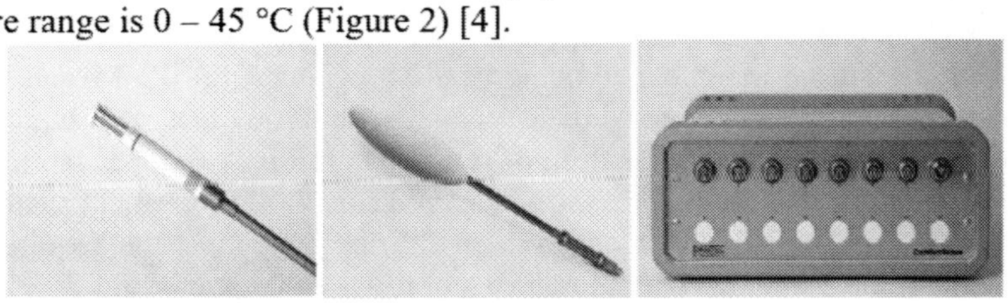

Fig. 2. Humidity probe, operative temperature probe and main frame.

3 ASHRAE STANDARD 55

ASHRAE Standard 55 (Thermal Environmental Conditions for Human Occupancy) is a standard that provides minimum requirements for acceptable thermal indoor environments. It establishes the ranges of indoor environmental conditions that are acceptable to achieve thermal comfort for occupants. Percent dissatisfied (PD) represent percentage of people predicted to be dissatisfied due to local discomfort. Predicted mean vote (PMV) is an index that predicts the mean value of the votes of a large group of persons on the seven point thermal sensation scale. Predicted percentage of dissatisfied (PPD) is an index that establishes a quantitative prediction of the percentage of thermally dissatisfied people determined from PMV [5-7].

4 Experimental setup

We have a laboratory of chemical analysis. This workplace has three devices, which have furnaces with heating up to 1 500 °C. We have several systems to ensure optimal working conditions. The laboratory is oriented to the South-East side. Roller shutters are installed on the windows to regulate the intensity of sunlight during the day when the laboratory is most used. Floor heating is used to provide heat, ceiling cooling to ensure the cold. The ventilation of the laboratory is ensured by circulating air through the ceiling spouts. We also use a ceiling air-conditioning unit to reduce room temperature. If all three devices are in full operation, room temperature rises rapidly. Workers are subject to varying temperature changes in a relatively short time in this laboratory. Laboratory space is 3 m high (Y), wide is 4.5 m (X) and long 8 m (Z). The basis of experimental verification of laboratory environment parameters will be the monitoring and graphical comparison of PMV, PPD, velocity, humidity and operating temperature. The standard STN 73 0540-3 determines the optimal values of environmental parameter according to the purpose of use. Setting the room measurement and parameters is in the Dantec program. It is possible to set the room size, location of the measuring probes in the space, display options. The natural state of the environment, the working status of the various analysers, the use of the air-conditioning unit, the draft and various weather conditions can be simulated using the laboratory's technological equipment.

5 Results of measurement

Measurement of environmental parameters was performed during simulated loads. The first measurement was in full operation (PP), load conditions were monitored at full operation of two machines without cooling by an air conditioning unit or natural ventilation. The second measurement was the condition using the air conditioning unit. In this measurement, the effect of air conditioning was measured at the maximum (C) and minimum (D) setting of the unit. The maximum temperature setting was 29 °C and minimum 22 °C. Measurement PP – ambient temperature 20 °C, relative humidity 37 %, wind velocity 10 km/h, measurement C and D – ambient temperature 11 °C, relative humidity 82 % and wind velocity 10 km/h. The graphical method is used to evaluate the PVM and PPD indices, which display individual experimental measurements and allow instant comparison with the permissible and optimal value. Evaluation of relative humidity - optimal value according to STN 73 0540-3 determines 30 – 70 %, we measured for PP measurement value 27.2 %, for C - 40.10 % and for D 34.14 %. Measurement PP does not meet the requirements in the standard. Evaluation of operative temperature - optimal value according to STN 73 0540-3 determines 20 – 24 °C, we measured for PP measurement value 25.08 °C, for C – 24.48 °C and for D 22.25 °C. Condition D complies with the norm. It is cooling by the air conditioner at 22 °C. Load PP and C - the measured values exceed the optimal value. Figure 3 shows the evaluation of the PPD and PMV parameters and their interdependence, which meets the requirements of the ASHRAE 55 Index Standards at the thermal comfort limits. For PPD 10 % (OPTIMUM value) and 20 % (POTENTIAL value) of dissatisfied persons in the given thermal state of the environment and PMV In the range of -0.5 <PMV <0.5 (OPTIMUM value) and -1 <PMV <1 (PERMIT VALUE).

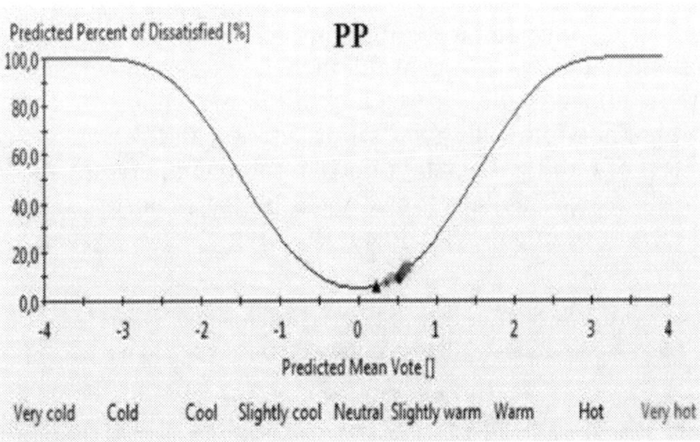

Fig. 3. Index PDD a PMV – measurement PP.

Figure 4 shows that load measurements "C" and "D" under the conditions of simulated load comply with the ASHRAE 55 standard for PMV and PPD.

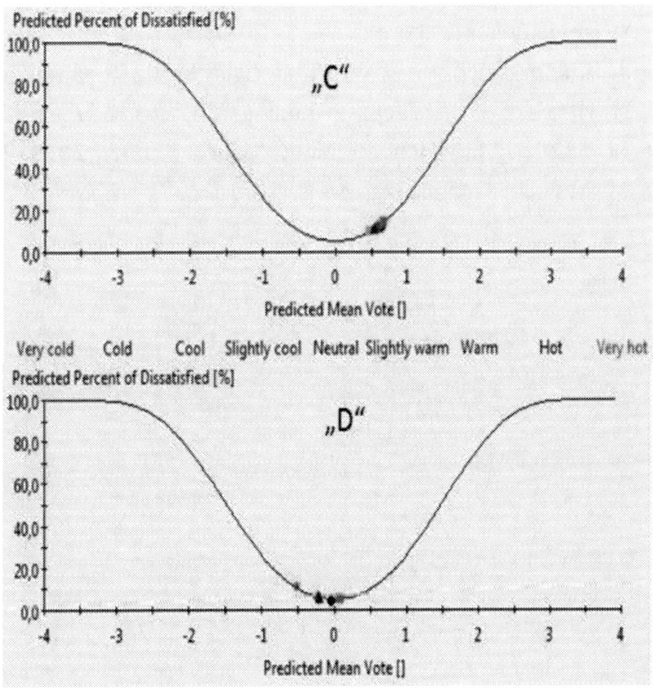

Fig. 4. Index PDD a PMV – measurement C and D.

6 Conclusion

Ensuring thermal comfort is a challenging task in an investigated laboratory. Analysers must be kept at a minimum temperature of 300 – 600 °C even out of operation. This means that the space is constantly heat-laden regardless of the working activity. The analysers are heated at 1 000 – 1 500 °C during operation. When the samples are incinerated, there is a smell that needs to be removed from the space. The sun shines until the afternoon on that side. The ventilation outlet is in the ceiling at the window. The temperature in the laboratory is kept constant at 24 °C regardless of the operation. We mainly ensure the safe operation of the instruments and a constant temperature in the room. The greatest load was caused by incorrectly adjusting the ceiling slats and lamellas on the ventilation unit.

We have optimized these settings using the Dantec system and we have eliminated such adverse effects on human health. The feeling of thermal comfort for all workers at the same time is difficult but possible with good set of systems and personnel arrangement. We will examine the effect of turning the outer blinds to room temperature and the effect on thermal comfort. We will create lamella settings based on the intensity of sunlight, which will be automatic and integrated into the Intelligent Building Control System.

The research is supported by the European Structural Funds – project ITMS: 26220220048 (50%) and by the project KEGA No. 042ŽU-4/2016" Cooling by physical and chemical processes".

References

1. K. Parsons, *Human Thermal Environmentals*, CRC Press, 635 p. (2014)
2. S. Carlucci, *Thermal Comfort Assessment of Buildings*, Springer-Verlag Mailand, (2013)
3. STN 73 0540-3:2012: *Tepelná ochrana budov. Tepelnotechnické vlastnosti stavebných konštrukcií a budov.* Časť 3: Vlastnosti prostredia a stavebných výrobkov
4. https://www.dantecdynamics.com
5. http://shop.iccsafe.org/media/wysiwyg/material/8950P219-sample.pdf
6. P. Vician, M. Holubčík, M. Palacka, J. Jandačka, J, *AeaNMiFMaE* 2016 7p., (2016)
7. P. Nemec, M. Smitka, M. Malcho, Scientific world journal, **2014** (2014)
8. P. Nemec, M. Malcho, J. Jandačka, *AIP Conference Proceedings*, **1558** (2013)

Conceptual design of the mobile granulation line for production fertilizers from digestates and ash mixtures

Marek Wróbel[1,*], *Marcin* Jewiarz, *Krzysztof* Mudryk, *Jarosław* Frączek, and *Krzysztof* Dziedzic

[1]University of Agriculture in Krakow, Department of Mechanical Engineering and Agrophysics, 120 Balicka st. 30-149 Kraków, Poland

Abstract. Paper presents a concept of the mobile line for the production of granulated fertilizer from digestates and ash mixtures. The main assumption of the designed line is its mobility which allows line work in the place of production of digestate. The premise for developing such model of line was the fact that the transport of highly hydrated digestate, as well as its dried form, is not advisable for economic reasons. More advantageous solution is to collect the digestate at the place of its production and it is processed into a fertilizer. Loose components, characterized by high bulk density, can be delivered to the place of production and the same means of transport can be used to transport the produced fertilizer to warehouse. This production model allows to process the digestate produced in several biogas plants, using one production line and reducing the cost of production. of fertilizer granules and the mobile line is transported to the next post-fermentation production point. Based on an analysis of the raw material characteristics and preliminary research on the raw materials mixture granulation process it was assumed that the stage of 3 hrs mixture stabilization time is required in the line.

1 Introduction

A digestate is generated in a biogas plant in anaerobic methane fermentation process. Substrates in the fermentation process is subject to the following transformations: reduction of content of organic substance, decomposition of odour generating compounds, solid particles defragmentation, improvement of dehydration ability and fractional hygienization with considerably low losses of fertilizer components [1, 2].

Remains after the process digestate includes: biomass of microorganisms which carry out the fermentation process, non-fermented organic compounds and mineral components. These substances are very precious from the fertilization point of view, because they include N, P, K in the mineral form directly available for plants. Moreover, the content of organic

* Corresponding author: marek.wrobel@urk.edu.pl

matter which is subject to mineralization is a source of nutrients and positively influences the physical and chemical properties of soils improving thus their productivity.

From manure and different types of biomass, content: general nitrogen – 3.4–4.0; N-NH4 – 2.4-2.9; P – 0.8–1.2; K – 2.7-3.1 – (values provided in $g \cdot l^{-1}$) [3].

From an economic point of view, transportation of non-processed digestate at the distance exceeding 10 km from the biogas plant is non-profitable if it is going to be used as fertilizer [4].

So another, way of using digestate for fertilization purposes is searched for. One of it is granular organic and mineral fertilizers based on digestates and mineral substance. Seems to the best component for this purpose it's fly ash from professional energy industry installations, where only biomass is combustion. This kind of ash may be used as an good fertilizer [5]. Positive fertilization effects may be also achieved using mixtures of ash with organic substances [6, 7].

Mentioned raw materials, due of his origin, have different and varied physical properties such as: size distribution, density, porosity which significantly influence on the quality of obtained granulates [8, 9].

In order to be able to produce granulated fertilizer from mixtures based on ash and digestate with the addition of other valorizing components, the designed technological line must be able comprehensive processing of raw materials into granules in the form of pellets. Due to the properties of raw materials, it must enable thorough mixing of raw materials in various proportions, stabilization of the obtained mixtures in order to equalize the moisture of raw materials and also agglomeration, stabilization, purification and packaging of the obtained granulate. Designed line is supposed to have the character of a mobile line. The work presents the concept of a line that meets the above assumptions.

2 Characterization of the raw materials and their mixtures

The basic materials will be ash from the combustion of biomass in industrial installations and the digestate from biogasification installations for agricultural biomass and residues from slaughter of animals. Sulfur, urea and phosphorite were selected as valorizing additives. Digestate share in the mixture will be 25 – 75 %, sulfur and urea up to 5 % and phosphorite up to 15 %.

The results of the tests carried out on laboratory scale have shown that the optimal moisture content of the raw material mixtures is at the level 18 %, and the acceptable level is 20 %. The working humidity of ash and valorizing additives is about 5 %. Therefore, to prepare mixtures with the assumed share of digestate and assumed moisture levels, digestate should have dedicated batch moisture (depending on the produced mixture between 21.6 – 45.7 %). So digestate moisture is much higher than the moisture of the target mixture. To make the components of the mixtures obtain the required moisture a stabilization during time of 3 hours is required (value obtained on the laboratory tests).

Raw materials have a granulometric composition, which does not cause problems during granulation, and therefore they do not require additional grinding.

3 Project assumptions

Knowledge of raw materials parameters and interactions between them and also knowledge of its agglomeration process allowed the following assumptions and design requirements have been made:

- Line should be built on a car trailer, on which all the elements of line, necessary for its work are transported: a mixture preparation unit, granulator with a cooler, conveyors, raw

material acceptance points and final product packaging unit, platforms, safety barriers, etc.
- Digestate drying and milling systems are not part of the line (it can be a stationary system, located in a biogas plant or a separately mobile system).
- The power supply system (power generator) is a separate mobile unit.
- Line equipment includes a forklift that supports the line during work and during preparation for and after work. It's transported on the trailer with the line.
- Estimated line efficiency is 0.6 - 08 t per hour of final granulate.
- The operation of the line is planned in a continuous mode, which forces the use of appropriate devices.
- Valorising components are supplied in individual packages stacked on pallets. The ash is delivered in a BIG-BAG packaging.
- The digestate with required moisture is delivered to the line acceptance point with the loader (equipment of the plant in which the production is carried out) or the conveyor directly from the dryer.
- The line should be protected against dust escaping from it and should not generate postproduction waste.
- The final product will be packaged in BIG-BAG.
- Prepared fertilizer mixture, before granulation requires at least 3 hour stabilization.
- The line should have automatic control with the ability of regulate the performance of line units.

Accordingly, the mobile line should consist of the following units:
- Transport - delivering the line to the workplace.
- Preparing the mixture - implementing processes dosing and mixing of components and mixtures stabilization.
- Agglomeration - granulation prepared and stabilized mixtures.
- Cooling and stabilizing - performing processes of screening, cooling, stabilization and packaging of produced fertilizers.

The line must be equipped with additional devices, such as:
- Transport systems for raw materials, mixture and final product.
- System for separating dust and fines from granulates and filtration of technological air.
- Platforms, ladders, railings, stands ect. for comfort, quality and safety of work.
- Protection of engines and gears against overload.

4 Concept of the mobile granulation line

Based on the assumptions, a flowchart of the planned line was developed (Figure 1.), which illustrates the operations of processing the raw material and flow direction. There are pointed places where raw materials are supplied to the line (ash, valorising components, and digestate). The scheme also includes a technological air filtration system, with an indication of point from which the polluted air is captured and which receives dust and fines. It was also marked which elements of line remain on the transport trailer during work and which are placed outside of it.

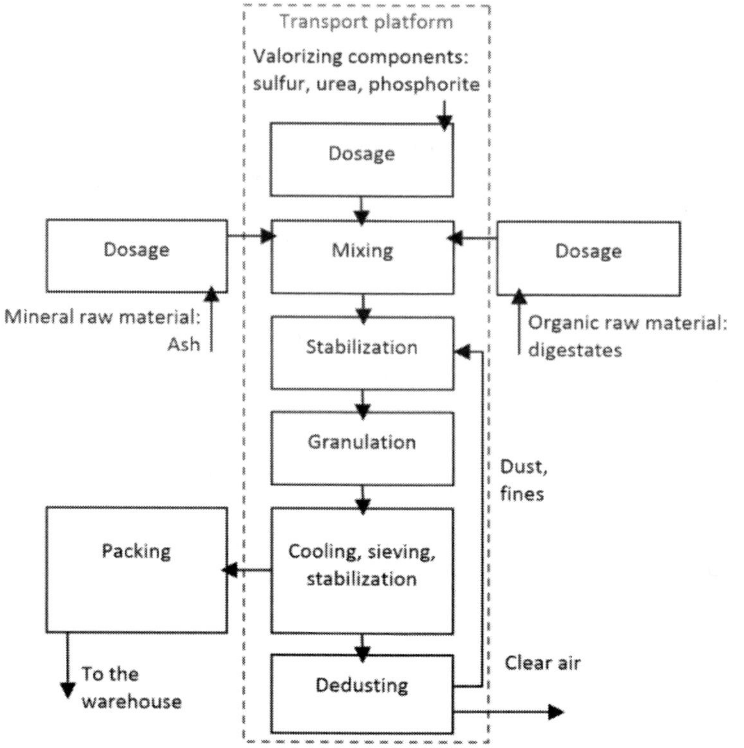

Fig. 1. Block diagram of the projected technological line.

The following technical solutions are proposed for each stage of granulated fertilizer production line (numbering in accordance with the designations in Figure 2 and Figure 3):

1. Ash in BIG-BAG unloading station.
2. Ash feeder – screw conveyor.
3. A set of tanks dispensing valorising components along with a conveyor supplying loose components to the mixer (No. 14).
4. Mixtures stabilization tank equipped with a screw conveyor feeding the mixture to the conditioner of the granulator (No. 7) and a screw system for uniform filling of the tank
5. Digestates feed station (hopper with horizontal screw conveyor)
6. Granulates coolers dedusting system (fan, cyclone, screw conveyor)
7. Ring granulator with conditioner.
8. A set of conveyors supplying raw granules into coolers (No. 9), conveyor with corrugated belt and horizontal belt conveyor with a material distributor for two coolers.
9. Pellet cooling system (2 drum-sieve cooler equipped with a horizontal screw conveyor collecting fines and feeding it to the dust extraction system (No. 6).
10. A set of conveyors with corrugated tape feeding granules to the BIG-BAG filling station (No. 11).
11. BIG-BAG filling station (two racks with BIG-BAG carrier, granulate distributor).
12. Transport trailer - folding forklift can be transported in a pallet box or hung at the rear of the trailer.
13. Digestates feeder – screw conveyor.
14. Unit for the fertilizer mixture preparation (twin-shaft mixer).
15. Mixture feeder to stabilizing tank (No. 4) – screw conveyor.

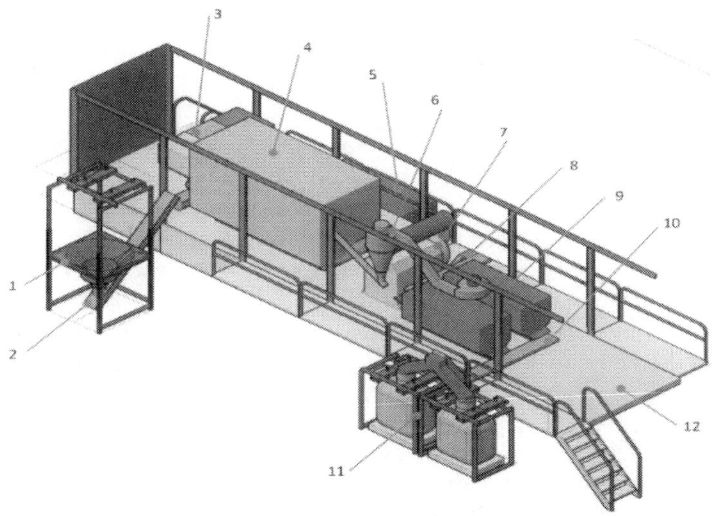

Fig. 2. 3D visualization of the designed line – view from the ash loading side.

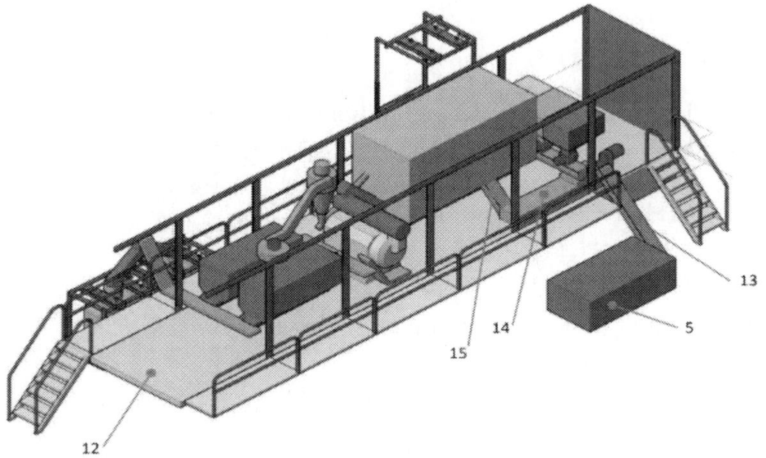

Fig. 3. 3D visualization of the designed line – view from the digestate loading side.

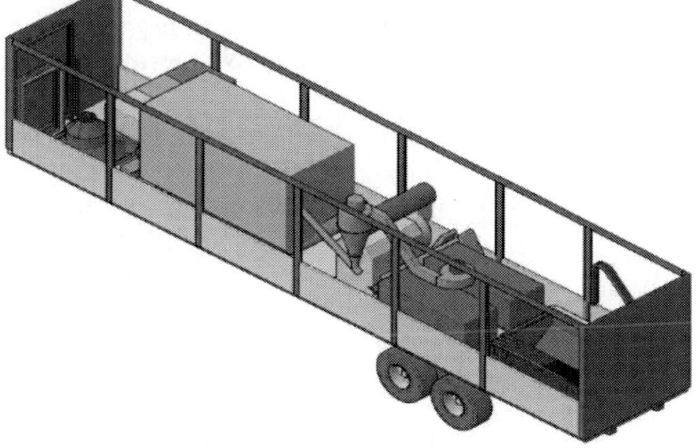

Fig. 4. 3D visualization of the designed line – view of transport position.

Figure 4 presents the 3D visualization of the line in the transport position with all line elements arranged on the transport trailer.

5 Summary

The presented concept of a mobile technological line for the production of granulated fertilizer from mixtures of ash, digestete and valorising components in form of pellets takes into account all assumptions and in complex way allows for processing of raw materials with different and variable input parameters (different morphology and granulometric composition, moisture). Proposed mixtures stabilizing step allows to equalize the moisture content of mixtures components, what improves the quality of the obtained granules. The air dedusting system allows to clean the air before leaving the line and extracted dust returns to the granulation, this solution doesn't generate production waste during the process. The main innovation of the line is its mobility that allows the digestate to be processed at the place of its generation.

Research was funded by the National Centre for Research and Development and the National Fund for Environment Protection as a part of GEKON program - project no. GEKON1/05/214543/38/2015 "Proecology production of organic and mineral fertilizers based on waste: by-porducts of combusiton and biogasification of biomass".

References

1. M. Bavutti, L. Guidetti, G. Allesina, A. Libbra, A. Muscio, S. Pedrazzi, Energy Procedia **45**, pp. 1344-1353 (2014) Doi:10.1016/j.egypro.2014.01.141.
2. G. Grinzi, L. Guidetti, G. Allesina, A. Libbra, P. Martini, A. Muscio, 20th European Biomass Conference and Exhibition (2012)
3. J. A. Alburquerque, C. Fuente, A. Ferrer-Costa, L. Carrasco, J. Cegarra, M. Abdad, M. P. Bernal, Biomass Bioenergy **40**, pp. 181-189 (2012)
4. H. Döhler, P. Schliebner, *Verfahren und Wirtschaftlichkeit der Gärrestaufbereitung*. Darmstadt: KTBL (2006)
5. X. J. Huang, N. He, M. D. Liu, D. Yang, Y. J. Wang, Applied Mechanics and Materials **535**, pp. 523-527 (2014)
6. L. C. Ram, R. E. Masto, Earth Sci. Rev. **128**, pp. 52-74 (2014)
7. T. Kumar, K. Tedia, D. Devdas, Journal of Progressive Agriculture **6**(1), pp. 78-82 (2015)
8. K. Mudryk, M. Wróbel, M. In *The acquisition and processing of biomass of Virginia mallow for energy purposes*. (ed. W. Piekarski). Wydawnictwo „Wieś jutra", Warszawa, ISBN 978-83-62815-10-4 (2011)
9. M. Wróbel, J. Frączek, S. Francik, Z. Ślipek, K. Mudryk, Engineering for Rural Development Jelgava, Available at:
http://tf.llu.lv/conference/proceedings2013/Papers/121_Wrobel_M.pdf (2013)

Digestate, ash and Trichoderm based fertilizer – production line concept design

Marcin Jewiarz[1,*], *Marek* Wróbel, *Jarosław* Frączek, *Krzysztof* Mudryk, *Krzysztof* Dziedzic

[1]University of Agriculture in Krakow, Department of Mechanical Engineering and Agrophysics, 120 Balicka st. 30-149 Kraków, Poland

Abstract. Paper presents design process of the prototype line, dedicated for production of fertilizer pellets from digestate-ash mixture. For enrichment of the nutritional characteristics of the final product additives were also used, like urea, elemental sulphur or phosphorite, and special fungal strains from Trichoderma spieces. During this process raw material will be dried (raw digestated), mixed, granulated and packed in one technological line. Presented concept, shows that there is possible to produce high quality fertilizer, from waste materials derrived form Green energy production, what should be one of very interesting ways to develop digestate and ash from biomass combustion.

1 Introduction

Production of the renewable energy, leads in many cases to waste production. The best examples of such technologies, are biomass combustion and biogas plants. In first case, there is produced ash, which could be reclaimed as valuable fertilizer. The Table 1 presents specific data of ash from biomass combustion

On the other hand, biogas plants produces large quantities of solid digestate, which in case of fermentation of agricultural residues can be also treated as valuable source of nutrients for plants. In many cases such highly watered digestate is deposed directly to land fields, but this method has some important disadvantages. First of all, usually, fresh digestate emits large amounts of odours, what limits its usage near household. What is also important, in many countries, there are limits concerning amount of digestate dropped on the field during year. The last one of the cons is profitability of the transportation of raw digestate, which in many cases is limited up to 15 km from the place of production [1].

Increase in energy share of biogas and biomass combustion, leads to higher amounts of this two waste-by-products, which needs to be processed in some way.

The aim of this paper is to present concept line design, capable of production of fertilizer mixture, with flexible composition. For the production will be used waste by-products form Renewable Energy production (ash form biomass combustion and solid digestate) with supplementation of sulphur, urea and phosphorite. What is also important to enhance

* Corresponding author: marcin.jewiarz@urk.edu.pl

assimilability of the nitrogen to the plants, part of digestate will be inculcated with special strains of Trichoderma [2, 3].

Table 1. Ash composition form biomass combustion [4].

Parameter	Value range
Moisture(%)	0.6 – 3.3
TC (%)	1.4 – 12.3
Cl^- (g/kg, DM)	6.2 – 254.2
$NO3^-$ (g/kg, DM)	0 – 2.2
$PO4^{3-}$ (g/kg, DM)	13.4 – 39.5
Na (g/kg, DM)	3.3 – 47
K (g/kg, DM)	47.2 – 570
Mg (g/kg, DM)	0.4 – 30
Ca (g/kg, DM)	5.4 – 316
Al (g/kg, DM)	0.01 – 19.1
Fe (g/kg, DM)	0.7 – 13.2

Increase in energy share of biogas and biomass combustion, leads to higher amounts of this two waste-by-products, which needs to be processed in some way [5].

2 Design Methodology

The design process was based on selection of unit operation best equipment by means of Scoring Design Method and derived from it [6–8]. Concept provides analysis of each unit operation by means of several key factors, selected at the first stage of design. What is important, each factor is described by importance multiplier, usually in rage from 0 to 1. Then selection of each apparatus is made by means of scoring each of key factors (scale 0-20 points). Sum key factors score multiplied by importance coefficient, is used to rank devices (apparatus with the highest score are selected), and based on the highest score in each unit operation groups

2.1 Installation requirements

Based on experience from earlier research during project, which were aimed on assessment of physical properties of the materials used in production of this fertilizer mixture, several key boundaries conditions for design process were stated.

- For stationary production, the facility should be able to work in continuous regime, to assure high profitability of final product.
- The installation capacity should be higher than 1Mg/h to provide profitable production
- As a feedstock for production will be used dewatered digestate, with moisture content below 75%, will be stored in bulk under shelter or roofed site
- The transport of the digestate to the infeed hopper of the dryer will be realised by multifunctional wheel loader, used also for other transportation procedures during production.
- The hopper should have capacity which will be sufficient for 1h of production (approx. 2Mg of digestate).
- The drying of the digestate, will be realized by means of drum or belt dryer.

- Additives: ash, elemental sulphur, urea and phosphorite will be delivered by tankers to silos located in the factory, the capacity of silos should be enough for 24h of continuous production.
- Based on the research made in previous parts of the Grant, the digestate and ash mixture should be stabilised up to 3 hour before granulation
- The composition of the fertilizer, should be flexible, as it will be set according to the current needs of market.
- The installation should be capable of production two types of pellets – form mixture digestate, ash and additives, and form digestate incubated with Trichoderma, as this fungi will not survive in the mixture with ash (to high pH level).
- The final mixture of pellets with and without ash should be prepared just before confection process.

2.2 Raw material description

Production of this type of fertilizer requires mixing of several raw materials of mineral and organic origin. That's why it was important during first stages of project to determine key quality parameters of this materials. The physical analysis of the raw materials used in pellet production are provided in Table 2.

Table 2. Basic analysis of physical quantities of ash and digestate.

Quantity	Digestate	Ash
Moisture (%)	73 – 75	1 – 3 %
Bulk density (kg/m^3, FM)	475 – 500	570 – 580
Specific density (kg/m^3, DM)	220 – 233	-
Absolute density (g/cm^3, DM)	1.85 – 1.91	2.46 – 2.48
Absolute porosity (%, DM)	92 %	76
Angle of repose (°, DM)	47 – 56	52 – 60

3 Design presentation

Results of the design process were placed in the warehouse with dimensions 46x19m and 7 m high (no 25 on Figure 1). Precise visualisation is presented of Figure 1. The nominal capacity of the whole line was set in the range 1.5 – 1.7 Mg/h (depends upon composition of the fertilizer mixture).

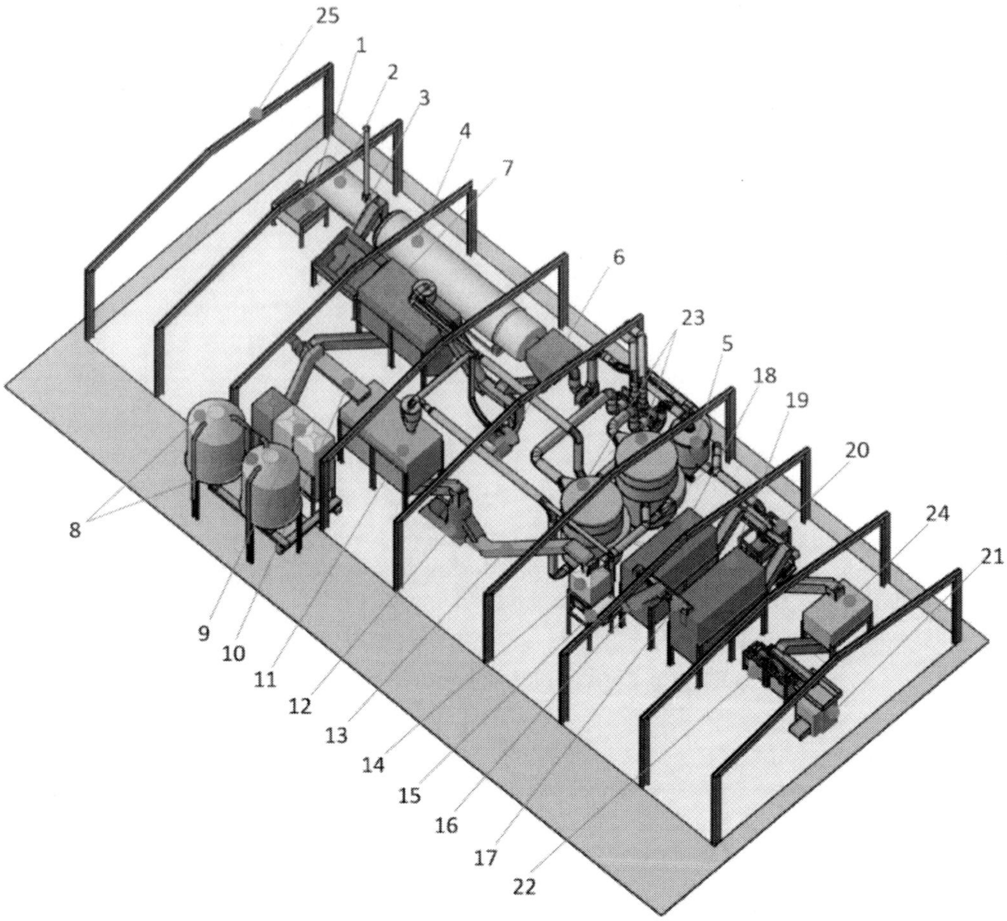

Fig. 1. Complete digestate-ash to fertilizer line (description of numbers in text).

3.1 Preparation of raw materials

3.1.1 Drying of digestate

Granulation of the organic materials, requires specified condition, one of them is moisture content of raw materials, which usually should be in the range 20 – 25 % to obtain satisfy quality product. As moisture content of the digestate could vary (even if it was initially dewatered), so there is need to take out the water excess by means of drying. In this case best solution (according to score method) is classic direct contact drum dryer (no 4). As source of drying agent (flue gases) biomass burner was selected (no 2) with 4.5 m^3 fuel tank (no 1). What is important, biomass combustion heater has a main advantage, as the ash produced during combustion, can be utilize on site, as feedstock for fertilizer production.

The heat output of the burner was estimated on approx. 1.5 MWth, and should be able to dry up to 2 Mg of digestate form moisture content 70 % to 30 %. Higher than range 20 – 25 % humidity of the raw material is connected with mixing it before granulation with ash (dry material). As the hopper for the raw digestate (no 3) should be able to supply line for 1 hour of raw material, the volume to it should be approx. 6 m^3. The fresh material will be supply to hopper by means of multifunctional loader. If the structure of the digestate requires fragmentation, the installation is equipped with hammer mill (no 6). When the digestate is free of big or fibrous particles, material after drying is transported directly to mixing stage. In other case movement of reversible belt conveyer transports the digestate to milling stage.

3.1.2 Storage of raw materials

The storage of the raw material is divided to two stages. In first are stored digestate, and digestate inoculated with Trichoderm. Both this processes are realized in roofed shelter, in case of Trichoderm there will be also polymer covers, to maintain best conditions for fungi. Both storage places should be equipped with waste effluent collection system, connected to waste water treatment facility. The second part of the raw material storage system, will be additives section: ash, sulphur, urea, phosphorite silos (no 8 and 9). For ash, as it share in the final product will be the highest thru all additives (up to 75 %), the total capacity of the ash silos were estimated on 67 m^3, what should suffice for daily production of the most ash consuming mixture. The rest of the additives will be stored in 12 m^3 tanks. Supply of the materials will be realised by means of pneumatic transport form the tankers.

3.1.3 Raw material mixing

Preparation of accurate mixture is one of the key operations to achieve repeatable quality of the fertilizer. Continuous production requires high capacity of mixers, to supply production process. For this raw materials, best method seems to be double shaft mixer (no 10) with capacity up to 6 m^3/h, mainly due to flexibility in raw material composition. After this operation mixture is transported to stabilization tank (no 11), with capacity 40 m^3. Recommended stabilization period is 3 h.

3.2 Granulation and confection

Production of granules will made by means of flat die pellet press (no 12), with output capacity up to 2 Mg/h. What is important, characteristics of the materials requires special geometry of the die channels. After granulation fertilizer pellets are transported by belt conveyers (no 13) to cooler (no 14), and next there is separation (no 15), crushed pellets and loose parts are recycled and high quality granules are transported to selected storage container by reversible conveyer (no17). To first goes granules with digestate, ash and supplements (no 18), to second digestate with Trichoderm (no 19).

3.3 Mixture preparation and confection

Final mixture is prepared in batch type single shaft mixer (no 20), with precise dosing cells. After mixing (no 24), product is packed by means of fully automated bagging machine (no 21), or packed to the Big-Bags (no 22).

3.4 Process gas cleaning

The production of pellets usually is connected with production of high amount of dust, which should be recycled to the installation. To omit emission of fine particles to air, mostly all devices (when it is possible) works in the slight vacuum, and all air channels are connected to the dedusting system. For rough cleaning of air, cyclones (no 5) were used, with recycle of particles form dust hopper. At the end of installation, there is section of bag filters and fans (no 23), which are capable to separate smallest particles. From this stage, dust is also recycled to the process line.

4 Summary

The presented in the paper concept of the stationary production line for production of multicomponent fertilizer, form ash, digestate, and other components, is a very complex system, involving several aspects presented in the assumption. The flexibility of the installation, gives opportunity to process materials with different materials with unstable properties (moisture, morphology, densities). The design is focused on near to zero waste production, as ash from drying air heater can be used directly to fertilizer production, and all dust separated form air stream is recycled to the granulation process. The innovation during production, is addition a Trichoderm fungi to digestate, what should increase nitrogen assimilability for the plants.

Research was funded by the National Centre for Research and Development and the National Fund for Environment Protection as a part of GEKON program - project no. GEKON1/05/214543/38/2015 "Proecology production of organic and mineral fertilizers based on waste: by-porducts of combusiton and biogasification of biomass".

References

1. J. Dahlin, M. Nelles, C. Herbes, Res., Cons. and Rec., 118, 27-38 (2017)
2. M.B. Rubio, R. Hermosa, R. Vicente, F.A. Gómez-Acosta, R. Morcuende, E. Monte, W. Bettiol, Front. In. Plant Sci., **8**, article 294 (2017)
3. A. Pertiwinin, E.C. Budyanto, M. Hidayat, R.Y. Soeherman, M.F. Habibi, Jou of Biol. Sci., **17**, 21-27 (2016)
4. C. Lanzerstorfer, Jou. of Env. Sci., **30**, 191-197 (2015)
5. J. Lindner, S. Zielonka, H. Oechsner, A. Lemmer, Bio. Tech., **178**, 194-200 (2015)
6. A. Jahan, K.L. Edwards, M. Bahraminasab, *Multi-criteria Decision Analysis for Supporting the Selection of Engineering Materials in Product Design*, (Butterworth-Heinemann, 2016)
7. J.W. Creswell, *Qualitative Inquiry and Research Design: Choosing Among Five Approaches*, (SAGE Publications, Inc, 2012)
8. J.R. Moore, N.R. Baker, IEEE Trans. on Eng. Man., **16**, 90-98 (1969)

Dynamics of flow in a branch of a branching channel

Václav Uruba[1,2,*], *Pavel* Procházka[1], and *Vladislav* Skála[1]

[1]Institute of Thermomechanics, ASCR, v.v.i., Dolejškova 5, Praha 8, CR
[2]UWB, Faculty of Mech. Eng., Depart. of Power System Engineering, Universitní 8, Plzeň, CR

Abstract. Flow in a branched channel is studied experimentally using the PIV technique. The branches are issuing from the main channel perpendicularly, all channels are of rectangular cross-section. The flow dynamics in one particular branch is investigated using the OPD method.

1 Introduction

The branched channels are applied in many practical situations. In general, the main goal of the branching is distribution of the flowing fluid across the various locations. However, the flow-rate distribution across the branches is irregular even if theirs geometry is similar.

There are extensive knowledge resources in available literature, many authors address this problem both from theoretical and practical point of view. In classical engineering book [1] there are a lot of variants of this configuration. Recently there are many numerical studies on the problem with detailed analysis of the flow-field available.

The presented study was motivated by geometry of cooling channels in rotor of a power generator.

2 Experimental setup

The experimental model was designed and fabricated from the Plexiglas to allow optical access to the flow. Experiments were performed using the time-resolved PIV technique.

2.1 Channel geometry

The main channel was of the cross-section 25 x 30 mm^2, 1 450 mm long with the dead-end. The 13 branches (A, B, …, M) of reduced cross-sections 38 x 4 mm^2 are distributed regularly along the main channel perpendicularly to the main channel axis, as shown in schematic picture in Figure 1.

* Corresponding author: uruba@it.cas.cz

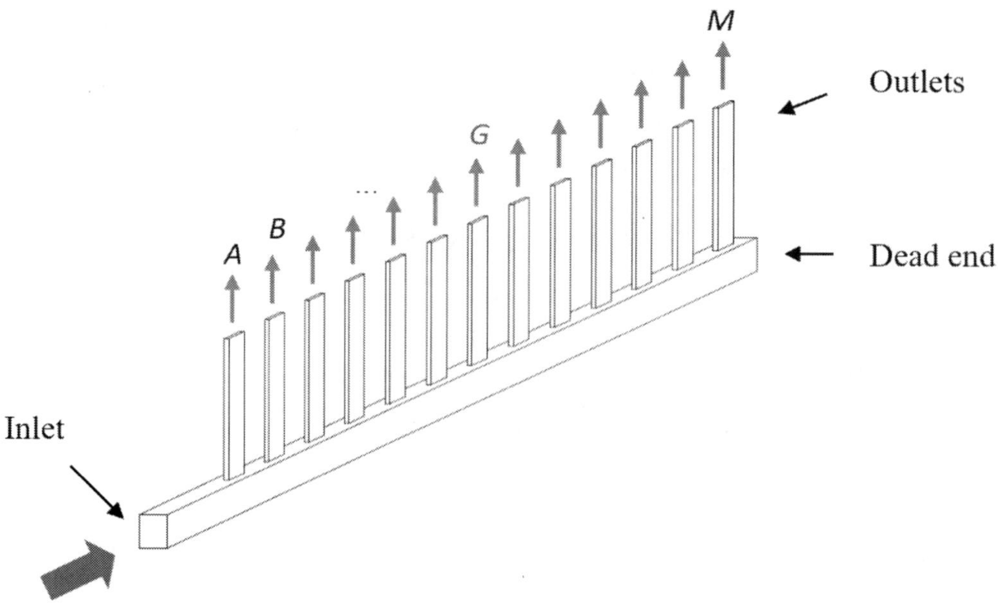

Fig. 1. Schematics of the channel.

In the inlet, fully turbulent and developed channel flow was present. The air flow in the main channel inlet is characterized by Reynolds number of 11 300, velocity of the air in the input is about 6.4 m/s. However, in the presented paper we are interested in the selected branch only, the branch G in the middle of the main channel. Flow rate in the selected branch G is about 3 10^{-4} m^3/s, maximal velocities 5 m/s.

Detailed description of the channel model used for the experiments is given in [3].

Schematics of the branch G under study is in Figure 2, the Cartesian coordinate system is introduced. The measuring plane was located in the mid-span of the channel. The area of the branch was divided into 2 overlapping parts GI and GII for technical reasons.

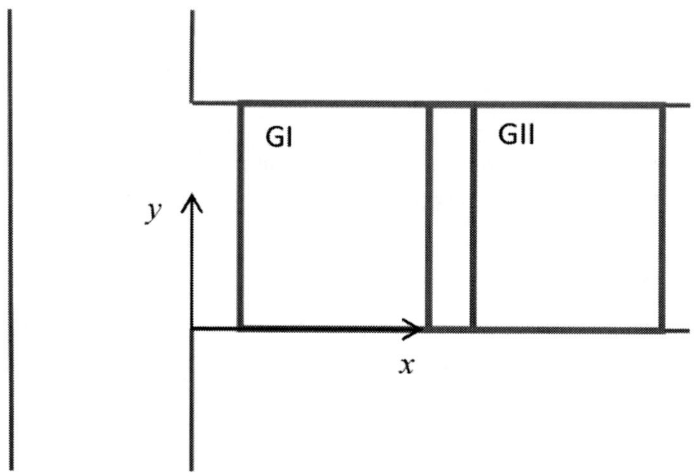

Fig. 2. Schematics of the branch G.

2.2 Instrumentation

The time-resolved PIV method was used for the experiments in the channel model. The measuring system DANTEC consists of the double-pulse laser with cylindrical optics and CMOS camera. The software Dynamics Studio 3.4 was used for velocity-fields evaluation. Laser New Wave Pegasus Nd:YLF, double head, wavelength 527 nm, maximal frequency 10 kHz, a shot energy is 10 mJ for 1 kHz (corresponding power 10 W per head). Camera Phantom V711 with maximal resolution 1 280 x 800 pixels and corresponding maximal frequency 3 000 double-snaps per second. For the presented measurements, the frequency 1 kHz and 4 000 double-snaps in sequence corresponding to 4 s of record for mean evaluation was acquired. As particles the oil droplets generated by SAFEX fog generator was used.

Detailed description of the measuring system could be find in [2, 4].

2.3 Analysis methods

Instantaneous velocity fields in the measuring plane have been evaluated using "Adaptive PIV" method. The averaged velocity field has been evaluated first to obtain the flow statistics and remove the mean flow. Mean flow structure is evaluated as well as velocity components variances and correlation coefficient distributions.

For study of dynamical properties of the flow-field the Oscillation Pattern Decomposition method (OPD) was adopted resulting in series of OPD mode. Each OPD mode is characterized by topology in complex form (consisting of real and imaginary parts), frequency and attenuation of the pseudo-periodic (oscillating) behaviour. Attenuation or amplitude decay is described by e-folding time representing mean time period of the mode amplitude decay by "e". The other decay characteristic is "periodicity" which express the e-folding time in multiples of periods.

The details on OPD method could be find in [5].

3 Results

The results are divided into two parts. The first part is oriented on time-mean results, while the second part characterizes the flow dynamics using the OPD method.

3.1 Time-mean

The averaged characteristics will be presented in each figure both for the part GI on the left- and the part GII on the right-hand side. The distributions of mean characteristics follow from one part to the other satisfactory.

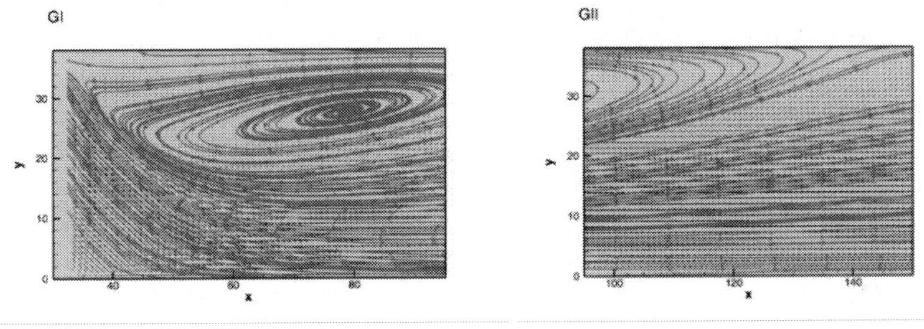

Fig. 3. Mean velocity vector fields.

Mean velocity vectors distributions for GI and GII areas are shown in Figure 3. For clarity the vector lines are added arbitrarily. The back-flow region associated with the vortex, centre located in position [79;26], is visible clearly.

To indicate regions with high dynamical behaviour the sum of variances x and y velocity components is to be presented next in Figure 4.

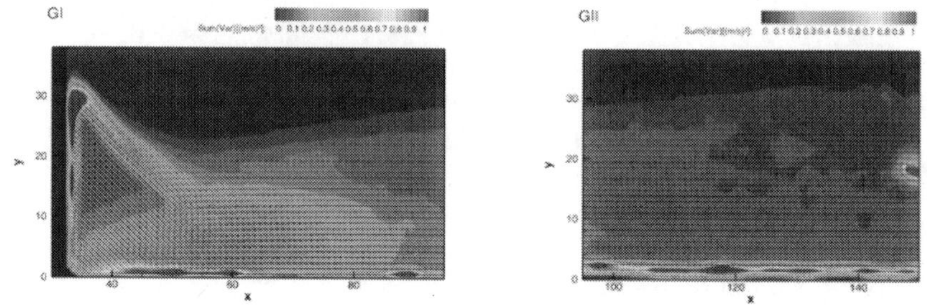

Fig. 4. Velocity components variances sum distributions.

In Figure 4 dark-blue colour represents low dynamical activity regions, red colour corresponds to excessive dynamics due to frequent errors in velocity evaluation using the standard PIV evaluation procedure in near-wall regions. Finally light-blue and green parts are those with high dynamical activity in the flow itself.

To estimate turbulence generation the correlation coefficient is evaluated in Figure 5.

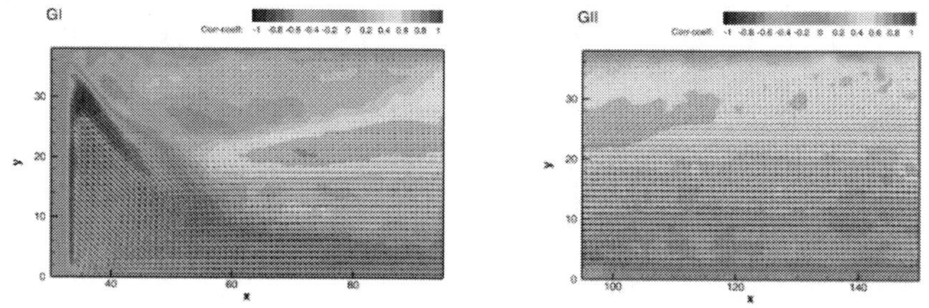

Fig. 5. Velocity components correlation coefficient distributions.

The correlation coefficient value close to 0 indicates low or even no turbulence generation, while negative (or positive) values approaching 1 indicate turbulence production regions.

3.2 Dynamics

The OPD method has been applied on the GI and GII zones respectively. The analysis of both parts is completely independent to each other and OPD modes from one part could not be linked directly to those of the other part.

In Figure 6 there are OPD modes frequency and periodicity combinations evaluated by the standard procedure. The results for the GI part correspond to blue points, while red points correspond to the part GII. Numbers of the modes ordered according to the e-folding time (descending order) are indicated within the points.

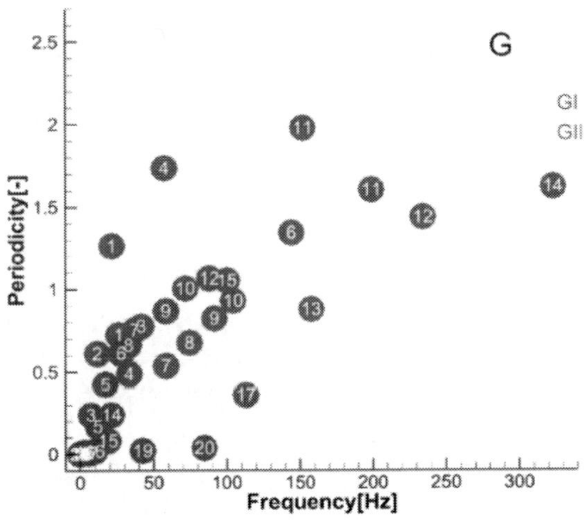

Fig. 6. OPD modes, frequency and periodicity.

A few modes have been selected with the highest value of periodicity, in particular modes 6, 11, 12 and 14 for the GI and 1, 4 and 11 for the GII. Those modes could be considered as oscillating or pseudo-periodic, the others decay too quickly. All parameters, of the selected modes, frequency, e-folding time and periodicity, are given in Table 1.

Table 1. Selected OPD modes.

	OPD mode	frequency [Hz]	e-folding [ms]	periodicity [1]
GI	6	144.2	9.32	1.34
	11	199.1	8.05	1.60
	12	234.5	6.13	1.44
	14	323.2	5.02	1.62
GII	1	21.5	58.71	1.27
	4	57.2	30.31	1.73
	11	152.0	13.00	1.98

Topology of the selected modes is to be shown next. Each mode topology consists of real and imaginary parts respectively. The real part corresponds to the phase angle 0, phase angle $\pi/2$ corresponds to the imaginary part, than phase π is characterized by negative real part and $3\pi/2$ by negative imaginary part. The process is pseudo-periodical with decaying amplitude.

Topology is shown as vector fields. For better clearness the vector lines are added arbitrarily, for the real part in red and for the imaginary part in blue colours. The GI OPD modes are to be presented first. In Figure 7 the GI OPD mode number 6 is shown, real and imaginary parts.

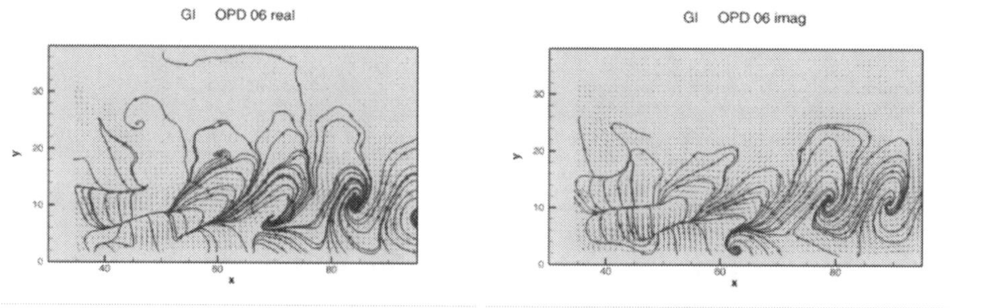

Fig. 7. Topology of the OPD mode 6, GI, real and imaginary parts.

The OPD mode 6 for the GI region is characterized by vortex train of 5vortice located in the lower part moving in the x direction, alternate orientation. Topology of the OPD modes 11, 12 and 14 is rather similar, so they are presented in one Figure 8.

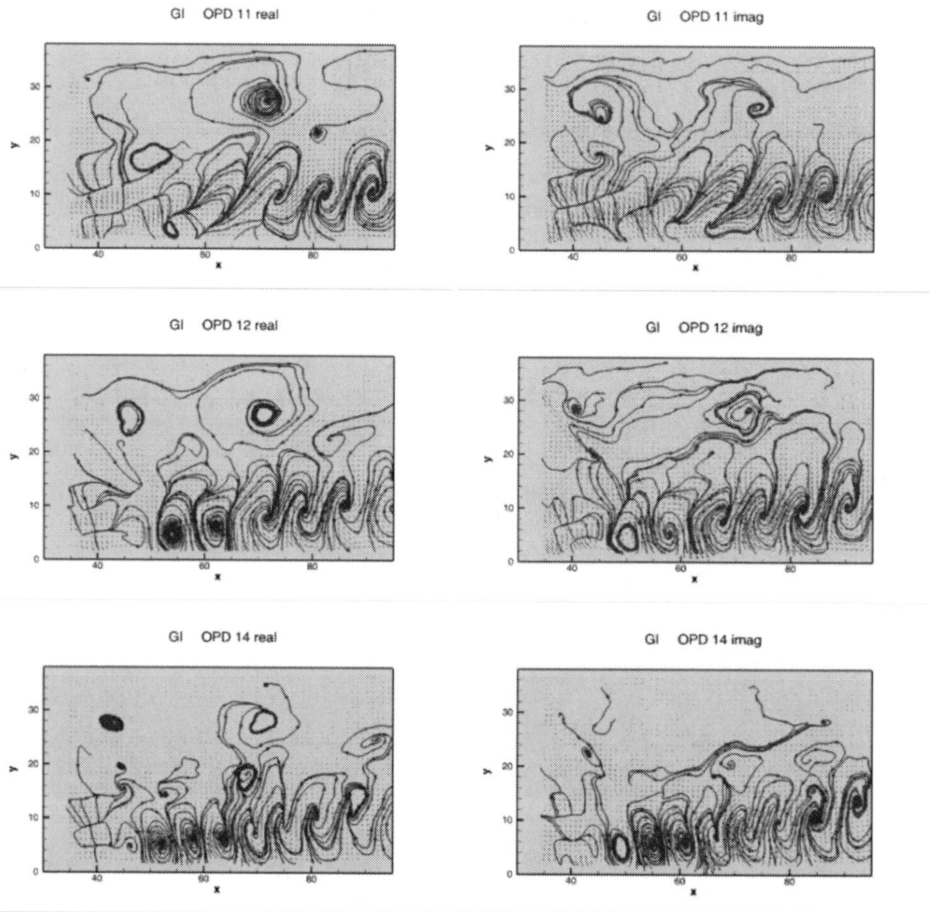

Fig. 8. Topology of the OPD modes 11, 12 and 14, GI.

The OPD modes 11, 12 and 14 are characterized by vortical structures in one row approximately in the x direction. They are of alternate orientation and they differ by number of vortical structures from 5 to 8. The frequencies are also different (see Table 1). Generally: higher number of structures, higher frequency of the mode indicating approximately constant structures propagation velocity, this velocity is close to the mean-flow velocity in the given area. All vortical structures are located within the shear layer between the back-flow (upper part) and forward flow (bottom) regions – compare with the mean velocity vector field in Figure 3.

Next, the OPD modes topology in the GII region is to be presented. Unlike in the GI region, the selected modes with high periodicity are of the three completely different types. The OPD modes 1, 4 and 11 are presented in Figure 9.

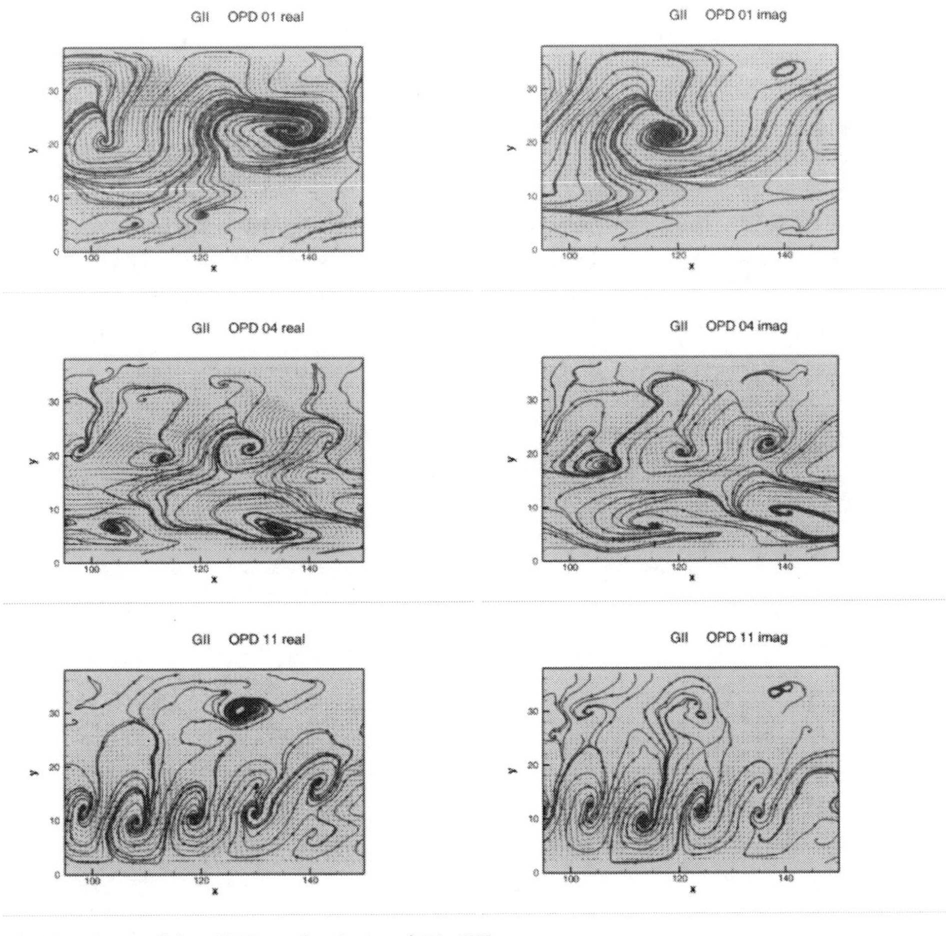

Fig. 9. Topology of the OPD modes 1, 4 and 11, GII.

The OPD mode 1 is characterized by very low frequency 21.5 Hz and the advected vortices are of the size of the whole domain.

The OPD mode 4 consists of the two rows of vortices in chequered configuration. The vortices are relatively small.

The OPD mode 11 is similar to the modes in the GI region presented above, it is formed by vortical structures in a single row in the streamwise direction located in the channel bottom.

Vortical structures in all presented modes propagates in the streamwise direction, as visible from the phase shift between the imaginary and real parts.

4 Conclusions

In the paper the flow structure of a selected branch of the branched channel is shown as a result of PIV experiments.

Both statistical methods based on averaging and special methods for vector field frequency analysis are applied. Typical pseudo-periodic dynamical structures are detected. The typical dynamical structures are trains of vortices of alternative orientation aligned in the x direction and propagating in the streamwise direction.

This work was supported by the Grant Agency of the Czech Republic, project No. 17-01088S.

References

1. I.E. Idelchik, Handbook of Hydraulic Resistance, *Jaico Publishing House* (2005)
2. P. Procházka, V. Uruba, Bistable flow occurrence in the 2D model of a steam turbine valve. *AIP Conference Proceedings*, Vol. 1889, Article number 020029 (2017)
3. V. Uruba, P. Procházka, V. Skála, Flow in Branched Channel, *Topical Problems of Fluid Mechanics*, pp. 299-306 (2018)
4. V. Uruba, P. Procházka, V. Skála, Coanda effect in valves. *EPJ Web of Conferences*, Vol. 143, Article number 02136 (2017)
5. V. Uruba, Near Wake Dynamics around a Vibrating Airfoil by Means of PIV and Oscillation Pattern Decomposition at Reynolds Number of 65 000, *Journal of Fluids and Structures*, 55, pp. 372-383 (2015)

Stereoscopic TR PIV measurement and POD analysis of flow behind the turbine prototype of model size

Darina Jasikova[1,*], *Michal* Kotek, and *Vaclav* Kopecky

[1]Technical University of Liberec, Studentska 2, 46117 Liberec, Czech Republic

Abstract. When measuring turbine prototype flow, a number of modes with different flow rates and cavitating vortex rope settings have been measured. The turbulent flow in the draft tube of diameter 240 mm reaches the Reynolds number of 10^5. The vortex and highly swirling flow is developed in the draft tube. The usage of kHz-order frame rates was necessary to resolve the velocity field. The investigated area in the mixing vessel was examined by the stereoscopic time-resolved PIV technique. Further mode evaluation allows comparison of cross-sectional velocities in the vector map. The Proper orthogonal decomposition analysis was applied on 3D vector maps, those including u, v, w velocity compounds. In the selected regimes – operating points with volume flow rate 97 l/sec and 114 l/sec, the greatest risk of transition to cavitation mode was identified, so they were studied in detail.

1 Introduction

The development of energy storage re-sources is a response to weather fluctuations and growth of green power plants (solar and wind power). There raises an urgent need for energy storage and its use depending on the immediate increase. It is also very closely connected with ensuring the safety and stability of the distribution system. The most effective functions are assured by pumped-storage hydroelectric power plants (PVE). The unique feature is their rapid response to the needs of the transmission network and the ability to ensure their re-commissioning in case of blackout. In Europe, most of the PVEs were built more than 20 years ago, i.e. before volatile renewable resource boom, where the requirements for dynamic PVE functions and bandwidth were less demanding than today. The number of PVEs installed in Asia and South America has been so far inadequate. The usage of conventional reverse turbines in off-optimal modes brings an unstable closed flow causing the vibration pulsation of the turbocharger. It forces the limitation of the working scope.

The progress in turbine optimization is in a design of new shape of the hydraulic parts of the reverse turbine, assuming the rotation of the fluid in the draft and suction tube. The design of impeller, draft and suction tube of reverse (pump) turbine with an extended operating band is based on computational modelling (CFD, mathematical optimization).

* Corresponding author: darina.jasikova@tul.cz

Validation of models can be performed by advanced experimental techniques (time-resolved PIV, tomography, high speed visualization).

The intention in progression is to suppress pulsations, which are the result of the cavity of the swirling head. The novelty of the reverse turbine design is in optimization of the plant as a whole unit (impeller plus suction unit) using computational modelling with help of advanced experimental techniques. The solution is based on a thorough theoretical analysis on stability and flow dynamics. The verification of the prototype runs in testing room on a reverse turbine prototype. This prototype meets requirements and demands of dynamic control function, thereby increasing the safety and resilience of the distribution network.

The validation of the numerical calculations should be done by the method that allows nonintrusive contactless measurement of fluid flow meeting the temporal resolution of an unsteady flow. The turbulent flow in the draft tube of diameter 240 mm reaches the Reynolds number of 10^5. The vortex and highly swirling flow is developed in the draft tube. The usage of kiloHertz-order frame rates was necessary to resolve the velocity field. Another condition influencing the selection of the appropriate measuring methodology was the 3D character of flow. The flow behind the turbine is anisotropic. The velocity component in the mainstream direction greater then swirl flow. The time-resolved PIV (TR PIV) can provide access to spatial - temporally varying flow fields at sampling rates of kHz. The mainstream time-mean velocity was calculated from volume flowrate as $W_0 = 2$ m/s and the velocities in swirl flow ware expected as U_0 and V_0 0.3 m/s. Measurement in two planes was impossible for spatial reasons and optical approach to the prototype. Therefore, a one-off measurement was provided to get information on all three velocity flow components. Confirming the expectations was to use stereo PIV system setup know from conventional PIV system. Here, we set the stereo TR PIV (3D TR PIV) system on Nd:YLF laser Litron running on maximal flashing frame 1kHz in double frame mode and two high speed cameras set in 3D geometrical arrangement using Scheimpflug mounts. The results of fluid flow presented in this paper are complemented by proper orthogonal decomposition calculated for three velocity components [1-7].

2 Experimental Method

The selection of the measuring methodology and experiment conditions on the turbine prototype were determined by the high Re number ($Re > 10^5$) and the anisotropic character of the fluid flow in the draft tube. The diameter of draft tube at the measuring plane was 240 mm. Average expected water flow through the cross section was 100 l/sec.

The time – resolved PIV extended with two high speed cameras to obtain stereo records of fluid flow was suitable solution for monitoring of three velocity component using two optical accesses. The experimental measurement was run in the test hall on the turbine prototype. Together with the design of the geometric arrangement of the cameras and the laser, it was necessary to design a suitable optical approach and to supplement the tube with the optic extensions. According to requirements of optical measuring methods, the prototype of suction and draft tube was manufactured with Plexiglas to ensure optical access also for purpose of visualizations. It was mainly the reduction of the distortion caused by the transition of light through various environments (water, Plexiglas, air) and curved walls of draft tube.

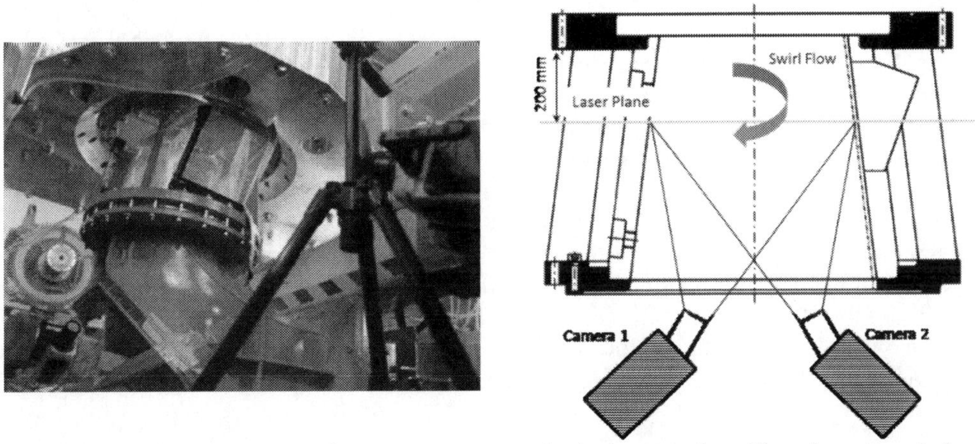

Fig. 1. Photo of TR PIV system layout - two cameras in the bottom view. There is seen a window for equalizing different refractive indices of the environment and curvature of draft tube. Scheme of draft tube geometry behind the impeller and location of the laser plane and stereoscopic TR PIV camera setup.

2.1 Measurement Technique Setup

The investigated area in the mixing vessel was examined by the time-resolved PIV technique. The stereoscopic TR PIV system was built on two high speed Speedsense cameras, complete with a sync module and a Litron Ld:Y300 pulse laser operation on the frequency 1 kHz. This kind of double cavity laser emits pulses of energy reaching 15 mJ in each pulse on wavelength 527 nm. The laser beam was extended into the vertical plane with cylindrical optics to reach the parameter of the planar laser sheet of thickness 1mm and spread into the 300 mm width. The investigated area's diameter was 240 mm.

The two high speed cameras SpeedSense working on frequency 1kHz with resolution of (1 280 x 800) px in double frame mode were equipped with low-passing filters to eliminated the backward flashes from the laser sheet that arises on the blades surfaces. The wavelength of the optical filter corresponds with the emitted light of the particles. The set of lens system Nikkon f/2.8 60 mm was also used. The cameras were fitted into optical amounts to reach Scheimpflug condition. Conventional particle image velocimetry uses a single camera oriented orthogonally to the illuminated plane. Stereoscopic simultaneous views from two different off-axis directions provide sufficient information to extract the out-of-plane and in-plane velocity components. We used the angular-displacement method in our experimental camera setup. That means that two cameras axes are not parallel, but rotated inward such that they intersect at the midpoint of the domain to be recorded. The object plane is not parallel to the lens plane Therefore it becomes difficult to obtain particle images that are well focused across the image plane. Most simple solution is to increase the depth of field of the recording optics. The depth of field $dof = 4(1 + M^{-1})^2 f^2 \lambda$, where M is the camera magnification, f is the f-number, and λ is the wavelength of the illuminating laser. A large depth of field can be obtained of increasing the f-number, implying that a smaller fraction of the light scattered by the particle will reach the camera. The second method to obtain images in focus over the entire image plane is to enforce the Scheimpflug condition. That requires the object plane, the lens plane, and the image plane are collinear. We used both methods with regard to insufficient saturation in our experiment.

The exposition time for both cameras operated on 330 usec. The increase of exposure time increases the particle density and illumination.

The laser and camera system was synchronized via timer box and controlled from the DantecStudio software. The expectation of the maximum frequency of the vortex structures was in tens of Herz. According to Nyquist criterion of sampling frequency it was set 500 Hz. The dataset of 5 000 images was captured. The raw images were also processed in this software.

The analyses of captured images run in three main categories. There were run pre – processing analysis, i.e image masking, image algorithm for blur and light reflection reduction. There was run the dewarping algorithm with the calibration function. The pre-processed images were analysed using standard cross-correlation method with interrogation area size (32 x 32) px and overlap 50 %. The raw vector maps were validated with peak and range validation methods to obtain correct dataset. For the purpose of overview of the complex flow behaviour the statistical evaluation and the turbulence index was calculated. The dynamically changing velocity field was analysed by the Proper Orthogonal Decomposition (POD) for identifying the energy fractions and partial vortex structures in the fluid flow.

2.2 Calibration

The stereoscopic TR PIV requires spatial calibration with a specially designed calibration target. The use of a calibration target with an accurate dot matrix also enables subsequent post processing of captured images and dewarping of local distortions.

Standard calibration of the stereo PIV system is performed by a traversing calibration target. The movement of the target in a closed floating waterway could not be realized, therefore a multi-level target method was chosen.

The calibration target was designed as a spatial matrix with regular point alternation in two planes. When designing a target, it was important to keep the square structure of both planes, which later defined the axial orientation. The designed round calibration target was manufactured using 3D printing method. This method enables precise two colour printing with high resolution. The appropriately designed target must have the following parameters: number of pixel points in the image 10 x 10, the ratio of the size of the main points to the image width of 1:50, the ratio of the main points to the central 1: 1.5, and the ratio of the main point size and 1: 1 layer spacing. The target was designed as white dots on a matt black background. The main dot size was 7 mm and the regular dot was 5 mm in diameter. The distance between dots was 20 mm and the plane distance was 4mm.

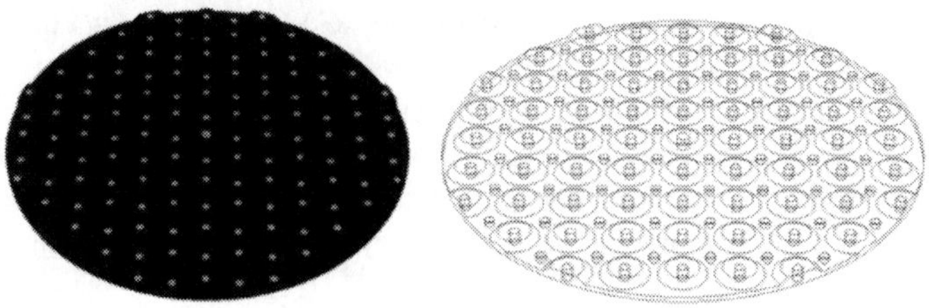

Fig. 2. Calibration target with a diameter of 240 mm for measurement in the experimental track.

2.3 Seeding

Solution for seeding was first supposed to use a natural flow of saturation. This was especially content of impurities and bubbles that have proven themselves in the past. The advantage of natural saturation was more considerate to the device. There was a legitimate concern about blade damage and the deposition of particles in the system. There was a problem with the seeding of the fluid flow due to the spatial layout of two cameras to the laser plane from experimental point of view. Each camera records the evaluated area at a different angle to the scattering angle on the particle. The problem was the diffusion of light at different angles. The bubbles show a backscatter of laser light. Solid particles, on the other hand, exhibit forward scattering. However, the natural impurity content was not sufficiently proportional for both cameras to record particles of sufficient intensity, notwithstanding the fact that two non-comparable seeding units (density, viscosity, reflection index) were captured. The stereoscopic TR PIV method is particularly sensitive to the quality of seeding, especially when it comes to high-speed recording with short exposure times and limited light power. These arguments outweighed concerns about the damage to the system, which was not proven later. The volume of 50 m^3 (whole volume in the turbine, pipe system and reservoir) of water was to be seeded with 50 μm PA particles.

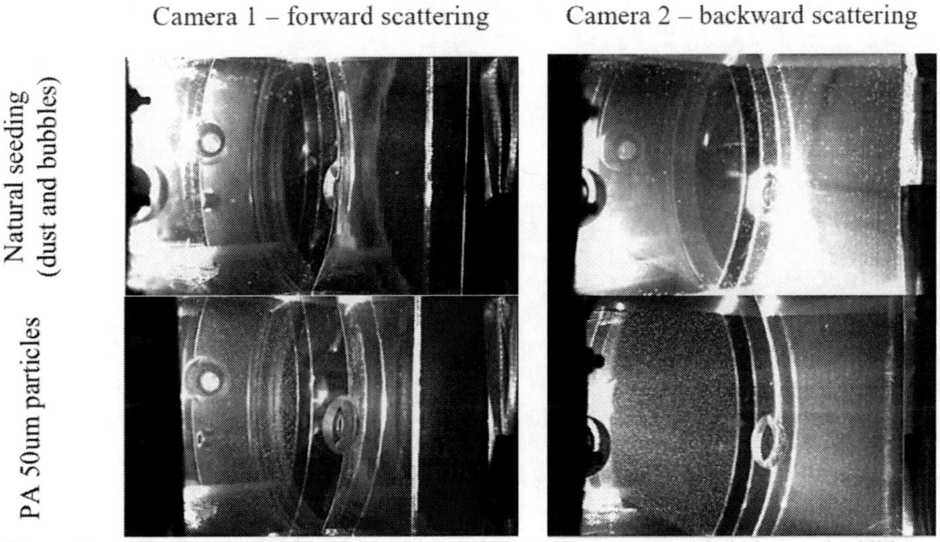

Fig. 3. Examples of seeding using presence of natural dust, impurities and bubbles, and addition of PA 50um particles.

3 Results and Discussion

When measuring turbine prototype flow, a number of modes with different flow rates and cavitating cavity settings have been measured. This article lists results from two modes that are most prominent. There is a risk of an early transition to the cavitating character of the fluid flow (due to under pressure) for both operating modes. The impeller optimization design was based on the suppression of the cavitating whirl. In the selected regimes – operating points marked as 293 and 298, the greatest risk of transition was identified, so they were studied in detail.

Mode 293 was running at fluid flow 97 l/sec with a slight overpressure of 20 kPa. In this mode, the cavitating rope did not go into the gaseous phase, so it was possible to perform PIV measurements. By lowering the pressure of 20 kPa under atmospheric pressure, the core of the cavitating whirl changed the phase to the gas occupies a large part of the evaluated area. Condition for PIV measurements are harder to reach due the reflections and lack of seeding inside the cavitating rope. Mode 298 is characterized by a flow of 114 l/s at atmospheric pressure. This mode was stable under current operating conditions.

The stereoscopic model used to reconstruct three component vector maps was loaded with a number of limits. First and foremost, due to the complexity of the scene and the views to the measured area, it was not possible to cover the entire cross-section of both cameras. In places where the view of a single camera is hidden, it is not possible to reconstruct the data properly. This leads to misinterpreted results, especially in some extreme areas. However, most of the cross-section is interpreted correctly and it is certainly possible to have the overall flow.

The basic evaluation of fluid flow behind turbine is time-averaged statistics (Figure 4). We captured in the following vector maps for following measurement points 293, and 298. The basic physical quantities of the operation were recorded at all of these measurement points.

a

b

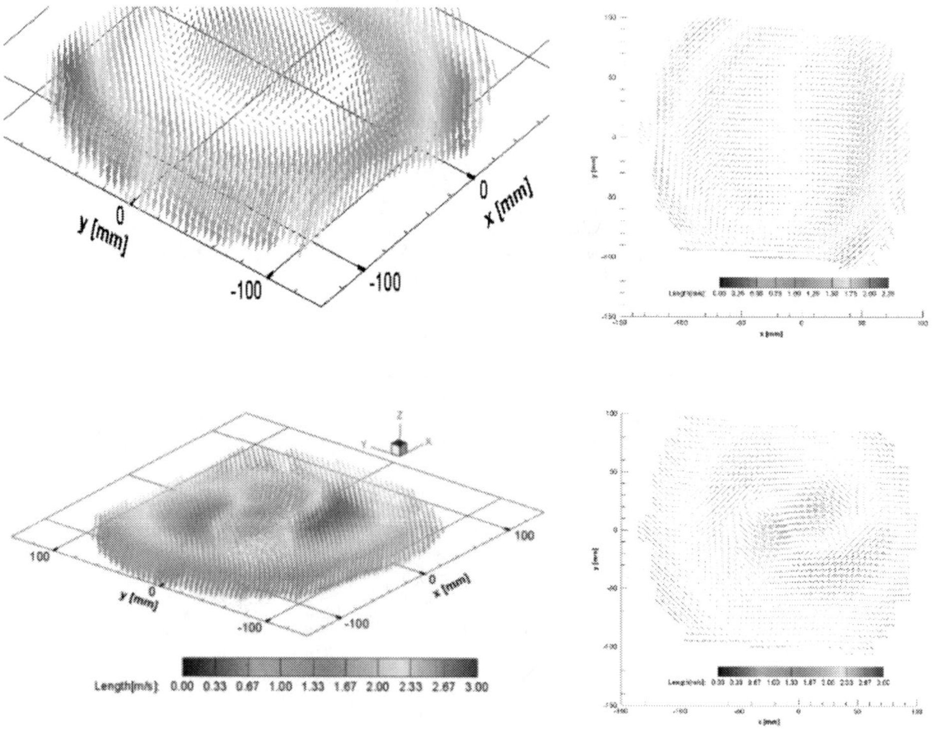

Fig. 4. Time-averaged statistics of fluid flow behind the turbine prototype processed as 3D and 2D vector maps for regime a) 293, and b) 298.

From the basic view of the vector maps, there is a difference in the mode at the measurement point 298. While in the other modes exhibits the dominant stream of water is in the peripheral regions, at the point 298 mode, the main stream with the vortex structure moves closer to the centre of the cross-section. The maximum velocity are higher in this mode, $w_{max} = 3m/s$. An interesting comparison of paragraphs 298 and 293 shows another difference in the characteristics of the modes.

There is a difference between rotations of the main current behind the impeller from this interpretation of the results (Figure 4). While in the modes at points 293 rotates the main current according to the direction of the impeller, in this case in the clockwise direction, the mode at point 298 rotates in the opposite direction of the impeller. This behaviour of the vortex structures was subsequently confirmed by high-speed visualizations in the presence of the gaseous component.

Further mode evaluation allows comparison of cross-sectional velocities in the vector map.

The Proper orthogonal decomposition analysis was applied on 3D vector maps, those including *u, v, w* velocity compounds. It is a mathematical method that gains and clarifies the flow structures based on extracting a basis for modal decomposition from an ensemble of signals. [8] We can reconstruct flow field and demonstrate the influence of different flow structures by the selection of main dominant POD modes. The major POD modes that effects the dominant flow is seen in Figure 5. The POD Snapshot is based on autocorrelation function and depends on the number of instantaneous acquired samples acquired. We can see in figure 5 two dominant Modes that are significant for flow regime 293. For comparison to this we chose the first dominant mode even for the regime 298.

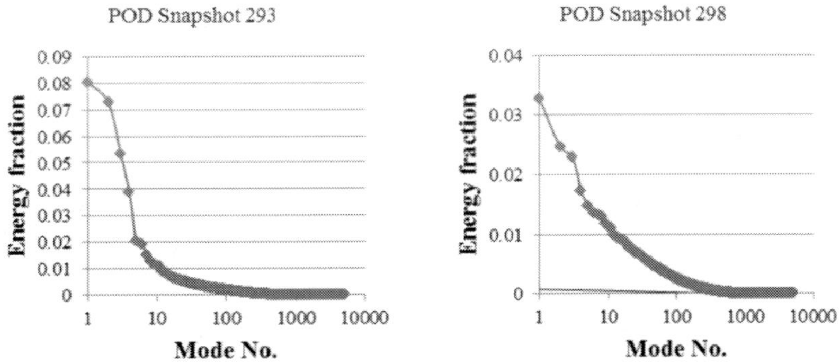

Fig. 5. POD Snapshots for regime 293, and 298 according to energy fractions.

The general aim of POD is to find the optimal representation of measured field realizations. The eigen functions are sometimes referred to coherent structures. The structures are highly correlated in an average sense with the flow field. The POD analysis was performed on 5 000 sequential 3D vector maps. The POD projection of Mode 1 and 2 for both regimes is seen in Figure 6 and Figure 7. These two dominant modes affect the resulting flow structure. The presentation of all three components of velocity, i.e. *u, v* in plane and *w* as vertical velocity component, is also evident in anisotropy and orientation of the main and swirl flow.

The interpretation of 3D vector velocity fields is difficult, especially if they are strongly anisotropic. In this experiment, the component of the vertical velocity *w* was predominantly in the mainstream direction, so it follows the main swirl flow. We focused on viewing this folder, when interpreting data.

In regime 293, the centrifugal flow orientation is predominant, resulting in a structure with a maximum velocity in the laser plane (*u, v* velocity components) – horizontal, near the walls. The rotational frequency of these structures is 7 Hz. The swirl structure and centred local velocity decrease is dominant in both POD modes. The return fluid flows in the centre time-averaged velocity statistics is created by averaging these vortex structures, which rotate with the main stream. The increase of velocity near the walls is well visible on the projection of 3D images in figure 6.

In regime 298 in POD mode 1, and in 2 prevail the vortex structures in the centre of swirl structure. Complex, irregularly structured vortex are projected into time-averaged vector statistics as two opposite vortex structures moving in counter clockwise direction of rotation. In averaged statistics, these partial vortex structures cause a uniform velocity profile as it is seen in figure 8. The occurrence of both POD modes is time-based as it is seen in figure 7. This local closure predicts the incidence of cavitation rope if pressure drop.

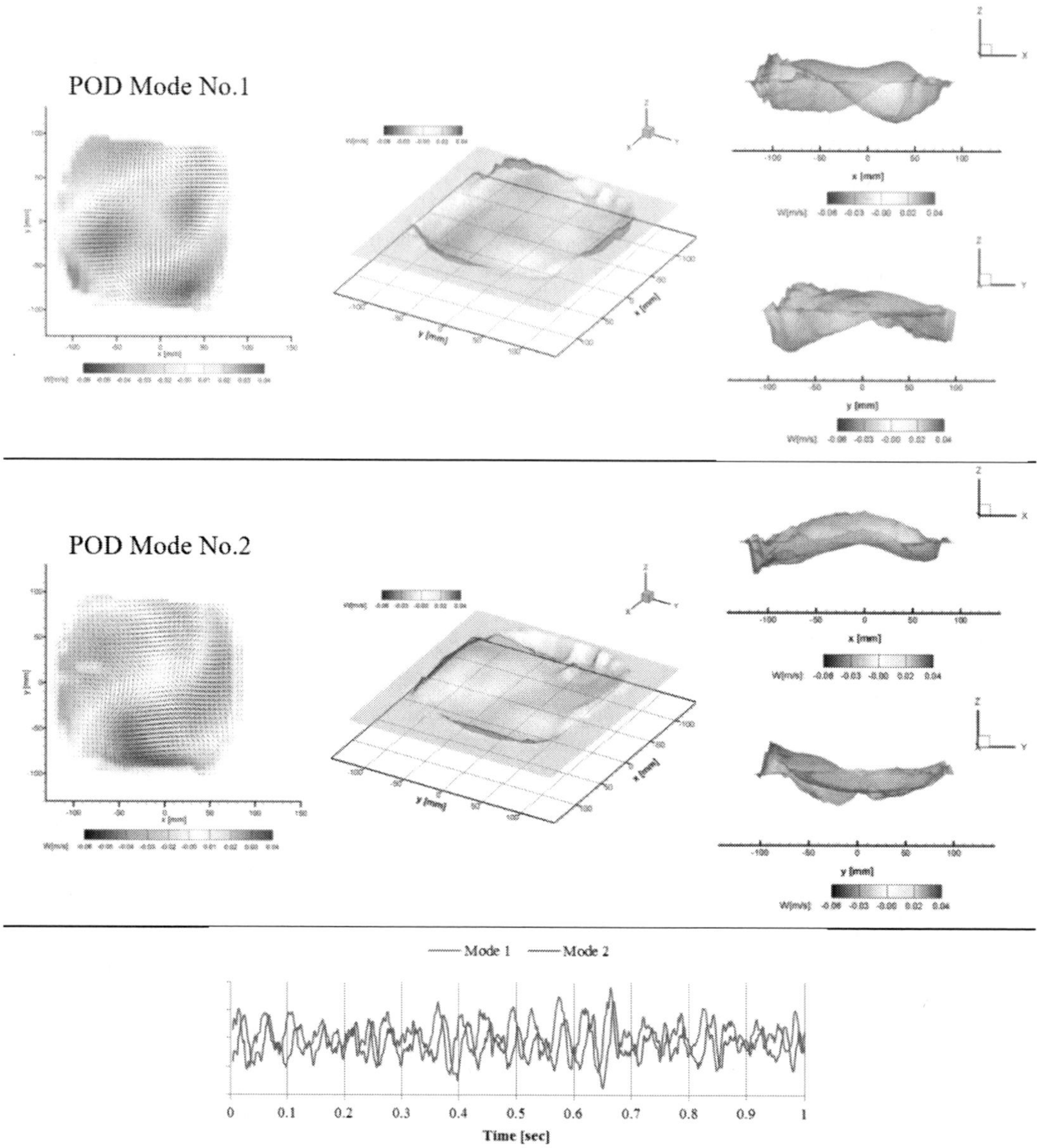

Fig. 6. POD 1st, and 2nd mode of flow structures in regime 293 and time history of POD coefficient for both modes.

There is seen significant vortex breakdown. This breakdown is characterised by decrease of axial velocity along the axis, leading to internal stagnation point or even backflow [7].

Two modes differing in flow, achieve significant different flow characteristics, which are also reflected in the speed profile evaluation that was taken across the transverse region passing through the centre of the diffuser. According to the velocity profiles for vertical velocity component *w* in Figure 8 can be watched the dominant trend of the flow reaches the maximum near the wall of the diffuser. Towards the centre of the diffusers, the velocity of the velocity decreases and there is a backward flow and significant closure in the horizontal plane.

Against this mode 298 towards the centre of the diffuser, the speed increases, including the rotary component, see vector map above. The small drop is due to the rage behind the impeller's charge.

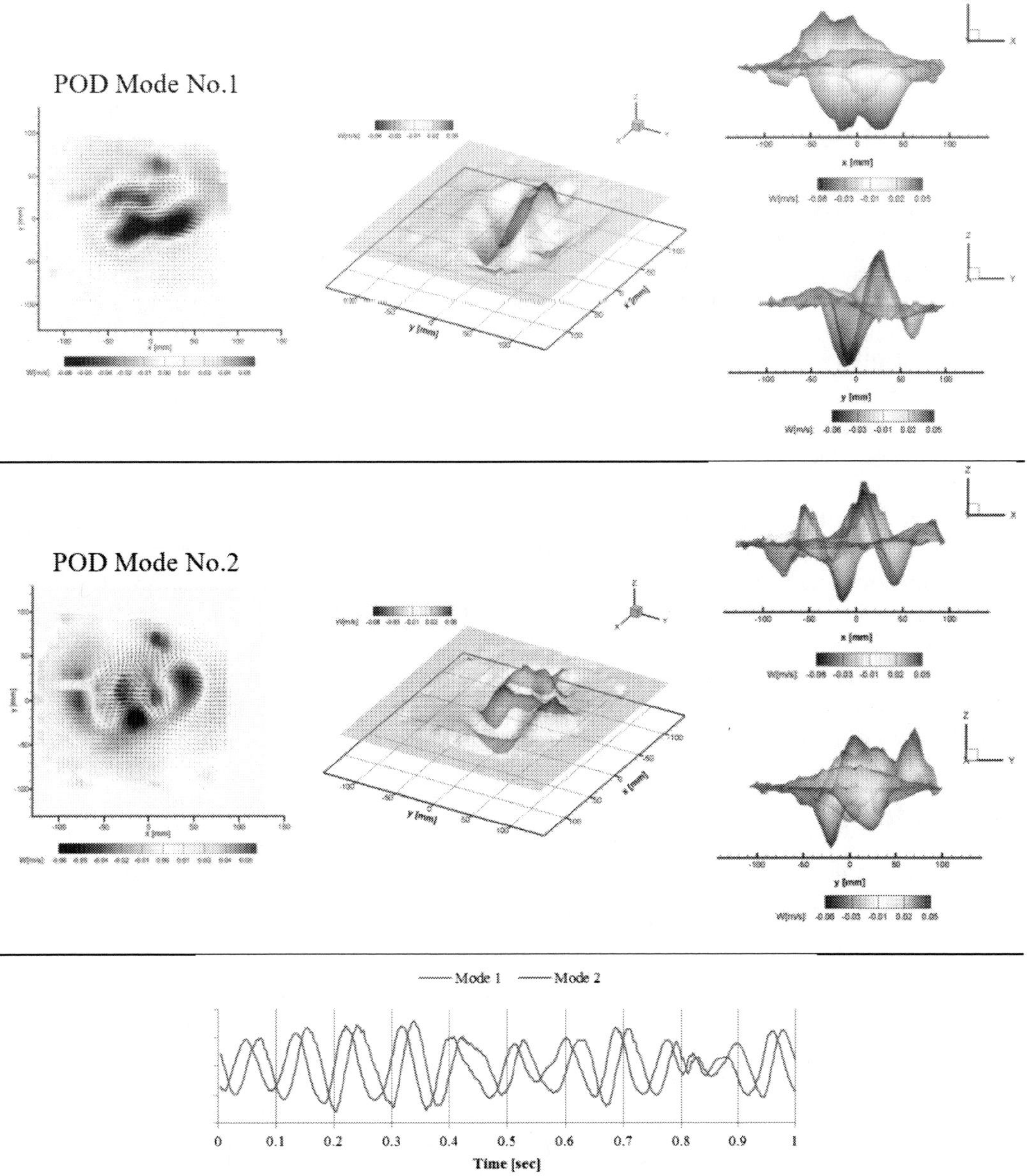

Fig. 7. POD 1st, and 2nd mode of flow structures in regime 298 and time history of POD coefficient for both modes.

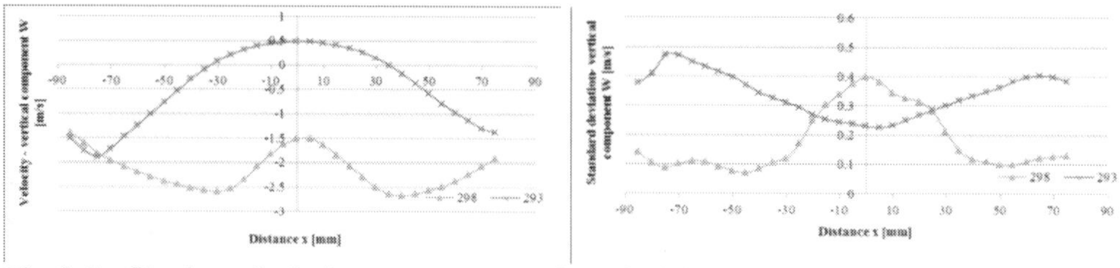

Fig. 8. Profile plots of velocity component *w* and standard deviation *Std w* taken across the measured area under the impeller.

4 Conclusion

The measurement in draft tube below the impeller was a technically demanding. In addition, it has to be solved with compensating optical elements. The distortion near the curved walls was too large to be able to measure the speed field in these areas. The evaluation of field speeds near the wall provides the most interesting results, especially in the area behind the rotating impeller, when the main current is closed.

Using optimized PIV stereo methods and high speed recordings, it was possible to monitor and evaluate this complex task as well. By application of the POD analysis, dominant vortex structures were detected, which were evaluated in the form of 3D graphs. The distinction between the two modes is reflected not only in the evaluation of the velocity profiles across the monitored area, but also in the resulting frequency analysis, where the dominant frequencies of the vortex structures can be detected. This measurement responds to the CFD simulation already conducted to optimize the turbine.

The results of this project LO1201 were obtained through the financial support of the Ministry of Education, Youth and Sports in the framework of the targeted support of the "National Programme for Sustainability I".

References

1. A. Upatnieks, K. Laberteaux, S. L. Ceccion, Experiments in Fluids **32** (2002)
2. S. Wang, Y. Chen, Y Z. Li, J Vis **21** (2018)
3. D. F. Kurtulus, F. Scarano, L. David, Exp Fluids **42** (2007)
4. L.L. Shi, Y.Z. Liu, J.J. Wan, Journal of Mechanical Science and Technology **24** (2010)
5. J. Basley, L.R. Patur, F. Lusseyran, Exp Fluids **50** (2011)
6. A. Villegas, F.J. Diez, Exp Fluids **55** (2014)
7. P. Rudolf, D. Štefan, IOP Conf. Series: Earth and Environmental Science **22** (2014)
8. L. Sirovich, Quarterly of applied mathematics **3** (1987)

Evolution of vortical structures behind an inclined flat plate

Pavel Procházka[1,*], *Václav* Uruba[1,2], and *Vladislav* Skála[1]

[1]Institute of Thermomechanics, ASCR, v.v.i., Dolejškova 5, Praha 8, CR
[2]UWB, Faculty of Mechanical Engineering, Department of Power System Engineering, Univerzitní 8, Plzeň, CR

Abstract. 2D3C TR-PIV technique was utilized to investigate streamwise-oriented vortical structures behind an inclined flat plate. The angle of attack was set to 7 deg, several fields of view in the wake were investigated. The instantaneous velocity vector fields were captured, dynamics of the flow was studied using POD method. The streamwise structures are determined by vorticity and low- and high-velocity streaks are defined. The acquired results are in a good agreement with the new hypothesis of a principle of flight.

1 Introduction

The invention of a wing is more than a hundred years old. Nevertheless, there are some doubts recently, how the lift force is actually generated. This experimental research is a reaction to the new ideas about principle of lift and flight by Hoffman and Johnson [1, 2]. This new hypothesis determine the streamwise vortical structures over and behind the airfoil as the main source of the lift force generation. This streamwise vorticity is developed as a consequence of the instabilities of the boundary layer developing under adverse pressure gradient of the suction side of the airfoil. Some experimental studies of flow instabilities and their consequences are given in [3, 4, 5].

During previous investigations the vortical structures topology within boundary layer was examined very close to the inclined plate surface [6]. Another research was focused on the wake behind the flat plate [7]. Presented study is dealing with the streamwise-oriented vortical structures and their identification using POD (Proper Orthogonal Decomposition) method from velocity vector maps located perpendicularly to the flow direction. Several parallel planes were investigated along longitudinal direction and the evolution of streamwise vorticity was observed.

2 Experiment description

A studied test airfoil is simple flat plate which is inclined at the angle of attack (AOA) 7° with respect to the incoming flow. The flow is coming from blow-down wind tunnel, cross-

* Corresponding author: prochap@it.cas.cz

section dimension 250 x 250 mm^2, uniform flow speed 5 m/s, low turbulence intensity level (around 0,2 %). The plate has semi-cylindrical leading and trailing edges, chord length c is 100 mm, thickness of 2 mm and span dimension is 300 mm. The corresponding Reynolds number is approximately 33 000. The schematic view of the experiment arrangement can be seen in Figure 1. The coordinate system was introduced so that x- and y-axis are in the plane of measurement and z-axis is in the direction of flow. Several fields of view in streamwise direction were measured at dimensionless positions: z/c = 1, 1.3, 1.5, 1.8, 2, 2.3 and 2.5. The position z/c = 1 corresponds to trailing edge location.

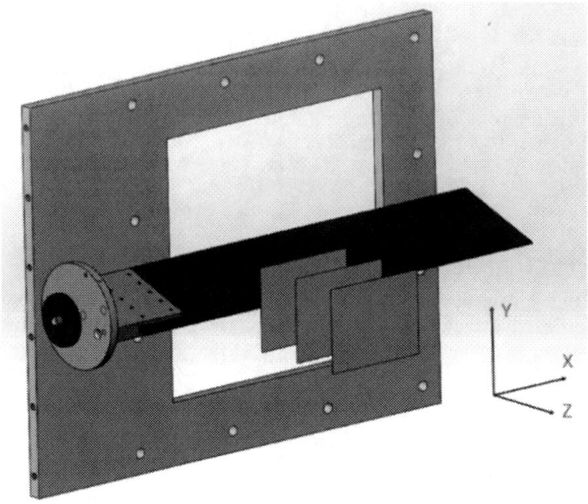

Fig. 1. The setup of inclined plate and wind tunnel orifice, coordinate system and measured planes z/c = 1.5, 2 and 2.5.

Stereo time-resolved PIV (Particle Image Velocimetry) was used to acquire the data in parallel planes, subsequent post-processing resulted in velocity vector map with all three velocity components in a distinct plane. The measuring technique is from Dantec company and consists of laser with cylindrical optics and two cameras. The laser New Wave Pegasus Nd:YLF has double head emitting light of wavelength 527 nm by maximal frequency equal to 10 kHz. A shot energy is 10 mJ for 1 kHz and corresponding power is 10 W per head. Two CMOS camera NanoSense MKIII have full resolution 1 280 x 1024 and maximal double-snap acquisition frequency is 500 Hz for full resolution. It was used half resolution in height which enabled to use double value of acquisition frequency.

Thus used camera resolution was 1 280 x 512 pixels. The field of view (FOV) dimensions were 60 x 25 mm^2. The correlation algorithm determined 68 x 29 vectors with maximal 10 % of substituted ones. Acquisition frequency was set to 100 Hz (10 sec of measurement) and to 1 000 Hz for the case of statistical quantities evaluation and for dynamical analysis, respectively. Proper Orthogonal Decomposition (POD) was applied to selected datasets. POD analysis is used here to extract deterministic structures from random turbulent flow field resulting in topological orthogonal modes. Each instantaneous PIV measurement is considered here as a snapshot. The mean flow field is calculated and it is subtracted from all snapshots – the calculation is further provided from velocity fluctuations. Further calculation (see e.g. [8]) results in series of POD modes (structures of the flow field) which are ordered according to kinetic energy content. This means that dominant structures are represent in the first modes.

3 Results

Although the investigated flow velocities were 5 and 10 m/s, the presented vector fields will be connected only with smaller velocity magnitude. Black vectors represent the velocity projected into the plane of measurement. The color represent a scalar value (velocity vector modulus, w-component of velocity, sum of velocity variances, z-vorticity). All figures are plotted in dimensionless coordinates x/c, y/c and z/c. The black dashed lines denote the position of trailing edge. Green lines are added either for z-vorticity distribution or to make visible the streamlines.

3.1 Mean flow field

Firstly, the mean flow field of the wake was evaluated. There is comparison of four distance from the trailing edge in Figure 2. The angle of attack is set to 7 and downwash effect of the inclined plate is well visible. The minimal velocities are located behind rather below the trailing edge for z/c = 1.3. With increasing distance in downstream direction, the position of local minima is shifted below and for the farthest measured plane is out of the FOV. The velocities of freestream are approximately 5 m/s. The minimal velocities were detected just behind the trailing edge for z/c = 1 (not plotted here). However, no streamwise-oriented vortical structures can be seen from time-mean images. The location of enhanced dynamical activity can reveal the distribution of velocity variances. The Figure 3a, b plotted for z/c = 1.3 and 1.5 imply that maximal values are located rather above the profile.

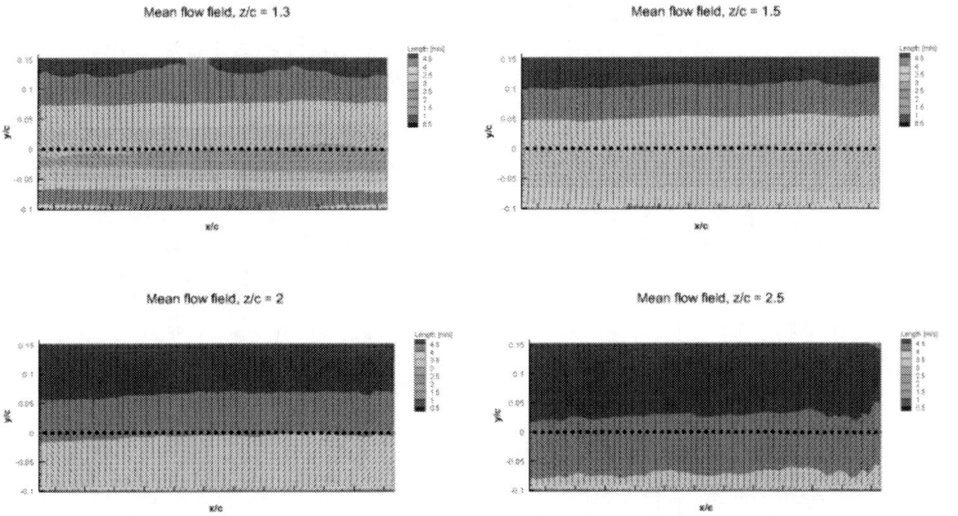

Fig. 2. a, b, c, d. Mean flow field for planes z/c = 1.3, 1.5, 2 and 2.5; distribution of velocity modulus.

3.2 Instantaneous flow field

The topology of instantaneous flow field is completely different. The irregular and perturbed wake behind the inclined plate is visible at four distinct positions at Figure 4. The extension of the wake is decreasing and its position is shifted below with longitudinal coordinate. From the vorticity distribution as well as from streamlines one can see that the

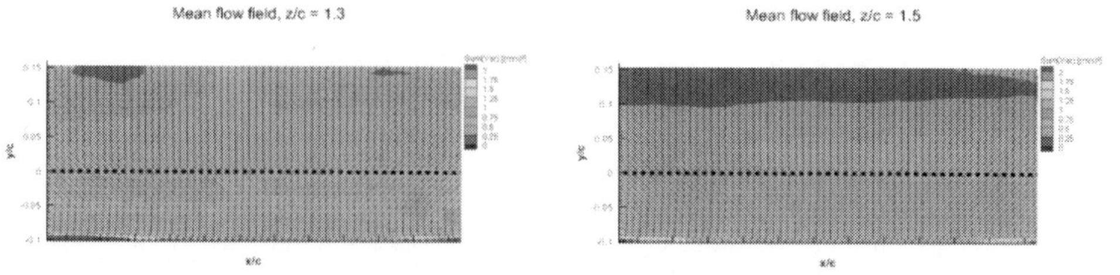

Fig. 3. a, b. Fluctuation activity for planes z/c = 1.3, 1.5; distribution of sum of vel. variances.

streamwise-oriented vortical structures gradually disappear. The last investigated plane (150 mm downstream the trailing edge) has negligible amount of vortical structures compared to the first mentioned measured plane. Flow patterns are distributed rather in the upper side of the profile. However with increasing distance in downstream direction, they are homogenously distributed throughout the FOV. The downwash effect is rather detectable in the farthest measured planes. The topology of flow field is obviously very dynamical and therefore the POD analysis was utilized to detect the deterministic structure from the random flow field.

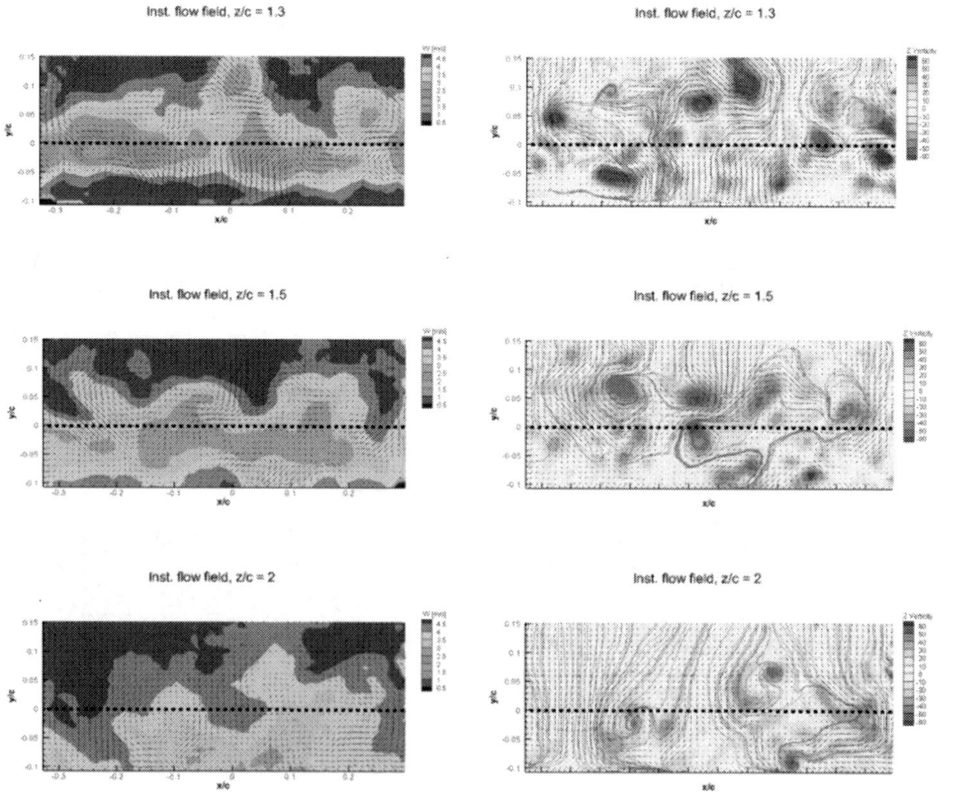

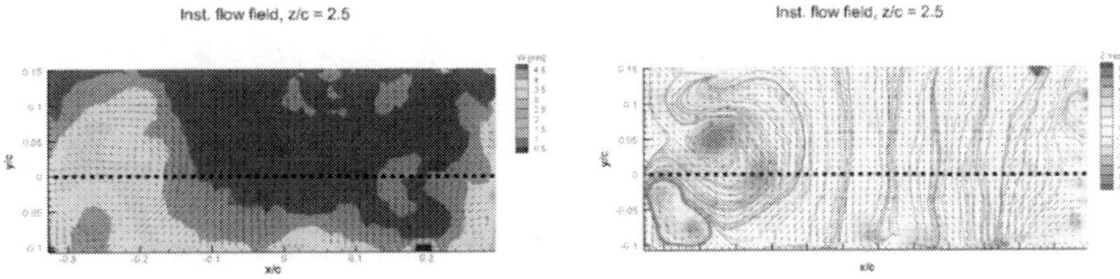

Fig. 4. a, b, c, d, e, f, g, h. Instantaneous velocity vector map for planes z/c = 1.3, 1.5, 2 and 2.5; distribution of streamwise velocity and z-vorticity with streamlines.

3.3 POD modes

As it was mentioned previously, the first POD modes are dominant concerning relative kinetic energy content which means that they consists of the biggest flow structures. The relative kinetic energy of the 30 POD lowest order modes is plotted in Figure 5. Only about 10 POD modes have relative kinetic energy higher than 2 % (the kinetic energy is calculated from velocity fluctuations). The kinetic energy of low-ordered mode is increasing with the streamwise distance. The blue line plotted for z/c = 1 implies that the flow field here consists of larger amount structures with lower energy contribution. On the contrary, the farthest plane z/c = 2.5 consists of rather bigger structures created by merging. The Figure 5b reveals that low-ordered POD modes contain bigger and more energetic structures for lower velocity 5 m/s compared to the double value.

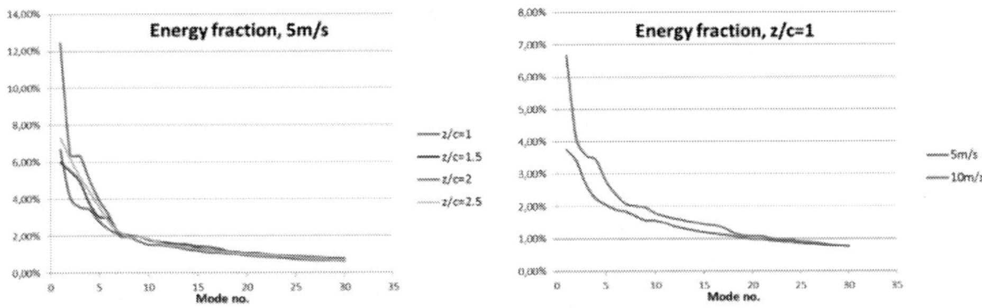

Fig. 5. a, b. The relative kinetic energy content of first 30 POD modes.

The topology of each POD mode will be given by vector distribution patterns in the plane of measurement. Color will denote the w-component of velocity in streamwise direction. Red color is used for positive velocity differences and hence high velocity region and blue color for negative velocity differences and low velocity region. Each individual vortex is identified by streamwise vorticity component isolines. In this case, the z-vorticity is calculated from velocity components in the plane of measurement. However, vorticity cannot distinguish between the swirling motion and shear but here no distinct shear regions are expected. The vortex core is located in the center of isolines and more isolines means more powerful swirling motion. Please note, that the vortices are of alternate orientation – vorticity is positive and negative.

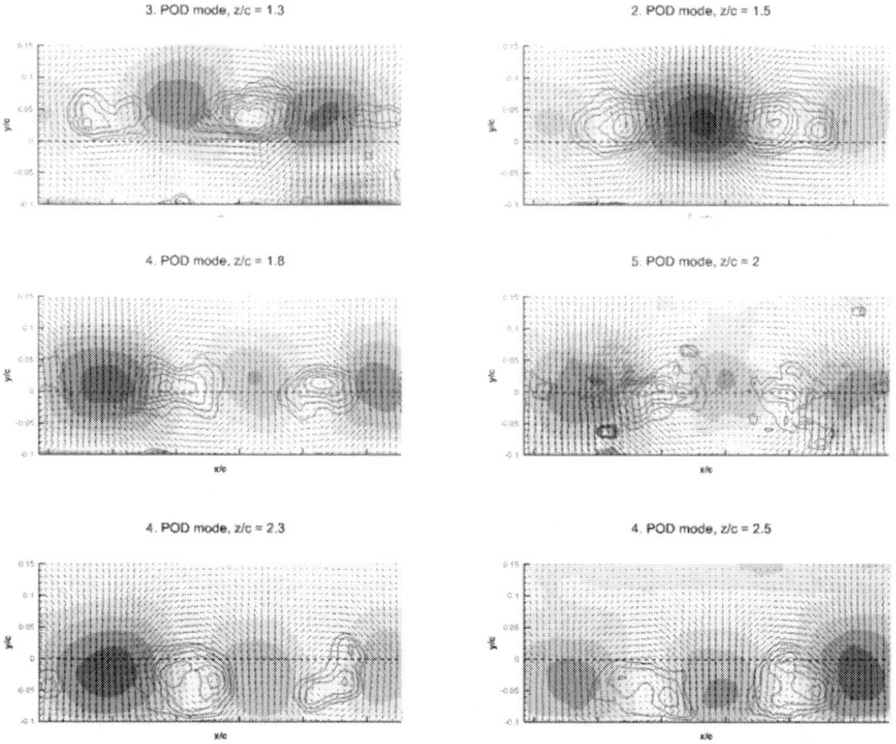

Fig. 6. a, b, c, d, e, f. POD modes of one pair of vortices for planes z/c = 1.3, 1.5, 1.8, 2, 2.3 and 2.5; distribution of fluctuation of streamwise velocity, isolines of vorticity.

There are several examples of the interesting POD modes in Figure 6-8 plotted for various planes. Again the position of trailing edges is marked by black dashed line. Note that for planes close to the trailing edge (z/c = 1.3 or 1.5) the dominant flow structures are situated above the profile in suction side. Due to downwash effect these structures are traversed rather below in more distant planes.

A large number of POD modes was detected. We focus here only on low-order modes containing interesting coherent structure. The topology of each mode consists of one pairs of counter-rotating vortices along the profile chord (Figure 6). It is visible the evolution of this pair in longitudinal direction. Although this structures was detected for distinct plane in another mode number (2nd, 3rd, 4th or 5th) it is obvious that it dominates as one of the most important. The region between streamwise vortices is created by the fluid of either high or low streamwise velocity. This distribution could indicate the existence of the streaks. Size of the vortical structures or streak region detected in these modes is not dependent on longitudinal direction in the studied range.

There is also no dependency on z-coordinates for other POD modes (Figure 7) which are characterized by two pairs of counter-rotating vortices. The size of these structures is significantly smaller and therefore they are connected with lower energy contribution to the mean flow (POD modes no. 6 and 7). The space between two adjacent vortices is again periodically filed with high or low streamwise velocity.

 https://doi.org/10.1051/matecconf/201816805003

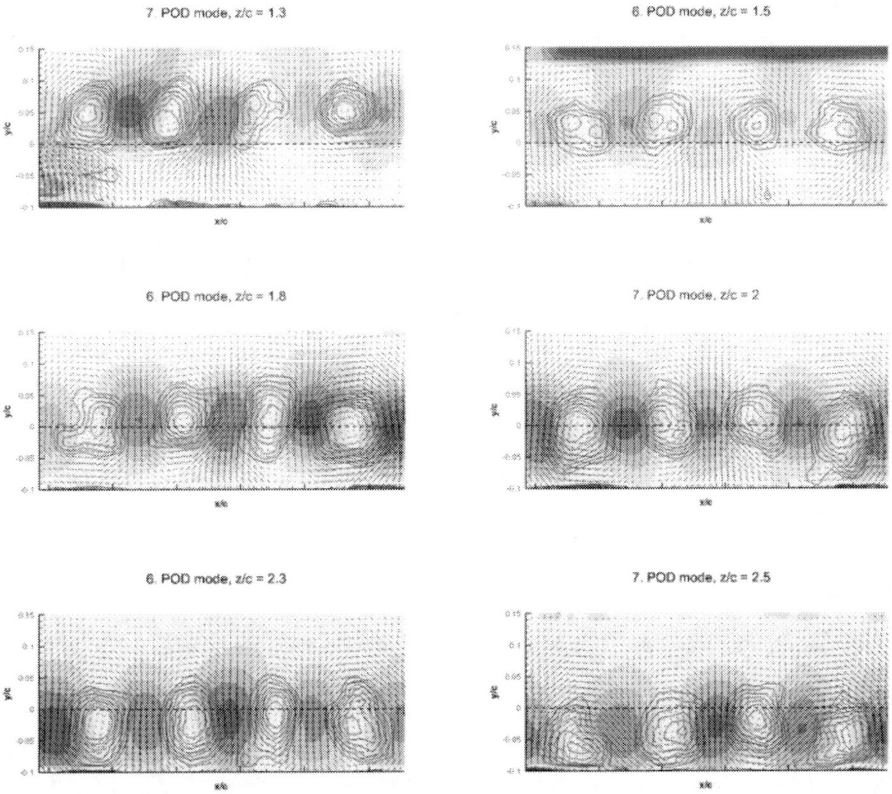

Fig. 7. a, b, c, d, e, f. POD modes of two pairs of vortices for planes z/c = 1.3, 1.5, 1.8, 2, 2.3 and 2.5; distribution of fluctuation of streamwise velocity, isolines of vorticity.

Generally it can be said that the lowest order POD modes contain long streamwise structures and positive or negative streak regions. With decreasing amount of kinetic energy of each mode the mode topology is more structured and complex (Figure 8). The existence of streamwise structures is apparent up to the farthest distance z/c = 2.5. However the role of streamwise vorticity is decreasing gradually with the longitudinal direction and there was not detected the boundary where this vorticity can be declared as negligible.

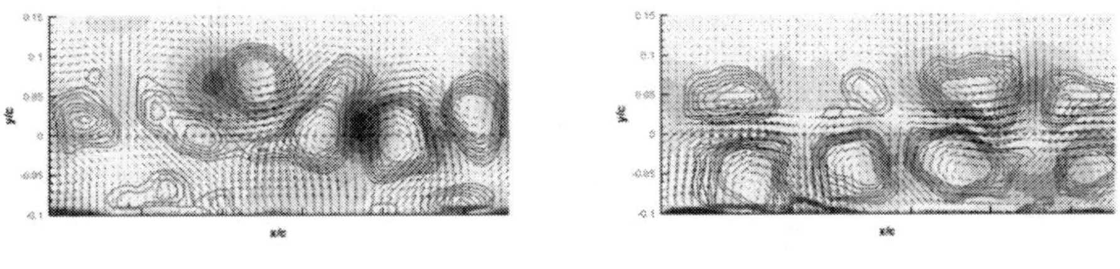

Fig. 8. a, b. POD modes of more complex topology for planes z/c = 1.5 and 1.8; distribution of fluctuation of streamwise velocity, isolines of vorticity.

4 Conclusions

This article was dealing with preliminary results of evolution of flow structures behind the inclined plate obtained using time-resolved stereo PIV technique. There is no information about streamwise vorticity from the mean flow field. However instantaneous vector maps and several POD modes have revealed the existence of streamwise vortices and probably streaks in the boundary layer region and in the wake behind a simple profile. Downwash effect of the inclined profile was presented. It was shown that low ordered modes are characterized by the existence of pairs of streamwise vortices. The topology of other modes with low energy content is more complex. The behaviour of the flow is fully dynamic. The instantaneous images have revealed that firstly smaller structures just behind the trailing edge merge into larger structures along longitudinal direction. More investigated planes should be incorporated in the future research to get more detailed information about streamwise vorticity evolution.

This experimental research created another piece that fully supports a new theory of flight and lift generation.

The authors gratefully acknowledge the support of the Grant Agency of the Czech Republic, project No. 17-01088S.

References

1. J. Hoffman, C. Johnson, The Mathematical Secret of Flight, *Normat 57*, **4**, 1-25, (2009)
2. J. Hoffman, C. Johnson, Resolution of d'Alambert's paradox, J. Math. Fluid Mech., **12(3)**, 321-334, (2010)
3. V. Uruba, Force Interaction of an Airfoil in Fluid-Flow, *15th conference on Power System Engineering, Thermodynamics and Fluid Flow – ES 2016*, June, 2016, Pilsen, Czech Republic, 7375, 7, (2016)
4. V. Uruba, On Aerodynamics Forces Physical Mechanism, *SKMTaT 2016*, Slovak University of Technology in Bratislava, 258-262, (2016)
5. V. Uruba, ON 3D Instability of Wake behind a cylinder, *AeaNMiFMaE 2016*, University of Žilina, Department of Power Engineering, 253-256, (2016)
6. V. Uruba, P. Procházka, V. Skála, On 3D Flow Structure of the Boundary Layer on the Suction Side of a Plate, Experimental Fluid Mechanics 2017, TU Liberec, 694-701, (2017)
7. V. Uruba, D. Pavlík, P. Procházka, V. Skála, V. Kopecký, On 3D flow-structures behind and inclined plate, EPJ Web of Conferences, **143**, 02137, (2017)
8. V. Uruba, Decomposition methods in turbulent research, EPJ Web of Conferences, **25**, 01095, (2012)

The PIV study of air flow past the counter-swirler 53983

Daniel Duda[1,*]

[1]Department of Power System Engineering, Faculty of Mechanical Engineering, University of West Bohemia in Pilsen, Univerzitní 22, 306 14 Plzeň, Czech Republic

Abstract. PIV (particle image velocimetry) measurement of the air flow past a counter-swirler 53983 (anticlockwise swirler surrounded by clockwise swirler) is performed. The measurement is focused to an area at the boundary between the inner swirling jet and the outer one rotating oppositely. The Reynolds number Re based on the inner swirler diameter ranged form $1.2 \cdot 10^3$ to $2.1 \cdot 10^4$. By using *band pass filtering* the shear layer and vortices in the contact region between counter-swirling jets is highlighted. The shear layer between these regions shortens and decays into vortices as Reynolds number increases.

1 Introduction

Free jet into a liquid usually produces turbulence due to the *Kelvin-Helmholtz instability* in the free sheer layer between the moving jet and the surroundings. In mixing studies, the jet can be *swirled* by using a fixed propeller in the nozzle, which widens the jet profile and speeds up the turbulence evolution [1]. This is in the literature refereed as a *swirling jet* [2]. If one would like to avoid net angular momentum in the downstream [3], then the counter-swirler consists of a inner-ring and a oppositely oriented outer-ring of blades resulting into a *pair of concentric jets* rotating oppositely.

From the physical point of view, this kind of flow is interesting because it combines the *Kelvin-Helmholtz instability* with rotation, or, from the second side, it can be seen as a *Taylor-Couette instability* between "*fluidic walls*".

The aim of this study is to perform a preliminary preview on this kind of flow generated by a mass-produced component number 53983 manufactured by the company "Lego". The geometry of the part 53983 seems to be far from ideal, Figure 1, and its planed improvement will be based on the here published results.

2 Experimental setup

A standard method of *Particle Image Velocimetry* (PIV) has been used; the method is based on suspending small particles (in this case of *oil*) into the studied fluid (in this case *air*) and believing that these particles follow the motion of the fluid [4]. These particles are

* Corresponding author: dudad@kke.zcu.cz

illuminated by a laser sheet and observed from direction perpendicular to this sheet. PIV does not allow direct determination of velocity of an individual particle, instead it provides a statistical estimate of the *Eulerian* velocity field in an arbitrary *interrogation areas* (IA) with a small number (~10) of particles. The velocity in one IA is calculated as the shift of a cross-correlation peak between two consecutive photographs of the particles illuminated by a laser. The time between these frames is set according to the velocity and, in this case, it ranges from 20 to 160 μs. As the computational algorithm, the function "Adaptive PIV" performed by the "Dantec Dynamic studio" has been used.

The flow is generated by the wind tunnel borrowed by the company "WT-tech", and the obstacle (counter-swirler 53983) is located at the beginning of the measurement volume of square cross-section of 125 mm side. The illuminated plane is oriented along the flow and crosses the center of the obstacle.

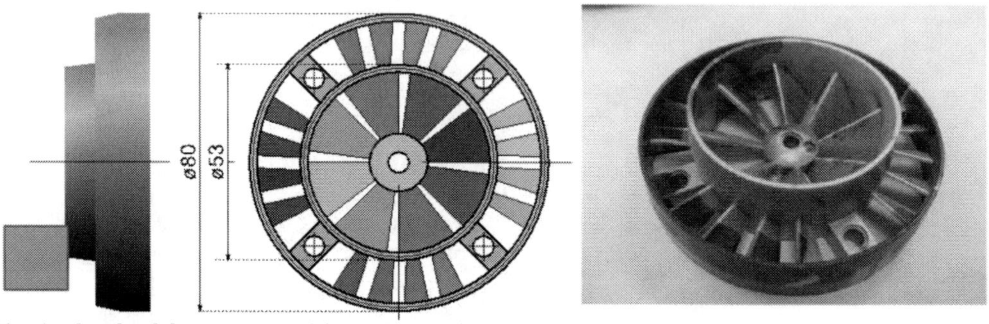

Fig. 1. Sketch of the counter-swirler 53983, units in mm. The green rectangle of size 17.4 × 17.4 mm highlights the smaller studied plane. Right: photograph of the device.

3 Observations

3.1 Large-scale overview

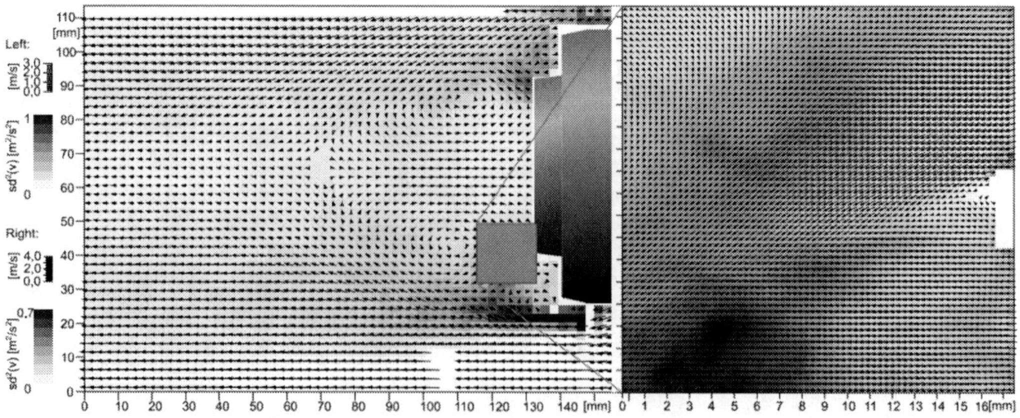

Fig. 2. Mean flow field past the obstacle (sketched in the figure). The color corresponds to the fluctuating energy content. Left: the large-scale view, right: the region of interest drawn as a green rectangle 17.4 × 17.4 mm.

First lets look at the large-scale structure of the flow past the obstacle, Figure 2-left. Scale of this measurement is 2.58 mm/IA showing the overview of the wake, but at this scale it is impossible to deal with the structures in fluctuating field. We can see a pair of big vortices (would be a big vortex ring in 3D) of recirculating fluid behind the obstacle. At this scale the flow is dominated by the regime "flow past a disk" and the counter-swirling character is *hidden* at larger distances from the swirler. In shorter distance, there can be observed the centrifugal flow widening [5]. Therefore, a smaller region of interest has been selected close to the obstacle just at the boundary between the clockwise and counterclockwise swirling regions, see figure 2-right. The fact the swirler is situated like a bluff obstacle is not a problem there. The corresponding scale is 273 μm/IA.

3.2 Detailed look

Figure 3 shows the dependence of *Intensity of turbulence* I_T usually defined as

$$I_T = \frac{\sqrt{\left\langle \left(\mathbf{v} - \langle \mathbf{v} \rangle_T\right)^2 \right\rangle_T}}{\left|\langle \mathbf{v} \rangle_T\right|} = \sqrt{\frac{E_{RR}}{E_{SP}}}, \quad (1)$$

on the Reynolds number *Re* based on the inner swirler diameter D_{in} = 53 mm. In equation (1) $\langle\cdot\rangle_T$ means the averaging in respect to *time*, which is in this case, due to quite small camera frame-rate, averaging over *independent* realizations. E_{RR} means the mean energy content of the fluctuating field obtained by the so called *Reynolds decomposition*: $\mathbf{v}'(t) = \mathbf{v}(t) - \langle\mathbf{v}\rangle_T$, and E_{SP} is the energy content of the *mean field*, of-course, they must fulfill that $E = E_{SP} + E_{RR}$, where E is the total kinetic energy content[†].

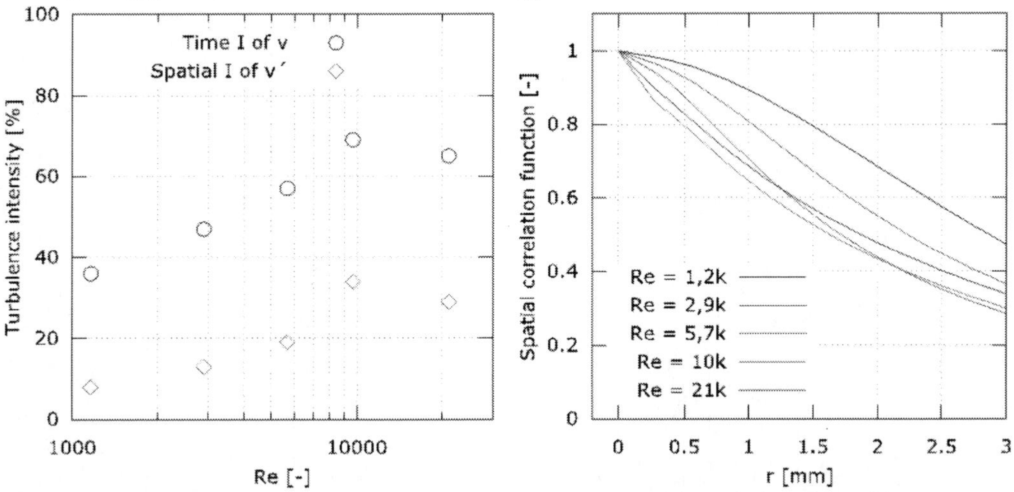

Fig. 3. Left: blue circles: the dependence of turbulence intensity, equation (1), on the Reynolds number *Re* based on the measured velocity and the inner swirler diameter D_{in} = 53 mm. Brown diamonds: spatial turbulence intensity in respect to 1 IA Gaussian filtering of the fluctuating velocity field, equation 3 i.e. a measure of loses by smoothing. Right: the spatial correlation function $R(r)$, equation (2), for different *Re*. Note the lack of inflexion point for higher *Re*. The value of $R(r)$ is normalized by its value at $r = 0$, which corresponds to the energy of fluctuating component E_{RR}.

In Figure 3-left one can see the drop of I_T at the highest *Re* point. This is probably caused by the insufficient spatial resolution, which misses lot of energy in fluctuating field because this energy is in length-scales smaller than is the current resolution. Note that PIV provides velocity in one IA, which is already averaged over this IA. This problem is alternatively illustrated in figure 3-right, where the *spatial correlation function* $R(|\mathbf{r}|)$ is plotted.

† Lets explain the strange unit [J m²/kg] used in this paper: $E = \frac{1}{2}\rho \int \mathbf{v}^2 \, dV$, where $dV = dx \cdot dy \cdot dz$ and only $\mathbf{v}$, dx and dy are measured, $\rho \approx 1.2$kg/m³ is the density of air and $dz \approx$ 1 mm is the thickness of the laser sheet, but these quantities are *not measured*, therefore we publish only *safe* numbers: $E = E_{real} / \rho / dz$.

$$R(\mathbf{r}) = \langle \mathbf{v}'(\mathbf{x}) \cdot \mathbf{v}'(\mathbf{x}+\mathbf{r})\rangle_S, \tag{2}$$

where the subscript S denotes *spatial* averaging and **v'** is the already Reynolds-decomposed velocity field. We can see that for larger velocities, the curve lacks inflexion point -- it is convex since the origin saying that parabolic region is smaller than one IA.

3.3 Looking for middle-scale structures

Figure 4-left shows the Reynolds decomposed velocity field convoluted with Gaussian with $\sigma = 1$ IA = 0.27 mm. The convolution with Gaussian averages out structures smaller than σ in order to kill the noise, which is uncorrelated, but on the other hand we loose physical information at this length-scale, which anyway is indistinguishable from the noise. The measure of loses caused by this procedure in sense of turbulence intensity is shown in figure 3-left as brown diamonds and calculated as

$$I_S = \frac{\sqrt{\langle (\mathbf{v}' - \mathbf{v}' * G_\sigma)^2\rangle_S}}{\sqrt{\langle \mathbf{v}' * G_\sigma^2\rangle_S}} = \sqrt{\frac{E_{1-G1}}{E_{G1}}}. \tag{3}$$

Highlighting the *middle-scale structures* can be achieved via the convolution with a *band-pass filter* [6] referred as *Agrawal decomposition*, which removes structures larger than its upper width, see figure 4-right. The filter is just difference of two Gaussians with widths σ_L and σ_H:

$$P(\mathbf{x}) = G[\sigma_L](|\mathbf{x}|) - G[\sigma_H](|\mathbf{x}|) = \frac{1}{\sqrt{2\pi}}\left(\frac{1}{\sigma_L}e^{\mathbf{x}^2/2\sigma_L^2} - \frac{1}{\sigma_H}e^{\mathbf{x}^2/2\sigma_H^2}\right). \tag{4}$$

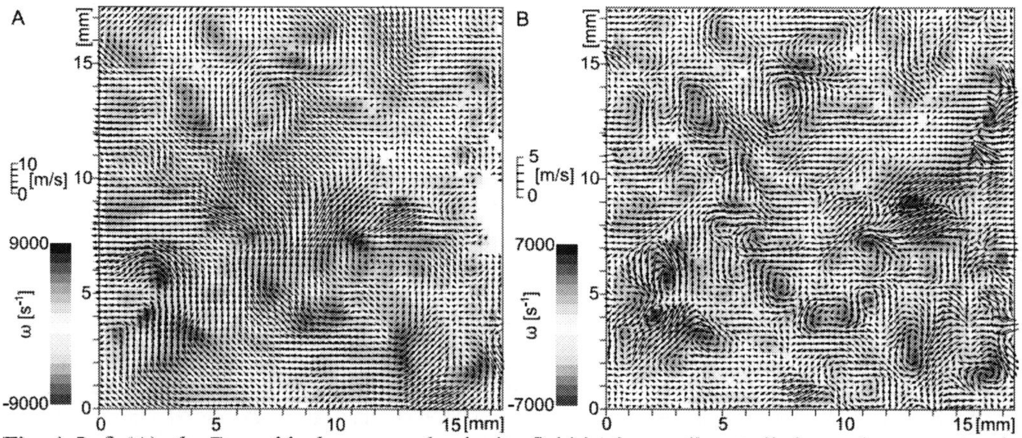

Fig. 4. Left (A): the Reynolds decomposed velocity field in the smaller studied area (green rectangle in figure 1) convoluted with Gaussian with $\sigma = 0.27$ mm. Right (B): The *Agrawal decomposition*, i.e. original velocity field convoluted with *band pass filter* with upper width $\sigma_H = 1.1$ mm. Both pictures show the same particular realization of the flow with $Re = 10^4$. Color in both cases corresponds to classical vorticity.

The particular similarity of both pictures suggests that the fluctuating component consist of such sized structures, but not only of them, as the energy content of the particular case in figure 4-left is $E_{RR} = 0.9$ mJ·m^2/kg, while for the band-pass filtered data it is $E_{Pas} = 0.13$ mJ·m^2/kg, for comparison, the energy of the mean field is $E_{SP} = 2.2$ mJ·m^2/kg. The value of the upper width of the band filter, which is in the presented figure $\sigma_H = 1.1$ mm, is justified by the Figure 3-right showing that the spatial correlation function crosses the value of $1/2^{1/2} \sim$ 71 % approximately at such distance. While the Agrawal decomposition shows nicely small vortices, the Reynolds decomposition in addition shows larger masses moving coherently, which is naturally responsible for some of energy difference between Reynolds and Agrawal decomposed fields.

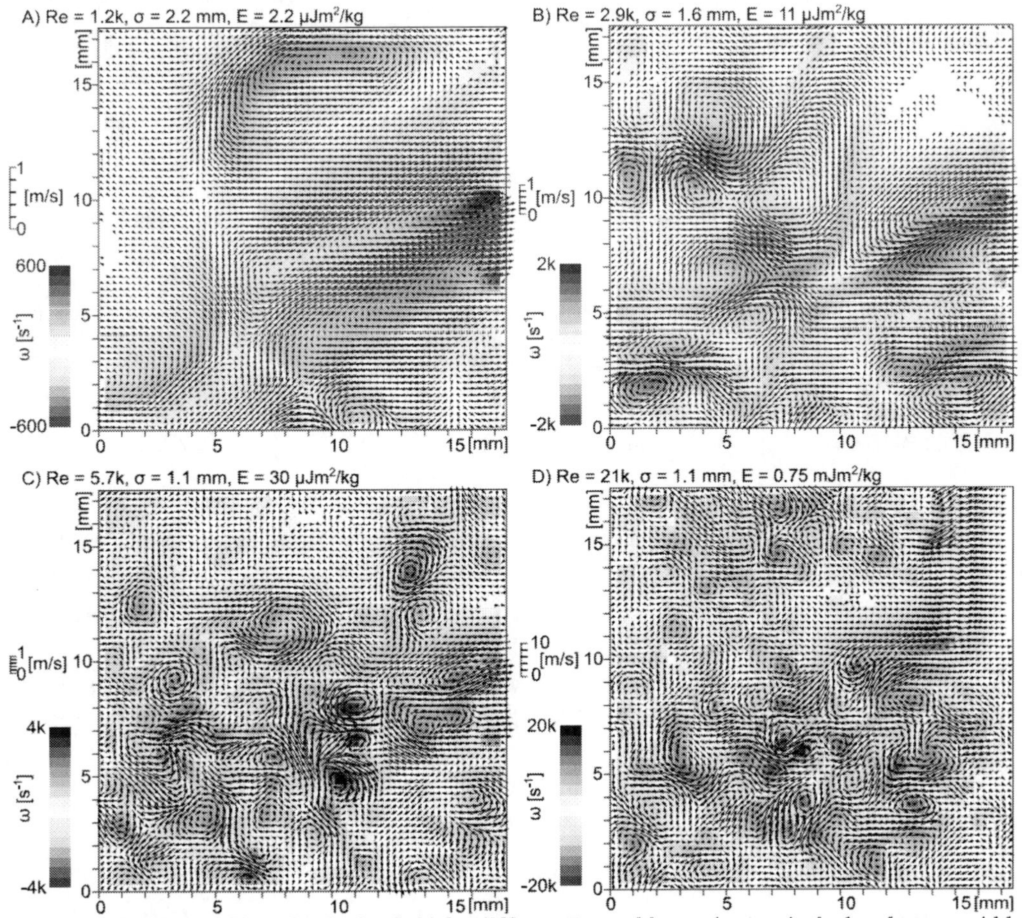

Fig. 5. The band-pass filtered velocity field for different Reynolds numbers. σ is the band upper width chosen according to the spatial correlation function.

Figure 5 shows band-filtered velocity field of other cases. For smaller *Re* the larger σ_H has been chosen according to the *spatial correlation function*, Figure 3-right. Lets note one of the disadvantages of the Agrawal decomposition -- it highlights the areas of large velocity gradient as if it would be shear layers, especially if these are larger than the band upper width. On the other hand, at higher *Re* and more intensive turbulence with many overlapping vortices of different sizes, the Agrawal decomposition makes visible also smaller vortices carried by larger ones [7], because it subtracts the correct local advection velocity and it is not needed to estimate it manually [8]. This makes this method better for the vortex identification, than is the standard Reynolds decomposition.

In Figure 5 we can see shortening of the shear layer between clock- and anticlockwise swirling region with increasing *Re*. At the smallest one, this layer is about 1 cm long, at Re = 2.9k it is about 7 mm and at *Re* = 5.7k it shortens down to 3 mm, while it is difficultly recognizable at larger *Re*. Behind this region, the shear layer decays into vortex tangle.

4 Conclusion

The flow field past a double concentric counter-swirler of type 53983 has been visualized by using a standard Particle Image Velocimetry (PIV) technique. The measurement was focused to a small area behind the swirler at the boundary between the inner swirling jet rotating anticlockwise and the outer swirling jet rotating clockwise. The Reynolds number Re based on the inner swirler diameter ranged from $1.2 \cdot 10^3$ to $2.1 \cdot 10^4$. By using *band pass filtering* suggested by Agrawal [6] the shear layer and vortices in the contact region between counter-swirling jets has been highlighted. The shear layer between these regions shortens and decays into vortices as Reynolds number increases. The *spatial correlation function* analysis states, that for the largest Re the used spatial resolution is inadequate and has to be increased for future experiments.

This work was supported by the project SGS-2016-045 (Increase of efficiency, reliability and a lifespan of power machines and devices 4). The author thanks Martin Duda working at WAT in Warsaw for providing the Lego part number 53983 used in this study.

References

1. M.R. Ruith, E. Meiburg, Journal of Turbulence **3**, 37 (2002)
2. F. Cozzi, A. Coghe, R. Sharma, Experimental Thermal and Fluid Science **94**, 281 (2018)
3. H. Werschnik, M. Schneider, J. Herrmann, D. Ivanov, H.P. Schiffer, C. Lyko, International Journal of Turbomachinery, Propulsion and Power **2**, 12 (2017)
4. V. Uruba, O. Hladík, P. Jonáš, Journal of Physics: Conference Series **318** (2011)
5. C.E. Cala, E.C. Fernandes, M.V. Heitor, S.I. Shtork, Experiments in Fluids **40**, 267 (2006)
6. A. Agrawal, A. Prasad, Experiments in Fluids **33**, 565 (2002)
7. A. Agrawal, Experiments in Fluids **39**, 836 (2005)
8. R.J. Adrian, K.T. Christensen, Z.C. Liu, Experiments in Fluids **29**, 275 (2000)

Efficiency calculation on 10 MW experimental steam turbine

Michal Hoznedl[1,2,*]

[1]Doosan Škoda Power s.r.o., Experimental Research, Tylova 1/57, 30128, Plzeň, Czech Republic
[2]University of West Bohemia, Department of Power System Engineering, Univerzitní 8, 30100, Plzeň, Czech Republic

Abstract. The paper deals with defining flow path efficiency of an experimental steam turbine by using measurement of flow, torque, pressures and temperatures. The configuration of the steam turbine flow path is briefly described. Measuring points and devices are defined. The paper indicates the advantages as well as disadvantages of flow path efficiency measurement using enthalpy and torque on the shaft. The efficiency evaluation by the help pressure and temperature measurement is influenced by flow parameter distribution and can provide different values of flow path efficiency. The efficiency determination by using of torque and mass flow measurement is more accurate and it is recommended for using. The disadvantage is relatively very complicated and expensive measuring system.

1 Introduction

Measurement of experimental turbine flow path efficiency is a rather complex, but at the same time a routine activity. However, attention must be paid to measuring all parameters influencing the efficiency calculation relation. It is necessary to adjust calculation relations so that they describe better the real behavior of the machine, and also to use more precise measuring equipment to calibrate individual sensors and to optimize the measuring chain in order to lower uncertainties and in some cases also measurement errors. Last but not least the experiment must be constructed so that the sensors, pressure samplings or probes are located in a suitable position to measure the required steam flow parameter typical for the whole measured cross section. It means, for example, that the thermometer in the blading outlet must be located at such a blade height and circumferential position where the expected temperature is very similar to the mean temperature in the whole cross section. It is obviously a very complex task not directly related with defining measurement uncertainties. The experiment designer will use his experience as well as some preparatory numerical simulations of the flow. On the other hand, the designer is limited by many construction limitations and a limited number of measuring points. These limitations often complicate data evaluation. At this moment numerical simulations must be used whose boundary conditions are more exactly set using the already measured values. A number of

* Corresponding author: michal.hoznedl@doosan.com

measuring places in the flow path serves only for verification whether at specific points the steam has the same parameters that were measured by the sensor. If so, then the CFD results can be taken as credible and usable for further analyses [1] serves as a suitable example.

Steam turbine efficiency can be defined by measured pressures and temperatures that can determine the steam enthalpy. This way of measurement is the only one possible when it is necessary to define efficiency of individual stages placed in one rotor. An example of such an experiment is shown in [2]. However, problems occur when wet water steam appears in the last stage outlet cross section or even before it. In this case the only possibility to determine flow path efficiency is to measure the mass flow through the flow path and the torque. Distribution of efficiency among individual stages is in all cases very complicated. This turbine configuration is often used when the last stage or group of stages is placed on a special rotor with a brake for defining only the torque of this last group of stages, e.g. [3, 4]. The design of flow path efficiency for the situation when wet steam is found in the last or the last four out of five stages is shown in [5].

The submitted paper aims to describe the flow path of a 10 MW experimental steam turbine as well as the measuring points and their positons. In the next part of the paper equations are described for defining efficiency using the flow, the torque and measurements of steam state. Equation members are briefly described. Advantages and disadvantages of efficiency determination using both types of equations are provided.

2 Description of experimental rig

Experimental measurement was carried out on the steam turbine with the configuration of the flow path named "Boiler Feed Pump Turbine (BFTP)". It is a special design of flow path, when the steam, after passing through the radial inlet sieve, turns to the axial direction and flows through both stages. Behind the second stage one half of the steam continues to the third (and at the same time the last) stage and the diffusor. The second half of the steam is diverted before the third stage and flows through a bypass to the fourth stage and the following diffusor.

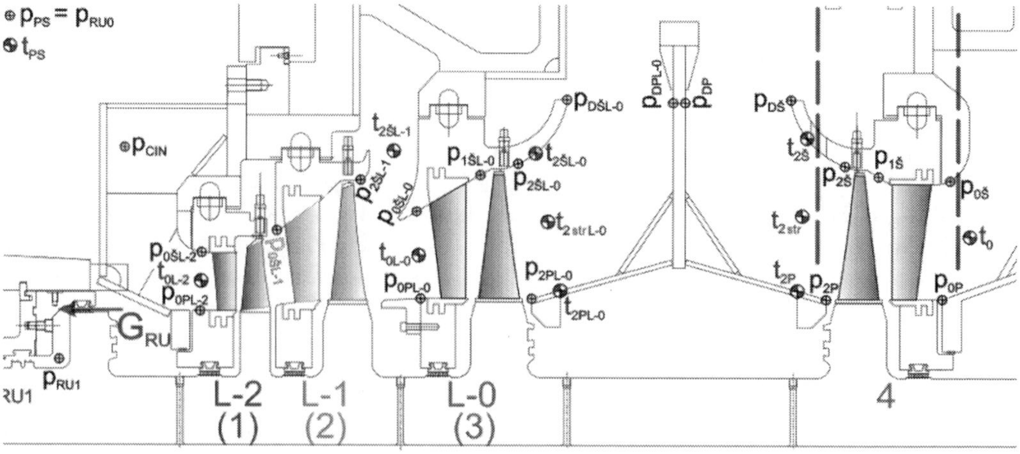

Fig. 1. Flow path with measuring places decription.

In the outlet casing the steam from the last two stages unites. In the flow part and on the outer gland casing a total of 54 static pressure sample collecting places are located. All the pressures are doubled. It means that in each particular place, e.g. on the tip limiting wall behind the stationary blade the pressure sample is on the left and the right side of the turbine about 10° below the dividing plane. In case of one measuring place malfunction, it is possible to use the second pressure. When both pressure pipes are working, the obtained data are averaged.

In the flow path also a total of 37 thermometers are placed. Again, the thermometers are circumferentially doubled. It is worth mentioning that outlets of both last stages are equipped with many measuring points, as it is necessary to measure the distribution of temperatures along the blade length for small turbine outputs, in the so called ventilation regimes.

In order to obtain complete knowledge of the flow path behaviour it is necessary to measure the torque twice (using the torque meter and the water brake) and the rotor speed. Last four important data are measurements of flow times of specific oil tank, main condensate and gland and vent steam condensates. Using known tank volume and water density it is possible to calculate mass flows.

Schematic section of the flow path is in Figure 1. All names of pressure samples are labelled with letter „p“. All temperatures are labelled with letter „t“.

The second scheme in Figure 2 shows basic steam or water flows important for calculation of efficiency using flow and torque measurement. During the measurement it is necessary to determine the steam flows using orifice of admission steam (flow G_{pc}) and gland steam (G_{vv}). The amount of water injected into the admission steam in order to regulate the inlet temperature G_{vv} is evaluated as well as the amount of water measured in the condensate tank G_k and the amount of water condensed after the passage through the gland system. Specifically, it is the amount that goes through the vent and gland steam condenser.

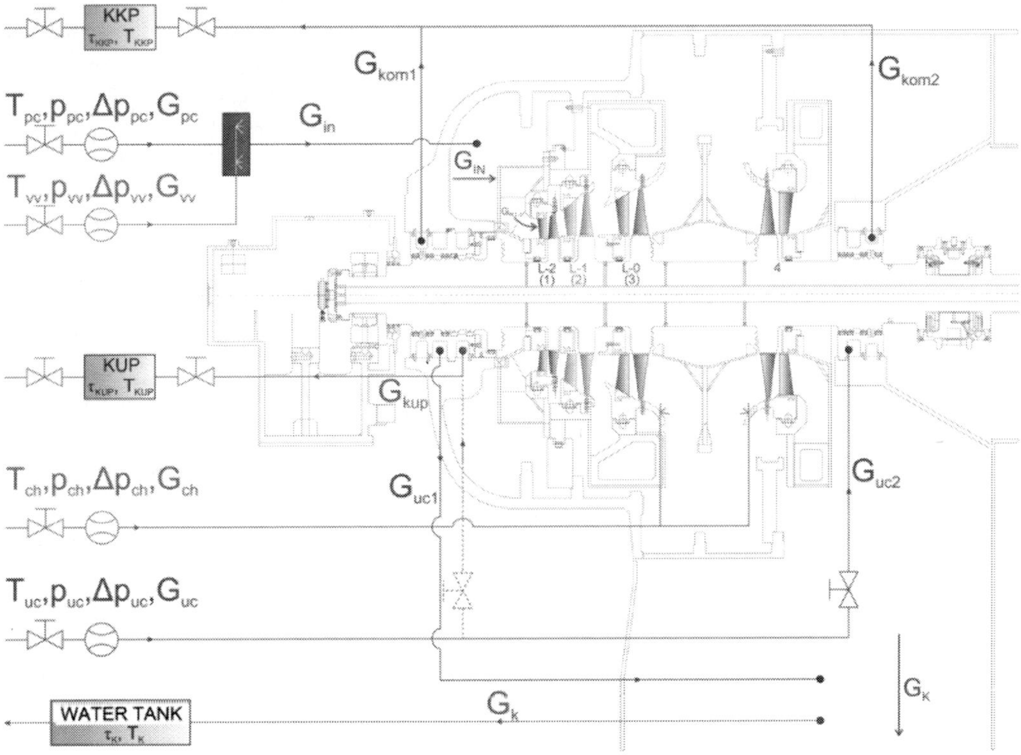

Fig. 2. Water and steam flow in the turbine.

3 Relationships for calculation

3.1. Calculation of efficiency based on temperatures and pressures

For calculations that use equation (1) it is expected that the whole flow path is located in the superheated steam area. In this case it is possible to define the relation for efficiency calculation static-static (in the inlet - index 0 – in the outlet – index 2 – from blading are considered static states) as follows:

$$\eta_{ss} = \frac{h_{0-2}}{h_{is0-2}} = \frac{i_0 - i_2}{i_0 - i_{2is}} \tag{1}$$

In practice the relation is considered with directly measured data:

$$\eta_{ss} = \frac{IAPWS\_97(p_0,t_0) - IAPWS\_97(p_2,t_2)}{IAPWS\_97(p_0,t_0) - IAPWS\_97(p_2,s_0)} \tag{2}$$

It is evident that for calculation of water steam qualities water steam tables are used in the industrial version IAPWS_97. Using these tables even entropy in the flow path inlet can be simply calculated as $s_0 = f(p_0, t_0)$.

It may seem that the easiest choice for efficiency calculation using temperatures and pressures are textbook relations. The static pressure in the blading inlet is measured on the tip and root limiting wall (altogether 4 samples). The temperature is measured in the middle of the blade length (altogether 2 temperatures). In both last stages the pressure is measured again on the tip and root limiting wall (altogether 4 samples) and temperature is measured even in the middle of the blades length (there are 6 temperature measurements behind each last stage).

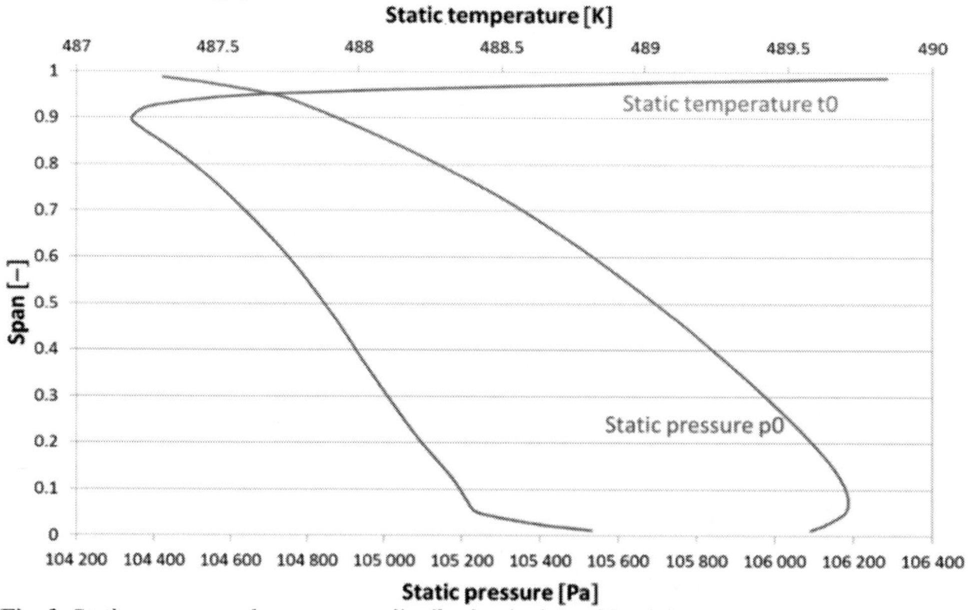

Fig. 3. Static pressure and temperature distribution in the turbine inlet.

Now it is possible to take a look at the results of numerical simulation of the flow with regard to pressure and temperature distribution along the blade height in the flow path inlet (in the plane labelled in relations by index 0). They are in Figure 3.

It is evident that both parameters are in no case evenly distributed. In Table 1 the calculation of flow part efficiency by various ways is shown providing the parameters in the blading outlet are constant and at the level p_2 = 0.22 bar and t_2 = 89 °C. All efficiencies are calculated using relation (2). It is in fact an expression of measurement error that can occur in an inappropriate location of the static pressure and temperature samples.

Table 1. Possibilities of efficiency calculation when changing inlet parameters.

Measured parameters	η_{ss}
Average value of static pressure (= 1.056 bar) and temperature (=215 °C) from CFD calculation	**0.818**
Static pressure value on the inner and outer limiting wall, thermometers are placed in the specific blade length 0.07 and 0.93 – i.e. about 10 mm in the steam flow.	**0.819**
Static pressure value on the inner and outer limiting wall, thermometers are placed in the specific blade length 0.15 and 0.85.	**0.819**
Static pressure value on the inner and outer limiting wall, thermometer is placed in the specific blade length 0.50. This configuration is used in the experiment	**0.819**
Static pressure value in the inlet is average from the probe measurement with step 0.1 dimensionless blade length; thermometer is placed in the specific blade length 0.50.	**0.818**

From Table 1 it is evident that the method of measurement of steam flow inlet parameters is suitable and even for an incorrect position of thermometers the efficiency values are not distorted. It is also obvious that it is not necessary to traverse the inlet pressure field distribution.

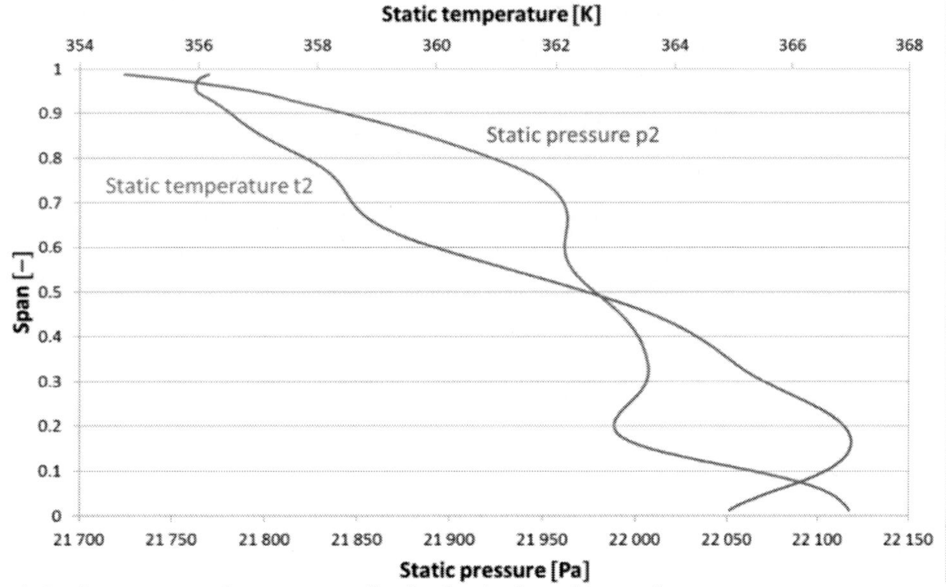

Fig. 4. Static pressure and temperature distribution in the turbine outlet.

The situation when there is constant pressure and temperature in the inlet (which is according to the data in Table 1 quite justifiable) and there will be changes in the pressure and velocity field behind the third stage) is presented in Table 2. In this case pressure $p_0 = 1.056$ bar and temperature $t_0 = 215$ °C. Distribution of steam flow parameters behind the 3rd stage is in Figure 4. It is evident that while the static pressure distribution is only with the difference of 400 Pa, the difference in minimal and maximal temperature in the flow field is up to 11 °C.

Table 2. Possibilities of efficiency calculation when changing outlet parameters.

Measured parameters	η_{ss}
Average value of static pressure (= 0.22 bar) and temperature (=89 °C) from CFD calculation	**0.815**
Static pressure value on the inner and outer limiting wall; thermometers are placed in the specific blade length 0.03 and 0.97 – i.e. about 10 mm in the steam flow.	**0.823**
Static pressure value on the inner and outer limiting wall; thermometers are placed in the specific blade length 0.05 and 0.95.	**0.822**
Static pressure value on the inner and outer limiting wall; thermometers are placed in the specific blade length 0.15 a 0.85.	**0.814**
Static pressure value on the inner and outer limiting wall; thermometers are placed in the specific blade length 0.03; 0,50 and 0.97. This configuration is used in the experiment	**0.819**
Static pressure value on the inner and outer limiting wall; thermometer is placed in the specific blade length 0.50.	**0.810**
Static pressure value in the inlet is average from the probe measurement with step 0.1 dimensionless blade length; thermometer is placed in the specific blade length 0.50.	**0.811**

From Table 2 it is evident that outlet parameters have considerably higher influence on efficiency than inlet parameters. At the same time the outlet flow field is very unbalanced and the location of thermometers positions is very important. In general, the longer the blade and the lower the steam parameters behind it, the higher a possible measurement error is caused by the influence of wrong location of measuring sensors.

3.2. Efficiency calculation based on flow and torque measurement

For the above mentioned reasons it is advisable to try to measure efficiency in a different way. It is necessary to measure also the flow through the blading and the torque. However, this is a significant technical complication for the whole experiment design and implication. The relation for the efficiency calculation is presented here:

$$\eta_{tt\_b} = \frac{3600 \cdot N_{nl}}{G_{in} \cdot \left(h_{1zp} + h_{is0-2}\right)} \tag{3}$$

Where N_{nl} is the output on turbine blades, G_{in} is the flow mass in the first stage inlet. Then h_{1zp} is enthalpy drop calculated from inlet axial velocity, density and flow cross section or using total pressure probe. Drop h_{celk_is} is the total drop between the static state in the inlet and the total state in the blading outlet.

It is necessary to realize that the output on blades N_{nl} is the sum of outputs (torques) measured by torque meter or on the brake, torque loss caused by discs' friction and cylindrical rotor surfaces, and in the bearings. The torque loss in bearings is measured using oil heating and flow through the bearings. Here thus fundamental measurement error cannot be expected. The problem is to determine the disc friction losses. There are many relations available [6]. In general, it can be said that for the given example of the flow part the friction losses are, because of low steam density and a small rotor diameter, at the level of 0.1 % of the total torque. More important errors can occur in the situation when low turbine outputs are measured with high pressure steam, only slightly higher than total losses in bearings and by friction.

The calculation of total iso enthalpy drop h_{is0-2} is done using the relation of denominator in relation (2). As mentioned above, to determine the value of the bracket in relation (2) should not be a basic problem.

However, it is complicated to define the steam flow to the turbine. It can be done in two ways. Firstly by measuring the admission steam flow G_{pc} through the steam orifice and using orifice for injected water G_{vv}. Secondly by measuring the condensate flow using the filling time of the specific tank G_k. Theoretically, both ways of measurement should provide the same flow amount. In practice total agreement was not reached.

The difference in flows can be caused mainly by the amount of steam passing to the steam pipes drainage between the orifice and the turbine. It is a flow that cannot be exactly measured. For the flow through the specific tank it is necessary to realize that the steam from the rear gland casing G_T gets to the tank, but this steam did not go through the flow part. Thus it must be deduced from the flow measured by the tank. The scheme of the rear gland casing is in Figure 5. On the other hand it is necessary to deduce from the flow measured by orifices the steam leaving the front gland casing G_{ru}. At the same time the flow through the front gland casing whose scheme is in Figure 6, drained to the gland steam condenser G_{kup} . The flow through the front part of the gland is regulated by gland steam G_{uc1}. A part of the steam is drained to the vent steam condenser. According to Figure 2 the whole system is a sophisticated complex of flow measurements with the assumption of its tightness.

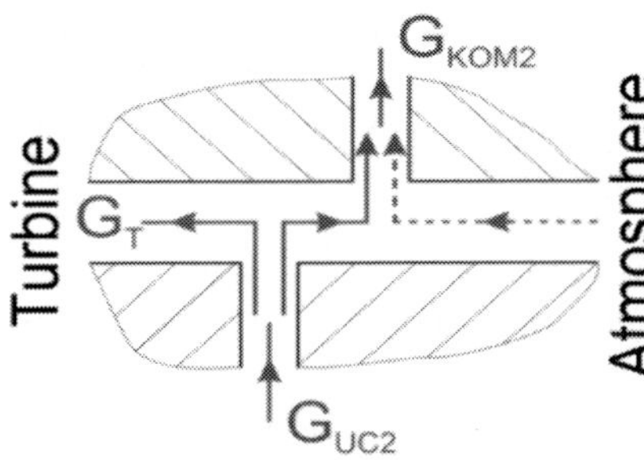

Fig. 5. Scheme of rear gland casing.

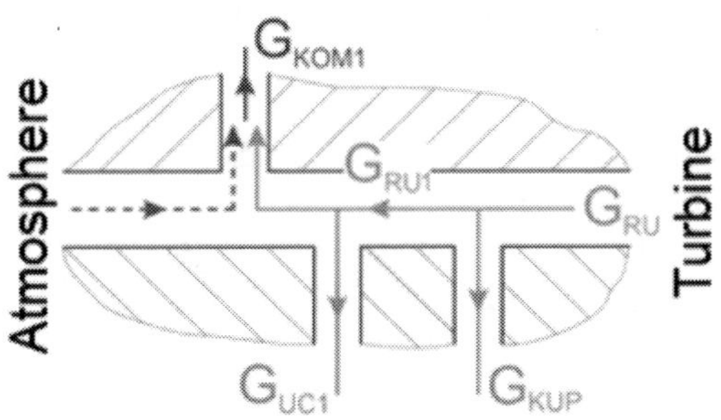

Fig. 6. Scheme of front gland casing.

From the flow balances it can be calculated that based on measured data G_{ru} (measured using flow through the gland), $G_{kom} = G_{kom1} + G_{kom2}$ (measured using vent steam condenser), G_{uc} (measured using gland steam orifice) and G_{kup} (measured using gland steam condenser), it is possible to derive the following equation:

$$G_{uck} = G_{ru} - G_{kom} + G_{uc} - G_{kup} \tag{4}$$

Amount G_{in} measured using condensate tank can be then defined as:

$$G_{ink} = G_{k} - G_{ch} - G_{uck} + G_{v} \tag{5}$$

In this equation the amount G_{ch} appears for cooling the turbine outlet parts and amount G_v, which represents the flow of saturated water steam to the vacuum pump.
For drop h_{is_celk} = 293 kJ/kg, inlet drop h_{Izp} = 4 kJ/kg and output 4 460 kW two values of flow were measured.

Table 3. Influence of flow measurement on efficiency.

Flow defining	Flow value [kg/h]	η_{ss}
By orifice	65 900	**0.820**
By tank	64 500	**0.838**

It is evident that the difference in efficiency calculation using two ways of flow measurement can be (and usually is) in per cents. It is again a measurement error. The efficiency measurement uncertainties are often lower: less than 1 %. When measuring efficiency using water brake according to relation (3) it is possible to define steam enthalpy i_2 in the case when wet water steam occurred in the outlet cross section. The relation for enthalpy i_2 calculation is as follows:

$$i_2 = i_0 - \eta_{tt\_b} \cdot h_{is0-2} \tag{6}$$

From the knowledge of enthalpy i_2 and pressure p_2 it is possible to determine outlet wetness x_2.

4 Conclusions

In the paper an outline is provided of measurement errors analysis when defining the flow path efficiency of BFPT turbine. Two relations are described for efficiency calculation. The first one is simpler, usable mainly in the situation when short blades are measured, without considerable changes in the temperature or pressure field along the blading height. This relation is also the only one that enables efficiency of individual stages in a multistage rotor to be defined.

The second relation is more complex. It defines efficiency using flow measurement through blading and using the torque. Flow measurement is done using condensate tanks or orifices. The common difference between defining efficiency from orifice or tank is 1 – 3 % of the efficiency value. The differences are caused mainly by a complicated assessment of supporting flows through the glands without a possibility to verify the tightness of the gland and vent systems.

In Doosan Škoda Power the value of efficiency defined using measurements from condensate tank flow is considered credible also because of the fact that it is possible to define the outlet wetness x_2 based on output measurement using water brake.

References

1. A. Živný, M. Hoznedl et al., *Proc. ASME TurboExpo 2017*, Numerical investigation and validation of the 1090 MW steam turbine exhaust hood flow field, GT2017-63576, doi:10.1115/GT2017-63576 (2017)
2. M. Hoznedl et al. *Proc. 11th European conference on Turbomachinery*, Experimental Research on Flow in a 5-Stage High Pressure Rotor of 1000 MW Steam Turbine. ETC2015-032 (2015)
3. K. Chandler, M. Melas, T. Jorge Proc. ASME TurboExpo 2015, A study of spontaneous condensation in an LP test turbine, GT2015-42458, doi:10.1115/GT2015-42458 (2015)
4. T. Eberle, M. Schatz et al., J. of Power and Energy, vol. 228, no. 1, pp. 97-106, doi: 10.1177/0957650913508119 (2014)
5. M. Hoznedl, L. Tajč et al. Proc. ASME TurboExpo 2018, Influence of wet steam on the five-stage steam turbine efficiency, GT2018-75049, (to be published)
6. V.M. Neumin, Thermal Eng. 61, **10** (2014)

Measurement of operating parameters of a hydrogen compressor using metal hydride materials

Tomáš Brestovič[1,*], *Natália* Jasminská[1], *Marián* Lázár[1], and *Romana* Dobáková[1]

[1]Faculty of Mechanical Engineering, TU Košice, Department of Power Engineering, Vysokoškolská 4, 042 00 Košice, the Slovak Republic

Abstract. The present article deals with the possibilities of using metal hydride materials in the construction of hydrogen compressors. For the purpose of continuous operation, a tandem tank connection is used. By increasing the tank temperature, hydrogen pressure increases as well, whereas the heat is transported between the tanks by means of a heat pump. Operating parameters are measured for the purpose analysing prevention of the device failures that occur when the maximum pressure behind the heat pump compressor is exceeded.

1 Introduction

Efforts to eliminate the drawbacks of high-pressure and cryogenic method of hydrogen storage lead to the use of metal hydride materials that facilitate hydrogen storage through absorption into the metal structure of the alloy. Hydrogen absorption into the alloy begins once the required pressure is achieved; at such pressure, a catalytic effect of the used metals facilitates dissociation of a hydrogen molecule and subsequent diffusion of hydrogen atoms into the intermetallic space of the lattice [1,2,3,4]. The absorption process is accompanied with a slow pressure increase up to the value of complete metal saturation. The plateau pressure, at which the absorption occurs, significantly depends on the alloy temperature [5,6,7,8]. The difference in plateau pressures of MH materials at the acceptable temperature change resulted in the effort to create a hydrogen compressor which would use the heat cycle of MH tank heating and cooling.

2 Description of the compressor´s temperature cycle

Figure 1 shows the basic temperature cycle of a compressor using the $La_{0.85}Ce_{0.15}Ni_5$ alloy that is capable of forming MH. The cycle is drawn in the measured PCI (Pressure Concentration Isotherms) graphs between the temperatures of 10 and 50 °C. The baseline point A represents the initial state of the alloy with the 0.16 % concentration of the absorbed hydrogen at the pressure of 0.12 MPa.

* Corresponding author: tomas.brestovic@tuke.sk

Hydrogen is absorbed into the alloy between points A and B and it is accompanied with the heat release. For the $La_{0.85}Ce_{0.15}Ni_5$ alloy, the amount of the released reaction heat is identified by experimental measurements as 1 $MJ \cdot m^{-3}$ of H_2 (measured at 20 °C). Following the alloy saturation up to the concentration of 0.66 %, the pressure increases up to 0.2 MPa. Further concentration increase is inappropriate, with regard to the increasing pressure gradient, as it would lead to the reduction of the compression ratio and to a useless increase of the operating pressure of the compressor (point C). Once the required concentration is reached, the alloy temperature must be increased by adding the heat [9,10,11,12].

$$Q_{BC} = \sum_i (m_i \cdot c_i) \cdot \Delta T + q_{des} \cdot \Delta V_{des\,H_2} \quad \text{(J)} \qquad (1)$$

whereas $\Sigma(m_i \cdot c_i)$ is the total heat capacity of the MH storage tank, including the integrated heat exchanger and the absorbed hydrogen ($J \cdot K^{-1}$), ΔT – the difference between the temperatures of working isotherms (K), q_{des} – reaction heat at hydrogen desorption ($J \cdot m^{-3}$), $\Delta V_{des\,H2}$ – amount of released hydrogen at the temperature increase in ΔT (m^3).

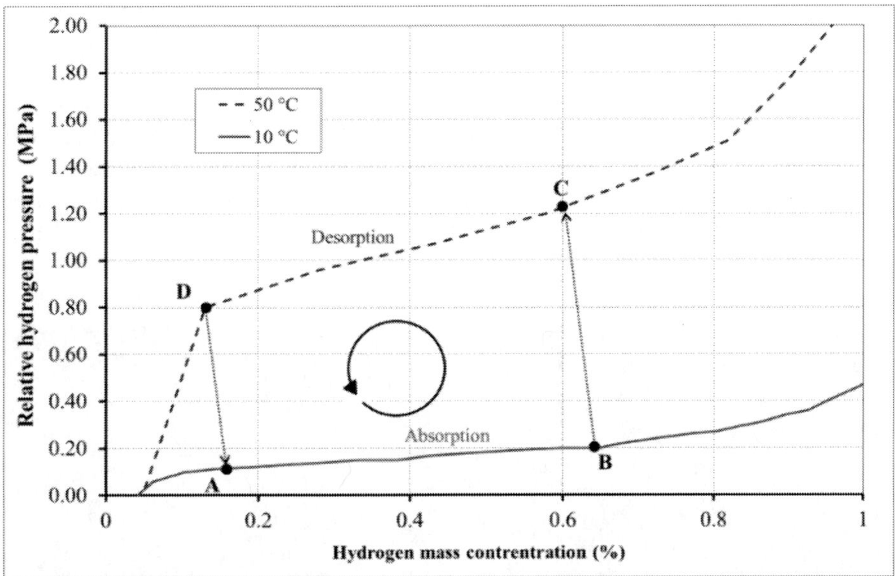

Fig. 1. Description of the temperature cycle of hydrogen compression using the $La_{14}Ce_2Ni_{84}$ alloy.

As a result of supplying the heat between points BC, the temperature increases in 40 °C, which is reflected in the pressure increase approximately up to 1.22 MPa. Due to the desorption effect accompanied with the pressure increase, hydrogen is released in a small amount and increases the pressure of gaseous hydrogen between the alloy grains; as a result, the shift of point C to the area with a lower concentration significantly depends on the grain size and on the volume of the free space between the alloy grains.

After the pressure is increased by heating the alloy, it is possible to start hydrogen desorption between points C and D that is accompanied with the consumption of the supplied reaction heat. Concentration decrease is only possible after achieving the value of 0.13 %, as at lower concentrations the pressure decreases very quickly. The pressure in point D reaches the value of 0.8 MPa; therefore, the compression ratio at hydrogen compression by the temperature cycle can be identified while using the formula

$$\varepsilon_k = \frac{p_D}{p_B} \quad (-) \qquad (2)$$

whereas p_D is hydrogen pressure in point D (Pa), p_B – hydrogen pressure in point B (Pa). During the operation of a cyclical compressor with the used $La_{0.85}Ce_{0.15}Ni_5$ alloy within the temperature range from 10 to 50 °C and the mass concentrations of 0.16 and 0.66 %, the compression ratio resulting from measurements of PCI curves is $\varepsilon_k = 4$.
To achieve the temperature and pressure of the baseline point A again, the storage tank must be cooled and the heat must be removed

$$Q_{DA} = -\left[\sum_i (m_i \cdot c_i) \cdot \Delta T + q_{abs} \cdot \Delta V_{abs\ H_2}\right] \quad \text{(J)} \qquad (3)$$

whereas q_{abs} – reaction heat at hydrogen absorption ($J \cdot m^{-3}$), $\Delta V_{abs\ H2}$ – amount of absorbed hydrogen at temperature decrease in ΔT (m^3). On the basis of the cycle, the basic concept of the compressor described in the next chapter was proposed.

3 Experimental measurements

Operation of a heat pump used in a hydrogen compressor significantly differs from the conventional heat pump use in heating. The difference consists mainly in jump changes in temperatures of liquids supplied into the condenser and evaporator when switching between the heating and the cooling modes. At the Department of Energy Technology in Košice there is a currently running development of a prototype of a hydrogen compressor using MH materials within the APVV project. For this purpose, a water-to-water heat pump with the heating power of 1.5 kW and the cooling power of 1.2 kW is being developed in cooperation with the TATRAMAT, s.r.o. company. R134a (weight of 700 g) is used as the coolant. The compressor has a standard operating input power of 400 W. The cooling power is optimised for a two-stage tandem compressor. During the measurement, the heat pump was connected as shown in the scheme in Figure 2.

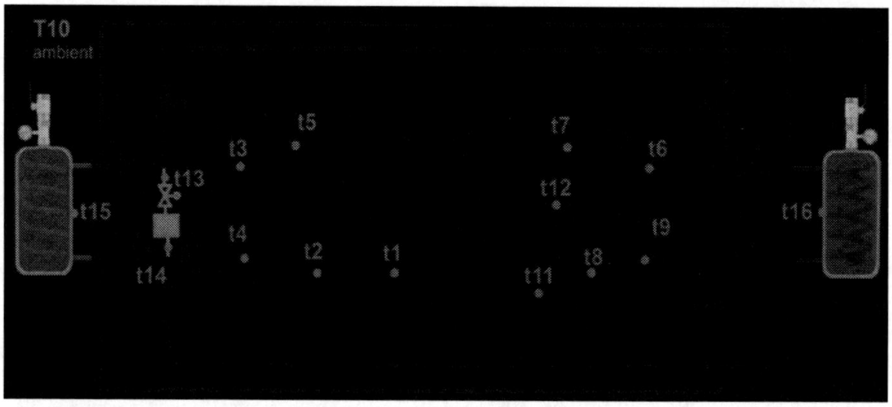

Fig. 2. Connection scheme of the hydrogen compressor using MH tanks with a heat pump.

The measurements were carried out using MH tanks of HBond 500 type with the $La_{0.85}Ce_{0.15}Ni_5$ alloy weighing 3.125 kg in a single tank.

Once the V_1 to V_4 valves are opened, water from the condenser flows into the left MH1 tank and water from the evaporator flows into the right MH2 tank. Once the V_1 to V_4 valves are closed and the V_5 to V_8 valves are opened, there is a switch between the cooling mode and the heating mode. Such switch invokes supplying the cold water into the condenser and hot water into the evaporator. The purpose of the measurements is thus provide a closer view of thermodynamic changes occurring when individual tanks are switched from the heating mode to the cooling mode (and vice versa) between the tank temperatures of 10 and 50 °C (Fig. 3).

Temperatures were scanned using electronic 1-wire sensors of the DS18B20 type with the measurement range from -55 to 125 °C. The temperature measurement accuracy is ± 0.5 °C. The pressure was scanned using Huba Control sensors with the range from 3 MPa (pressure behind the compressor) to 1.6 MPa (pressure behind the evaporator). The maximum measurement accuracy is 0.5 % of the sensor's range. The output current signal from pressure sensors is 4 to 20 mA, conversed by a resistor into the voltage signal. The values were recorded using the Raspberry Pi PC with additional cards for 1-wire sensors and the A/D converter.

As for heat pumps, the thermal power removed from the condenser is always higher than the cooling power received by the evaporator. Therefore, an additional air cooler is placed on the condenser's side.

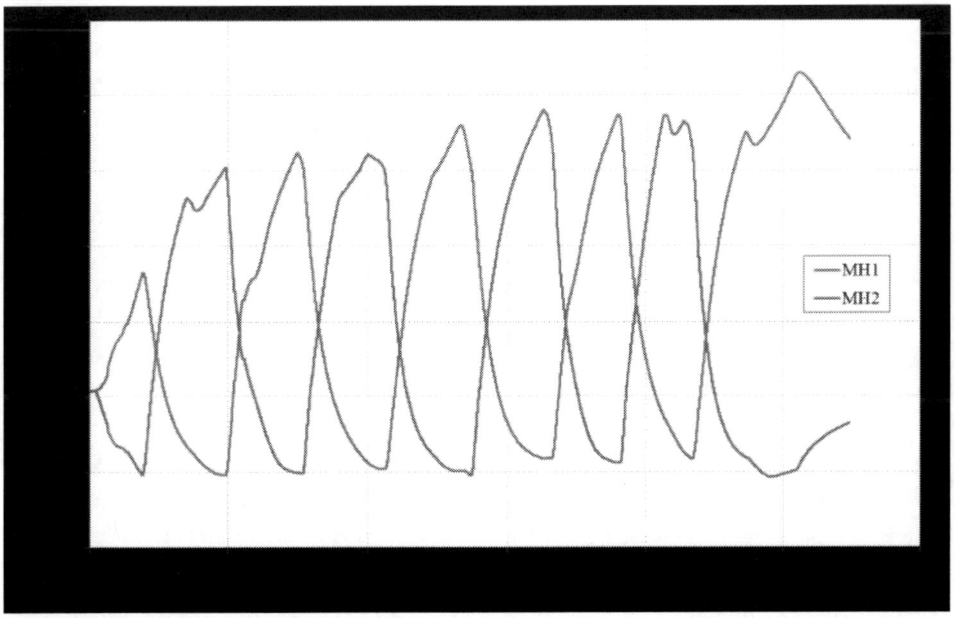

Fig. 3. Curves of surface temperatures of MH tanks during eight switches between the heating mode and the cooling mode.

The curve of coolant pressures before and behind the compressor, including the measured active electric input into the compressor is presented in Figure 4. The vertical dash line represents the time of switching from the cooling mode to the heating mode. With regard to the fact that warmer water is supplied in jumps into the evaporator, such switch is followed by a gradual increase in the coolant temperature before the compressor, leading also to the consequent increase in the relative pressure behind the compressor. The value of active electric input into the compressor is increased as well, decreasing thus the COP of the heat pump.

When the heat pump operation is not interfered with, emergency stop of the device frequently occurs due to exceeding the maximum permissible pressure behind the compressor which is detected by a safety pressure switch with the preset stop pressure of 2.3 MPa. To prevent such heat pump stoppage and achieve a continuous operation of the device, an air cooler (Ch) was turned on before switching between the heating and the cooling of MH tanks.

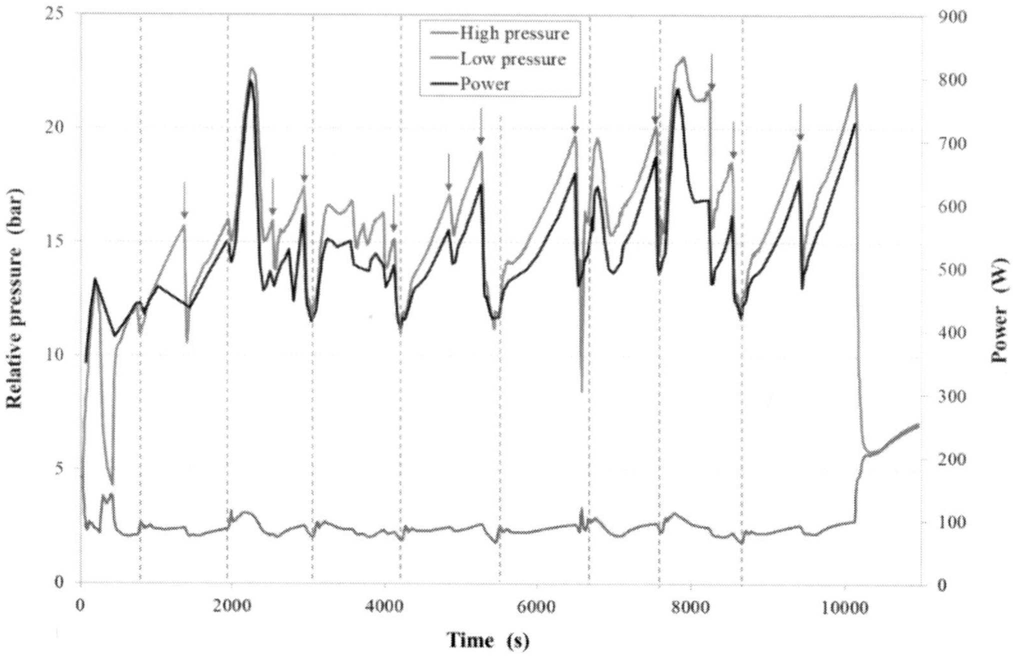

Fig. 4. Curves of coolant pressure before and behind the compressor, including the measured active electric input into the compressor.

MH1 and MH2 tanks contain tubular spiral heat exchangers with a rather small internal diameter of pipes; this significantly decreases the water flow rate and facilitates low COP values during the hydrogen compressor operation. This drawback can be significantly eliminated by creating a second-stage of the compressor, i.e. adding another tandem pair of MH tanks to the first stage.

4 Conclusion

The present article deals with the possibility of using a heat pump in the construction of a hydrogen compressor using the heat cycle at hydrogen absorption and desorption into metal hydrides. We carried out the measurement with cyclical switching between the heating and cooling of MH tanks, while maintaining a continuous operation of the heat pump in the event that before the switch the water in the condenser was cooled below 45 °C, which decreased the subsequent heating up of the evaporator. By measuring the operating parameters and subsequent adjustment of the compressor operation, a stable and continuous heat pump operation may be achieved. Also, the minimum time for changing the tank temperature between the temperatures of 10 and 50 °C was determined as 700 s in the existing conditions. The compressor operated as described above must also be subjected to energy optimisation because the average COP is approximately 1, whereas in a simulated run of heating and cooling of tanks with water the average COP was 3.3. The solution consists in concurrent addition of the second stage of the hydrogen compressor, using one more tandem pair of MH tanks containing the alloy with the composition corresponding to the required pressure behind the second stage of the compressor.

This paper was written with the financial support of the granting agency APPV within the project solution No. APVV-15-0202, of the granting agency VEGA within the project solution No. 1/0752/16 and of the granting agency KEGA within the project solution No. 005TUKE-4/2016.

References

1. A. Sarkar, R. Benerjee, International journal of hydrogen energy **30**, 8 (2005)
2. S. Bouaricha, J. Huot, D. Guay, R. Schulz, Hydrogen Energy **27**, 9 (2002)
3. W. Grochala, P. Edwards, Chem Rev **104** (2004)
4. H. Imamura, K. Masanrik, M. Kosuhara, H. Katsumoto et. all, J Alloys Compds **381**,1-2 (2005)
5. A. Kapjor, T. Gressak, J. Huzvar, AIP Conference Proceedings (2014)
6. M. Vantuch, J. Huzvar, A. Kapjor, EPJ Web of Conferences **67** (2014)
7. G. Popeneciu, V. Almasan, I. Coldea, D. Lupu et all, Journal of Physics **182**, (2009)
8. P. Mlynár, M. Masaryk, BME Budapest (2012)
9. K. Ferstl, M. Masaryk, Prenos tepla (Heat transfer) (2011)
10. M. V. Lototskyy, V. A. Yartys, B. G. Bollet, Boweman R. C, International journal of hydrogen energy **39**, 11(2014)
11. N. Jasminská, T. Brestovič, Ľ. Bednárová, M. Lázár, R. Dobáková, International Journal of Engineering Research and Science **3**, 9 (2017)
12. Ľ. Bednárová, N. Jasminská, T. Brestovič, M. Lázár, R. Dobáková, International Research Journal of Advanced Engineering and Science **2**, 4 (2017)

Evaluation of the energy production efficiency of CHP systems during the summer and winter season operation on the cogeneration unit

Miroslav Rimár[1], *Milan* Abraham[1,*], *Marcel* Fedak[1], *Jakub* Váhovský[1], and *Andrii* Kulikov[1]

[1]Technical university of Kosice, Faculty of Manufacturing Technologies with a seat in Prešov, Bayerova 1, 080 01 Prešov, Slovak Republic

Abstract. The main idea of using combined heat and energy production is to reduce usage of primary fuels and with that connected reducing of pollutions due to the integration of renewable energy sources and with regard to the natural environment. Presented contribution is dedicated on issue on the operation of a cogeneration unit integrated in central heating system. The purpose of currently contribution is to analyse and compare effectivity of cogeneration unit in each summer and winter season.

1 Introduction

Term combined heat and power supply nowadays is becoming and important issue for power generation management. Integration of cogeneration systems and renewable energy for heat generation on the basis of act no. 309/2009 [1] is becoming an important part of the energy generation. This integration into the national heat and power generate network may be either as separate sources or as a part of central systems for the power and thermal energy generation. Presented contribution is oriented to describe installed cogeneration unit driven by natural gas functioning on diesel principle which produces heat and electric energy. The main objective is in describing and comparing effectivity in individual summer and winter seasons. In addition, the conventional cogeneration system has been set optimally according to economic and energy requirements with respect to environmental criteria and controlled automatically through microcontrollers. The main idea on installing cogeneration is to reduce cost of energy and reducing pollution due to the integration of renewable energy sources such as biogas etc. Several existing installed systems demonstrated that there is many useless heat in summer seasons and that is reason for building accumulation or water storage tanks to save heat and use it for example in peaks when people need the hot water. When implementing energy sources, it is necessary to

* Corresponding author: milan.abraham@tuke.sk

consider with implementation directive of the European parliament and of the Council of EU 2012/27/EU on energy efficiency [1, 2].

2 Term cogeneration

Heat supply nowadays requires part of primary energetic source and easiest way is to burn right fuel and by transfer send to heat customer. Effectivity of this transformation is high effective (80 – 100 %) but it loses exergy of fuel (ability to do mechanical work with the released heat). Energetically much more superior way to obtain heat is by using cogeneration, which represents combined production of electrical energy and heat. The main part of cogeneration unit is modified engine which are powering usually asynchronous generator. Engine modification consists of replacement of classic carburettor by mixing device for air and natural gas. Thermal energy from cooling engine and heat from exhausted gases for heating is collected by a system of heat exchangers. Thermal energy from engine is being transferred by water at temperature 70/90 °C. Engine with generator are connected in one unit using the elastic silent block attached to the frame as shown at Figure 1. [1, 2].

Fig. 1. Cogeneration unit engine with generator.

The production of heat and its delivery as final product is a sector whose operation is important not only from the point of view of industrial generation processes, but also from a social point of view. Presented cogeneration unit heats and prepares hot water for about 1 600 apartments, elementary school and shopping centers. Cogeneration unit itself consists of gas combustion engine MWM/TCG 2016 V12C with generator which provides 600 kW of electrical rated power and 652 kW of thermal nominal power.

3 The efficiency of the cogeneration unit

The electrical efficiency of this cogeneration unit is about 39 %. To the quantity of consumed fuel, we add calorific value which is around 10.761 $kWh.m^{-3}$ and we will get theoretical energy included in that amount of fuel [8]. By summing each electric energy and thermal energy and dividing it by theoretical energy included in fuel we get % effectivity of cogeneration unit. Average fuel consumption in year 2017 is around 81 500 m^3 of gas. Monthly fuel consumption is shown in Figure 2. Graphics values shows that cogeneration unit in summer season is very ineffective and in winter season effectivity is around 82 – 87 % which is still a little. Low values can cause overexposure system. From the point of view of the generation of thermal energy for the final customer the advantage of the central heat supply compared to other forms are relatively low prices for heat, in connection with the increase of the usage of cogeneration [7, 8].

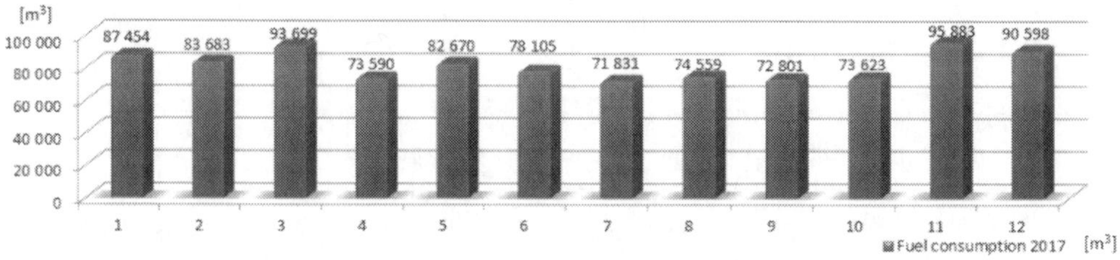

Fig. 2. Fuel consumption of cogeneration unit in 2017.

Monitored value of produced electric energy and heat is shown in Figure 3 separately. Total average effectivity of monitored cogeneration unit is around 87 %. The priority on the production of cogeneration unit is in thermal energy as shown in the figures. Overall power of cogeneration unit was not recorded, the only recorded values are shown at figures.

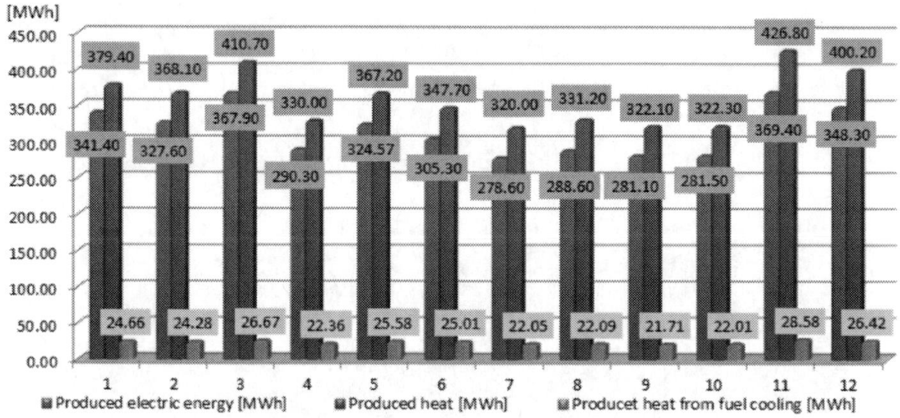

Fig. 3. Comparison of produced electric energy and heat in 2017 [MWh].

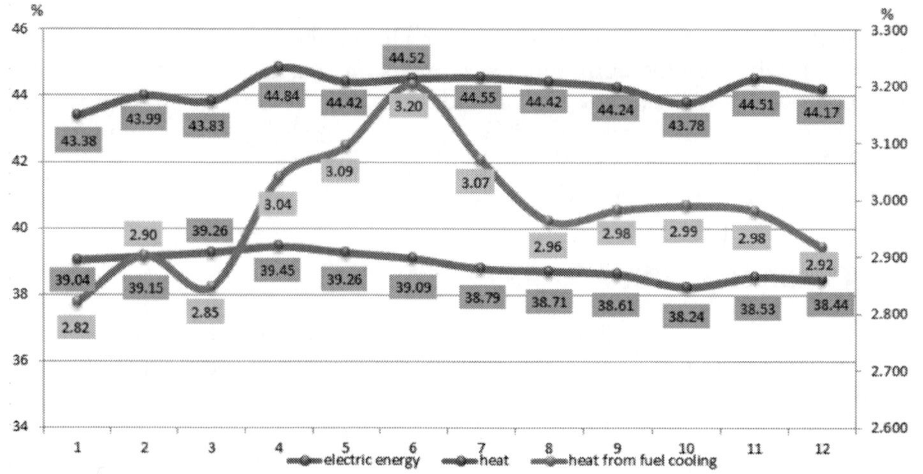

Fig. 4. Effectivity of each energy production separately [%].

The theoretical and measured values lead to knowledge which you can read above that allows understanding the cogeneration unit effectivity and compares the effectivity in summer and winter seasons. Figure 4 describe effectivity of each produced energy in percentage. Measured values shows that there are not big differences in produced electric energy. The largest differences in produced energy are in heat from cooling fuel as shown in Figure 4 [4].

4 Conclusions

Since natural gas is the dominant fuel for the production of heat, greater emphasis need to be placed on combined energy production methods and directly connected to cogeneration. As shown, there is a direct correlation between the generated electrical energy and the heat in combined heat and energy production. Figures confirm problems with non-economical usage of cogeneration unit and shows that whole system is ineffective in summer season, in winter season efficiency of cogeneration unit is around 87 %. Compared to conventional thermal energy and electric energy production cogeneration is effective but there are still many others systems that could be more effectively. Heat from fuel cooling has higher value in summer seasons compared to heat from engine, due to operating hours and performance value of running cogeneration unit.

This work was supported by the Slovak Research and Development Agency under the contract No. APVV-16-0192

References

1. Act No. 309/2009 *Coll. on promotion of renewable energy sources and high efficiency combined generation, and on amendments and supplements of certain Acts in the version of latter provisions*
2. Directive 2004/8/EC *of the European Parliament and of the Council of 11 February 2004 on the promotion of cogeneration based on a useful heat demand in the internal energy market and amending Directive* 92/42/EEC
3. E. Dvorský, P. Hejtmánková, *BEN - Technická literatúra* (2005)
4. J. Krbek, L. Ochrana, B. Polesný, *Heat supply and cogeneration* (1999)
5. M. Rimár, P. Skok, *Thermodynamics* (2013)
6. M. Struž, *Utilization of CHP in Slovakia and its further outlook* (2014)
7. M. Rimár, M. Fedak, *Combustion processes* (2014)
8. *Hodnoty vážených priemerov kvalitatívnych parametrov zemného plynu, ktorý na území Slovenskej republiky distribuuje spoločnosť SPP - distribúcia, a. s. Bratislava*
   http://www.spp.sk/sk/velki-zakaznici/zemny-plyn/o-zemnom-plyne/emisie/

The Effect of Moisture Content of Black Locust Wood on the Heating in the Saturated Water Steam during the Process of Colour Modification

Ladislav Dzurenda[1,*]

[1]Department of Wood Working, Technical University in Zvolen, Zvolen, Slovak Republic

Abstract. The effect of moisture content of black locust wood on the heating of sawn timber with the thickness of h = 32 mm in saturated water steam during the processes of colour modifications by modes: mode I – temperature of saturated water steam t = 112 ± 2.5 °C for τ = 5.5 hours, mode II – temperature of saturated water steam t = 127 ± 2.5 °C for τ = 6.5 hours, and mode III – temperature of saturated water steam t = 137 ± 2.5 °C for τ = 7.5 hours is analysed in the paper. The increase in moisture content affects the heating time, it lasts longer. When the moisture content of black locust wood ranges from W = 60 – 65 %, the heating time by modes I – III did not last longer than the time $\Delta\tau \leq 0.5$ min. and no negative impact on the final colour modification is observed.

1 Introduction

Wood placed in the environment of hot water, saturated water steam or saturated humid air is getting warmer and its physical, mechanical and chemical properties changes. Mentioned facts come in useful in the technology of wood steaming and boiling used in the producing veneers and plywood, bent furniture or pressed wood. Processes of wood heat treatment in saturated water steam result not only in changes of physical, mechanical and chemical wood properties but also in changes of colour. In the past, colour modification, especially wood darkening, was used to remove undesirable differences in colour of lighter sapwood and darker heartwood, or to remove wood stains resulting from steaming or moulding. Recently, the issue of colour modification of various wood species has become of great importance [1-6].

Firstly, the wood is heated to the required temperature in the shortest time. Initial moisture content of modified wood must be taken into account during manufacturing processes. Various moisture contents of analysed wood species can affect physical properties of wood, such as density, heat capacity, thermal conductivity. Process and the length of non-stationary temperature distribution during heating are affected by mentioned physical properties directly.

* Corresponding author: dzurenda@tuzvo.sk

The aim of the paper is to analyse the effect of initial moisture content of black locust wood ranged from w = 40 – 100 % on the heating time of sawn timber with the thickness of h = 32 mm in saturated water steam during the processes of colour modification by modes: mode I – temperature of saturated water steam t = 112 ± 2.5 °C for τ = 5.5 hours, mode II – temperature of saturated water steam t = 127 ± 2.5 °C, for τ = 6.5 hours, and mode III – temperature of saturated water steam t = 137 ± 2.5 °C for τ = 7.5 hours.

2 Model to Calculate the Heating Time of Sawn Timber in Saturated Water Steam

Sawn timber heating by saturated water steam in an autoclave is described by Fourier's equation of heat conduction, one-dimensional, steady-state conduction, i.e. amount of heat transmitted under the given temperature and boundary conditions.

$$\Theta = \frac{t_x - t_p}{t_0 - t_p} = \sum_{i=1}^{\infty} \frac{2.(-1)^{i+1}}{\beta_i} \cos\left(\beta_i \frac{x}{b}\right).e^{-\beta_i^2 .Fo} \quad (1)$$

Lykov [7] published a nomogram of the relation **θ** = ***f(x/b, Fo)*** to determine dimensionless heat θ of the infinite long and wide sawn timber b << y<< z for wood heating at the high value of heat transfer coefficient α→∞. Heating of both surfaces x/b = 1 is as follows:

$$Fo = -0.419 \cdot ln(\theta) + 0.0769 \quad (2)$$

Time necessary for heating the sawn timber can be defined following the Fourier number by the equation:

$$\tau = \frac{Fo.b^2}{a} \quad (3)$$

Fo is Fourier number, b is 1/2 of the thickness of a board, a is a heat transfer coefficient.

2.1 Technological Parameters of Wood Heating by Saturated Water Steam

Heating time of black locust wood with the thickness of h = 32 mm in saturated water steam by individual modes of colour modification process is analysed for initial values of moisture content ranged: W = 40 ÷ 100 %. Physical properties of heated black locust wood: wood density ρ_w, specific wood heat capacity c_w, heat transfer coefficient perpendicular to the fibres in tangential direction $\lambda_{w\perp}$ and thermal conductivity of wood a_w for individual moisture content W and the temperature calculated t_{vyp} are shown in Table 1. Temperature calculated in the process of non-stationary temperature distribution during heating by individual modes of colour modification are determined using the equation:

$$t_{vyp} = \frac{t_0 + (t_x + t_p)/2}{2} \quad (4)$$

t_0 is a temperature of sawn timber at the beginning of heating, t_x temperature of heated sawn timber in the 1/2 of thickness, t_p is a temperature on the sawn timber surface.

Temperature in the centre of the sawn timber thickness t_x is lower Δt = 2.5 °C in comparison to the temperature on the sawn timber surface.

$$t_x = t_p - 2.5 \quad (5)$$

t_p is a temperature of the sawn timber surface.

Table 1. Parametres of physical values of black locust [8-12].

Temperature	Parameter	40 %	50 %	60 %	80 %	100 %
calculated t_{vyp}	ρ_w [kg.m^{-3}]	875	937	1.000	1.125	1.250
mode I t_{vyp} = 60 °C	c_w [J.kg^{-1}.K^{-1}]	2.738	2.869	2.984	3.175	3.328
	$\lambda_{w\perp}$ [W.m^{-1}.K^{-1}]	0.452	0.491	0.532	0.611	0.685
	a_w [m^2.s^{-1}]	$0.188\cdot10^{-6}$	$0.182\cdot10^{-6}$	$0.178\cdot10^{-6}$	$0.171\cdot10^{-6}$	$0.163\cdot10^{-6}$
mode II t_{vyp} = 68 °C	c_w [J.kg^{-1}.K^{-1}]	2.770	2.901	3.016	3.207	3.360
	$\lambda_{w\perp}$ [W.m^{-1}.K^{-1}]	0.467	0.508	0.549	0.631	0.705
	a_w [m^2.s^{-1}]	$0.192\cdot10^{-6}$	$0.186\cdot10^{-6}$	$0.182\cdot10^{-6}$	$0.174\cdot10^{-6}$	$0.168\cdot10^{-6}$
mode III t_{vyp}= 73 °C	c_w [J.kg^{-1}.K^{-1}]	2.797	2.928	3,043	3.235	3.388
	$\lambda_{w\perp}$ [W.m^{-1}.K^{-1}]	0.473	0.515	0.557	0.641	0.725
	a_w [m^2.s^{-1}]	$0.193\cdot10^{-6}$	$0.187\cdot10^{-6}$	$0.183\cdot10^{-6}$	$0.176\cdot10^{-6}$	$0.171\cdot10^{-6}$

Wood temperatures during the heat treatment process by individual modes of colour modification of black locust wood are shown in Table 2.

Table 2. Temperatures of sawn timber during the process of heat treatment.

Mode	Temperature of sawn timber		
	t_o	t_p	t_x
I	10 °C	t_p = 112.5 °C	t_x = 110 °C
II		t_p = 127.5 °C	t_x = 125 °C
III		t_p = 137.5 °C	t_x = 135 °C

3 Result and discussion

Time of non-stationary temperature distribution during heating of black locust sawn timber with the thickness of h = 32 mm in saturated water steam by individual modes of colour modification is illustrated in Figure 1.

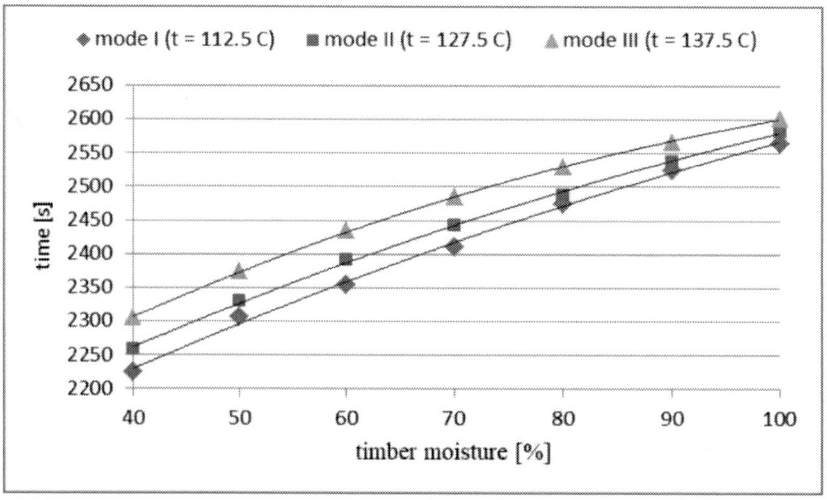

Fig. 1. Heating time of black locust sawn timber depending upon the moisture content of wood and mode of colour modification.

Dependences of heating time of black locust sawn timber with the thickness of h = 32 mm on the moisture content of wood W = 40 – 100 % to the required temperature in individual modes of colour modification are shown in Table 3.

Table 3. Dependences of heating time of black locust sawn timber on the temperature of the saturated water steam.

Temperature of saturated water steam	**Functional dependence**	**Coefficient of determination**
$t'' = 112.5$ °C	$\tau = -0.0214 \cdot w^2 + 8.6286 \cdot w + 1{,}918.4$	$R^2 = 0.9974$
$t'' = 127.5$ °C	$\tau = -0.0243 \cdot w^2 + 8.7143 \cdot w + 1{,}951.7$	$R^2 = 0.9990$
$t'' = 137.5$ °C	$\tau = -0.0350 \cdot w^2 + 9.8000 \cdot w + 1{,}970.4$	$R^2 = 0.9998$

The increase in the moisture content of black locust wood affect the heating time of sawn timber to the required temperature, it prolongs. The change in mode I is from $\tau_{w=40\%} = 37.08$ min to $\tau_{w=100\%} = 42.75$ min, it means the heating time is $\Delta\tau = 5.67$ min longer.
The average moisture content of modified sawn timber in operating conditions is 60 % and moisture content variation is $\Delta W = \pm 5$ %. Table 4 shows the values of prolonged heating associated with the moisture content variation $\Delta W = \pm 5$ %, as well as extreme values $\Delta W = \pm 10$ % for individual modes of colour modification of black locust sawn timber with the moisture content W = 60 % and the thickness of h = 32 mm.

Table 4. Effect of increased moisture content of black locust wood on the prolonged heating to the required temperature.

Mode	**Heating time of sawn timber h = 32 mm thick, w = 60 %**	**Prolongation of heating time of sawn timber because of higher moisture content ΔW = +5 % and ΔW = + 10 %**		**Relative prolongation of heating time of sawn timber** $\Delta\tau = \frac{\tau_{60\%} - \tau_{60\%+\Delta W}}{\tau_{60\%}} \cdot 100$ %	
Mode I	$\tau = 39.25$ min	$\Delta W = +5$ %	$\Delta\tau = 0.46$ min	$\Delta W = +5$%	$\Delta\tau = 1.17$ %
		$\Delta W = +10$ %	$\Delta\tau = 0.91$ min	$\Delta W = +10$ %	$\Delta\tau = 2.28$ %
Mode II	$\tau = 39.85$ min	$\Delta W = +5$ %	$\Delta\tau = 0.43$ min	$\Delta W = +5$%	$\Delta\tau = 1.08$ %
		$\Delta W = +10$ %	$\Delta\tau = 0.87$ min	$\Delta W = +10$ %	$\Delta\tau = 2.12$ %
Mode III	$\tau = 40.57$ min	$\Delta W = +5$ %	$\Delta\tau = 0.42$ min	$\Delta W = +5$ %	$\Delta\tau = 1.03$ %
		$\Delta W = +10$ %	$\Delta\tau = 0.85$ min	$\Delta W = +10$ %	$\Delta\tau = 2.04$ %

Due to the increase in the moisture content of black locust wood, the heating time of sawn timber to the required temperature is longer. Mentioned prolongation of heating time depending upon the temperature of saturated water steam by the mode of colour modification at the standard moisture content variation $\Delta W = \pm 5$ % is no longer than $\Delta\tau \leq 0.5$ min. and does not affect the process in negative way. Mentioned findings are confirmed by determining the colour of black locust wood [13].

4 Conclusion

The findings of the analyses of the effect of moisture content of black locust wood on the time of non-stationary temperature distribution during heating by saturated water steam during the process of wood colour modification are presented in the paper. The increase in moisture content can cause the changes in physical properties of wood, such as wood density, specific heat capacity, heat transfer coefficient, thermal conductivity. Subsequently, the heating time of wood to required temperature is longer. The time of non-stationary temperature distribution during heating of black locust sawn timber with the thickness of h = 32 mm, with the moisture content W = 60 % in saturated water steam with the temperature of $t'' = 112.5$ °C by mode I is $\tau = 39.25$ min, in saturated water steam with the temperature of t'' = 127.5 °C by mode II is τ = 39.85 min and in saturated water steam with the temperature of t'' = 137.5 °C by mode III is τ = 40.57 min. When the moisture content of black locust wood ranges from W = 60 – 65 % the heating time by any mode of colour modification did not prolong the time $\Delta\tau \leq 0.5$ min and did not have a negative impact on colour modification.

This paper was prepared within the grant project: VEGA–SR No: 1/0563/16, as the result of work of author and the considerable assistance of the VEGA–SR agency.

References

1. L. Tolvaj, R. Nemeth, D. Varga, S. Molnar, Drewno. **52 (181)**, 5-17, (2009)
2. I. Klement, P. Marko, Wood research **54 (3)**, 45-54, (2009)
3. L. Dzurenda, Acta Facultatis Xylologiae Zvolen, **56 (1)**, 13-22, (2014)
4. Š. Barcik, M. Gašparík, E.Y. Razumov, Wood Research. **60 (3)**, 385-396, (2015)
5. J. Barański, I. Klement, T. Vilkovská, A. Konopka, BioResources, **12 (1)**, (2017)
6. L. Dzurenda, BioResouces, **13 (1)**, (2018)
7. A.V. Lykov, *Teplo i masoobmen v processach sushki*. Moskva. Gosenergoizdat (1956)
8. B.S. Chudinov, *Teorija teplovoj obrabotki drevesiny*. Moskva. Nauka (1968)
9. N. Deliiski, *Transient Heat Conduction in Capillary Porous Bodies*, (InTech 2012)
10. K. Radmanović, I. Dukić, S. Pervan, Drvna industrija **65 (2)**, 151-157, (2014)
11. R. Hrčka, M. Babiak, Intech Europe, Rijeka, Croatia. (2017)
12. N. Deliiski, L. Dzurenda, D. Angelski, N. Tumbarkova, Acta Facultatis Xylologiae Zvolen. **60 (1),** 101-112, (2018)
13. L. Dzurenda, Acta Facultatis Xylologiae Zvolen. **60 (1),** 61-70, (2018)

Control and regulation systems of energy networks in buildings

Zuzana Kolková[1,*], *Peter* Hrabovský[1], *Jozef* Matušov[1], *Martina*[1] Antošová and *Michal* Holubčík[2]

[1]University of Žilina, Research centre, Univerzitná 8215/1, 010 26 Žilina, Slovakia
[2]University of Žilina, Department of Power Energy, Univerzitná 8215/1, 010 26 Žilina, Slovakia

Abstract. Measurement, regulation and control systems offer direct savings and reduce energy consumption, regulating heating, cooling, ventilation or lighting in the intelligent buildings. They provide quick and accurate information on the status of regulated systems or possible malfunction. Systems can use the special meteorological stations to get information about wind velocity and direction, outdoor temperature, intensity and direction of sunlight. They respond flexibly to changes in external parameters. Intelligent buildings combine architecture and aesthetics of the construction, safety, comfort and quality of the living. These buildings are productive, energy efficient and environmentally acceptable. Intelligent buildings combine internal and external intelligence building, intelligence used materials and constructions. The most important aspect is the cooperation of people with those systems. Intelligent buildings should be permanent, healthy, technologically advanced.

1 Introduction

The intelligent building control is based on communication with all embedded system devices in the form of data messages sent by the data cable. Sensors, thermostats, motion sensors, fire detectors are designed to monitor the situation and then send the measured data bus. The bus will send a command that accepts and executes the requested operation. Dimmers, binary outputs, power switches are among the actuators. Changes to the control system or its deployment can be realized without the necessity of reworking the project or installing new cables. Changes to an existing system can be made after several years [1].

Communication between the devices involved in the control system with each other and with the control system ensures its proper functioning. This connection can be secured by cable or wireless. The benefits of cable connection include high speed, cost and availability of the devices needed for the connection. It is the most used connection because of the connection of electrical equipment to the devices [2, 3].

Intelligent buildings should be permanent, healthy, technologically advanced. They should meet the requirements and needs of the people and should be flexible to the changes you want. The role of intelligent buildings is also using its built-in system to create

* Corresponding author: zuzana.kolkova@rc.uniza.sk

high-quality internal and external environment while meeting environmental and effective values. This means that the design, construction, technical equipment and building management should be on one level. The most important aspect is the cooperation of people with these systems. System thinking is necessary in the design and management of buildings with the talent that create, innovate and remain mostly practical and is always focused on maximum simplicity. The most important function of the building is to create a suitable environment for humans [4, 5, 6].

2 Energy systems

Heating and cooling are used to create thermal comfort in the building. For its function using thermal and cold heat sources, which are mostly electric heat pumps ground – water, air – water and gas heat pump air – water (Figure 1.). The main objective is to understand the functioning of all intelligent building systems and ensure optimal use of all energy sources.

Fig. 1. Gas and electric heat pumps, gas boiler.

Heat pumps use heat from the environment either through boreholes from the ground or from the air. The intelligent building has 10 boreholes. Each borehole has 100 meters. Heat can also earn seasonal rooftop solar tube collectors. In the extreme case is used gas boiler. Cold is obtained directly from boreholes. If this is not enough, system can use chillers on the roof. In the future, the heat source will be used a natural heat storage. Heat and cold are distributed as needed to the building from the utility room in the basement. Distribution is ensured by controlling heating and cooling branches. Areas of the buildings are supplied with a hot and cold individual with a room-based accuracy based on user requirements. Set temperature can be changed directly in the room ± 3 °C or centrally via building management. Technological elements of heating and cooling are managed centrally at the resources. Boreholes are closed independently, thus it is possible to change their order and count in the required pumping capacity cold around 8 °C or heat through the heat pump (water-glycol media). Research centre areas of the building are in normal operation priority supplied heat through underfloor heating and cooled ceilings.

Cooling and heating are possible in all the settings for research purposes. Section of the light and blinds is also used to create thermal comfort of people in the building. LED light sources are used, part is bionic. Blinds eliminate excessive dazzle in conjunction with lighting. The desired light intensity can be adjusted separately for each room. Lighting in hallways and bathroom is automatic. Classical switches are also used in technical premises (Figure 2). Section of ventilation and air conditioning provides ventilation in the building. Three air handling units (on the roof) using for this function. Regulation is centralized and automated. It is possible to control the flaps to the nearest room. Ventilation in sanitary facilities is an automatic under pressure.

Cooling of server rooms and DAT Centre is independent and automatic (5 units). Through central control and monitoring intelligent building can monitor and control individual components collect data in real time and from a database. It serves as a support for projects and research tasks. Management is based on software tools that are an extension of the management processes of the building. Point of contact is a SCADA or OPC server. Display information about experiments, availability of equipment, laboratories are located in front of each room.

Fig. 2. Humidity sensor, CO_2 sensor, regulator of lighting, temperature and blinds.

Additional technologies include elements built into the construction of buildings. Pressure, temperature, humidity, radiation and wind speed are part of these technologies. Intelligent building have also camera system, sensor of lighting in the room and the presence of persons in the building, fire detector FES, fire button FES, and fire alarm FES.

3 Control system

Control system of intelligent building should reduce building energy and maintenance costs compared to a non-controlled building. A demonstration of the control of one floor is shown in Figure 3.

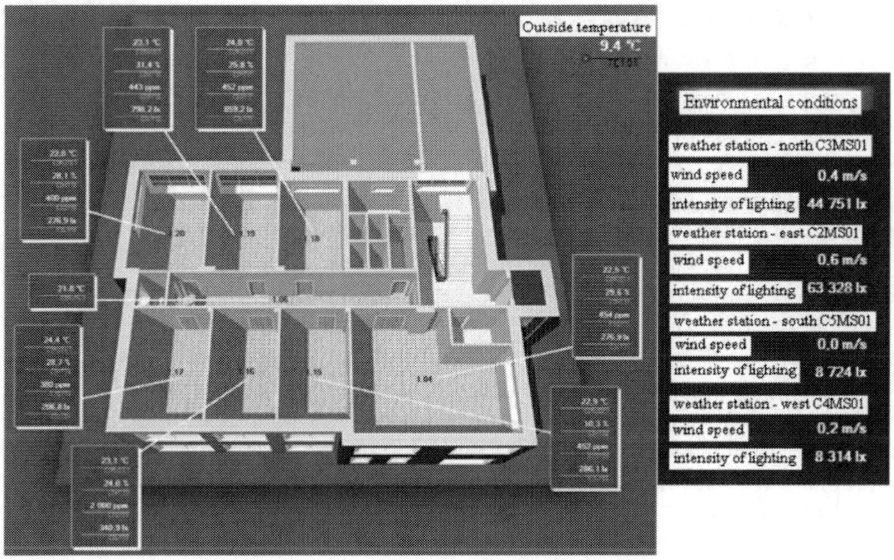

Fig. 3. Control system for one floor.

Central management programs run on internal servers and workstations of the university, data repositories, and in an internally doubled separated LAN network with a connection to the external communication and computing infrastructure of the school.

The system is controlled through the BACnet (Building Automation Control Network) open communication standard. Regulators and automation stations communicate with this interface. External environmental conditions are recorded by the weather station, which is located on the building. It is possible to set individual parameters for each room individually or for one floor or for the whole building.

The systems for heating, ventilation, cooling and heating of hot water are controlled in the technical room according to Figure 4.

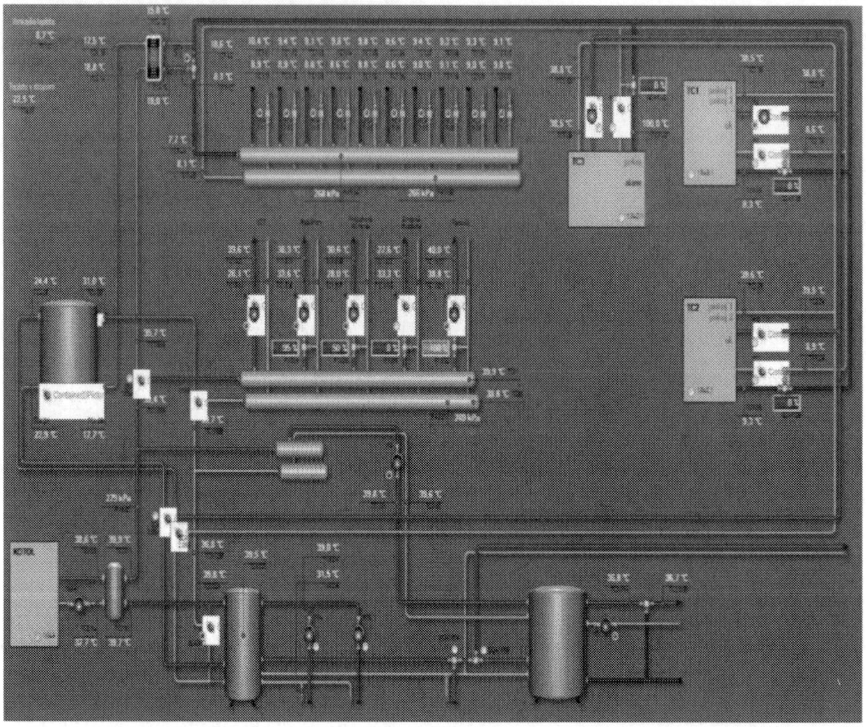

Fig. 4. Control in the technical room.

Our climatic conditions do not allow designers to use only one type of renewable energy source. The combination of several types of energy sources must be used to provide the necessary energy. Currently, the building is set to heat by electric heat pumps. In case of higher energy consumption, the gas condensing boiler is used. The solar system is primarily designed for hot water heating. In the next step, it charges the storage tank. After securing these features, it is used to charge the ground thermal storage. In this case, the heat is deposited with a volume of 770 m^3 in the rock.

Ground thermal storage has three heating wells with water. Heating is provided via the solar system. The rock is heated by the wells and the heat accumulates during the summer. Ground thermal storage has earth needles that serve to recover heat in the winter. The perimeter walls and the top wall are insulated. The bottom wall is not. The location of the thermal ground storage is on the subsoil, which can isolate itself, or losses to the environment are acceptable. If the temperature falls below 17 °C for 2 hours, the heat pumps will automatically switch on – heating mode. If the temperature is more than 20 °C 2 hours, the cooling mode switches.

The cooling ceiling system is used for cooling. The laboratories also have air conditioning units to ensure an optimal operating environment. Ventilation is controlled by air breathing sensors in every rooms.

The control of the ventilation is individual for each floor, because the first floor contains chemical laboratories and the offices are on the other floors (Figure 5).

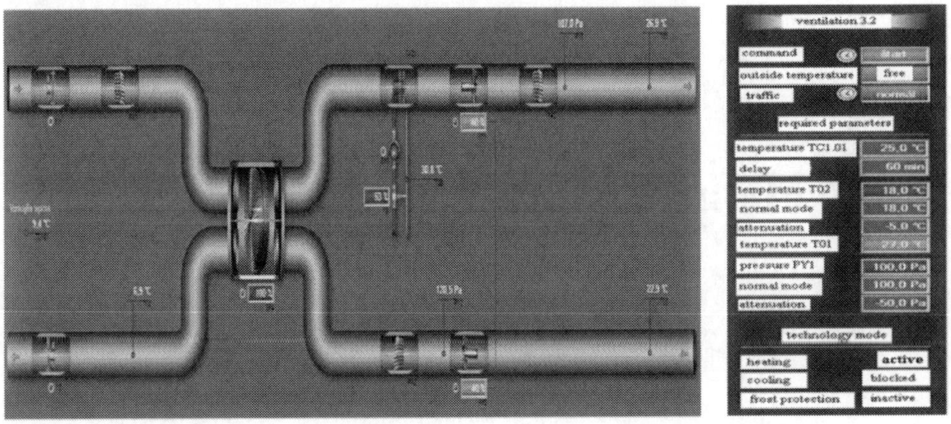

Fig. 5. Control and monitoring system for ventilation.

4 Conclusion

Our aim is to create conditions for the implementation of applied research in the field of low-energy construction and technology for the construction of low-energy buildings, high-efficiency energy sources using renewable energy sources and fossil fuels, the optimal control, optimal control of energy consumption in buildings and premises suitable choice of hot and cold sources, ensuring optimum energy consumption, safety and comfort in intelligent passive buildings, reducing the emission load environment.

Our main priority is to establish a functioning, modern system and achieving economic, environmental and proper functioning of several systems. Information on the possibility of heat accumulation in the ground thermal storage will be very interesting from an energy and economic point of view. Currently, we are working on optimizing the management program, which still needs certain programmatic corrections. We solve and eliminate problems with synchronization of individual technologies.

The research is supported by the European Structural Funds – project ITMS: 26220220121 (50%) and KEGA 046ŽU-4/2016 "Unconventional systems using renewable energy".

References

1. S. Wang, *Intelligent Buildings and Buildings Automation* 64 (2009)
2. D. Clements-Croome, *Intelligent buildings: An introduction* (2011)
3. M. Jobb, Ľ. Kosa, R. Nosek, M. Malcho, AEaNMiFMaE (2016)
4. B. Garlík, *Inteligentní budovy* (2012)
5. S. Števo, *Čo tvorí inteligenciu budov?* (2012)
6. R. Lenhard, T. Puchor, *EPJ Web of Conferences*, **143** (2017)
7. P. Nemec, M. Smitka, M. Malcho, *Scientific world journal*, **2014** (2014)
8. P. Nemec, M. Malcho, J. Jandačka, *AIP Conference Proceedings*, **1558** (2013)

Measurement and Calculation of Explosion Pressure for Hydroxyl Substituted Hydrocarbons Mixture

Jan Skřínský[1,*]

[1]Energy Research Centre, VSB-TU Ostrava, 708 33 Ostrava - Poruba, Czech Republic

Abstract. Methanol as a biofuel requires information about the explosion parameters measured in air under various conditions in different volumes. Forty explosion pressures owing to the methanol-air mixture have been measured in the temperature range 298 – 423 K between Φ = 0.65 – 2.65 in 1 m^3 explosion chamber. This work supports the systematic investigation owing to 0.005 m^3 and 0.020 m^3 volumes.

1 Introduction

Today, usually fossil fuels are used for fuel production. The reserves of these fossil-based fuels are being rapidly depleted. Besides that, when these fuels are used in internal combustion engines, they produce air pollutants such as CO, HC, NO_x and particulate matter. Alternative sources of energy are needed in order to replace the non-renewable resources and to improve air quality [1]. Fundamental combustion research can help the clean and efficient utilization of alcohols. Lower alcohols (≤C4) with high octane number and relatively low cost have been successful in the practical use as the gasoline additives. However, low alcohols still have their disadvantages such as low energy content and high hygroscopicity and corrosivity. While, higher alcohols (≥C5) have disadvantage in poor ignition [2].

It has been well established that studies concentrated on the low alcohols (ethanol, 1-butanol, 1-pentanol, 2-pentanol and 3-pentanol) in small stainless cylinder vessel with a diameter of 180 mm and length of 210 mm [3]. An experimental study of the explosive combustion of methanol-air mixtures at various initial conditions (temperatures, pressures and fuel/air ratios within the explosion limits) was performed in a small 0.005 m^3 spherical stainless steel with central ignition [4]. The most recent study determined the maximum explosion pressures, the maximum rates of pressure rise, the severity factors and the maximum explosion delay time in two spherical vessels (volume: 0.005 m^3 and 0.020 m^3) for various ethanol/air mixtures (3.5-20.0 vol%) at various initial pressures (10.1 – 101 kPa) and temperatures (298 – 373 K) [5]. In the present preliminary study, explosion severity indices of the methanol, CH_4O, have been measured in the temperature range 298 – 423 K and over stoichiometric concentrations 0.65 – 2.65, in 1 m^3 chamber, respectively.

* Corresponding author: jan.skrinsky@vsb.cz

2 Experiment

Two experimental setups have been used in the present study. One employed 1 m^3 chamber located in Energy Research Centre, VSB-TU Ostrava to record explosion severity indices. Measurements of the ultra-high speed digital images were obtained using high-speed camera from the Faculty of safety engineering, VSB-TU Ostrava.

2.1 High-speed camera

The high-speed camera Photron FASTCAM SA-Z (speed > 1M fps; resolution > 12k fps) has been used to capture high-resolution digital images at ultra-high speeds. The high-speed camera has been installed in a chamber opening and focused on ignition source. FASTCAM Viewer tools allow image calibration and measurement of angles and distances from explosion image data. Schematic representation of the camera and the whole system is shown at Figure 1.

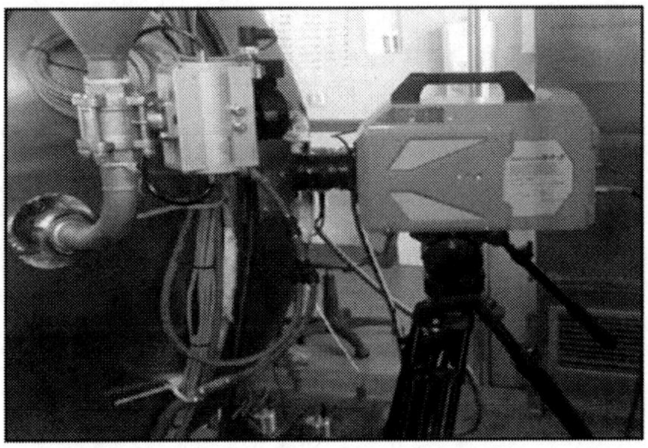

Fig. 1. High-speed camera system.

2.2 Explosion

Experiments were done in 1 m^3 chamber made by OZM Research in accordance with EN 1839:2013 and EN 15967:2012standards [6, 7]. The chamber is a vessel of a spherical shape with an inside diameter of 1 240 mm. The vessel is provided with an opening of an inside diameter of 800 mm, which is provided with a flange and a lock for locking the door. The material and components comply with the use of equipment for experiments at atmospheric initial pressure and initial laboratory temperature (15 – 200 °C). There is one liquid dispersion system installed at the chamber. The system consists of a liquid reservoir with a capacity of 0.3 dm^3. Reservoir is designed for operating pressure up to 0.5 MPa and testing pressure 10 MPa. Length of the reservoir to its diameter is 3 to 1. The reservoir is equipped with a lid, which allows for pouring the sample liquid by syringe. Special dispersion nozzle and homogenisation were used for all experiments. The reservoir is equipped by pneumatic valve for pressurization and by remote pressure sensor for control of dispersing pressure. Reservoir´s installation contains also manual ball valve for manual release of overpressure from the reservoir in case of system malfunction. The explosion chamber is equipped with measurement of time depended dynamic pressure using two pressure sensors by Kistler type 701A calibrated for the range up to 2.5 MPa. Natural frequency of the sensor is 70 kHz. Data are recorded be four-channel data card with a sampling rate 50 kS/s/channel. PLC Siemens Simatic 1215 connected to PC used as a user interface automatically control whole procedure including fast acting valve timing.

Electric discharge was used as ignition source [8]. The whole system is schematically introduced at Figure 2.

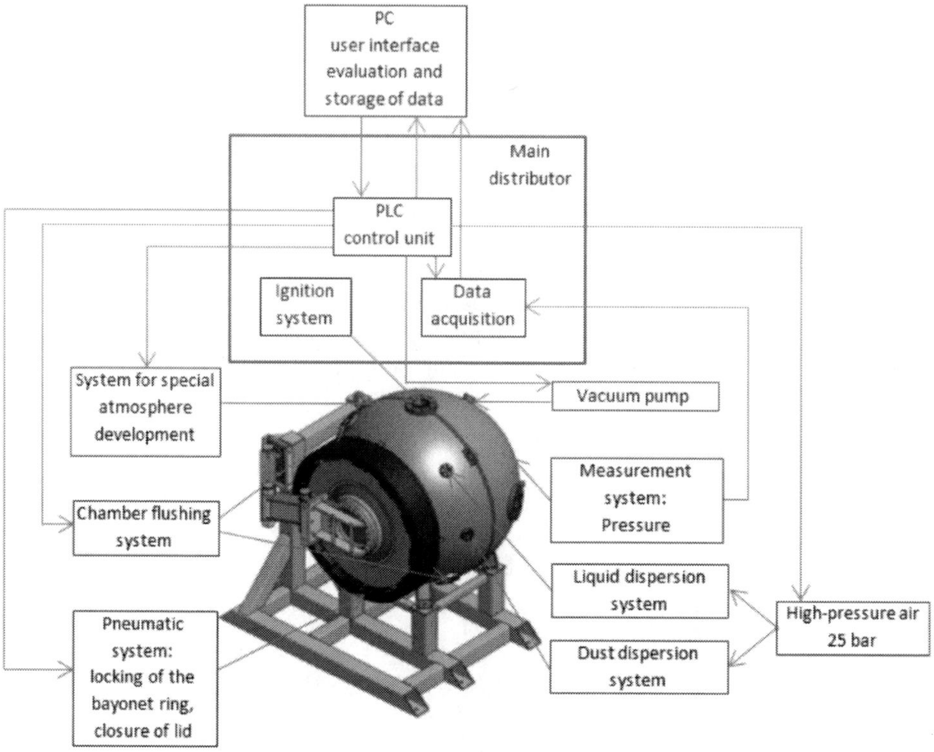

Fig. 2. Experimental set-up for liquid-air explosion parameters measurement.

2.3 Procedure

Procedure for vapour explosion starts with evacuation to pressure less than 0.89 bar to leave a space for the vapour. For example, the use of methanol concentrations of 5 vol. % and 10 vol. % require the partial pressures of methane to be added are 0.05 bar and 0.10 bar respectively. Then the evacuation pressures should be 0.84 bar and 0.79 bar respectively. PLC starts the experiment where 600 ms is counted as ignition delay time, and then the gas is mixed by blowing dispersion air into the methanol-air mixture inside the chamber. After that the mixture is ignited by the electric discharge. The explosion indices are measured and calculated [8].

3 Predictions

Method used for calculating equilibrium compositions at a specified temperature based on the minimisation of free energy technique (NASA method). The detail description of the equations and computer program for computations involving chemical equilibria in complex systems is given in [9]. These calculations provide us with the predictions for experiments at a defined temperature and constant volume for assigned thermodynamic states. The calculations considers condensed species as well as gaseous species. This approach represents ideal deflagrations in closed systems and gives the highest possible attainable explosion pressures. The present numerical method allows the determination of the most conservative values of adiabatic pressures and were used as suitable for prediction of measured experimental data.

4 Results

4.1 High-speed images

Figure 3 illustrates the real non-adiabatic explosion. Figure 3 illustrates the nearly laminar spherical flame that have been produced during the explosion. Beyond in hundreds of ms the spherical flame became turbulent and the flame speed increases monotonically with increasing turbulences and fuel air mixing.

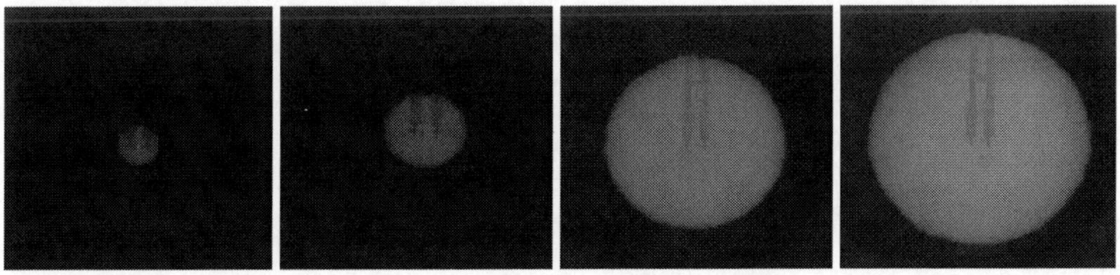

Fig. 3. High-speed images of methanol -air spherical flames ($T_0 = 298$ K, $\Phi = 1.15$, $P_0 = 0.1$ MPa).

4.2 Explosion indices

The results are very similar to the our recent gas-air investigation published in [10, 11] with exception in 298 K. Figure 4 is a plot of the normalized peak explosion pressure, P_{ex}/P_0, and adiabatic explosion pressure, Pad, versus equivalence ratio at atmospheric pressure and four initial temperatures. The P_{ex}/P_0 explosion pressure has the maximum value of explosion pressure, at $\Phi = 1.15$ as indicated in Figure 4. With the increase of temperature, the peak explosion pressure is decreased due to the decrease of the total fuel mass in gas/vapour cloud (turbulences) and shifted to the fuel rich region. This is because the species in flames at elevated temperatures are more reactive and taking into account the effect of density (point concentrations) and soot's formation combined with the wall effect. The adiabatic explosion pressure, P_{ad}/P_0, gives the peak value around $\Phi = 1.0$.

Leaner or richer mixtures make the drop of pressure between the real and adiabatic pressure. The reason is, that the real combustion is not under the absolute adiabatic conditions. With the mixture becoming richer, more soot formed in the flame, causing larger difference between P_{ad}/P_0 and P_{ex}/P_0. This is in agreement with [3].

As shown in Figure 5, $(dP/dt)_{ex}$ and K_G give an approximate value at varied initial temperatures, indicating that $(dP/dt)_{ex}$ and K_G are sensitive to the variation of temperature. Rate of pressure rise is affected by both flame speed and heat release (turbulences). Flame speed increases monotonically with the increase of temperature, but the decrease of total fuel mass results in the decreased heat release and offsets the effect of flame speed on the explosion pressure.

The main result is the behaviour of the vapour-air mixture bellow the Antoine's limit, i.e. saturated vapour pressure. These correlations describing the relation between vapour pressure and temperature for pure components limit the evaporation rate of methanol. For methanol is saturated vapour pressure at 298 K equal to 0.167837 bar [12]. It is not possible to evaporate more methanol under the given temperature and pressure conditions. Therefore, the P_{ex}/P_0, became constant. To overcome the 16.0 vol. % limit we have to increase the temperature up to 338 K at which is the saturated vapour pressure equal to 1.02432 bar. Both, 298 K (over 16.0 vol. %) and 323 K (over 55.0 vol. %) real P_{ex}/P_0, $(dP/dt)_{ex}$ and K_G values are affected by this limit.

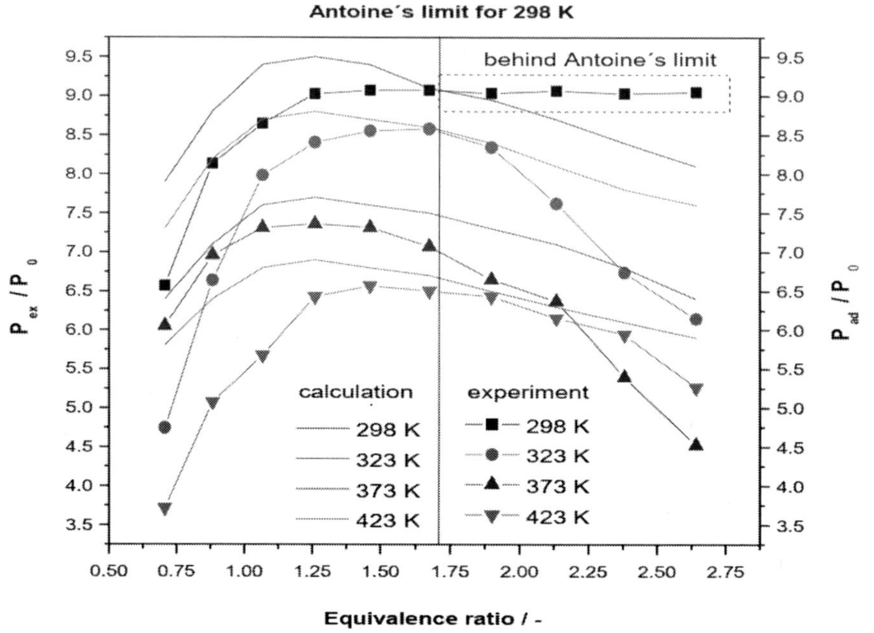

Fig. 4. P_{ex}/P_0 and P_{ad}/P_0 at $\Phi = 0.65 - 2.65$, $P_0 = 0.1$ MPa and $T_0 = 298 - 423$ K for methanol-air mixtures.

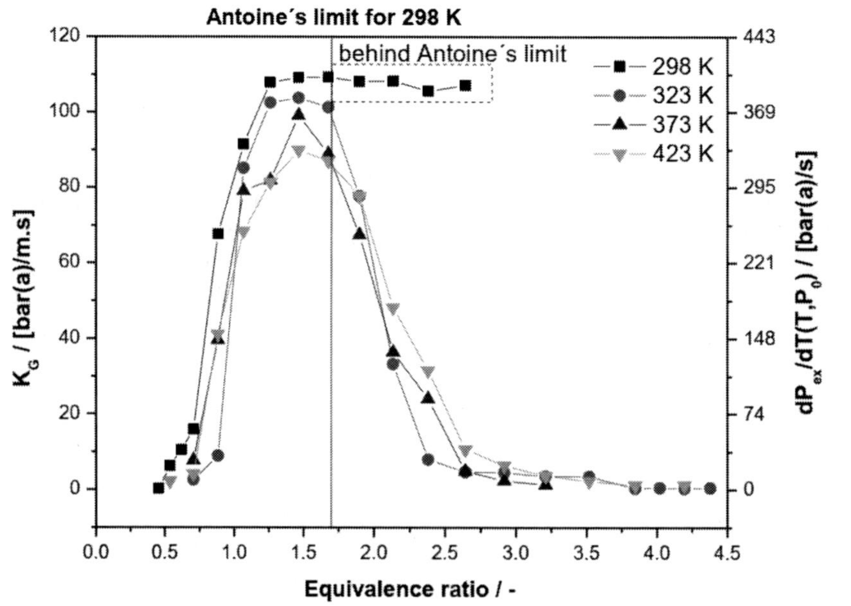

Fig. 5. (dP/dt)ex and K_G at $\Phi = 0.65 - 2.65$, $P_0 = 0.1$ MPa and $T_0 = 298$-423 K for methanol-air mixtures.

5 Conclusion

Explosion indices data are essential for a quantitative risk assessment of explosion hazard associated with the use of methanol gas. The present work is the preliminary study to investigate the influence of various initial temperatures and stoichiometric concentrations on explosion characteristics of methanol/air mixtures.

The explosion pressure, the maximum rate of pressure rise and the deflagration index are the fundamental explosion characteristics. We have described, that they have decreased with the increase of initial temperature. In all cases, the turbulent gas flow and point concentrations play crucial role in the promotion of the whole reactions due to fuel air mixing and hence the faster flame speed, and heat release.

The effect of saturated vapour pressure has been characterized in the gas phase and has been applied for methanol vapour in thermodynamic equilibrium with its condensed phases for given temperature in a closed system.

This work started a systematic comparison among the $C_nH_{2n+2}O$ analogues (where n=1-5) carried out in heated 1 m^3 explosion apparatus built by OZM Research s.r.o. at Energy Research Centre, VŠB - Technical University of Ostrava.

This work was prepared within the project „Innovation for Efficiency and Environment – Growth", identification code LO1403 with the financial support from the Ministry of Education, Youth and Sports in the framework of the National Sustainability Programme I. J.S. is thanked P. Lepik for providing the high-speed camera and B. Janovsky for helpful discussions about Antoine's.

References

1. M.B. Çelik, B. Özdalyan, F. Alkan, Fuel, **90**, 1591-1598 (2011)
2. J. Skrinsky, J. Veres, V. Peer, P. Friedel, AIP Conference Proceedings, **1745**, art. no. 020057 (2016)
3. Q. Li, Y. Cheng, Z. Huang, Journal of Loss Prevention in the Process Industries, **37**, 91-100 (2015)
4. M. Mittu, E. Brandes, Fuel, **158**, 217-223 (2015)
5. M. Mittu, E. Brandes, Fuel, **203**, 460-468 (2017)
6. EN 1839:2013. Determination of explosion limits of gases and vapors. Berlin: Beuth Verlag (2013)
7. EN 15967:2012. Determination of maximum explosion pressure and the maximum rate of pressure rise of gases and vapours. Berlin: Beuth Verlag (2012)
8. B. Janovsky, J. Skrinsky, J. Cupak, J. Veres, Draft-paper in 12th International Symposium on Hazards, Prevention and Mitigation of Industrial Explosions Kansas City, USA - August 12-17 (2018)
9. S. Gordon, B.J. McBride, Computer program for calculation of complex chemical equilibrium compositions and applications, NASA Reference Publication 1311
10. J. Skrinsky, WSEAS Transactions on Environment and Development, **13**, 421-430 (2017)
11. J. Skrinsky, J. Veres, J. Travnickova, Materials Science Forum, **844**, 65-72 (2016)
12. DDBST Dortmund Data Bank Software & Separation Technology GmbH, DDB Explorer Version (2015)

Introductory measurements on the physical model of the VVER 440 nuclear reactor II. fuel cell assembly

Zdenko Závodný[1], *Peter* Mlynár[1], *František* Urban[1], *Ján* Hollý[1], *Zoltán* Fuszko[1], and *František* Világi[1]

[1]Slovak University of Technology in Bratislava, Faculty of Mechanical Engineering, Institute of Energy Machinery, Nám. slobody 17, 812 31 Bratislava, Slovakia

Abstract. Qualitative and quantitative analysis of the relationship between the coolant temperature in the fuel cell assembly outlet and the mean coolant temperature profile in the thermocouple plane is required for safe and effective loading of nuclear fuel cells. Physical model of the VVER 440 nuclear reactor fuel cell assembly serves to analyze the influence of the coolant mass flow on the coolant velocity and temperature profiles at the plane of the thermocouple position in the fuel cell assembly.

1 Introduction

During the operation of the VVER 440 pressurized water reactor, uneven development of the heat in the assemblies occurs because of the burn-out and the placement of the individual fuel cell assemblies in the active zone of reactor. Heat generation is also uneven in the fuel rods, therefore an uneven temperature profile of the coolant is present at the fuel rods outlet. The coolant flows further through the alignment grid. In addition to the change in the shape of the cross-sectional area from the hexagonal to the circular, the size of the flow area also varies at the outlet from the assembly. The temperature profile of the coolant flow is equalized. Further mixing of the coolant takes place during its flow in the inlet of the protective tube block. The coolant temperature at the fuel cell assembly outlet is measured by a thermocouple which is located in the fuel cell assembly axis about 300 mm from the end of the fuel rods. Qualitative and quantitative analysis of the relationship between the coolant temperature in the fuel cell assembly outlet and the mean coolant temperature profile in the thermocouple plane is required for safe and effective loading of nuclear fuel cells [1].

2 Physical models of VVER 440 fuel cell assembly

In the laboratory of the Institute of Energy Machinery of the Faculty of Mechanical Engineering STU in Bratislava are the experimental models of I. and II. fuel cell assembly of VVER 440 nuclear power plant. The dimensions of the available production drawings

* Corresponding author: peter.mlynar@stuba.sk

and documentation of the reactor V 213 of the nuclear power plant with the pressurized water reactor VVER 440 were used to create the fuel cell assembly models. The physical models are used to analyze the water flow in the outlet portion of the fuel cell assembly of the nuclear reactor.

2.1 Physical model of VVER 440 I. fuel cell assembly

The fuel cartridges are replaced with 8 mm diameter aluminium and copper pipes in the experimental models. The diminished physical model of the real object – fuel cell assembly of reactor V 213 of nuclear power plant VVER 440 – was created in scale 1:1.125. In this case we are talking about the geometric similarity between the model and the real object. The heated water in the fuel cell assembly flows around the fuel cartridges. The fuel cartridges in the physical fuel cell model are formed of 117 aluminium tubes with closed sides alongside with 3 triple copper tubes. With these triple copper tubes, cool water can be fed to the model. In the plane of the end of fuel cartridges the mixing of heated and cold water takes place. In this way, the temperature field is modeled at the outlet of the fuel cartridges.

Temperature and velocity profiles are measured with 2 combined probes located in 3 planes at the outlet of the fuel cell assembly physical model (Figure 1):

- The plane 1 is situated between the end of the fuel rod bundle and mixing grid.
- The plane 2 is situated behind the mixing grid and the ending of the central tube at the beginning of the cylindrical head of fuel cell assembly outlet.
- Plane 3 is at the position of thermocouple of the nuclear reactor fuel cell assembly [2, 4].

Fig. 1. The outlet of I. fuel cell assembly physical model with combined probe traverse devices located in planes 1, 2 and 3.

2.2 Physical model of VVER 440 II. fuel cell assembly

Based on the analysis of the measurement result of coolant mixing in the outlet portion of the I. fuel cell assembly and the published CFD simulations [5, 6] it has been shown, that the dependence of the water flow through a centralized tube on the temperature and velocity field of the coolant in the plane 3 has to be discussed in detail. For this reason, an experimental device with the II. fuel cell assembly physical model of the nuclear reactor VVER 440 was designed, manufactured and presented. The II. physical model is simplified. It consists of a circular section of the fuel cell assembly of the reactor VVER 440. The casing of the fuel cell assembly with inner diameter Ø 81.7 mm is fitted with 30 fuel cartridges Ø 9.53 mm, 6 fuel cartridges Ø 6 mm at the casing itself, and with a Ø 11 mm central tube (Figure 2). The fuel cartridges of 2 770 mm length are vertically positioned using 11 simplified spacing grids. The central tube is closed from the bottom side. The holes in the central tube are used to hold the spacing grids and at the same time the water is pumped through them to the fuel cell assembly. The mixing grid is mounted to the upper side of the central tube (Figure 3). The dimensions of the II. fuel cell assembly physical model are scaled in 1.068 compared to the fuel cell assembly of the reactor VVER 440.

The experimental device with the II. fuel cell assembly physical model consists of a pump with frequency converter, a fuel cell assembly, an inlet and drainage pipe and a circuit for adding cold water to the water stream flowing through the fuel cell assembly model. Hot and cold water is fed to the experimental device from a public water supply network. The velocity profile in plane 0, located 2 900 mm from the pump, on the physical model inlet is measured by two Pitot probes. In planes 1, 2 and 3 are 8 combined probes located and used to measure temperatures and water flow velocities. There is a set of pressure transducers, resistive thermometers and water flow diaphragm available.

Fig. 2. Outlet of the II. fuel assembly physical model – upper end of the fuel cartridges, central tube and combined probe in plane 1.

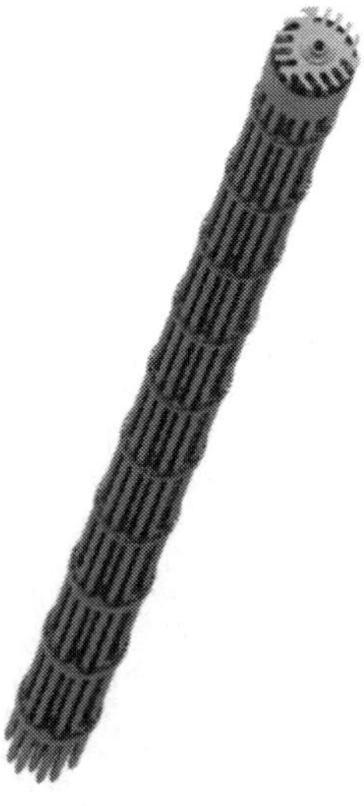

Fig. 3. Fuel cartridges, spacing grids, alignment grid and the central tube of the II. fuel cell assembly model.

3 Physical similarity of water flow in physical models of fuel cell assemblies I, II. and the fuel cell assembly VVER 440 reactor

Cooling water mass flow through the fuel cell assembly located in the VVER 440 nuclear power plant is 26.31 $kg.s^{-1}$. The coolant flow at the fuel cell assembly outlet has the temperature 300 °C. In the plane of the thermocouple location this flow can be characterized by Reynolds number Re $4{,}735x10^6$ (Table 1).

Both fuel cell assembly physical models can be operated with water temperatures up to 60 °C. This is due to the way of heating the water in the condensing boiler for the physical model of the I. fuel cell assembly and the centrally prepared hot water for the physical model of the II. fuel cell assembly. High values of the dynamic viscosity η of water flowing in the I. and II. fuel cell assembly physical models. (6.530x10^{-4} Pa.s and 4.664x10^{-4} Pa.s) reduce the value of the Reynolds number *Re* (2.979x10^5 and 3.033x10^5, Table 1.). When analysing the results of measurements on the I. and II. fuel cell assembly physical models, we focus at determining the influence of water flow rate m_{cl}, flow velocity v, dynamic viscosity η and relative roughness of the pipeline *k/d* on the Reynolds number *Re* and on the velocity and temperature profiles in plane 3 at the outlet of the assembly, at the same simulated inequality temperature profile in plane 1. In Figure 4 is a Moody diagram showing the values of the friction loss coefficient λ depending on the Reynolds number *Re* and the relative roughness *k/d* of the VVER 440 nuclear reactor fuel cell assembly (point 1), the physical model of fuel cell assembly I. (point 2) and the physical model of fuel cell assembly II. (point 3). For the strictly turbulent region, the friction loss coefficient λ depends only on the relative roughness rather than value of the Reynolds number. The lower the relative roughness, the lower the Reynolds number. The values *Re* of the Reynolds number and the friction loss coefficient λ for the real fuel cell assembly are located in the strictly turbulent region (point 1). Measurements of velocity and temperature profile in the I and II. fuel cell assembly models are carried out near the automotive flow region where the friction loss coefficient λ is close to the constant value (points 2 and 3). As a consequence of keeping the dimensional and physical similarities between the VVER 440 reactor fuel cell assembly and the fuel cell assembly models I. II. the results of measurements in the model are in accordance with the real flow of the coolant at the fuel cell assembly outlet [7].

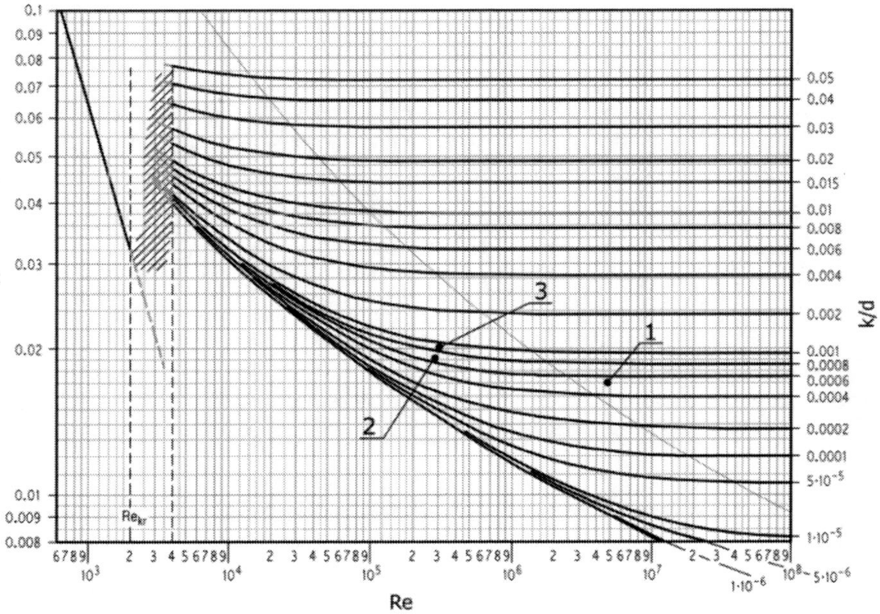

Fig. 4. Friction loss coefficient λ according to Reynolds number *Re* and relative roughness *k/d*
1 VVER 440 real fuel cell assembly,
2 I. fuel cell assembly physical model,
3 II. fuel cell assembly physical model.

Table 1. Values of the relative roughness *k/d* and Reynolds number *Re* measured in plane 3 of VVER 440 real fuel cell assembly and the I. and II. fuel cell assembly models.

Fuel cell assembly	p	t	ρ	η	d	$m\ cl$	v	Re	k/d
	(MPa)	(°C)	(kg.m^{-3})	(Pa.s)	(m)	(kg.s^{-1})	(m.s^{-1})	(-)	(-)
VVER 440	12.3	300	720.17	8.734E-05	0.081	26.31	7.09	4.735E+06	0.00056
Physical model I	0.3	40	992.31	6.530E-04	0.072	11.00	2.72	2.979E+05	0.00063
Physical model II	0.3	60	983.30	4.664E-04	0.054	6.00	2.66	3.033E+05	0.00083

4 Initial measurements on the II. fuel cell assembly physical model of the VVER 440 nuclear reactor

Experimental device with II. fuel cell assembly physical model should be designed considering the balance of velocity profile in the plane 0 at the fuel cell assembly inlet. The initial measurements are aimed at detecting the velocity profile measured by Pitot probes A0 and B0 in the range of water flow rates m_{cl} of the assembly. It is also necessary to determine the dependence of the pressure loss Δp_{kaz} of the II. fuel cell assembly physical model from the water flow rate m_{cl} measured by flow diaphragm. In the next phase of experiments, coolant mixing in the output part of the II. fuel cell assembly physical model will be analyzed. By pair of combined probes, the temperature and velocity profiles will be measured in plane 1: after the end of the fuel rods, plane 2: behind the alignment grid and plane 3: in the plane of the thermocouple placement in the VVER 440 reactor fuel cell assembly. Figure 5 is an example of the measured values on the computer during the initial measurement on the fuel cell assembly physical model of nuclear reactor.

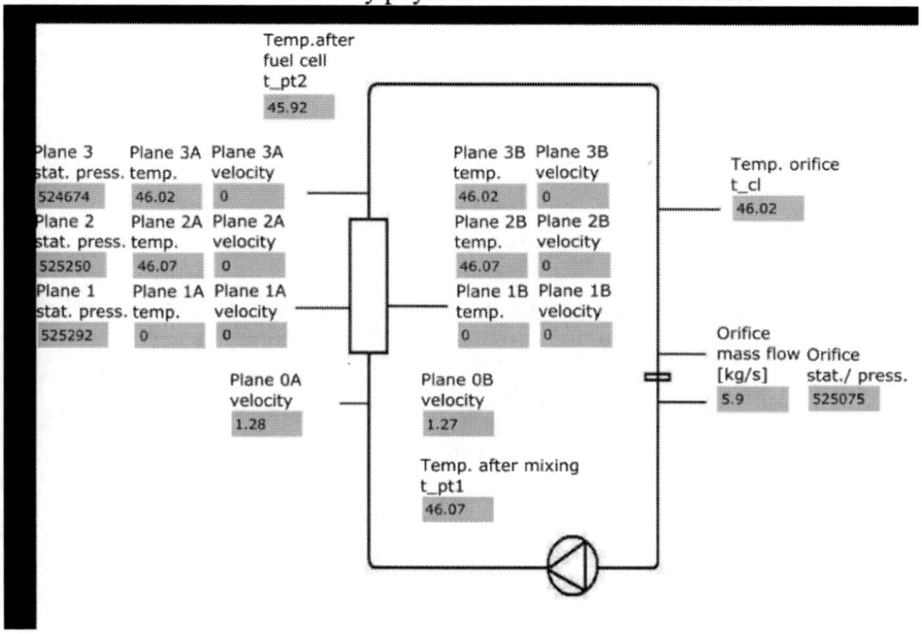

Fig. 5. Representation of the measured values on the computer screen during the initial measurement on the II. fuel cell assembly physical model of the VVER 440 nuclear reactor.

4.1 Pressure loss of the II. fuel cell assembly physical model

The pressure loss of the II. fuel cell assembly physical model is measured as the difference between two static pressures before and behind the assembly. Sampling is carried out on pipes of the same diameter Ø 82 mm. At water flow rates from 1.91 $kg.s^{-1}$ to 5.44 $kg.s^{-1}$, the pressure loss was between 10.297 kPa and 65.879 kPa (Figure 6).

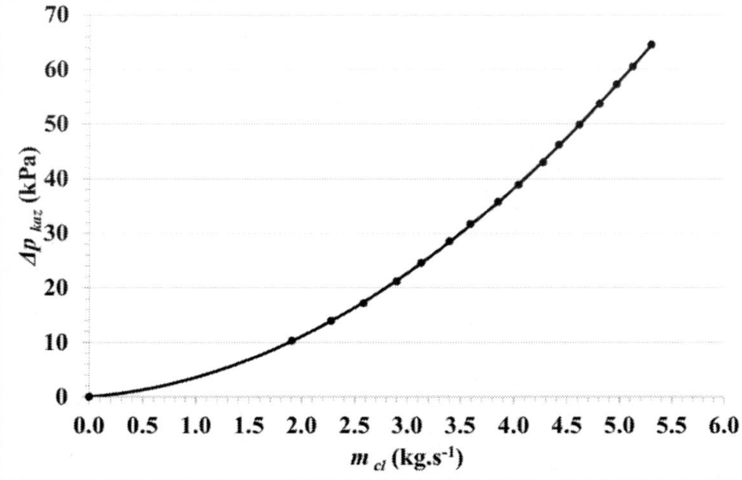

Fig. 6. Pressure loss Δp $_{kaz}$ of the II. fuel cell assembly physical model depending on the water flow rate.

4.2 Velocity profile in plane 0 at II. fuel cell assembly physical model water inlet

The velocity profile in plane 0 of inlet to the fuel cell assembly physical model is measured by two Pitot probes A0 and B0 that are installed in the traversing devices. The water flow velocity $v_{SR\,i}$, measured by probe S (S = A for probe A0, and S = B for probe B0); in the plane R (R = 0) at probe position i is calculated from the dynamic pressure $p_{d\,i}$ measured at point i and the density of water ρ_v by this equation:

$$v_{SRi} = \sqrt{\frac{2 \cdot p_{di}}{\rho_v}} \tag{1}$$

Velocity profiles of the water flow in the pipe (Φ 82 mm) at the inlet to the II. fuel cell assembly model which were measured by probes A0 and B0 are shown in Figure 7 up to Figure 10. Water flow at the inlet of the assembly can be evaluated as turbulent, axially symmetrical.
By integrating the velocity profiles measured by probes A0 and B0, the flows were determined

$$m_{traverzA0r\,\varepsilon\langle 0,|-41|\rangle},\; m_{traverzA0r\,\varepsilon\langle 0,41\rangle},\; m_{traverzB0r\,\varepsilon\langle 0,|-41|\rangle},\; m_{traverzB0r\,\varepsilon\langle 0,41\rangle}. \tag{2}$$

From these four values, an average value of the flow rate was calculated $m_{traverz}$.
Flow rate $m_{traverzA0r\,\varepsilon(0,41)}$ is calculated by equation:

$$m_{traverzA0r\,\varepsilon\langle 0,41\rangle} = \sum_{i=2}^{21} 2.\pi.\rho \int_{ri-1}^{ri} v_{A0i}(r).r.dr = \sum_{i=2}^{21} 2.\pi.\rho \int_{ri-1}^{ri} (\alpha_{v1\,i}.r + \alpha_{v0\,i}).r.dr, \tag{3}$$

points $v_{SR\,i-1}$ $(r = r_{SR\,i-1})$ and $v_{SR\,i}$ $(r = r_{SR\,i})$ was used to determine the equation of a line

$$v_{SR\,i}(r) = \alpha_{v1\,i}\cdot r + \alpha_{v0\,i} \tag{4}$$

After the adjustment we get

$$m_{traverzA0\,r\in\langle 0,41\rangle} = 2.\pi.\rho\sum_{i=2}^{21}\left[\frac{\alpha_{v1\,i}\cdot r^{3}}{3}+\frac{\alpha_{v0\,i}\cdot r^{2}}{2}\right] \tag{5}$$

In the same way, we calculate the flows

$$m_{traverzA0\,r\in\langle 0,|-41|\rangle},\ m_{traverzB0\,r\in\langle 0,|-41|\rangle},\ m_{traverzB0\,r\in\langle 0,41\rangle}.$$

Table 2 shows water flow rates $m_{traverz}$ at the inlet of the II. fuel cell assembly model, calculated by integrating the velocity profiles measured by Pitot probes A0 and B0. The relative deviation of the flow rates $(m_{cl} - m_{traverz}) / m_{cl}$ from the flow rates m_{cl} measured by the flow diaphragm was -2.21 % to -6.06 %.

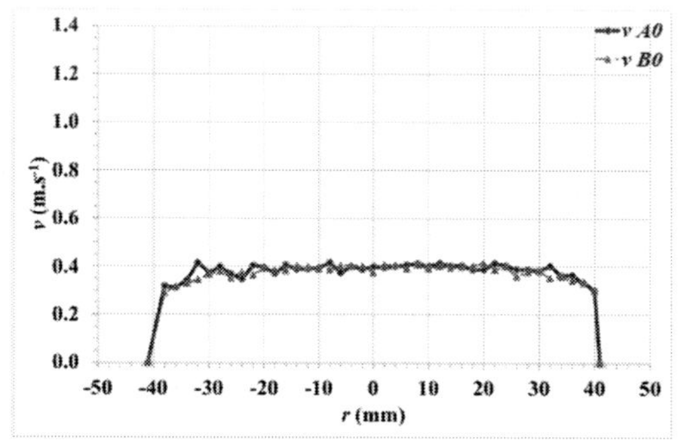

Fig. 7. Velocity profile in plane 0, water flow rate 1.75 kg.s^{-1}.

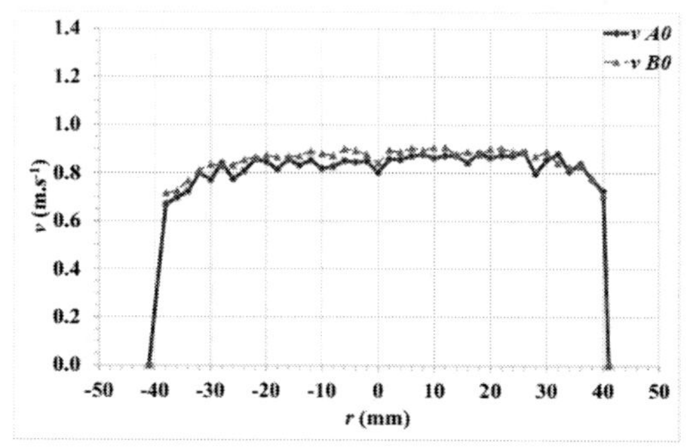

Fig. 8. Velocity profile in plane 0, water flow rate 4.03 kg.s^{-1}.

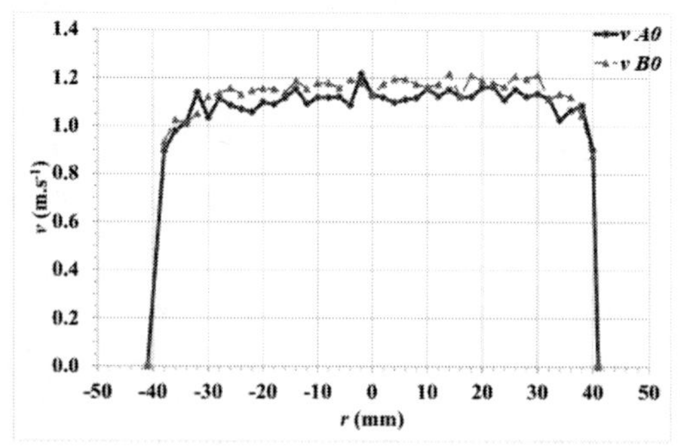

Fig. 9. Velocity profile in plane 0, water flow rate 5.23 kg.s^{-1}.

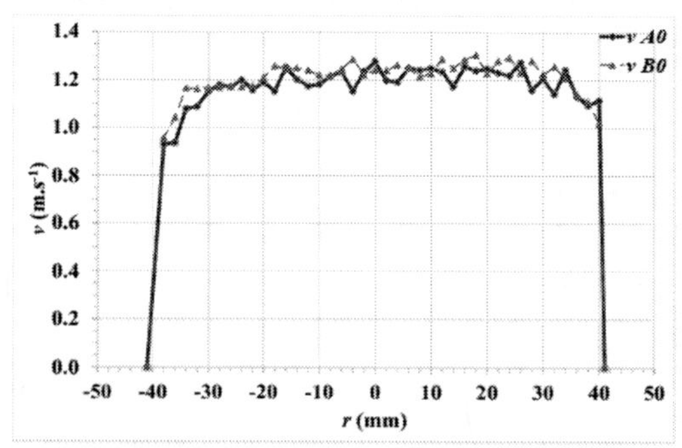

Fig. 10. Velocity profile in plane 0, water flow rate 5.74 kg.s^{-1}.

Table 2. II. fuel cell assembly physical model water flow rate comparison $m_{traverz}$ and m_{cl}.

$m_{traverz}$	(kg.s^{-1})	1.85	4.13	5.46	5.87
m_{cl}	(kg.s^{-1})	1.75	4.03	5.23	5.74
$(m_{cl} - m_{traverz}) / m_{cl}$ * 100	(%)	-6.06	-2.51	-4.22	-2.21

5 Conclusion

By measuring the velocity profile in plane 0 at the inlet of the fuel cell assembly it was confirmed, that the water flow is turbulent and axially symmetrical.
As a consequence of keeping the dimensional and physical similarities between the VVER 440 reactor fuel cell assembly and the fuel cell assembly models I. II. the results of measurements in the model are in accordance with the real flow of the coolant at the fuel cell assembly outlet.
For the future variants of the water mixing measurements at the output part of the II. physical model the CFD simulation will have to be performed to simulate the water flow.
The task will be to select a suitable method for CFD simulation to model the turbulence and generate a computational mesh with respect to measurement results on the physical model. After that the CFD calculations can be adjusted in such a manner that the measured temperature and velocity profiles in the plane 1 can be replaced by the profiles calculated for the real distribution of the heat output of the individual fuel rods in the VVER 440 nuclear reactor fuel cell assembly.

Európska únia

This contribution was created on the basis of the project "Research centre ALLEGRO" (ITMS project code: 26220220198), supported by Operational Programme Research and Development funded by the European Regional Development Fund.

References

1. L.L. Kobzar, D.A. Oleksyuk, *Experiments on Simulation of Coolant Mixing in Fuel Assembly Head and Core Exit Channel of Assembly Head (*AER on VVER Reactor Physics and Reactor Safety, 2006)
2. J. Bereznai, F. Urban, Ľ. Kučák, Engineering Mechanics, **20**, 195-204 (2013)
3. F. Urban, Ľ. Kučák, J. Bereznai, M. Pulmann, J. Tihányi, Communications **14**, 42-46 (2012)
4. F. Urban, Ľ. Kučák, J. Bereznai, Z. Závodný, P. Muškát, American Institute of Physics Publising LLC **1608** 249-252 (2014)
5. S. Tóth, A. Aszódi, *Detailed Analysis of Coolant Mixing in VVER-440 Fuel Assembly Heads* (AER on VVER Reactor Physics and Reactor Safety, 2008)
6. J. Jakubec, V. Kutiš, E. Mojto, *Study of mixing grid positioning in fuel assembly of VVER-440*, (International Scientific Conference EEE **1** 103-106 2014)
7. M. Čarnogurská, *Základy matematického modelovania v mechanike tekutín a termodynamike* (Vienala, Košice, 2000)

Research of material and power of photovoltaic panels of various types of degradation in operating conditions

Peter Tauš[1,*], *Jana* Tomčejová, *Marcela* Taušová, *Ľubomíra* Gabániová, *Dušan* Kudelas, *Matúš* Jeňo, and *Daniel* Šlosár[1]

[1]Institute of Earth Resources, Faculty of Mining, Ecology, Process Control and Geotechnologies, Technical university of Kosice, 04001, Kosice, Slovakia

Abstract. The paper deals with the results of research on the influence of material and operating parameters on the performance characteristics and efficiency of photovoltaic panels. The disadvantage of PV panels is that their performance depends on a number of factors such as the intensity of the sun's radiation, the outside air temperature, the surface temperature of the photovoltaic panel, its pollution, shielding, etc. In the paper is analyzed the model photovoltaic power plant, from which data were obtained during the monitored period and subsequently the progress of individually measured parameters was evaluated.

1 Description of the analyzed PV power station

PVE is located on an area of 30,000 m^2, 75 % of which was used to install the facility. The construction of the PV power plant was carried out in 2010 from September to November, at the end of November of the same year it was put into the testing operation. In December it acquired a license from URSO and a license to join the distribution system ZSE Distribúcia, a.s. They launched regular operation of the facility in March 2011. The PV system was designed using PC software PVsyst, weather data was used from Meteonorm software. The optimal inclination at the site was set at 35°. Additional parameters are shown in Table 1.

Table 1. Parameters of PVPP.

Total incident radiation over the year on a horizontal plane	1 360 kWh.m^{-2}
Total installed PVPP power	828 kWp
Specific amount of electricity produced	1 124 kWh.kWp^{-1}.rok^{-1}
Total estimated minimum amount of electricity produced (annually)	931 MWh
Total available land area	30 000 m^2

* Corresponding author: peter.taus@tuke.sk

1.1 Photovoltaic panels

We used 3600 polycrystalline PV panels from the REC Solar manufacturer, specifically the REC 230AE type with a maximum power of 230 Wp, which the company acquired directly from the manufacturer. It provides the quarantee of 90 % of power production for photovoltaic panels after 10 years and 80 % production after 25 years. The relative spacing of the individual rows with the PV panels is 8.61 m with fixation on fixed structures of the MSP-AL type with orientation 0° to the south [3, 4].

1.2 Other PVE components

Three Vacon (RS-IYZ-0460-3-250 kWp) inverters are installed to connect to the distribution network and are located in the LN substation next to the transformer station. In connection with the construction of a photovoltaic power plant, it was necessary to build a new 22 kV cable connection and a transformer station for power output from PVP. To supply electricity to the ZSE, a. s. a 1 000 kVA transformer was used. The measurement of the consumed and supplied energy is carried out in the switchgear of the transformer station.

2 Analysis of the operating parameters of the PV power station

At the PV power plant measurements of selected quantities related to the power and material properties of PV panels and cells were realized. These are the following parameters:

• power of global radiation P_{GZ} (W), operating temperature of panel T_{PVP} (°C) air temperature T_{vzd} (°C), produced electricity EP_{PVPP} (kWh), power of P_{PVP} PV panels (kW) and efficiency of panels f_{PPP} (%).

The data gathered for the years 2012 to 2015 with blackouts in 2014 due to a power plant failure. Blackouts are included as unavailable data in statistical analyzes, or as extremes - by data type. For the statistical evaluation, the 12th hour data was used with the highest average global radiation value during the year. The measured values of the inverter power outputs and the theoretical efficiency of the PV panels were calculated.

2.1 The course of operating parameters for the period under review

The average monthly maximum of 0.91 kW.m^{-2} was reached for the reference period in April 2015, the average monthly minimum was 0.22 kW.m^{-2} was in January 2012. The following graph (Figure 1) presents a comparison of the average monthly GZ in of the years examined.

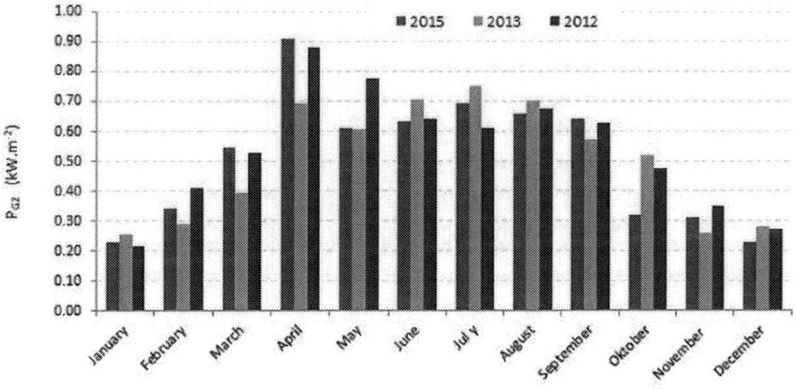

Fig. 1. Comparison of P_{GZ} in individual years.

The highest temperature of the entire monitored period of 35.8 °C was reached on 01.07.2012 and the lowest temperature was -4.92 °C on 13.01.2013. The following graph (Figure 2) presents a comparison of the average monthly temperatures of the PV panels in the individual years under review.

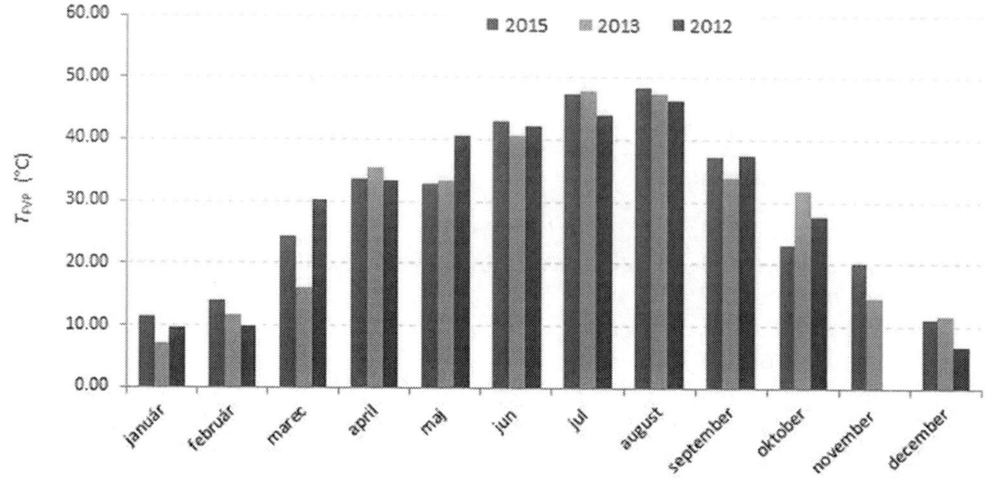

Fig. 2. Comparison of the average monthly temperature of PV panels in individual years.

The highest average monthly P_{PVP} is in the range of 624 to 403 kW between April and September. The lowest average monthly P_{PVP} was recorded from 259 to 159 kW between November and January (PVE was out of service in November 2012). The average monthly maximum of 624 kW was reached for the reference period in April 2015, with an average monthly minimum of 164 kW in January 2012. Figure 3 shows the comparison of the average monthly PV power outputs in the years under review.

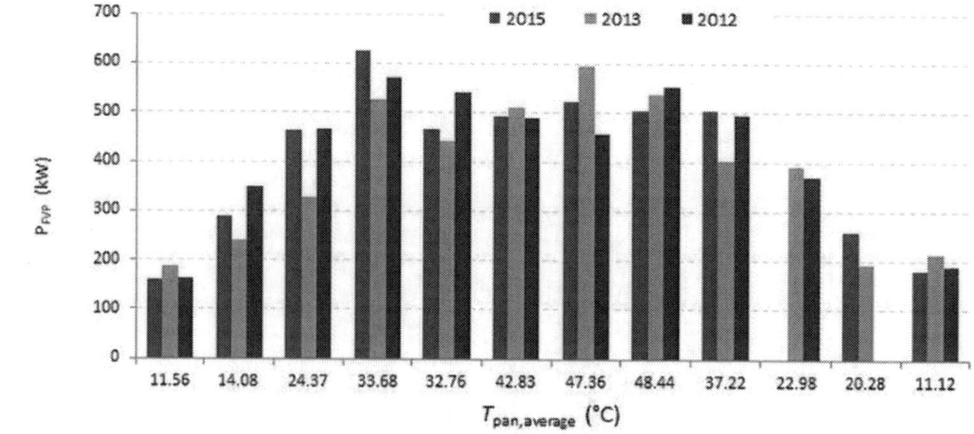

Fig. 3. Comparison of average monthly PV power over the reference period.

2.2 Evaluation of the interdependencies of the operating parameters of the PV power plant

The individual dependencies were evaluated on the analyzed PVP for a given reference period using correlation and regression analyzes of the measured values.

2.3 Dependence of the amount of energy produced on the output of global radiation

Figure 4 shows the relationship between the values of the PV energy produced and the power of the incident GR. As can be seen, there is a high direct correlation between these variables ($r = 0.8734$). The power variation of the incident GR(GR=GZ) explains the variability of the produced PV energy from 76.21 % ($r^2 = 0.7621$).

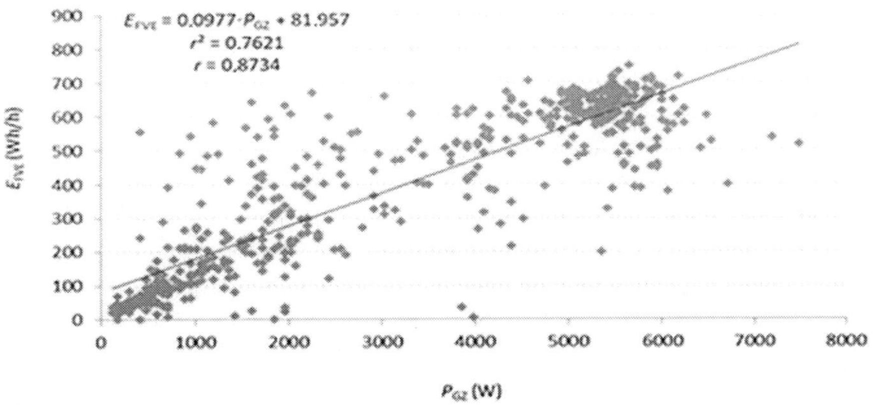

Fig. 4. Dependence of the generated PV power on GZ power over the period under review.

As can be seen from the analysis, the dependence of the electricity production from the PV panels on the incident GZ is confirmed. However, a considerable scatter of dependence is confirmed, which indirectly affirms the influence of temperature on the efficiency of PV panels. Most of the deviations from the average value are due to the temperature of the PV panels associated with the air temperature. Spatial values above the average line at low GZ values are recorded in the winter period, while values below the line GR at high GR show PV panels with elevated temperature during the summer period.

It follows from the above that, in the correct configuration of the PV field, we can maximize the power plant production for the different conditions of use or location of the installation.

2.4 Air temperature dependence on global radiation

As can be seen in Figure 5, there is a direct correlation relationship between the T_{vzd} air temperature and the daily totals of the GZ-incidence but significantly lower than in the previous case ($r = 0.5674$). The scattering of the correlation field points suggests that T_{vzd} is also dependent on atmospheric conditions such as wind velocity or hydrometeorologal conditions, in addition to incident GZ radiation, over the year. The variability of incident radiation GZ explains the variability of T_{vzd} from 32.19 % ($r^2 = 0.3219$). The confidence value is $r^2 = 0.3219$, that is, the 32.19 % increase in the operating air temperature is dependent on the increase in the power of the incident GZ.

This is, of course, the one-year period to consider when planning a PV plant or plant. Especially for seasonal PV plants that are currently on the rise.

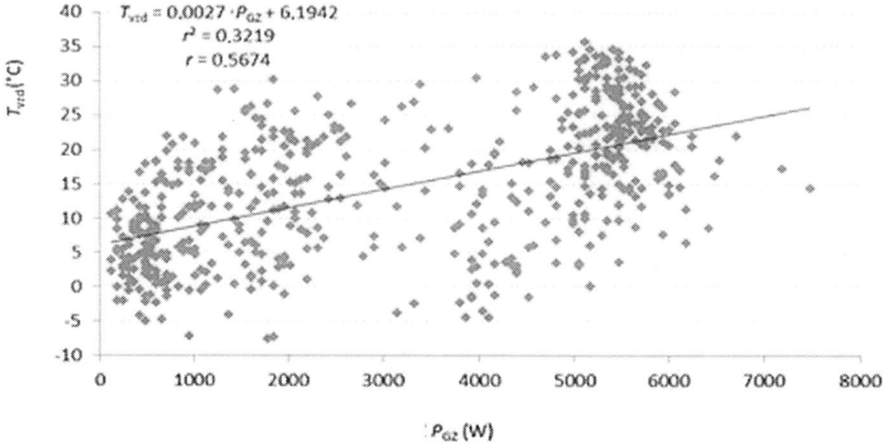

Fig. 5. Air temperature dependence on GZ performance over the period under review.

2.5 Temperature dependence of photovoltaic panels on the output of incident global radiation

There is a relatively high correlation between the temperature of the T_{PVP} photovoltaic panels and the output of the GZ incident, as evidenced by r = 0.7750. The variability of incident radiation GZ explains the variability of T_{PVP} from 60.06 % ($r^2 = 0.6006$).

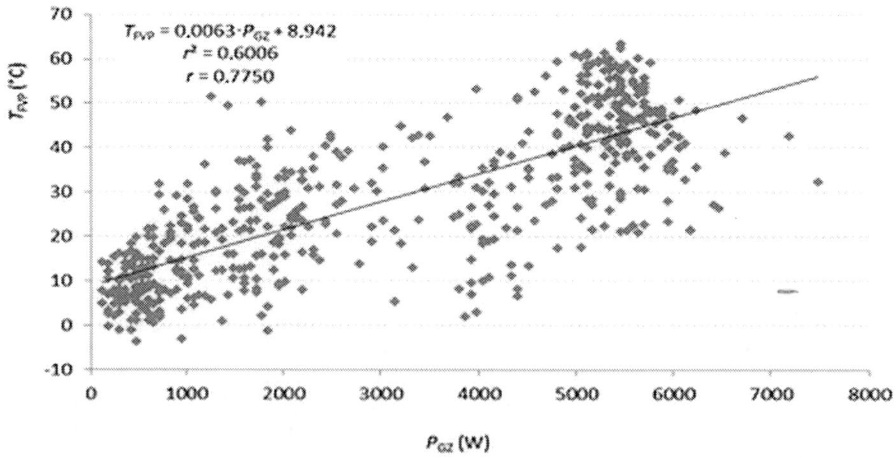

Fig. 6. The dependence of the PV panel temperature on the GZ output over the period under review.

All of the aforementioned dependencies show a relatively wide dispersion despite confirmed correlation dependence. This confirms the influence of the weather, respectively. Of the annual period to the power parameters of the PV panels.

2.6 Temperature dependence of photovoltaic panels on air temperature

As can be seen in Figure 7, there is a high direct correlation dependence (r = 0.9185) between the T_{vzd} air operating temperature and the temperature of the T_{PVP} panel. The scattering of the correlation field points suggests that T_{PVP} is also dependent on other atmospheric conditions such as GZ intensity and wind velocity. The T_{vzd} air temperature variability explains the T_{PVP} variability from 84.37 % ($r^2 = 0.8437$).

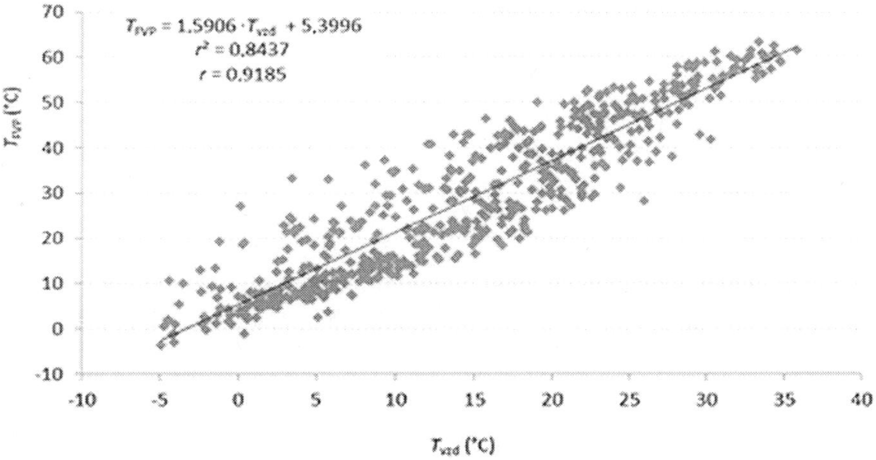

Fig. 7. Dependence of the PV panel temperature on air temperature over the period under review.

Unlike Figure 4, the FV panel temperature dependence on air temperature is more pronounced, which implies inter alia that there are no faulty PV panels in the power plant. Furthermore, this dependence is confirmed by the fact that neither the influence of GZ nor the temperature difference of PV panels and ambient air is affected.

2.7 Dependence of power of PV panels on performance of GZ

Graphic dependence on Figure 8 shows the relationship between PV panel power outputs and GZ output power.

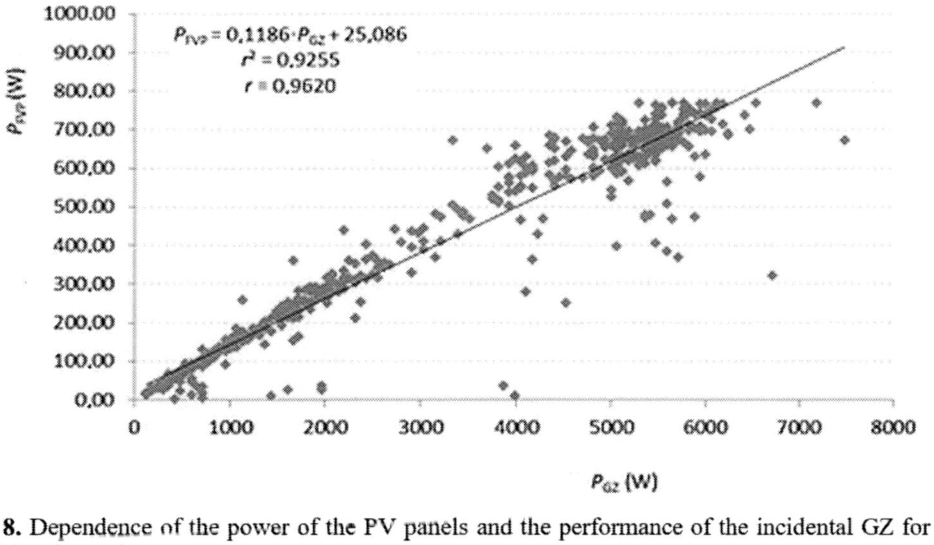

Fig. 8. Dependence of the power of the PV panels and the performance of the incidental GZ for the period under review.

From the correlation coefficient value, there is a relatively high correlation coefficient $r = 0.9620$ between these parameters. The confidence value is $r^2 = 0.9255$, that is, 92.55 % of the

PV power output is dependent on the increase in the power of the incident GZ. Similarly, this is also true for the next analysis, where the PV power performance of the PV panels is clearly confirmed. For most of the extremes shown in the graph in Figure 8 occurred during the summer period, which confirms material wear / fatigue due to increased temperature.

3 Conclusion

Many photovoltaic power producers often do not address the type of PV panels that are more advantageous for them in terms of quality, as well as the realization of their future power generation. The primary criterion for decision-making at the time of the largest PV boom was clearly the price of these panels, as there were large volumes of supply. In addressing this issue, the FVE investor has leaned to acquire quality panels from REC SOLAR. This manufacturer of PV Panels has a lot to do with the quality of its products from both material and power point of view. From the point of view of lifespan, as evidenced by the analysis of the measured data in PVP. Within the measurements, the dependencies between the temperature of the PV panels, the ambient air temperature, the efficiency of electricity generation and the incident solar radiation were evaluated by correlation analysis.

The results of the research show that the material used in the manufacture of the studied PV panels does not show the expected degradation. Positive is that the degradation of the PV panels is significantly slower than expected. Also, the direct dependence of the influence of temperature on the efficiency of PV panels was proven to be lower than the theoretical. This is a positive phenomenon from the point of view of the plant operator, because the efficiency of the panels, depending on their temperature, has fallen more slowly than the value mentioned. The results of the analyzes showed that in the winter the temperature influence on the efficiency of the PV panel is not relevant. The average efficiency of PV panels may be up to 1 % higher than the efficiency declared by the manufacturer during this period. On the contrary, in the summer, the average efficiency is 1.5 % lower than the declared efficiency.

This research has proven that RECOLAR PV panels of the REC230AE type are an appropriate choice for installation in our climatic conditions in terms of their tolerance to temperature and optimum performance. The manufacturer shall declare a temperature coefficient of 0.0046 $°C^{-1}$ for these panels. The calculated value for the reference period is about 0.0026 $°C^{-1}$ lower.

As a result, these panels are also designed for areas with higher hot summer days. As well as in terms of temperature, it is important to choose the correct type of PV panel also from the viewpoint of other climatic conditions, relative air humidity, pollution factor, etc. After careful consideration of these factors, it is possible to choose the correct PV panel or a situation where the power and material degradation of this component is minimized (taking into account the correct techniques and installation procedures). The recommendation for the photovoltaic electricity producer is to consider the possibilities and choose the right PV panels with the relevant characteristics for the given climatic conditions and location based on existing analyzes in similar locations.

References

1. J.Tomčejová, *Výskum materiálovej a výkonovej degradácie fotovoltaických panelov rôznych typov v prevádzkových podmienkach*, Písomná časť dizertačnej skúšky. s. 26 (2013)

2. J. Tomčejová, *Analýza vplyvu zmeny vstupných podmienok FVE na jej prevádzkové parametre na základe dopytu prevádzkovateľa*, Diplomová práca. s. 99 (2011)
3. P. Inn Inneichen, Th. Lejeune, PC software PVsyst Version 5.4. University of Geneva. (2009)
4. E. Skoplaki, J.A. Palyvos, On the temperature dependence of photovoltaic module electrical performance: A review of efficiency/power correlations. *Solar Energy*, **83**, s. 614-624 (2009)
5. V. Leitman, Identifikace a detekce poruch FV system. [online]. Bakalárska práca. VUT v Brně. [cit. 2018-02-18]. Dostupné na internete: <https://www.vutbr.cz/www_base/zav_prace_soubor_verejne.php?file_id=84404> (2013)
6. J. Hofierka, J. Kaňuk, Assessment of photovoltaic potential in urban areas using open source solar radiation tools. *Renewable Energy*. **34**, s. 2206-2214 (2009)
7. J. Hofierka, M. Šúri, The solar radiation model for Open source GIS: Implementation and applications. *Proceedings of the Open source GIS - GRASS users conference 2002*. 19 s. (2002)
8. M. Libra, V. Poulek, Fotovoltaika: teorie i praxe využití solární energie. 165 s. (2010)
9. M. Malcho, J. Jandačka, R. Lenhard, S. Gavlas, T. Ochodek, *Communications : scientific letters of the University of Žilina*, **18**, s. 115-121 (2016)
10. A. Klenovčanová, T. Brestovič, Možnosti využitia fotovoltických článkov na výrobu elektrickej energie v oblasti Košíc. *Acta Mechanica Slovaca*, **4** s. 511-516 (2007)
11. P. Bača, Posibilities of electric power storage from renewable sources. *Acta Montanistica Slovaca*, **15**, s. 100-104 (2011)

Boiling heat transfer augmentation on surfaces covered with phosphor bronze meshes

Lidia Dąbek[1], *Andrej* Kapjor[2], and *Łukasz J.* Orman[1,*]

[1]Kielce University of Technology, al. 1000-lecia P.P.7, Kielce, Poland
[2]University of Žilina, Faculty of Mechanical Engineering, Univerzitna 1, 010 26 Žilina, Slovakia

Abstract. The paper discusses the issue of boiling heat transfer augmentation on phosphor bronze wire mesh coatings during nucleate boiling heat transfer for distilled water and ethyl alcohol under ambient pressure. A significant enhancement of heat flux has been recorded for such treated surfaces in comparison to the smooth reference surface. The obtained results have been discussed and compared with models and correlations available in literature.

1 Introduction

Boiling heat transfer is highly efficient in dissipating significant heat fluxes at small temperature differences. Considerable improvement in the value of heat fluxes can be achieved by the application of heat enhancing microstructures on heat exchangers such as metal wire meshes, capillary – porous structures, sintered powders and other specially treated surfaces.

Poniewski [1] provided an overview of microstructures used for boiling heat transfer augmentation. According to the author, heat flux enhancement is caused by such a change in the heater structure that increases a number of active nucleation sites and ensures their stabilised operation. Boiling heat transfer enhancing structures are, among others, industrial microstructures, sintered or thermally sprayed powder structures as well as metal fibrous, wire mesh and combines structures [1].

Highly efficient are wire metal structures, which are the focus of the presented paper. They are cheap to produce and easily available on the market due to their other application in engineering. For example, they can be used for the production of heat pipes [2].

Sasin et al. [3] wrote one of the first papers on boiling heat transfer on mesh coated heaters. Water, ethanol and ether boiling was investigated on heat pipes of 300 mm and 470 mm length. It was found that at a given number of mesh layers there is an optimal aperture for which the maximal value of the critical heat flux can be obtained. Smirnov et al. [4] considered boiling of water and ethanol under different saturation pressures within the range of 0.1 bar to 1 bar. Copper and brass meshes were used. Generally, heat flux was independent of the structure height (namely, the number of layers) and weakly dependent on the structural material. Rannenberg and Beer [5] presented results of boiling of refrigerants R-11 and R-113 at ambient pressure on a horizontal copper

* Corresponding author: orman@tu.kielce.pl

surface covered with 2 – 9 layers of stainless steel and bronze meshes. Basing on the test results no impact of the structure height (the number of mesh layers) was concluded.

Tolubinskij et al. [6] experimented on water boiling under pressures ranging from 0.02 – 0.5 MPa on a surface covered with a stainless steel mesh of different aperture (0,04 – 3 mm). The coating was mechanically attached to the heater. According to the authors, heat transfer on single mesh layers is influenced by the value of the mesh aperture in relation to the bubble departure diameter. The paper by Tsay et al. [7] gave the test results of water boiling on a horizontal smooth and rough surface covered with a single stainless steel mesh layer. Basing on the visualisation studies it was concluded that the number of bubbles grown on the surface with the coating was higher than on the smooth surface.

Brausch and Kew [8] described the heat transfer phenomena in microstructures basing on the test results of water boiling on a heater surface covered with 1, 3 and 5 layers of mesh. It was reported that a single layer enhanced boiling heat transfer in comparison to the smooth surface at superheats up to 15 K, which the authors explained by the creation of additional nucleation sites. The visualisation studies proved an increased rate of bubble production at superheats up to 10 K on surfaces with a single layer. At higher superheats the mesh made the removal of bubbles from the structure more difficult.

Orzechowski [9] presented results of boiling heat transfer of water on a heater covered with 1, 2 and 3 copper mesh layers of wire diameter 0.18 mm and aperture 0.55 mm. Basing on the test results it could be concluded that the application of the coating enhances heat transfer in comparison with the smooth surface. Li et al. [10] listed four reasons of heat transfer enhancement in the porous layers, among others: the reduction of heat flux on the heater through additional surface extension and the presence of contact points which join the microstructure with the heater; this effectively reduces the process of vapour film forming on the surface. Liou et al. [11] investigated multi-layer copper meshes sintered to the heat pipe surface. Different combinations of structures of the mesh were used to form coatings of the height 0.26 mm – 0.8 mm. The authors concluded that with rising heat load, water film receded to form corrugated menisci in the analysed layer and evaporation resistance decreased till partial dry out occurred.

Since boiling on enhanced surfaces offers considerable advantages for industrial applications (for example in refrigeration or air conditioning systems, high performance cooling devices, etc.), more information is needed in terms of both experimental database and physical phenomena occurring during boiling of different working agents on various microstructure coated heaters.

2 Material and method

The considered material for the meshes was phosphor bronze. The analysed meshes had different aperture (distance between the neighbouring wires) of 0.32 mm and 0.40 mm but the same wire diameter of 0.20 mm. The experiments were carried out for distilled water and ethyl alcohol under the nucleate boiling regime and ambient pressure. The samples were prepared in such a way that the meshes were sintered to the copper discs in the reduction atmosphere of hydrogen and nitrogen to prevent oxidation.

The experiments were performed on the stand, whose main element – the heater – has been presented in Figure 1. The produced samples (copper discs with sintered meshes) were soldered to the copper heating block. The electric heater provided the heat flux which was conducted to the samples (the block was insulated from the surrounding with high temperature insulation). The power of the heater was increased during the measurements using the autotransformer with given steps to provide data points which enabled to draw the boiling curves.

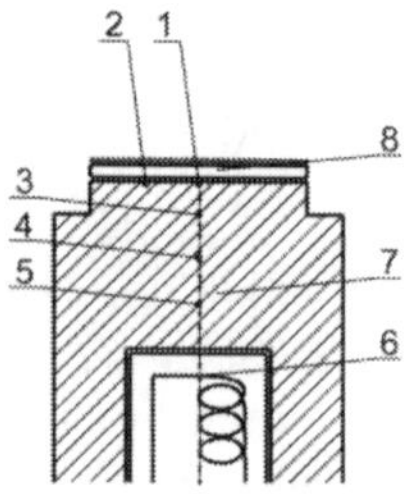

Fig. 1. Main unit of the experimental set-up: 1, 2 – thermocouples under the sample, 3, 4, 5 thermocouples in the axis of the heating block, 6 – electric cartridge heater, 7 – copper block, 8 – sample with the mesh [12].

The boiling curve is a dependence of heat flux and superheat (defined as the difference between the wall temperature and the saturation temperature of the liquid). The heat flux transferred to the samples was calculated using temperature readings recorded in the axis of the copper block. The temperatures under the sample were also recorded. As a result, the boiling curves could have been drawn. Boiling occurred in the thermally – resistant glass vessel located above the sample on the teflon plate. The generated vapour underwent condensation in a condenser located above it and was returned to the vessel gravitationally so that the liquid level above the sample was kept constant. The liquid temperature in the vessel was measured with a thermocouple. In the tests all thermocouples were of K type.

3 Test results and discussion

Two liquids have been considered, namely distilled water and ethyl alcohol of very high purity (over 99 %). The results obtained for the smooth surface (without any additional covering) have also been presented for reference purposes.

The influence of the aperture of the mesh layer on boiling heat transfer of distilled water and ethyl alcohol has been presented in Figure. 2 and 3, respectively. The wire diameter of the phosphor bronze meshes was 0.20 mm (which relates to the height of the microstructure of 0.40 mm). Two values of the aperture were considered in the experimental testing, namely 0.32 mm and 0.40 mm. The surface porosity amounted to 38 % and 44 %, respectively. The data on the smooth surface and the mesh of aperture 0.40 mm has been presented by the author in [12].

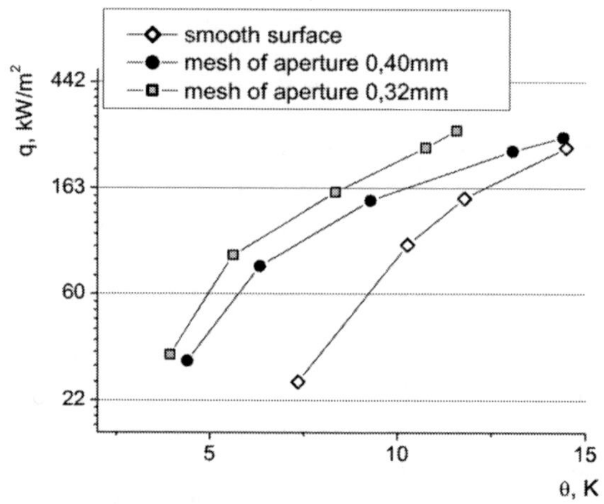

Fig. 2. Distilled water boiling curves for phosphor bronze meshes of different apertures.

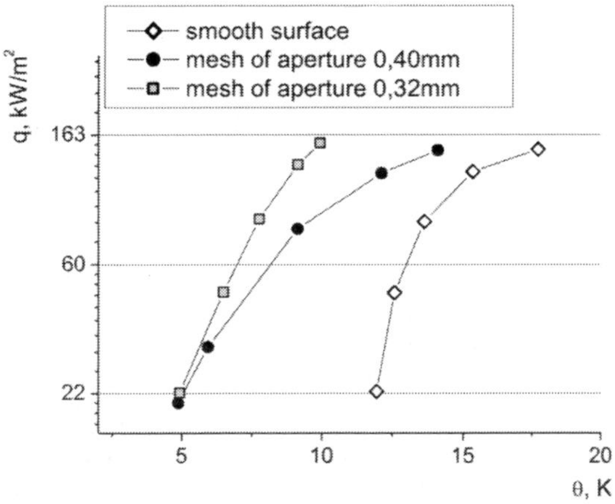

Fig. 3. Ethyl alcohol boiling curves for phosphor bronze meshes of different apertures.

For the analysed samples the best enhancement of heat transfer was provided by the mesh of the smaller aperture (as seen in figures below). It is especially evident in the case of water. For ethyl alcohol the results produced with the mesh of 0.32 mm aperture are also better, but the impact is much more significant for high superheats. For low temperature differences of ca. 5 K the results are very similar.

It seems that for smaller apertures the conditions of boiling heat transfer are improved. It can be related to the fact that the areas between the wires act as nucleation sites, where vapour bubbles are created and grown. As the aperture becomes smaller the density of active nucleation sites on the surface increases. Thus, heat transfer is enhanced. However, data from literature indicates that for very small apertures the heat transfer conditions are actually hampered. Consequently, this phenomenon needs to be considered taking into account the diameter of the bubbles (smaller bubbles are produced with ethanol boiling than with distilled water boiling.

The obtained test results also indicate that the highest influence of the meshes occurs for small superheats. For larger values data points might approach the values recorded for the smooth surface. It may be related to the fact that due to increased vapour generation within the coating there is a difficulty to sustain the process of efficient vapour removal from the microstructure and liquid transport to it. From this point a rapid transition to film boiling can occur. The details of the enhancement caused by the application of the mesh of aperture 0.32 mm have been presented in Figure 4.

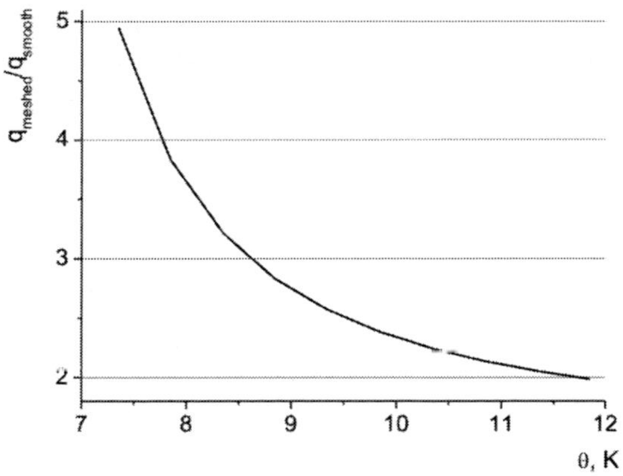

Fig. 4. Enhancement ratio for the mesh of 0.32 mm aperture for water boiling.

The enhancement ratio given in the figure above considers the heat flux dissipated from the meshed surface to the heat flux removed from the smooth reference surface for the same superheat value. The necessary data have been derived from Figure 2 for distilled water boiling. The calculations involved generating the polynomial fits of the second order of both the test results (the meshed and smooth surface) to produce a curve presented below.

The use of the mesh proved to be most efficient for lowest superheats where the heat flux for the meshed surface was almost five time higher than for the smooth surface. For larger superheats the effect diminished.

A different problem in the analysis of boiling heat transfer is modelling of this phenomenon. It is necessary for a proper design of heat exchangers. Data presented for the mesh of aperture 0.32 mm has been compared with selected correlations from literature, namely those developed by Smirnov et al. [4, 13, 14], Nishikava et al. [15] as well as Xin and Chao [16].

In order to make calculations according to the model presented by Xin and Chao, it was necessary to introduce certain modifications as for the geometrical parameters used in the formulae. The width of a single cell had to be considered as the total of wire diameter and aperture and the width of the tunnel as aperture. The results of the calculations have been presented in Figure 5 for ethyl alcohol as the boiling liquid.

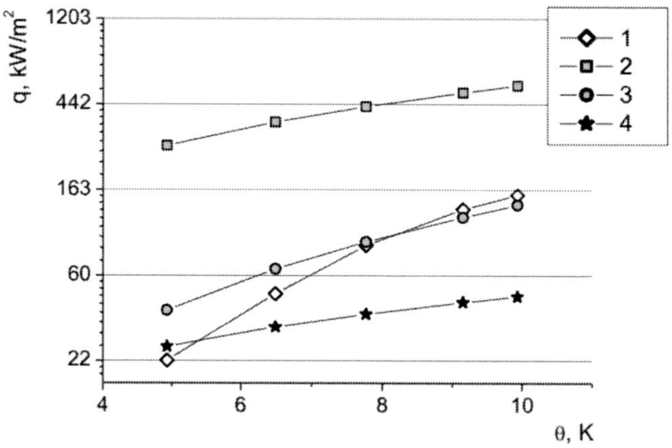

Fig. 5. Comparison of experimental data with the models for the mesh 0.32 mm x 0.20 mm (ethyl alcohol); 1 – experimental results as in Figure 3, 2 – calculation results according to Nishikava et al. correlation, 3 – calculation results according to Xin and Chao correlation, 4 – calculation results according to Smirnov et al. correlation.

The analysis of the produced results indicate that the best congruence of experimental data for the considered meshed surface was observed for Xin and Chao model, although the congruence is less evident for small superheat values (below about 7 K). The least efficient for the analysed case was the model by Nishikava et al., which gave much higher results than those obtained from the experimental analysis.

4 Conclusion

The application of the meshes can lead to significant enhancement of boiling heat transfer. Dissipated heat fluxes can be several times higher in comparison to the smooth surface without any additional coating. This might be linked to increased density of active nucleation sites on the heater. It might also be the reason why meshes of smaller aperture (0.32 mm) performed better than those of larger aperture value (0.40 mm) – both for distilled water and ethyl alcohol.

The comparison of the experimental data for boiling heat transfer on the phosphor bronze meshed surface of the mesh aperture of 0.32 mm with models from literature proved that the most precise was the model proposed by Xin and Chao with certain modifications, while the least accurate was the one presented by Nishikava et al. It should also be noted that microstructural coatings can also be applied for flow boiling [17, 18] and not only for pool boiling as discussed in the present paper. They can be efficiently used for heat pipes, which are highly efficient heat exchangers [19].

Work in the article was funded by statutory research of the Faculty of Environmental, Geomatic and Energy Engineering of Kielce University of Technology, No. 05.0.07.00/2.01.01.01.0033 MNSP.IKSI.16.003.

References

1. M.E. Poniewski, *Wrzenie pęcherzykowe na rozwiniętych mikropowierzchniach* (Wydawnictwo Politechniki Świętokrzyskiej, Kielce, 2001)
2. M. Vantúch, J. Jandačka, M. Malcho, R. Kiš, A. Kapjor, H. Smatanová, RESpect 2013 Proc of VIII Int. Conf. (2013)
3. V.Ja. Sasin, V.N. Fedorov, A.Ja. Sorokin, Doklady Naučno-Techn. Konf. po Itogam Naučno-Issled. Rabot za 1968-69, MEI (1969)
4. G.F. Smirnov, A.L. Coba, B.A. Afanasiev, AIAA Paper (1978)
5. S. Hasegawa, R. Echigo, S. Irie, J. of Nuclear Science and Technology, **12**, 11 (1975)
6. V.I. Tolubinskij, V.A. Antonenko, G.V. Ivanenko, Heat Transfer-Soviet Research, **21**, 4 (1989)
7. J.Y. Tsay, Y.Y. Yan, T.F. Lin, Heat and Mass Transfer, 32 (1996)
8. A. Brausch, P.A. Kew, The effect of surface condition on boiling heat transfer from mesh wicks, Proc. of 12th Int. Heat Transfer Conf. (2002)
9. T. Orzechowski, *Wymiana ciepła przy wrzeniu na żebrach z mikropowierzchnią strukturalną* (Wydawnictwo Politechniki Świętokrzyskiej, Kielce (2003)
10. C. Li, G.P. Peterson, Y. Wang, J. of Heat Transfer, 128 (2006)
11. J.H. Liou, C.W. Chang, C. Chao, S.C. Wong, Int. J. of Heat and Mass Transfer, 53 (2010)
12. Ł.J. Orman, Proc of Int. Conf. "Experimental Fluid Mechanics 2013", Czech Republic (2013)
13. G.F. Smirnov, Teploenergetika, **9** (1977)
14. G. Smirnov, B.A. Afanasiev, VI Int. Heat Pipe Conf. Proc. in *Advances in Heat Pipe Technology* (London, 1982)
15. K. Nishikawa, T. Ito, K. Tanaka, Heat transfer – Japanese Research, 8 (1979)
16. M.D. Xin, Y.D. Chao, Chem. Eng. Comm., 50 (1987)
17. M. Piasecka, Heat and Mass Transfer, **49**, 2, 261-275 (2013)
18. M. Hożejowska, M. Piasecka, M. Poniewski, Int. J. of Thermal Sciences, **48**, 6, 1049-1059 (2009)
19. Z. Kolková, M. Malcho, Structure and Environment, **5**, 4, 37-40 (2013)

Modeling the Production of Natural Gas Hydrates in Laboratory Conditions

Ján Siažik[1,*], *Milan* Malcho[1], and *Stanislav* Gavlas[1]

[1]Department of Power Engineering, Univerzitná 1, 010 26 Žilina, Slovakia

Abstract. Natural gas hydrates are in nature sources of primary energy. They are possible to created in laboratory conditions also under certain temperature and pressure conditions. With determining whether founding suitable conditions is closely linked to the creation of a physical model. The article deals with the model of natural gas hydrates.

1 Introduction

Energy storage is nowadays an interesting phenomenon that tends to move forward. There are many options to store energy, for example water energy – construction of dams and pumping tanks or capture of tidal wave energy. Also solar energy and its conversion to electrical energy and storage in accumulators, etc. natural gas or methane reserves are stored in the form of hydrates in nature. The hydrate can be produced also under laboratory conditions. Accumulation of energy into the form of hydrate is of great importance in terms of storage capacity, safety, transport to the place of consumption and coverage of unexpected energy peaks in the technology industry. The hydrate is generally a substance comprising water and hydrocarbons (or other gases). The storage of natural gas in hydrates (especially methane) is particularly advantageous in terms of storage capacity but also in terms of gas storage safety, which is possible at higher temperatures and lower pressures compared to other storage technologies. The natural gas hydrate is in a solid state, which accounts for about 85 % of water and 15 % of the gas bound in the clathrate lattice. Hydrates also have some uncertainty about ongoing research and their use as potential sources of income. The gas hydrate can be a great source of energy that begins to be considered for natural gas supplies for the coming decades. It is estimated that 99 % of the global stock of gas hydrates is present in marine sediments under the corresponding temperature and pressure conditions prevailing in the top ten to hundreds of meters of seabed, in sediments at depths of 300 m and up to 4000 m. Hydrates also occur on continent, but only in areas of permafrost ie in long-term freezing areas with temperature all year below 0 °C. Storing energy in a convenient form for safe storage and its subsequent release is a current challenge in processes that use the need for energy to accumulate. The use of methane hydrates in practice could be for example, ensuring peak coverage in cogeneration electricity and heat generation by natural gas released from hydrates [1-4]. In Figure 1 is a burning gas hydrate sample.

* Corresponding author: jan.siazik@fstroj.uniza.sk

Fig. 1. Burning sample of gas hydrate.

2 Modelling hydrate creation

This part of article describe a part of chosen parameters necessary to determination model of hydrate creation. Injection capacity Q_D [$m^3.s^{-1}$] is one of the important parameters. It is calculated according to the following relationship [5]:

$$Q_D = 5.4.10^{-6}.\left(\rho_w.d_0^3.\frac{v_0^3}{\sigma}\right).\left(\frac{l_0}{d_0}\right)^{0.75}.(\sin\alpha)^{-0.2} \quad (1)$$

Where ρ_W – density of liquid [$kg.m^{-3}$], d_0 – diameter of nozzle [$4.1.10^{-4}$ m], v_0 – average velocity of liquid at outlet was calculated as 16.8 [$m.s^{-1}$], σ – surface tension of water [0.065 $N.m^{-1}$], l_0 – lenght of nozzle [0.05 m], α – nozzle spray angle [0.96 rad].

As example was used volume flow rate of nozzle size 0.4 which at pressure 70 bar is 7.6 $l.h^{-1}$.

Velocity v_0 was calculated as ratio volume flow rate of nozzle [$m^3.s^{-1}$], and nozzle cross-sectional area S [m^2]. Injection capacity in our case is $8.96.10^{-7}$ $m^3.s^{-1}$.

Next parameter is Sauter mean diameter, calculated as follows [5]:

$$SMD = 4.4ST^{0.6}\mu_L{}^{0.16}\rho_L{}^{-0.16}Q_L{}^{0.22}\Delta P_L{}^{-0.43} \quad (2)$$

Where ST – surface tension [$kg.s^{-2}$], Q_L – volume flow rate of liquid [$m^3.s^{-1}$], Δp_L – fluid pressure [Pa], μ_L – dynamic viscosity [$kg.m^{-1}.s^{-1}$]. Sauter mean diameter is in our case $1.6.10^{-5}$ m.

For the gas-water transmission properties have a fluence diffusion coefficient. The diffusion coefficient [$cm^2.s^{-1}$] of the gaseous components in water was calculated as follows for the hydrate-forming components using the Wilke-Chang correlation [5]:

$$Dc = \frac{7{,}4x10^{-8}\,(\varepsilon.MW_w)^{0{,}5}T}{\mu_w.V_m^{0{,}6}} \quad (3)$$

Where ε – is an asociation factor of solvent (for water ε = 2.26), MW_w – molecular weight of water [$g.mol^{-1}$], μ_w – dynamic viscosity [$kg.m^{-1}.s^{-1}$], Vm – molar volume of gas [$dm^3.mol^{-1}$].

Also for transfer properties of the gaseous mass in the water during spraying influence the transfer properties. The weight transfer coefficient of gaseous components in water droplets was calculated on the assumption that the droplets from the spray nozzle were well mixed [6]. A model designed by Srinivasan and Aiken [7] based on Levich Theory [8] and was used to determine the weight transfer factors of water components in water. The weight transfer factor is as follows [5]:

$$K_l = \frac{0.16D_c}{d_p} Sc^{0.5} We^{0.5} Re^{5/16} \tag{4}$$

Where Dc – diffusion coefficient, dp – charakteristic dimension of dropplet [m], σ – surface tension [J.m^{-2}], ρ_w – water density [kg.m^{-3}], v_p – flow rate of the particles [m.s^{-1}], Sc – Schmidt number, We – Weber number, Re – Reynolds number.

$$S_c = \frac{\mu_W}{\rho_W . D_c} \tag{5}$$

$$W_e = \frac{v_p^2 \rho_W d_p}{\sigma} \tag{6}$$

$$R_e = \frac{d_p v_p \rho_W}{\mu_W} \tag{7}$$

The average weight transfer coefficient was obtained for the gas by normalizing the individual mass transfer coefficients of hydrate-forming components based on the composition of the gas-forming gas hydrate in the natural gas mixture. The initial particle velocity v_0 [m.s^{-1}] emerging from the nozzle is given by the value [9], where φ is the nozzle geometry taken as = 1 [5].

$$v_0 = \varphi \sqrt{\frac{2.\Delta.p_n}{\rho_p}} \tag{8}$$

The resistance coefficient is calculated from the following correlation as described in [10]:

$$c_d = \left(\frac{24}{R_{ep}}\right).\left(1 + 0.14.R_{ep}{}^{0.7}\right); 0.1 < R_{ep} < 1000 \tag{9}$$

Where R_{ep} is Reynolds number of particles given by relation [5]:

$$R_{ep} = \frac{SMD.v_0.\rho}{\mu} \tag{10}$$

It is also possible to calculate the feed rate in reactor r, provided that the sprayed particles were evenly deposited on the bottom of the vessel [5]. Where ρ_h^b is volume density of the hydrate [kg.m^{-3}] and D_v is the inside diameter of the container [m] taken as 0.18 m.

$$r = \frac{Q.\rho_W}{\rho_h^b.\pi.(D_v/2)^2} \tag{11}$$

Concentration of gas at the gas-liquid interface under pressure $Csol$ [mole.m^{-3}]:

$$C_{sol} = X.\frac{\rho}{M} \tag{12}$$

Where X – molar solubility of gas in water X (p, T); p = 1.10^7 Pa, T = 275.15K ρ – water density [kg.m^{-3}], M – mole mass water [kg.mol^{-1}], M = 0.018 kg.mol^{-1}. In our case, the concentration is 1388 mol.m^{-3}.

The gas constant is equal to R = 8314 J.kmol^{-1}.K^{-1} and the gas concentration in the fluid c_b [mol.m^{-3}] at the hydration pressure and equilibrium temperature is calculated as [5]:

$$C_b = \frac{p}{R.T}.1000 \tag{13}$$

The ratio of gas dissolution in r_1 [mol.s^{-1}] is calculated as follows:

$$r_1 = k_l . A_g . (c_{sol} - c_b) \tag{14}$$

Where: k_l – diffusion coefficient of gas in water [m.s^{-1}], D – methane coefficient in water at barometric pressure and temperature 25 °C = 0.0000149 [cm^2.s^{-1}], Ag – surface area of bubbles [m^2]. In our case, its value is 1.41.10^{-14}.

Hydrate formation in the water spray system in the gas environment would be subject to the process of heterogeneous nucleation, which occurs at the gas-solution interface. Areas of n-shaped units of hydrated lens are shown in Figure 2 together with "humidification angles" (θ and θ_0 equation 15, 16) with respect to each phase where σ is the specific surface energy of the interface solution-hydrate, σ_{sg} is the specific surface energy of the interface-solution gas, σ_{hg} is the specific surface energy of the hydrate-gas. In the case of lens-type clusters, the "humidification angles" are shown in [5, 11] as follows:

$$cos\theta = \frac{(\sigma_{sg}^2 + \sigma^2 - \sigma_{hg}^2)}{2\sigma_{sg}\sigma} \tag{15}$$

$$cos\,\theta_0 = \frac{(\sigma_{sg}^2 - \sigma^2 + \sigma_{hg}^2)}{2\sigma_{sg}\sigma_{hg}} \tag{16}$$

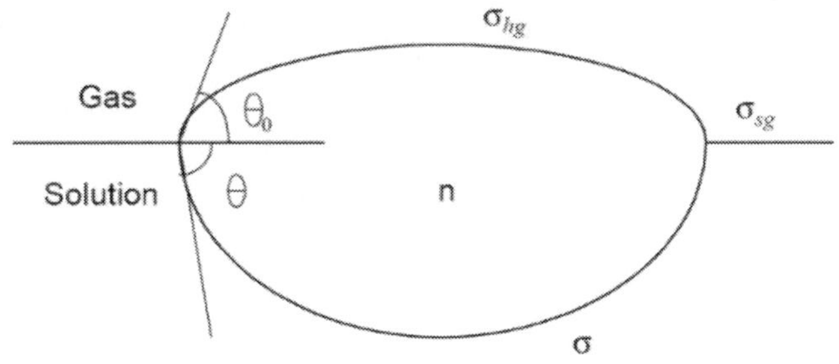

Fig. 2. Schematic diagram n shaped units at interface solution-gas.

The values of the specific surface energies used in the above calculation are founded in Table 1.

Table 1. Chosen nucleation parameters of hydrate.

Characteristic quantity	**value**
Surface energy solution/gas, σ_{sg}	70 mJ.m^{-2}
Surface energy hydrate/gas, σ_{hg}	80 mJ.m^{-2}
Surface energy solution/hydrate, σ	15 mJ.m^{-2}

At the stage of the solution and the finishing phase there is an experimental device, shown in Figure 3. The device will serve to create hydrates under laboratory conditions and in the following to compare actual consumption of energy consumed by the device with theoretical values [5-14].

Fig. 3. Experimental device in the finishing process.

3 Conclusion

Artificially created hydrates appear to be a prospective means of energy accumulation in the 21st century. Storage of gas in the form of hydrates is possible at relatively high temperatures and low pressures compared to other gaseous hydrocarbon storage technologies. At the same time, this method of energy storage appears to be advantageous both in terms of storage capacity and security. The article describes a modeling of hydrate formation using selected parameters related to the experimental device that is in the finishing phase.

Work on the article was financially supported by VEGA 1/0738/18. Optimalizácia energetických vstupov pre rýchle generovanie hydrátov zemného plynu a biometánu na akumuláciu vysokopotenciálnej primárnej energie

References

1. A. Demirbas, *ECM*, **51** (2010)
2. V. Foltin, J. Rajzinger, J. Juriga, Š. Guželľa, In *Vykurovanie*, SSTP, Bratislava (2013)
3. S. Čičmanec, Perspektívy využívania zemného plynu v SR (2004)
4. S. Depowski, In Kotliński, Szamałek (red.) *Surovce minerálne mórz i oceanów* (1998)
5. N. Gnanendran, R. Amin, *Chemical engineering science* **59**, 3849-3863 (2004)
6. A. Dimiccoli, M. Di Serio, E. Santacesaria, *Ind.& Eng. Chem. Res.* **39** (2000)
7. V. Srinivasan, C.R. Aiken, *Chemical Engineering Science* **43** (1988)
8. E.G.V. Levich, *Physioch. Hydrodynamics*, Prentice-Hall, Englewood Cliffs, NJ. (1962)
9. H.A. Lefebvre, *Atomization and Sprays*, Hemisphere, New York (1989)
10. R. Perry, *Perry's chemical engineers' handbook*, 7th ed., McGraw-Hill Comp. (1997)
11. D. Kashchiev, A. Firoozabadi, *Journal of Crystal Growth*, **243** (2002)
12. M. Holubčík, J. Jandačka, Š. Papučík, P. Pilát, *Manuf. Technology*, **15** (2015)
13. J. Jandačka, et al., *Emisné zaťaženie životného prostredia*, Edis (2015)
14. P. Ďurčanský, et al., *Scientific World Journal*, **2014** (2014)

Extrusion in Hydroenvironment in Laboratory Conditions

Ján Moravec[1,*]

[1]University of Žilina, Department of Technological Engineering, Univerzitná 8215/1, 010 26 Žilina, Slovakia

Abstract. The application of liquid environment in forward extrusion is described in this paper. Theoretical part states findings from fluidic mechanics applied in the experimental part. Design solution of the tool is also presented. The experiments were done in laboratory conditions. The findings are stated with their limitations.

1 Theoretical findings

According to Figure 1 points *N* and *N′* define the axis of vertical cylinder. Vertical height difference between points *N, N′* is *h*. All pressures perpendicular to the surface, here vertical, form a neutral resultant. The weight ρdSh contained in the cylinder operates in vertical direction, then pressures $p_{N'}dS$ and $p_N\, dS$. First two forces operate down, the third operates up [1, 2, 3].

Liquid in the cylinder is in balance according to: $\rho dSh + p_{N'}dS - p_N\, dS = 0$. When divided by surface element *dS, we get:* $p_N = p_{N'} + \rho h$. (1)

The difference of the specific pressures of two different points inside the liquid equals to the weight of vertical column of liquid, which has a unit of surface as a base and a difference of height contour of both points. The basic law of hydrostatics applies in any values of pressure $p_{N'}$ a p_N in equation (1). If pressure p_N rises by value Δp, $p_{N'}$ rises also by Δp. The increment of pressure is spreading evenly in all directions, because both considered points do not have to be on a common perpendicular [4, 5].

* Corresponding author: Jan.Moravec@fstroj.uniza.sk

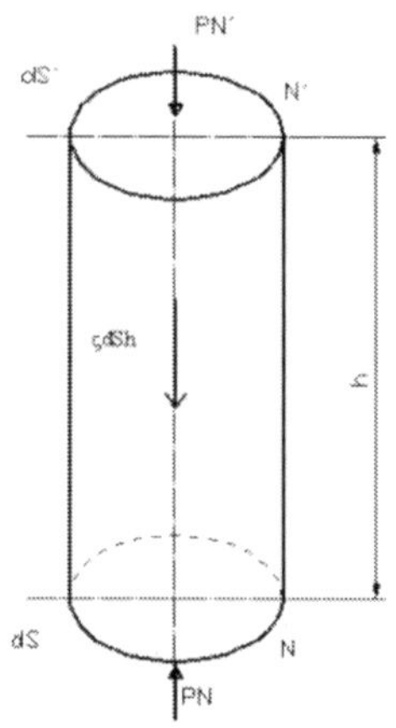

Fig. 1. Proportions in the vertical cylinder.

Let *M* be a point of surface of solid planar wall, which the fluid is touching in balance according to Figure 2. We set out an area element *dS* around point *M*. The liquid effects the wall with pressure force *dP*, solid wall reacts to liquid with force, which is equal, but opposite direction. Next we consider elementary cylinder with base *dS, dS′= dS* with surface perpendicular to the wall. In M′ point as a centre of surface *dS′* and infinitely close point *M* let p be the pressure of liquid. If the outer forces affecting the liquid contained in the elementary cylinder project themselves into the axis of the cylinder and neglecting volume forces, which are infinitely small, third order against the surface forces, we get: $pdS - dP = 0$, $dP = pdS$. (2).

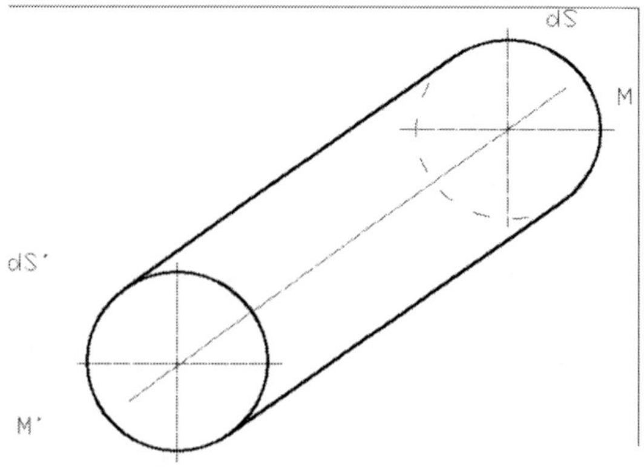

Fig. 2. Contact of the elementary cylinder with the planar wall.

Compressibility of the liquid is in general considered null. Although the molecules of the liquid move easily, liquids are noticeably compressible only by high pressures and we can say, that liquids are almost inelastic. The change of volume relative to original volume is proportional to change of pressure: $\Delta V/V \sim \Delta p_p$. If: *p*- change of pressure affecting the liquid, β – compressibility of liquid, ΔV – change of volume with rising pressure, *V* – volume of the liquid, then applies: $\Delta V = \beta \Delta p V$ (3).

Using appropriate experimental methods, we can specify tensile-deformation condition in any place (point) of the deformation zone. Integral outputs generate energo-force parameters, which are used in dimensioning the tool and determining size of the forming machine. The degree of plastic deformation (reduction, ductile ratio, removal of material, etc.), can be defined as a sum of small intensities of deformation, by which the considered element of formed material goes through in its flow – movement through focal spot of deformation. It can be expressed with equation: $\lambda = E\,dt$, (4), where: λ – degree of shear deformation in time frame of $0\text{-}t$, E – intensity of shear deformation velocity.

$$E = \sqrt{J_2(D_{\xi})} = \sqrt{2/3}\,\sqrt{(\xi_x - \xi_y)^2 + (\xi_y - \xi_z)^2 + (\xi_z - \xi_z)^2 + 3/2\,(\tau^2_{xy} + \tau^2_{yz} + \tau^2_{zx})} \qquad (5)$$

In equation (5): ξ_x, ξ_y, ξ_z – elements of velocities into linear deformations,
$\tau_{xy}, \tau_{yz}, \tau_{zx}$ – elements of shear deformations velocities.

For the flow line – axial, internal, surface it is necessary to count in the size of λ and E in between $t = 0$, to $t = t_K$, therefore from the entry of the line into the focal point of deformation. We know the tensile condition of formed object when it is possible to determine elements of tensions $\sigma_x, \sigma_y, \sigma_z, \tau_{xy}, \tau_{yz}, \tau_{zx}$, or the main tensions $\sigma_1, \sigma_2, \sigma_3, (\sigma_1 \geq \sigma_2 \geq \sigma_3)$ in the considered element.. Any expression containing these elements of tension is called tensile status indicator. *Nádai – Lode* coefficient of tension is preferred. It was found by verification, that this coefficient does not affect the limit plasticity λ_p very much. A more appropriate dependence from shape indicator $k_t = \sigma_{okt} / \tau_{okt}$ was determined experimentally. In equation: σ_{okt} is octaedric perpendicular tension, $\sigma_{okt} = 1/3\,(\sigma_1 + \sigma_2 + \sigma_3)$.$\tau_{okt}$ is octaedric shear tension $\tau_{okt} = 1/3\,\sqrt{(\sigma_1 - \sigma_2)^2 + (\sigma_2 - \sigma_3)^2 + (\sigma_3 - \sigma_1)^2}$.

Generally, every process of plastic deformation can be defined as a constant process of formation, spreading and closing (for example by welding) of submicro and micro defect. If φ is function of plasticity reserve, which has a starting status of material = 0, according to Kolmogorov [2]:

$$\varphi = \int_0^t B(t)\frac{E(t)}{\lambda_D[k_z(t)]}dt \quad \text{(a)} \qquad \varphi = \int_0^t B(t)\frac{E(t)}{\lambda_D[k_z(t)]}dt < 1 \quad \text{(b)} \qquad (6)$$

In relations $B(t)$ the quantity is dependent on the monotony of forming process. In monotonous processes, respectively almost monotonous, $B(t) = 1$. In strongly monotonous processes, $B(t) > 1$. $E(t)$ is function of time. In case of (3b), there will be no material breach, when shearing $\varphi = 1$ is desirable [6, 7, 8, 9].

2 Construction of tool

Using findings in the previous paragraph, a forming tool for forward extrusion was constructed – pic. 3. Basically it is a full metal tool, therefore by well-established criteria it should present a conventional forming technology. When replacing the extruder by liquid body, it is no longer a full metal tool. We can consider this „modified" tool as an unconventional extruding construction. Active part (extruder) function was taken over by liquid body, and the extruder became a piston. Although the liquid is incompressible (as is generally assumed), it is still a hundred times more compressible than steel. Compressibility of liquid can be described by a volume compression coefficient β_V or by elasticity modulus $k = 1/\beta_V$.

The volume conversation act applies in comparison of compressibility of water and steel at $t = 20$ °C.

Water:

$$\Delta V = -V\frac{\Delta p}{K} = -V\frac{\Delta p}{\frac{1}{100}E} = -100V\frac{\Delta p}{E} = 100\Delta V \quad (7)$$

Steel: *Δ V = - V Δp /E,* where E = 2.15 . 10 Pa. *k = 1/100 E.*

Active tool parts are made of class 12 steel without heat treatment. The tool was used in laboratory conditions, so following experimental verification respects this fact fully.

2.1 Strain on cavity and extrusion die body

We can consider the extrusion die as a thick walled vessel – Figure 3. Radial and peripheral strain is present.

$$G_r = \frac{pa^2 - p_o b^2}{b^2 - a^2} - \frac{(p - p_o)a^2 b^2}{(b^2 - a^2)r^2}$$
$$G_t = \frac{pa^2 - p_o b^2}{b^2 - a^2} + \frac{(p - p_o)a^2 b^2}{(b^2 - a^2)r^2} \quad (8)$$

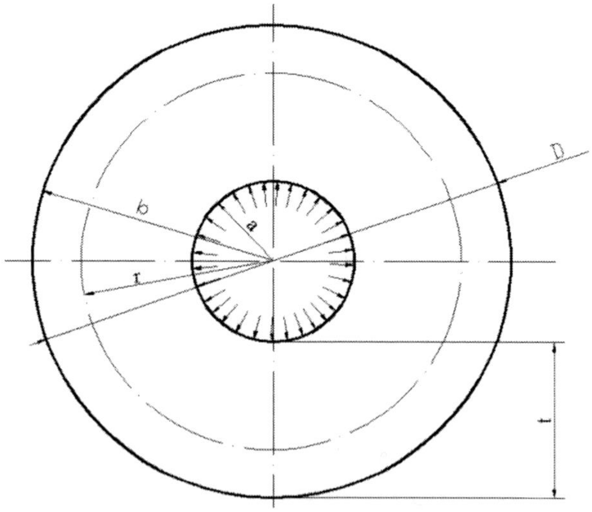

Fig. 3. Diagram for derivation of extrusion die strain.

The pressure in liquid column on the inside surface is: $p = \gamma . H$ (*H* – height of liquid column). Overpressure causes peripheral strain on the inner surface, for which the active parts needs to be dimensioned:

$$\sigma_{t\,max} = \frac{p(a^2 + b^2) - 2p_o b^2}{b^2 - a^2} \leq \sigma_{tdov} \quad (9)$$

if $p_o = 0$, we can determine necessary outer radius of the part: $b = a\sqrt{\frac{\sigma_{dov} + p}{\sigma_{dov} - p}}$

These realities need to be considered with concrete constructional solution of the real extrusion tool for hydroforming.

3 Experimental work

The tool for forward extrusion was verified in laboratory conditions, as shown in Figure 4. In first verification, a model material was extruded using classic technology, without application of liquid for second experiment a liquid (oil) was poured between inserted intermediate material and a face of the extruder (piston) and using liquid body the model material was formed. 5+5 specimens were made.
We can determine the forming pressure for forward extrusion according to [10]:
$p = k_1 - \sqrt{}\, k_2\, (k_3 - q_2)$ (N/mm^2), where k_1, k_2, k_3 constants dependent on total carbon percentage in steel,
$q_2 = d^2 \text{-} D^2 / d^2\ 100\ \%$ (cross reduction of area), d – beginning, D – terminal diameter.

According to stated relation and considering real process with application of liquid, it is quite obvious that in this method a maximum quality seal is necessary. The technological process in the second method differs from the conventional one by filling the cavity with pressure liquid, thus inserting intermediate material into the cavity od extrusion die, filling the cavity with liquid and inserting the extruder, the main course of process and after its completion, pulling out the extruder and removing the liquid follows.

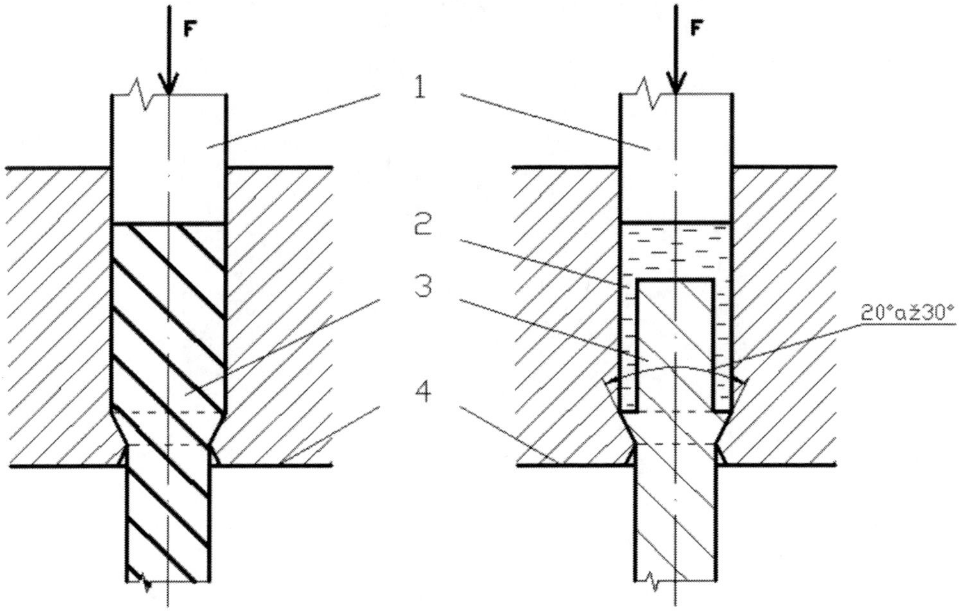

Fig. 4. Two conditions of the experiment: without and with the application of liquid.
1 – extruder (piston), 2 – liquid, 3 – formed material, 4 – extrusion die.

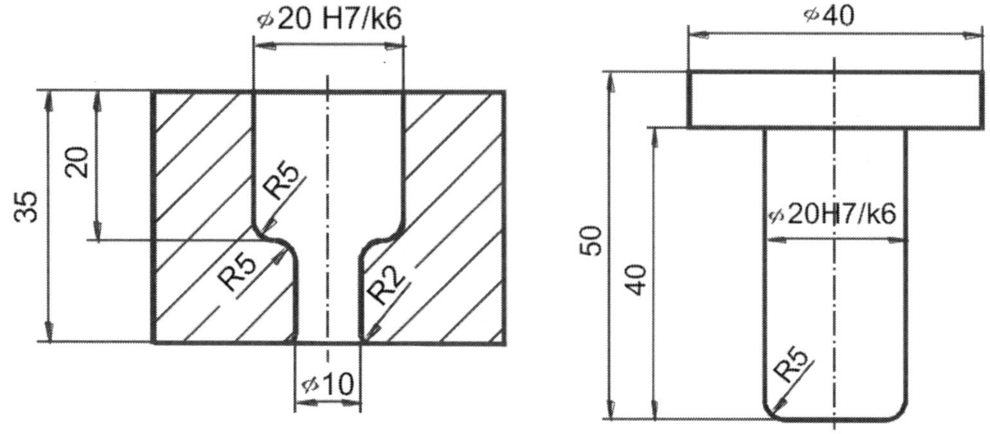

Fig. 5. Extrusion die (left) and extruder.

The photograph of the tool is in Figure 6.

Fig. 6. Experimental tool.

4 Results and findings

We can summarize them into these points: with pressing the extruder on the oil, process friction is reduced, angle of the extrusion die is lesser and quality of the product is far superior.

5 Discussion

Essential condition of optimal usage of formed material in the forming process is the forming tool, which should provide increased limit metal plasticity [11, 12]. It is about configuration of the forming process the way this condition is maximally satisfied. Limit plasticity of specified state and kind of material is not constant quantity, but is dependent on tension status, deformation velocity and temperature of the process. Forming processes can be deliberately influenced for:

- *preventing state of limit plasticity in critical spots of deformation area, focus of deformation,*
- *achieving maximal usage of plasticity in forming process,*
- *designing forming process from the perspective of optimal plasticity usage of formed metal.*

The diagram of hydrostatic extrusion through conical extrusion die is on pic. 7. In this method, the deformation focus is concentrated into the conical part of the extrusion die, where the variable factors are: extrusion ratio - $\kappa = D^2/d^2$, angle of the extrusion die – 2α and extrusion process velocity. If we consider the minimal extrusion pressure p as an indicator, then $\partial p / \partial \alpha = 0$, it is possible to transfer the dependencies into charts p- α. With p_{min} and $2\alpha = 45°$, with increasing extrusion ratio was transferred to higher α $(2\alpha = 70°)$. When extruding materials with lower plasticity (dural, molybdaen, high speed steel), the optimal pressure angle of the extrusion die does not guarantee compact products (transversal or tree fractures going deep are visible on the surface). By decreasing the deformation degree and increasing α leads to formation of internal axial fractures. Courses of λ and k_t are important and crucial for assessing the level of tool construction from the perspective of product quality. The value of λ from the entry to the deformation focus to its exit rises, this rising is almost linear for lower values of κ, especially with fibres around the wall of the extrusion die. With higher values of α and $\kappa > 5$, λ significantly rises in the centre of the deformation focus. Course of k_t is similar, but at higher κ a α we get to area of $k_t > 0$, because in this condition λ is higher. From equation (6) comes the probability of fracture formation closely under the surface.

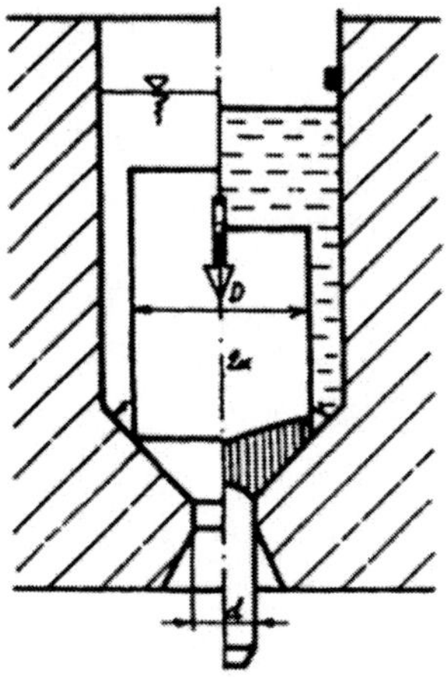

Fig. 7. Diagram of hydrostatic extrusion with focus of deformation shown.

6 Conclusion

Optimisation of forming processes based on their mathematization is the carrying science – technologic trend of following research in forming. Modern, computerised forming files with optimized technology and integrated operation can be built on these principles. Analyses and their implementation into the wide spectrum of forming problematics is a necessary premise for successful research. Physics as a science field of activity is an inexhaustible source of information, from which we can draw almost without limits.

References

1. W.F. Hosford, *Metal Forming, Mechanics and Metallurgy*, (2011)
2. V. Boljanovic, *Sheet Metal Forming Processes and Die Design*, (2004)
3. S. Kundar, *Technology of Metal Forming Processe*, (2008)
4. Z. Marciniak, *Theory of sheet metal forming*, (1964)
5. Z. Marciniak, et al. *Mechanics of Sheet Metal Forming*, (2005)
6. R. Pearce, *Sheet Metal Forming*. (1991)
7. Metal Forming Handbook, Schuler GmBH Gőppingen, (2009)
8. J. Moravec, *Unconventional metal forming technologies*, (2011)
9. J. Moravec, *18th Symp. on Anemo.*, (2003)
10. J. Moravec, *Contribution to the problem of hydroforming,* HaP, (2006)
11. J. Moravec, J. Bílik, (2017) *Forming machines and tools,* (2017)
12. J. Moravec, *The construction of a hydrostatic molding tool*, Technolog IV. (2012)

Methods of filling the heat pipes

Roman Banovčan[1,*], *Marcel* Novomestský[1], *Martin* Vantúch[1], *Andrej* Kapjor[1], and *Patrik* Nemec[1]

[1]University of Zilina, Department of Power Engineering, Univerzitná 1, Žilina, Slovakia

Abstract. The paper deals with individuals methods of filling heat pipes methods (filling by evaporation of the working fluid, filling by vacuuming of working fluid, freezing of working fluid – vacuuming). The second part of the paper is devoted to the comparison of the power of heat pipes after filling by individual methods.

1 Filling of heat pipes

The heat pipe may be filled in several ways to compare the extent to which the way the heat pipe function is affected. In all filling methods, distilled water is used as the working fluid, due to its availability and good thermophysical properties. For each filling method, the tube will be filled three times [1].

1.1 Filling by evaporation of the working fluid

This method of filling is followed by the total weight of the heat pipe.

• Check the weight of the heat pipe in the empty state, calculating the total volume according to the relationship (1), and determining the required quantity of the working fluid from it. Usually it is 15 % to 25 % of the total volume [2].

$$V = \pi . r^2 . v \tag{1}$$

• Fill the tubes with a syringe at a higher value than required.

• Verify weight after filling. The difference in weight before and after is the amount of working medium in the tube.

Fill of heat pipe:

Required quantity of working fluid

16.25 ml – 25 %

The tube placed on the weight is filled to:

24 ml – 37 % ↔ $m_{37\%} = 532.4$ g

Weight of heat pipe after filled to 25 %

$V = V_{37\%} - V_{25\%} = 25 - 16.25 = 7.75$ ml ↔ $m_0 = 7.75$ g

$m_{25\%} = m_{37\%} - m_0 = 532.4 - 7.75 = 524.65$

* Corresponding author: roman.banovcan@fstroj.uniza.sk

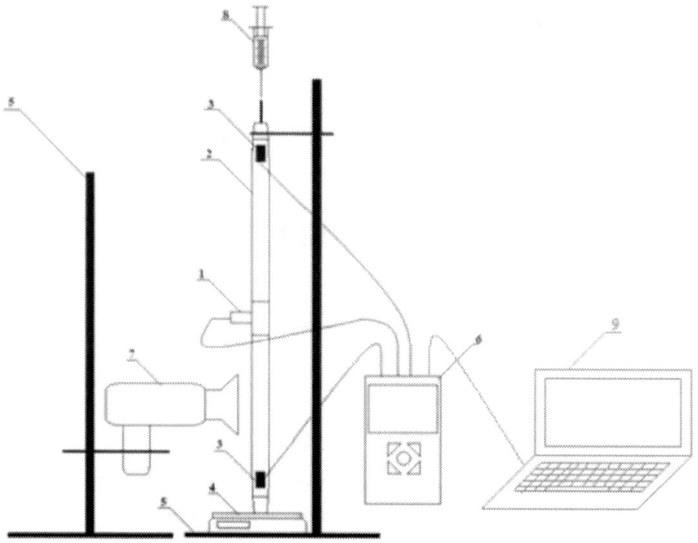

Fig. 1. Evaporation filling scheme 1) ALMEMO pressure sensor, 2) heat pipe, 3) temperature sensor, 4) weight KERN KB, 5) stand, 6) data logger ALMEMO 2890-9, 7) hot air gun, 8) Syringe, 9) PC.

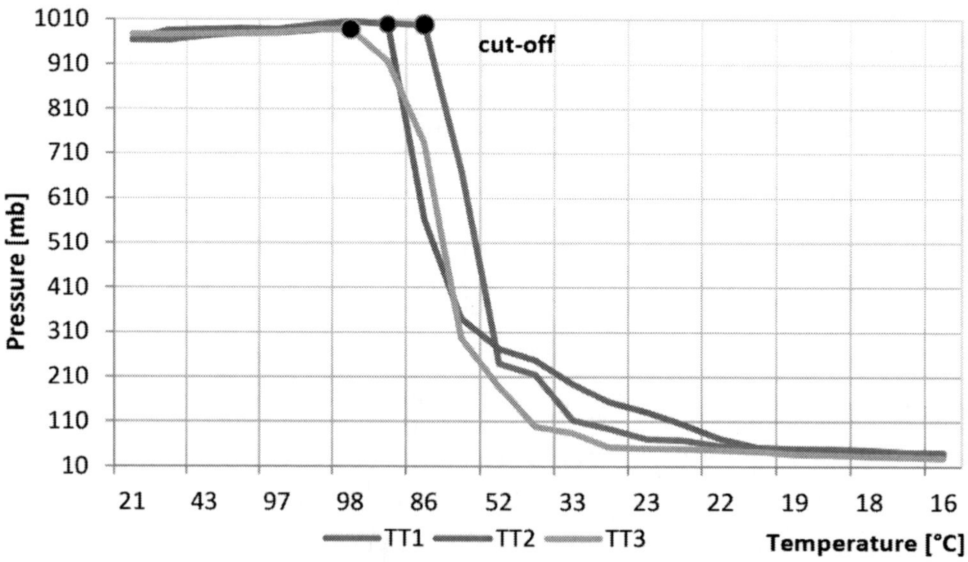

Fig. 2. Pressure in heat pipe during filling.

During filling TT1, the pressure dropped to 41 mb, TT2 to 32 mb and TT3 dropped to 21 mb. The evaporation and subsequent closing times vary depending on the amount of working medium in the tube, the heating intensity and the room conditions.

1.2 Filling by vacuuming of working fluid

This method is based on the principle of extracting air from the heat pipe by rotary vacuum pump via the container with liquid nitrogen [2].

Fill of heat pipe:

• The heat pipe is connected via a three-way valve to the metering vessel with the working fluid and the pump. The exchanger is connected to a container with liquid nitrogen to which a pressure meter is attached.
• The three-way valve is fully open. The pressure on the pressure gauge is monitored when it drops to a value close to 0 hPa. The three-way valve is switched so that the paths between the tube and the measuring vessel are permeable. Fill in the required amount that you calculate in the same way as in the first method.
• After filling, the filling tube is pressed and begins to start.

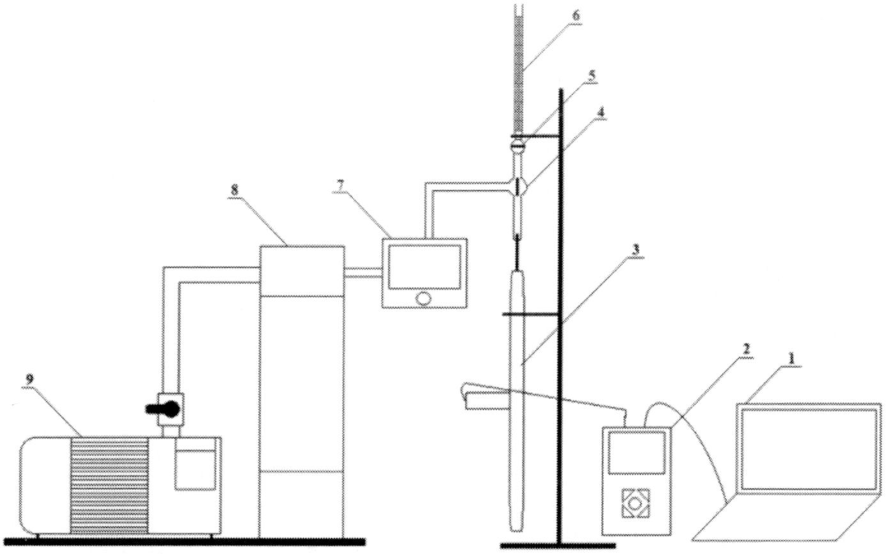

Fig. 3. Scheme of filling by vacuuming of working fluid 1.) PC, 2) data logger ALMEMO 2890-9, 3) heat pipe, 4) three Way Valve, 5) valve, 6) measuring Tank, 7) pressure gauge, 8) container with liquid nitrogen 9) vacuum pump.

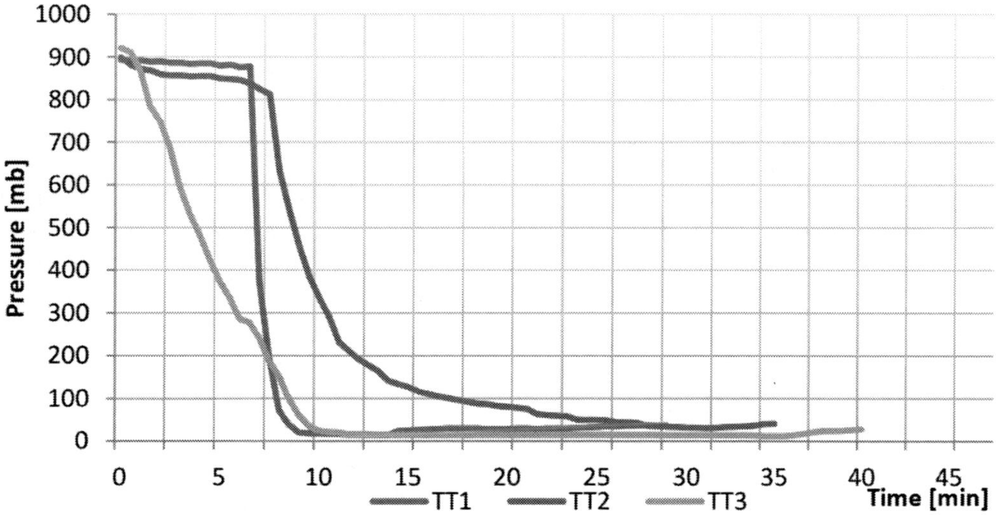

Fig. 4. Pressure in heat pipe during filling.

The vacuuming time for each tube varied between TT1: time - 32min, pressure - 36mb, TT2: time - 36min, pressure - 42mb and TT3: time - 41mi, pressure - 28mb. Room temperature at 18 °C.

During the measurement, the vacuum pump was cleaned, exchange the starter capacitor, the overload coil and the winding of the motor were changed. Despite these measures, we did not achieve the required pressure in the tubes, the pump overheated and lost the pull during suction.

1.3 Freezing of working fluid - vacuuming

It works on a similar principle as the "filling by vacuuming of working fluid", with the difference that the tube is filled with the exact amount of the working fluid. The working fluid is freeze and the air is pump out of the tube [2].

Fill of heat pipe:

• The tube is filled to the required amount, calculated as in the previous methods.
• Using liquid nitrogen is a cooled thermo-isolating vessel, after cooling, returns the nitrogen to the dewar vessel. Put the tube into the thermal insulator, lock it off with a 20 minute insulating plug.
• The pipe connects to the pump as in the second method, pump out the air and closed.

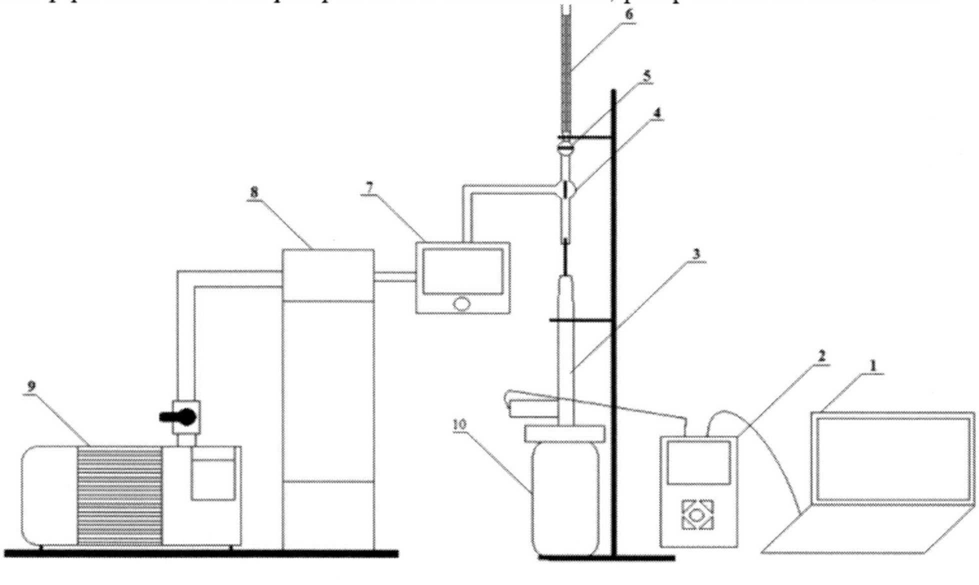

Fig. 5. Scheme of working fluid – vacuuming 1) PC, 2) data logger ALMEMO 2890-9, 3) heat pipe, 4) three-way valve, 5) valve, 6) measuring vessel, 7) Pressure gauge, 8) container with liquid nitrogen 9) vacuum pump, 10) thermo-isolating vessel.

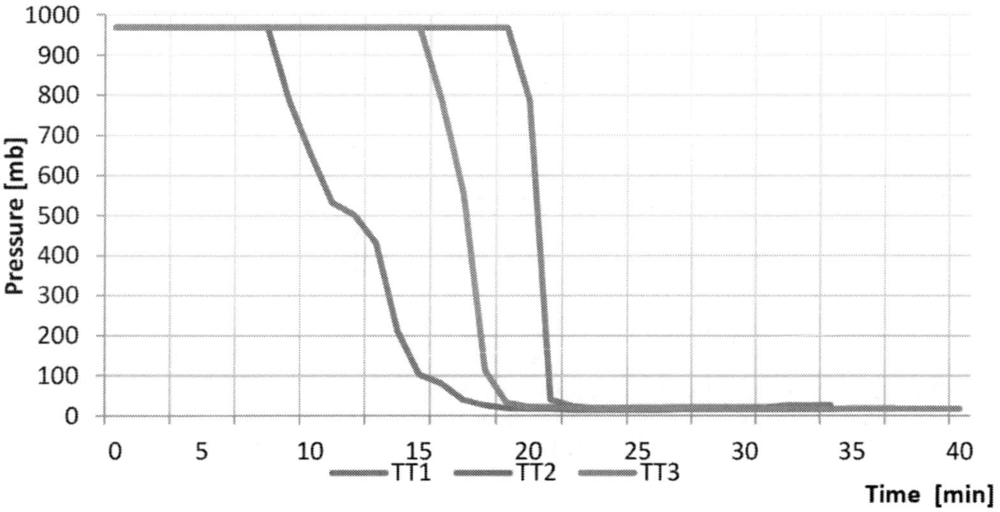

Fig. 6. Pressure in heat pipe during filling.

The time of filling of the suction tubes was different TT1: 32 min, pressure-26 mb, TT2: 36 min, pressure 18 mb, TT3: 40 min, 17 mb. Room temperature during filling of 19 °C.

2 Testing the power of different types of heat pipes

Power measurement was implemented at 0°, 15°, 30°, 60°, 90°, and it was repeated several times in order to achieve the most accurate results.

Measurement procedure:

• Set operating position 0°.
• Set JULABE to 80 °C.
• Record input and output temperatures along with flow rates every 10 seconds for 20 minutes (heating and cooling).
• When the measurement is complete, the device turns off. Close the hot water supply and open the cold water inlet. Cool the system to 30 °C.
• After cooling, a new working position of 15° is set and the measurement is repeated from point 2.

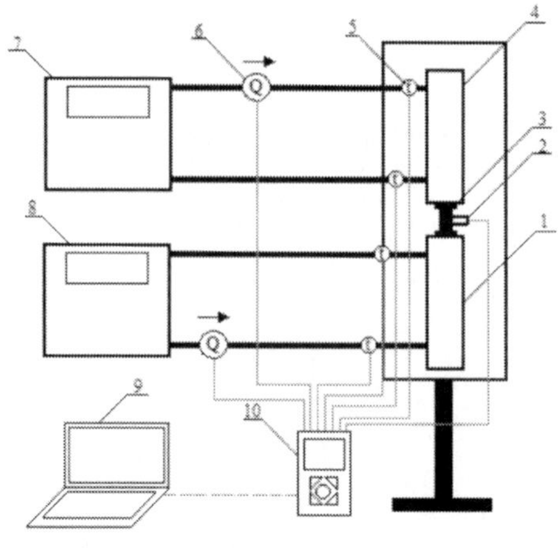

Fig. 7. Scheme for testing power of heat pipes. 1) heating part, 2) pressure sensor ALMEMO, 3) heat pipe, 4) cooling part, 5) temperature sensor, 6) flow meter ALMEMO, 7) JULABO for cooling 20 °C water, 8) JULABO heating 80 °C water, 9) PC, 10) data logger ALMEMO 2890-9.

In the first method filling (Figure 8) heat pipe TT1 reached higher performance than heat pipe TT2 and TT3, which have about the same power range. The TT2 and TT3 were measured on a modified measuring device, while the TT1 tube on the original device. By more accurate results, the measurement device passed through the adjustment. This is also seen when comparing TT1 with TT2 and TT3. Several measurements were made on the original device to confirm that the device must be edited. Only one of these measurements was selected to indicate the problem [3].

The device has been modified to prevent flow changes due to hose wounds when adjusting work positions and to allow better temperature sensing. The treatment concerned the heating section, where hoses and valves were replaced by water-jet galvanized fittings.

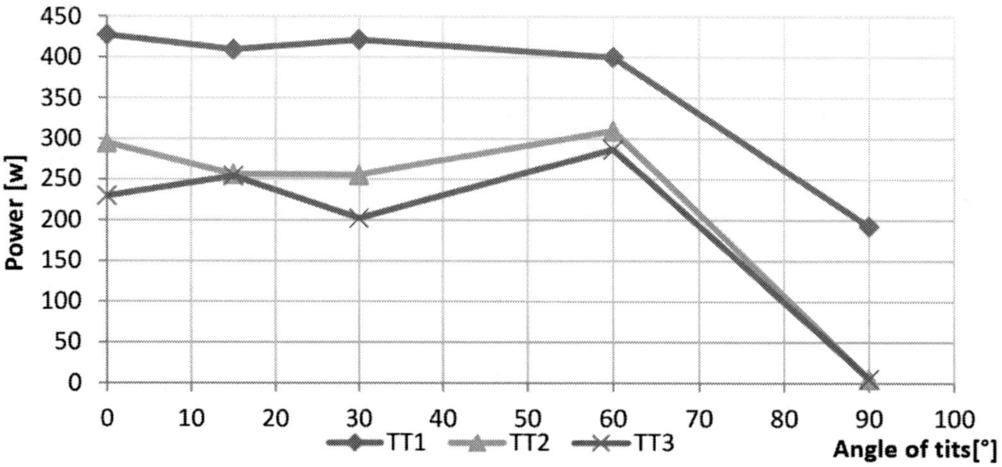

Fig. 8. Power of heat pipes filled by first method.

The TT1, TT2 and TT3 heat pipes of the second fill method (Figure 9) had similar performance ranges. In the TT2 heat pipe, the power dropped sharply at 15° from 214 W to 136W.

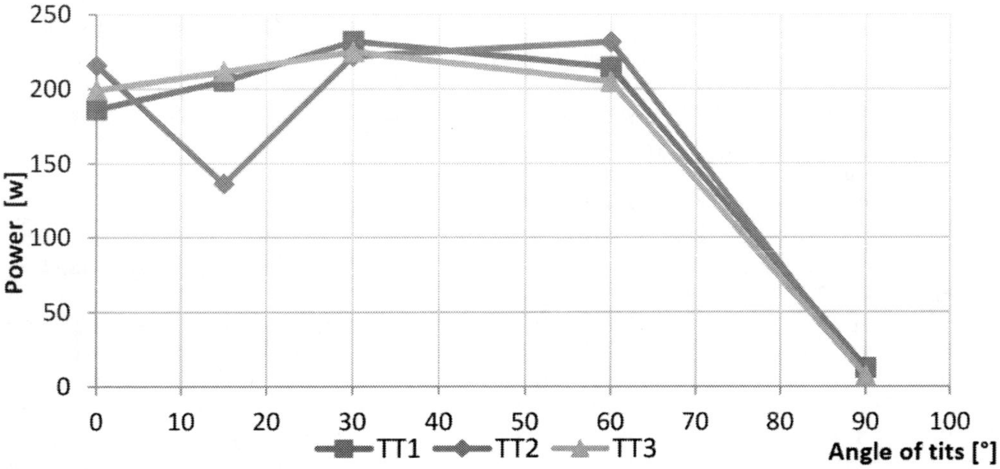

Fig. 9. Power of heat pipes filled by second method.

The TT3 heat pipe curve for the third filling method (Figure 10) has a decreasing character all the time, the performance achieved is significantly lower than that of the TT1 and TT2 heat pipes even though the best parameters have been reached during the TT3 heat pipe.

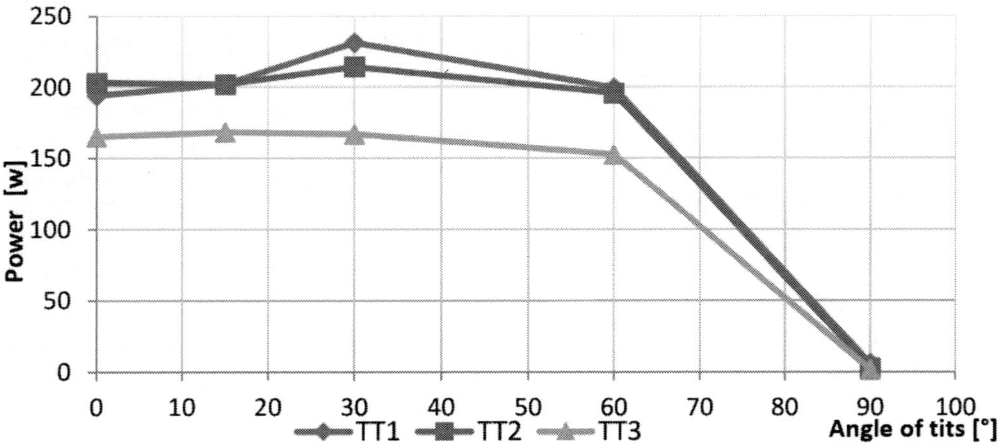

Fig. 10. Power of heat pipes filled by third method.

3 Evaluation of results

First method of filling (Filling by evaporation of the working fluid): The average pipe outputs are: TT1 = 413.5 W, TT2 = 279 W and TT3 = 243 W. The heat pipe TT1 was measured on the original measuring device. It points out that the measuring device has to be adjusted to achieve the most accurate results. The heat pipe TT2 had a higher power output of 36W even though the initial pressure was higher by 10 mb than heat pipe TT3.

Second method of filling (Filling by vacuuming of working fluid): In this method of filling, the average power of all three heat pipes was approximately equal to TT1 = 209.4W, TT2 = 201.3 W and TT3 = 210 W their outputs varied by an average of 5.2 W. In heat pipe TT1 and TT3, the smallest difference in power of 0.6 watts occurred with an initial pressure difference of 8 mb.

Third method of filling (Freezing of working fluid - vacuuming): In the third method of filling, the smallest performance was achieved. At heat pipe TT3, the lowest power was achieved at 163 W on average. These low power values can be caused by air bubbles created in the tube when they freeze the working fluid, which after thawing, has increased the internal pressure.

Table 1. Comparison of the power of heat pipes after filling by individual methods.

TT	1. filling method			2. filling method			3. filling method		
	TT1	TT2	TT3	TT1	TT2	TT3	TT1	TT2	TT3
0°	423.4	294.6	229.6	186.2	215.7	199.1	194.0	203.1	165.3
15°	409.4	256.3	253.8	205.0	136.2	211.4	202.0	201.9	168.5
30°	421.3	255.3	202.0	232.0	221.8	224.8	231.2	214.2	167.0
60°	399.9	309.7	286.6	214,4	231.4	204.7	199.4	195.6	152.7
90°	192.1	5.1	5.2	12.3	7.9	7.1	5.8	2.6	1.5

4 Conclusion

The filling technology also has a big influence on the heat pipe performance. Based on the different filling methods (filling by evaporation of the working fluid, filling by vacuuming of working fluid, freezing of working fluid – vacuuming) and the number of repetitions of each filling method, it was found that the best values of the transmitted power were achieved in the first filling method (filling by evaporation of the working fluid). The advantage of this method is that it is not as technologically demanding as other methods of filling. In the second filling method more accurate results would be obtained in the case of a better vacuum pump condition.

Article work was financially supported by the APVV-15-0778 project - The limits of control and convection cooling through phase changes of the working fluid in the thermosyphon LOOP heat pipe.

References

1. R. Lenhard, M. Malcho, J. Jandačka, Heat transfer engineering. **40**, 3 (2019)
2. V. Hlavačka, *Tepelné trubice v elektrotechnice*, SNTL, (1990)
3. R. Lenhard, J. Jandačka, *11th ICNAAM*, (AIP 2013)
4. Z. Kolková, M. Malcho, *European International Journal of Science and Technology*, (2014)
5. R. Lenhard, K. Kaduchová, P. Papučík, *EPJ Web of conferences 67*, (2014)

Application of mathematical modelling when determining the parameters effect of biomass densification process on solid biofuels quality

Peter Križan[1,*], *Michal* Svátek[1], *Miloš* Matúš[1], and *Juraj* Beniak[1]

[1]Institute of Manufacturing Systems, Environmental Technology and Quality Management, Faculty of Mechanical Engineering, STU in Bratislava, Nám. Slobody 17, 81231 Bratislava, Slovakia

Abstract. The main aim of this paper is to present the design of experiment (DOE) and evaluation methodology for this experimental plan in order to determine the parameters effect of biomass densification process on final solid biofuels quality. One of the recovery possibilities for waste biomass raw materials is production of solid biofuels. Using a variety combination of influencing variables can be improve the final quality of solid biofuels. Raw biomass material variables influence, especially (type of raw material, particle size, moisture content, compression pressure and compression temperature) can be recognized during the production of solid biofuels. Their effect can be seen through the quality indicators; especially mentioned variables significantly influence the mechanical quality indicators of solid biofuels. In this experimental research authors would like to investigate properties and behaviour of wood raw waste biomass during densification. This contribution discusses the analysis and design of experimental process, its individual steps and their subsequent DOE leading to the development of a mathematical model that will describe this process. This paper also presents the research findings regarding the effect of influencing variables on final density of solid biofuels during densification. Aim of the experimental process is to determine the mutual interaction between solid biofuels density and influencing variables during densification. Effect of compression pressure, compression temperature, moisture content and particle size on solid biofuels density from wood sawdust was determined.

1 Introduction

The densification process is a very interesting, but very complicated process, because many parameters effect the process and also the quality of the final solid biofuels. Solid biofuels quality is given by EU standards [1], and is evaluated by mechanical and chemical-thermic indicators of quality [2]. The properties of the raw material as well as technological demands are both very important during the solid biofuels production process [3]. The input raw material needs to be reduced to the optimal particle size, to dry to optimal moisture content level and we have to provide optimal technological parameters throughout the process of densification [4]. Each type of raw material requires an independent approach. Each small

* Corresponding author: peter.krizan@stuba.sk

change in the properties of the raw material can influence the final quality of the solid biofuels [5, 6]. Different raw material properties cause different conditions during the densification process [7]. The quality of solid biofuels is determined by the end-user's requirements on the heating system and the handling properties [8]. Low-quality pellets can cause operational problems in combustion systems, including undesired effects in the equipment such as slugging, fouling or corrosion, and may originate substantial amounts of gaseous and particulate matter (PM) emissions [9, 10]. In this contribution variables influencing the slid biofuels quality were specified and assigned.

The general purpose of this paper is to design a series of experiment and their evaluation methodology that can allow to determine the influencing variables effect. When the correct methodology and methods are used the increase of solid biofuels quality can be ensured. According to designed experimental plan we will be possible to develop a mathematical model which will describe a related variables interaction and variables effect on final biofuels quality. This mathematical model can be useful at optimizing of densification process related to an input raw material. Obtained research findings can increase the environmental responsibility with the environment protection.

2 Experimental plan

Design of experiment should to clearly define the general aim of experiment, experiment conditions, controllable variables, experiment levels, treatment, noise, replication and evaluation [11].

2.1 Experiment conditions

The main aim of our experiment is to determine the effect of raw material variables on final density of solid biofuels. The best way is to choose such variables which have during the densification process continuous characteristics. As an inputs variables is necessary to choose those variables in where the effect on outputs are expected. This may not be clear at the beginning of the experiment and then is needed to realize a screening experiment. In our case, according to the recent published research [12, 13] works and our experience determination of the mutual interaction between biofuels density (ρ), compression pressure (P), compression temperature (T), particle size (L) and moisture content (Wr) of raw sawdust used for densification were chosen.

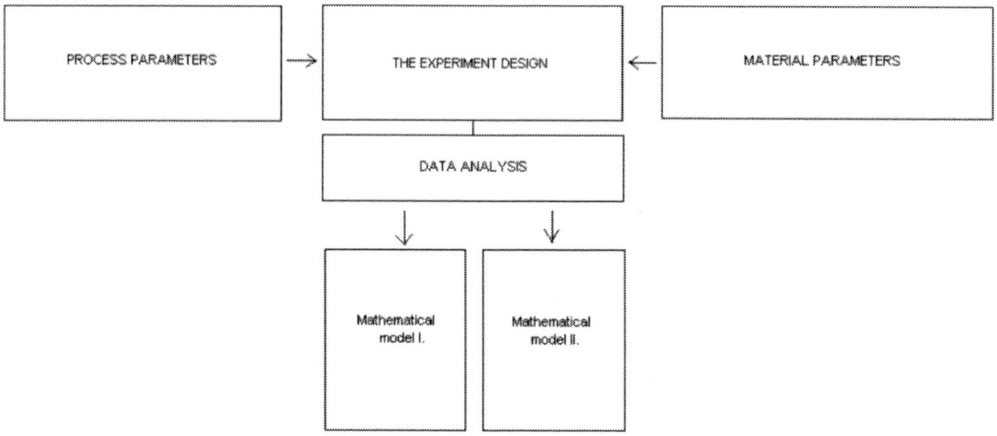

Fig. 1. A simplified diagram of the whole process [3].

With respect to the complexity of the presented process and the number of considered variables, 2 parameters of the process itself and 2 main material parameters, we have used second-order nonlinear model [11]:

$$w_i = a_0 + \sum_{j=1}^{k} a_i.t_{ji} + \sum_{j=l}^{k=j} \sum_{l=j+1}^{k} a_{ji}.t_{ij}.t_{ii} + ... + \sum_{j=1}^{k} a_{jj}.t_j^2 \qquad (1)$$

where
w_i is i-times measured value,
a_0, a_i .... a_{kk} are unknown parameters,
t_{1i}, t_{2i}, ..., t_{ji} are individual variables or their interactions.

Second-order models are used if the models of first-order do not describe the examined object sufficiently. Compared to the model of first-order, the experiment with second-order model is more complex, involve fact on more levels and therefore also higher number of measurements. Each variable within the second-order experiment must be changed on at least three levels. Currently there is a high number of second-order proposals based on various criteria and presenting various characteristics.
If we consider a system that has 4 parameters, the output variable can be considered as a function with 4 variables $\rho = f(P,T,Wr,L)$. Then we will consider this function in shape [3]:

$$\rho = b_0 + b_1.P + b_2.P^2 + b_3.T + b_4.T^2 + b_5.Wr + b_6.Wr^2 + b_7.L + b_8.L^2 \qquad (2)$$

where
b_i are regression parameters,
P, T, Wr, L are controllable variables.

Then for the estimation of regression function coefficients is valid [3]:

$$\hat{b} = (x^T x)^{-1} x^T \rho \qquad (3)$$

where
$\hat{b}$ is a estimates vector of unknown parameters of polynom,
x is the matrix of the plan,
ρ is the vector of the measured density values.

In this relation (3), the weight of individual measurements is not taken into account, so we extend the relationship by a covariance matrix. The previous relation (3) goes into the shape [3]:

$$\hat{b} = (x^T U^{-1}(\rho)x)^{-1} x^T U^{-1}(\rho)\rho \qquad (4)$$

where the vector $\hat{b}$ has the form [3]:

$$\hat{b} = \begin{pmatrix} b_0 \\ b_1 \\ . \\ . \\ b_8 \end{pmatrix} \qquad (5)$$

The design matrix x has the form [3]:

$$x = \begin{pmatrix} 1 & P_1 & P_1^2 & T_1 & T_1^2 & Wr_1 & Wr_1^2 & L_1 & L_1^2 \\ 1 & P_2 & P_2^2 & T_2 & T_2^2 & Wr_2 & Wr_2^2 & L_2 & L_2^2 \\ \cdot & \cdot & \cdot & \cdot & \cdot & \cdot & \cdot & \cdot & \cdot \\ \cdot & \cdot & \cdot & \cdot & \cdot & \cdot & \cdot & \cdot & \cdot \\ \cdot & \cdot & \cdot & \cdot & \cdot & \cdot & \cdot & \cdot & \cdot \\ 1 & P_m & P_m^2 & T_m & T_m^2 & Wr_m & Wr_m^2 & L_m & L_m^2 \end{pmatrix} \tag{6}$$

where the m determines the number of measurements.

Vector ρ has form: $$\rho = \begin{pmatrix} \rho_1 \\ \rho_2 \\ \cdot \\ \cdot \\ \rho_m \end{pmatrix} \tag{7}$$

The $U(\rho)$ covariance matrix has the form [3]:

$$U(\rho) = \begin{pmatrix} u^2(\rho_1) & u(\rho_1,\rho_2) & u(\rho_1,\rho_3) & \cdot\ \cdot & \cdot & u(\rho_1,\rho_m) \\ u(\rho_2,\rho_1) & u^2(\rho_2) & u(\rho_2,\rho_3) & \cdot\ \cdot & \cdot & u(\rho_2,\rho_m) \\ \cdot & \cdot & \cdot & \cdot\ \cdot & \cdot & \cdot \\ \cdot & \cdot & \cdot & \cdot\ \cdot & \cdot & \cdot \\ \cdot & \cdot & \cdot & \cdot\ \cdot & \cdot & u(\rho_{m-1},\rho_m) \\ u(\rho_m,\rho_1) & u(\rho_m,\rho_2) & u(\rho_m,\rho_3) & \cdot\ \cdot & u(\rho_m,\rho_{m-1}) & u^2(\rho_m) \end{pmatrix} \tag{8}$$

The result of composing the previous equations is calculation of the regression function parameters, with which we are able to gain the final shape of the regression function for further work.

2.2 Design of experiment (DOE)

In order that the experiment can be effectively and correctly realized, various tools are required that support both the creation and the manipulation of design-space descriptions [13-17]. Design-space descriptions are therefore usually based on design of experiments. Densification process is often dependent on a relatively large number of variables. It is practically impossible to test every combination of variables during process development in order to determine the relevant correlations between the individual variables. DOE uses a minimal number of experiments to provide an empirical process model for the interrelationship between the control and disturbance variables in the process and the resulting product and process characteristics [15]. On the Figure 2 in general the methods of statistical design of experiments are displayed [16]. Fractional factorial designs (screening designs) provide the possibility of significantly reducing the number of experiments. Results from screening designs can also be transferred to a subsequent series of experiments with less investigated factors, known as response surface designs [15, 16, 17]. Response surface designs are used to determine and then optimize non-linear interrelationships [17].

In the following Table 1 the chosen levels for each variable (raw material and process parameters) can be seen. Mentioned monitored variables (outputs) according to the technical standards and experimental device possibilities should be chosen. Outputs should represent quality indicators and these variables should have also a function of comparative criterions.

In our case, more than 2 levels of variables were chosen, than we have to use the Central Composite Design (CCD). Second but also very important reason for choosing of CCD in our case is volition to use surface response method, where 3 levels of variables is necessary condition for surface response modeling. The surface response method is a tool to investigate the response of a variable to changes in a set of design or explanatory variables and helps to find the optimal method for the response as a measurable output of the our interest [15,16].

Table 1. Input controllable variables of the process [3, 13].

Number of levels	Levels according CCD plan	Variables			
		Compression pressure - "P" (MPa)	Compression temperature - "T" (°C)	Particle size - "L" (mm)	Moisture content - "Wr" (%)
1	-α	63	55	0 – 0.5	5
2	-1	95	85	0.5 – 1.0	8
3	0	127	100	1.0 – 2.0	10
4	1	159	115	2.0 – 4.0	12
5	α	191	130	4.0 <	15

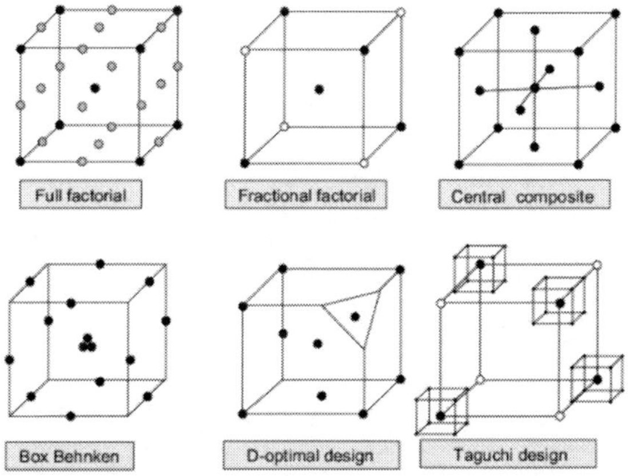

Fig. 2. Methods of statistical design of experiments [16].

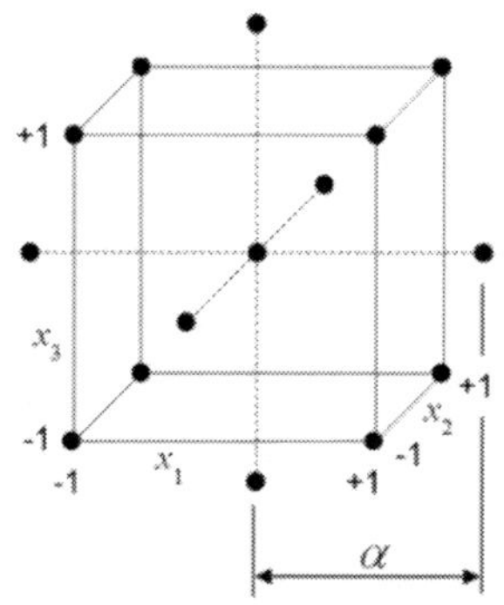

Fig. 3. Scheme of CCD plan [16].

The base of our CCD plan is two-level full factorial plan which is complemented by central and axial points. Value of number α and a number of central points on the base of other requirements (orthogonality, rotatability, number of variables, etc.), [11] related the CCD plan have to be calculated. On the Figure 3 can be seen the method by which can be CCD realized. In our case, due to the high number of examined variables and number of their levels we have decided for CCD of experiment consisting of three parts:

<u>Core of design</u> – is formed by factorial design 2^k. If we want to reduce the number of measurements, the core can be used as reduced experiment 2^{k-p}. For number of variables $k<4$ only complete experiment 2^k can be used as a core. For number of variable $5 < k < 7$ we use

the complete experiment 2^k or the reduced experiment 2^{k-p} as a core. For number of variables k > *7* also reduced experiment 2^{k-2}.
Axial points – points that are on the axes in distance of $\alpha > 0$ from the centre of design. Their number is $n_s = 2k$.
Central points – points in the centre of the design. There are $n_0 > 0$ of central points. By suitable selection of "α" and the number of central points - "n_0," the proposals characteristics can be influenced. CCD for our case is presented in following Table. 2.

2.3 Experiment realization

A set number of samples for testing will be produced according to the designed experimental plan shown in Table 2. The testing and sample's production process should be randomize order to avoid the effects of systematic errors or we can increase the number of replications for a given setting of variables. Raw material for experiment has to be treated. Wood sawdust have to be separate for obtaining the given particle sizes and also have to be prepared for achieving the given moisture content level. Final samples were produced by experimental pressing stand . In the final, the outputs variables were measured by appropriate measuring equipment for final density obtaining. The values of measured parameters can be calculated as ρ_1, ρ_2, ..., ρ_j, where $j = 1, 2, ..., r$ (in our case, r = 7), for experiments 1 t- 16. When evaluating, only measured values which characterize the core of the experiment for the centered composite design $2^4 = 16$ are used. From these measured values, the selection averages are calculated.

Table 2. Centered composite design with specific value of examined variables [3].

i	P (MPa)	T (°C)	Wr (%)	L (mm)
1	95	85	8	0.5 – 1.0
2	159	85	8	0.5 – 1.0
3	95	115	8	2.0 – 4.0
4	159	115	8	2.0 – 4.0
5	95	85	12	0.5 – 1.0
6	159	85	12	0.5 – 1.0
7	95	115	12	2.0 – 4.0
8	159	115	12	2.0 – 4.0
9	95	85	8	2.0 – 4.0
10	159	85	8	2.0 – 4.0
11	95	115	8	0.5 – 1.0
12	159	115	8	0.5 – 1.0
13	95	85	12	2.0 – 4.0
14	159	85	12	2.0 – 4.0
15	95	115	12	0.5 – 1.0
16	159	115	12	0.5 – 1.0

In this par, the processing methods and evaluation of the experiment in general are described. Initial information about the effect of a particular variable is obtained mostly with the aid of the multivariation diagram. In a multivariation diagram, the maximum and minimum values of a certain group are shown. The variability in one group is characterized by a certain range that corresponds to the length of the segment connecting both points [14]. This diagram allows us to monitor level changes and variability at the same time. When the multivariation diagram indicates differences in averages and variability on several levels, it is possible to use statistical tests for further verification. Evaluating the divergence of averages in groups makes it possible to use procedures derived from the control chart – average analysis (ANOM) or variance analysis (ANOVA) [16]. Both these methods assume normal distribution of the observed character and the same variability for some groups, which indicates an insignificant difference between selection distribution and range. The second of these assumptions verifies whether the Bartlett's test, typically used in relation with ANOVA or by the control diagram of range (R-diagram) or for variance (s^2-diagram), correspond by their character to ANOM.

Results of the experiment realized according to designed experimental plan also surface response can be used. Surface response will represent a group of points which forms a continuous surface when each axes in orthogonal view represents variables which influences the process and also the output variable [15, 17]. Influence of two input variables are related to the third axes, which mainly represents the output variable. With this type of diagram can be monitored a behavior of output variable according to the changes of influencing (input) variables. On the Figure 5 can be seen a result response surface with values of variables. Color scaling is helpful and is used for better orientation when the maximum, minimum or transition is wanted.

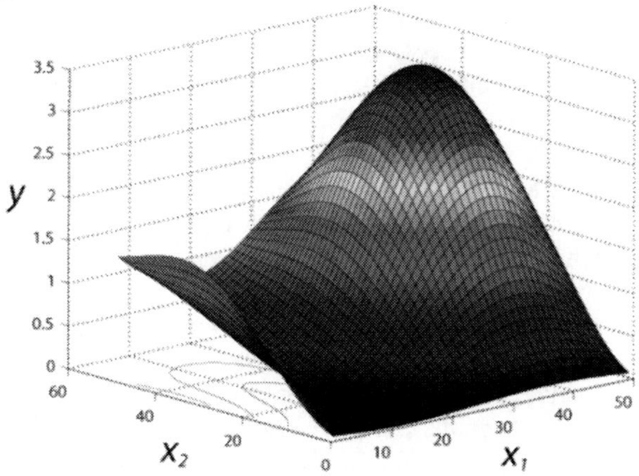

Fig. 4. Experiment result displayed through the surface response [15].

To we know create a surface response, it is necessary to have a "prescription" under which the points in a coordinate system will be bring out, in other words, a mathematical model that given us a three-dimensional function displayed as surface response will create. In nowadays for creation of mathematical models are various mathematical - statistics software widely used, which enable comfortable testing of hypothesis, testing of suitability of obtained data (values) and testing of model quality itself. With mathematical - statistics software can be generated many mathematical models, which described the process with some precision and can be chosen that model which suits us from variables composition and required precision points of view. We can find the optimal mathematical model from described points of view.

In our case multi-factorial analysis of variance (ANOVA) was used, because we have several input variables which influencing the process. ANOVA is used to compare and evaluate mean differences between two or more groups of variables on a single variable. Widely used mathematical - statistics software for all necessary analysis are e.g. JMP, Stat graphics, etc. This software can be used also in our case. Result of the mathematical - statistics software is following model with factorial design

$$y_{ijkl} = \mu + \alpha_i + \beta_j + \gamma_k + (\alpha\beta)_{ij} + (\alpha\gamma)_{ik} + (\beta\gamma)_{jk} + (\alpha\beta\gamma)_{ijk} + \varepsilon_{ijkl} \tag{9}$$

where:
μ - is level constant,
α_i - is share of i-th level of variable x_1,
β_j - is share of j-th level of variable x_2,
γ_i - is share of k-th level of variable x_3,
$(\alpha\beta)_{ij}$ - is combined share of i-th level of variable x_1 and j-th level of variable x_2 (interaction),
$(\alpha\gamma)_{ik}$, $(\beta\gamma)_{jk}$, $(\alpha\beta\gamma)_{ijk}$ - represents individual combined shares of given variables (their mutual interaction),
ε_{ijkl} - is share of l-th observation (error).

3 Evaluation and results

The first step in evaluation is to perform an average analysis and test the variance equality hypothesis using Bartlett's test [18]. This test can be employed when differences between selection variances s_i^2 or the range R_i are insignificant. The results of Bartlett's test obtained using the program *STATGRAPHIC S Plus* is the value *P* – the level of calculated significance, according to which the hypothesis H_0 is accepted or rejected on the level of specified significance $\alpha = 95$ %.

This means that, in order to accept the variance equality hypothesis, a value of *P* larger than 0.05 is necessary [18]. The analysis results returned values of P = 0.55001, which indicates that the hypothesis is acceptable. In this way, we satisfy one of the conditions for additional variance analysis using the method ANOVA. The selection can be regarded as homogeneous with respect to variability. This fact, based on the results, was also confirmed by the value of Cochran's tests.

The affecting factors are seen through the significant differences between selection averages [18, 19]. This means that the fluctuation of the selection averages over the total average is larger than the fluctuation of each result within the selection. Variance analysis, contrary to selecting averages, compares two variance estimates describing the variability type. For this analysis, the same assumptions apply for the averaging analysis. The role of the variance analysis is to find the effect of one or more factors on the average value of independent, normally divided, random variables with the same variance. In the table for the ANOVA method (F values), for the defined significance α, a certain number of degrees of freedom and critical value $F_{1-\alpha}$, are found. In another step, the ratio S_A^2/S_e^2 is compared with the critical value $F_{1-\alpha}$. If the value of the ratio S_A^2/S_e^2 is larger than that of the critical value of $F_{1-\alpha}$, the effects of this factor are considered to be significant [18, 20].

Table 3. Variance analysis table– results.

Variability source	**Sum of squares**	**Degrees of freedom**	**Average square**	**Ratio F**	**Value P**	**Significance**
A – pressure „P“	0.0210523	1	0.0210523	3.40	0.0923	3.
B – temperature „T“	0.13857	1	0.13857	22.37	0.0006	1.
C – moisture „W$_r$“	0.0709649	1	0.0709649	11.46	0.0061	2.
D – particle size „L“	0.0107137	1	0.0107137	1.73	0.2152	4.
Residual	0.0681389	11	0.00619445			
Sum	0.30944	15				

For this analysis and the determination of each factor's significance, the program *STATGRAPHIC S Plus* was used once again. The analysis results are listed in Table 3. The effect of each parameter or its significance is given by the value of *P* – level of calculated significance, which we compared. The effect of the parameter is considered important at a significance level of P^* (95 %); when the value of P is less than $(1-P^*)$, this means $P < 0.05$. From this, and based on the analysis results listed in Table 3, it is implied that the order of significance for each of the parameters are as follows: compression temperature, raw material moisture content, compression pressure, particle size.

For a closer determination, not only of the effect of each parameter but also of their mutual interactions, a factor effect method can be used. The degree of influence of a factor on an observed variable is known as the factor effect, which is defined as the magnitude of response change caused by a change in the factor level. It is often referred to as the main factor effect, unlike the interaction effect. The magnitude of the main effects and their interaction are estimated. Estimated (calculated) values of each of the effects are shown in Table 4, and Figure 5 illustrates the effect differences.

Table 4. The effect value of each parameter and their interaction.

B	C	BC	A	D	ABC	AC	AB
0.0465	0.0333	0.0254	0.0181	0.012938	0.012881	0.012065	0.008587
BCD	**AD**	**ABCD**	**CD**	**ABD**	**ACD**	**BD**	
0.00387	0.002897	0.002367	0.002005	0.000963	0.000661	0.000396	

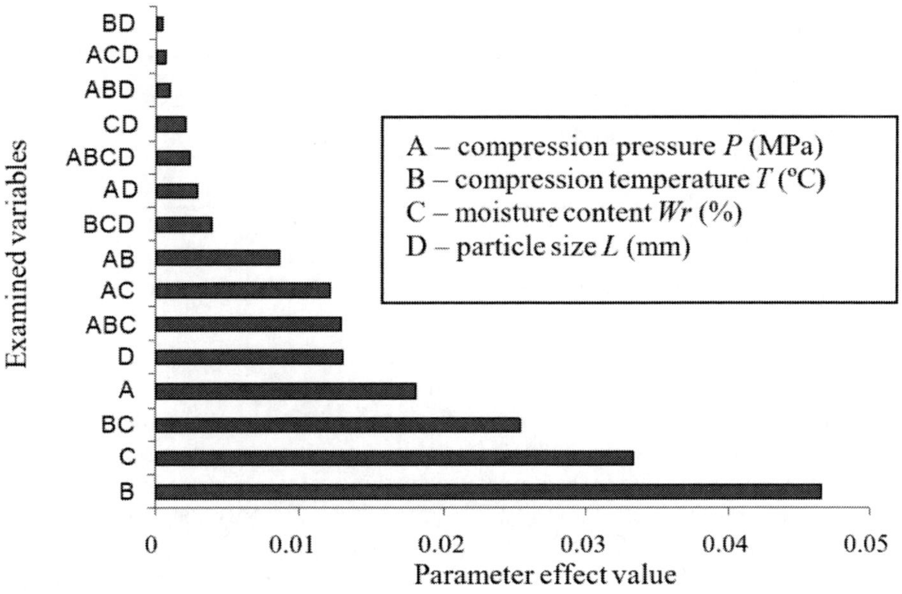

Fig. 5. Paret´s effect diagram [21].

We supposed that the dependency is given by a continuous function of four variables $\rho=f(P,T,Wr,L)$, which is expanded into a Taylor's series. A mode, which includes linear variables and the interaction of independent linear variables, is assumed. The model can be simplified into a functional dependence:

$$\rho = f(A,B,C,D,AB,AC,AD,BC,BD,CD,ABC,ABD,ACD,BCD,ABCD) \quad (10)$$

For a closer description of designing the model from the measured data, it is necessary to decide which of the variables in the relation (10) should be included in the model and which can be disregarded. Some variables have a negligible effect on the density (Figure 5) and can consequently be disregarded. A valid criterion is necessary to select use for the evaluation and comparison of variables in the model and regression model, respectively. Currently no theories exist that provide a precise specification of variables in a model. Nevertheless, there are several criteria that are suggested for use when selecting variable specifications for evaluating and comparing several regression models.

The next step was execution of selection of variables for mathematical model. For this selection we used three widely known criterions: index of multi-launching determination (R^2), Akaike's criterion (AICc) and Root mean squared error criterion (RMSE) [3]. Based on the three previously described criteria, we will compare and evaluate the designed models with the variables. For the selection and evaluation of models, the software JMP 8 is used. This software calculates the values for each of the described criteria for each designed mode. With the aid of JMP 8, we have calculated the values for the observed criteria (Table 5). These three criteria indicate that Model 3 is the most accurate.

Table 5. The best models calculated with JMP 8 on the basis of the evaluating criterion.

Model	**Parameters**	**Number of parameters**	**R²**	**AICc**	**RMSE**
1	B,C,BC,A	4	87.9 %	-30.1238	0.058448
2	B,C,BC,A,D,ABC,AC	7	96.8 %	-24.8346	0.035139
3	**B,C,BC,A,D,ABC,AC,AB**	**8**	**99.3 %**	**-32.7996**	**0.017764**
4	C,AB,AC,BC,ABCD	5	94.3 %	-35.5702	0.041982
5	B,C,AB,AC,BC, ABCD	6	96.1 %	-33.239	0.036413
6	B, C, BC, A, D	5	91.3 %	-28.8272	0.05183

We choose model with containing of all parameters and also with their most important mutual interactions (relation 11). This model had the best results after selection by criterions.

$$\rho = f(A, B, C, D, AB, AC, BC, ABC) \qquad (11)$$

In the next phase we were able to design the mathematical model with help of software SAS and JMP 8 and also we were able to calculate or estimate the regression parameters values [21]. The final designed form of mathematical model you can see in following relation 12, where the "ρ" is solid biofuel density. This model is valid only for densification of pine sawdust and is valid only by variables levels listed in Table 1.

$$\rho = e^{\left(\begin{array}{l} 4.98371-0.0261781p-0.0410292T-0.620594w_r-0.015446L+0.000228845p.T+ \\ 0.0031851p.w_r+0.00528717T.w_r-0.0000273004p.T.w_r \end{array}\right)} \quad (\text{kg.dm}^{-3}) \qquad (12)$$

Then, using the designed mathematical model, according to (12), the density values at levels according to the experimental plan in (Table 2) were calculated and compared with the densities of the solid biofuels obtained from experiment at these configurations.

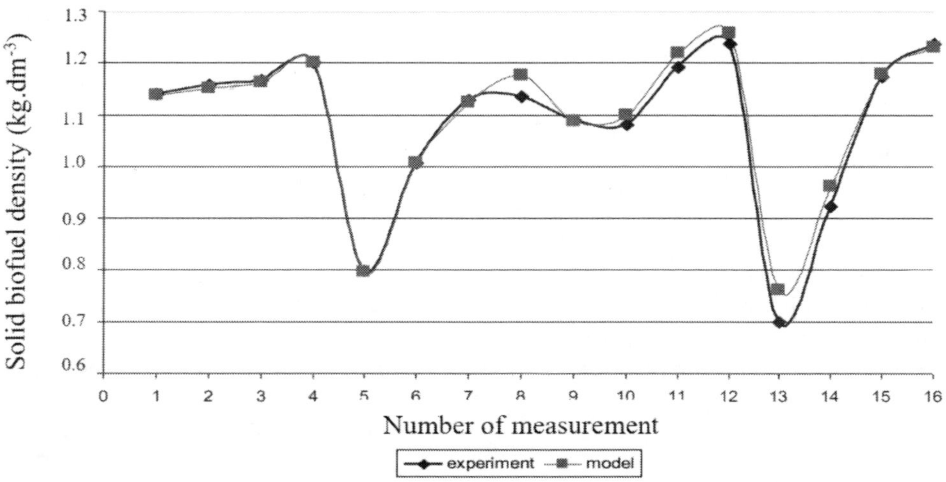

Fig. 6. Graphic comparison of solid biofuels densities obtained from experiment and calculated from the regression model.

If we want to look closely on the change of solid biofuel density, we have to use the "density response surface". Response surface is a great and useful tool for presenting the process in maximum amplitudes of individual parameters [21]. The result of experimental research can be presented in 2-axis interdependencies, but it is limiting to present the functionality with fixation of several parameters. Using the response surface we can get clear presentation of the process within the given extend by the means of surfaces [21, 22]. These represent set-up levels. The following figure shows the effects of interacting compression pressure and temperature on the density of the biofuels from pine sawdust at various moisture content levels.

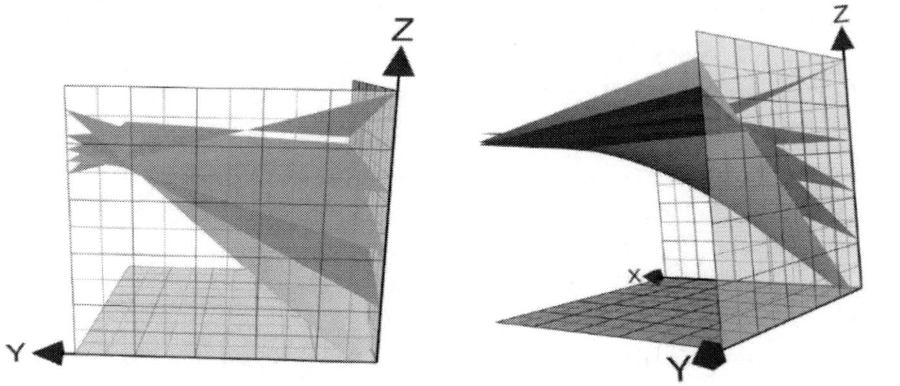

Fig. 7. Surface response of the density of biofuels from pine sawdust at a constant particle size / Labeling: x-axis → compression pressure (MPa); y-axis → compression temperature (°C); z-axis → biofuels density (kg.dm^{-3}); colour legend → levels of raw material moisture content: red = 5 %; green = 8 %; purple = 10 %; blue = 12 %; yellow = 15 %/.

4 Conclusion

Design of experimental plan, methodology for evaluation of experimental research and research findings regarding the important variables effect during densification of pine sawdust was presented in this paper. Authors presented in detail process and methodology of

DOE a evaluation process when the mathematical model have to be the final result. Presented mathematical-statistic methods are widely known, but the usage and application in this area is not usual. Presented research findings relates to the effect of material and densification process variables determination on final solid biofuels density. With application of mathematical and statistical methods we were able to design mathematical model of single axis densification of pine sawdust.

The main conclusions that can be drawn from this study are as follows:

• Applicability of mathematical and statistical methods was proven.

• All of the investigated input variables have an effect on monitored outputs.

• Compression temperature, raw material moisture content and mutual interaction of these two variables are most significant variables at densification of pine sawdust.

• The final solid biofuels density can be increased by a sensitive levels selection of influencing variables.

• Designed mathematical model can be used for prediction of solid biofuels density.

The paper is a part of the research done within the project VEGA 1/0420/16 – "Determination and research of parameters influence on final briquettes quality at waste biomass densification" funded by the Ministry of Education of Slovak Republic and to the Slovak Academy of Sciences, and also the paper was supperted by the project APVV- 16-0485 "Biomass compacting tools cast from the advanced wear-resistant cast irons" funded by the Slovak Research and Development Agency."

References

1. EN ISO 17225-2:2014, Solid biofuels – Fuel specifications and classes – Part 2: Graded wood pellets, (2014)
2. R. Nosek, M. Holubcik, J. Jandacka, BioRes., **11**, 44-53, (2016)
3. P. Križan, *The densification process of wood waste*, (De Gruyter open, 2015)
4. C. Serrano, E. Monedero, M. Lapuerta, H. Portero, F.Pro.Tech., **92**, 699-706, (2011)
5. S. Mani, L.G.Tabil, S. Sokhansanj, Bio.Bio., **30**, 648-654, (2006)
6. N. Kaliyan, R. Vance Morey, Bio.Bio, **33**, 337-359, (2009)
7. P. Lehtikangas, Bio.Bo., **19**, 351-360, (2001)
8. M. Baláš, M. Lisý, O. Štelcl, A.Pol., **52**, 7-11, (2012)
9. P. Kracík, M. Balas, M. Lisy, J. Pospíšil, Mat. Teh., **50**, 939-944, (2016)
10. M. Holubcik, J. Jandacka, M. Palacka, N. Kantova, E. Jachniak, P. Pavlik, Com.-Scie.Let.Uni.Za, **19**, 94-100, (2017)
11. B. Wortman, *The Certified Six Sigma Black Belt Primer*, (Quality Council of Indiana, 2001)
12. P. Križan, M. Matúš, Ľ. Šooš, J. Beniak, Ener., **8**, 6382-6398, (2015)
13. M. Matúš, P. Križan, Apl.-J.App.Math., **3**, 87-96, (2010)
14. T.P. Ryan, *Statistical Methods for Quality Improvement*, (John Wiley and Sons, New York, 2011)
15. S.R. Schmidt, R.G. Launsby, *Understanding Industrial Designed Experiments*, (Air Academy Press, Colorado Springs, 1994)
16. A.J. Duncan, *Quality Control and Industrial Statistics*, (Richard D.Irwin, Inc., Reno, 1986)
17. T.H. Wonnacott, R.J. Wonnacott, *Statistika pro obchod a hospodářství*, (Victoria Publishing, a.s. Praha, 1993) (in Czech)

18. R. Palenčár, I. Janiga, A. Horníková, J.M. Ruiz, *Štatistické metódy v metrologických a skúšobných laboratóriách*, (Publisher Ing. Peter Juriga, Bratislava, 2001) (in Slovak)
19. V. Chudý, R. Palenčár, E. Kureková, M. Halaj, *Meranie technických veličín*, (STU Bratislava, 1999) (in Slovak)
20. R. Palenčár, N. Jankura, *Proceedings of 5th International Conference on Measurement 2005*, (Smolenice, Slovak Academy Sciences, 2005)
21. P. Križan, Ľ. Šooš, M. Matúš, J. Beniak, M. Svátek, W.Res., **60**, 301-315, (2015)
22. P. Križan, M. Svátek, M. Matúš, M., *Aplimat 2011. Proceedings.* (Bratislava, FX s.r.o., 2011)

High-temperature reaction of sodium vapour with quartz glass

Radovan Nosek[1,*], *Zuzana* Kolkova[2], *Libor* Trško[2], *Tatiana* Liptáková[3], *Jan* Procházka[4], *Ondřej* Libera[4], and *Petra* Bublíková [4]

[1]University of Žilina, Faculty of Mechanical Engineering, Department of Power Engineering, Univerzitná 8215/1, 01026 Žilina, Slovakia

[2]Research Centre of the University of Žilina, Univerzitná 8215/1, 01026 Žilina, Slovakia

[3]University of Žilina, Faculty of Mechanical Engineering, Department of Material Engineering, Univerzitná 8215/1, 01026 Žilina, Slovakia

[4]Research Centre Řež, Hlavní 130, 250 68 Husinec, Czech Republic

Abstract. Heat pipes are one of the most efficient ways to transfer thermal energy hundred times more efficiently than copper. In this work, we present the investigation of sodium terminal velocity inside a glass pulsating heat pipe. Velocity measurements were conducted under operating conditions within the range of 500 – 1 100 °C. Since sodium is generally able to etch glass and ceramic materials, its presence resulted in glass reduction and sodium oxidation. From the XPS analysis of specimens extracted from a glass pipe after sodium explosion, it can be concluded that the reaction products are sodium oxide Na_2O and a thin layer of carbon, which is localized on the SiO_2 glass. The character of damage induced by chemical reactions is analysed in this manuscript.

1 Introduction

Primary aluminium is the material obtained from electrolytic cells or pots during an electrolytic reduction of metallurgical alumina (aluminium oxide). Therefore, it excludes alloying additives and recycled aluminium. The electrolytic cell uses electrolysis to produce aluminium in metal form. It means, that large currents must be used in order to refine the aluminium oxide to the pure aluminium. Unequal current distribution inside aluminium electrolytic cells causes energy losses, a phenomena impacting both the economy and the environment. Around 50 % of the supplied energy is the lost as thermal waste. The heat losses are necessary to maintain a frozen side ledge to protect the side walls of cells. As the energy costs account for 40 % of the total production costs, it is clear that the heat recovery optimization offers a significant economical end environmental potential in the industry.

Although waste heat is abundant in primary production plants, possibilities of thermal integration remain limited. Some of the main reasons impeding waste heat recovery and thermal integration in smelters are the low quality of the waste heat (i.e., low temperature)

* Corresponding author: radovan.nosek@fstroj.uniza.sk

and large distances between heat sources and locations where heat could be used (herein called heat sinks). Extracting the waste heat (capturing it in a heat transfer fluid) also requires adapted heat-exchanger designs. The technology based on the use of heat pipes has been developed for utilization of energy from the waste heat produced in the electrolytic process. The heat recovery system can control the ledge thickness and thermal energy can be utilized by other energy demanding processes while increasing the aluminium production by 20 % and still achieve thermal balance. With the heat recovery system, 10-12 % of the total energy consumed during the process can be recovered [1].

The project aims was to develop an innovative heat-to-electricity converter without any moving parts, placed on the pulsating heat pipe (PHP). The device can generate electric power by flowing fluid interacting with the magnetic field. The project aim was to investigate the fluid velocity within the heat pipe during the flowing.

2 Technology of efficient energy utilization

The investigated technology for the energy utilization of waste heat consists of PHP, which is a two phase heat transfer device with outstanding effective thermal conductivity. It is a vacuum tight device consisting of an envelope (here made of quartz glass) and flowing fluid (liquid sodium), which transfers heat under isothermal conditions at vaporization [2]. The fluid used in this study can operate in the range of 500 – 1 100 °C. Due to its electron configuration with a single electron in the valence shell, sodium is a very reactive element with one of the lowest electronegativity. It was expected, that its high chemical reactivity will be mitigated by high vacuum [3]. The evaporation part of the PHP was heated up to 750 °C which resulted in immediate explosion of the device (Figure 1).

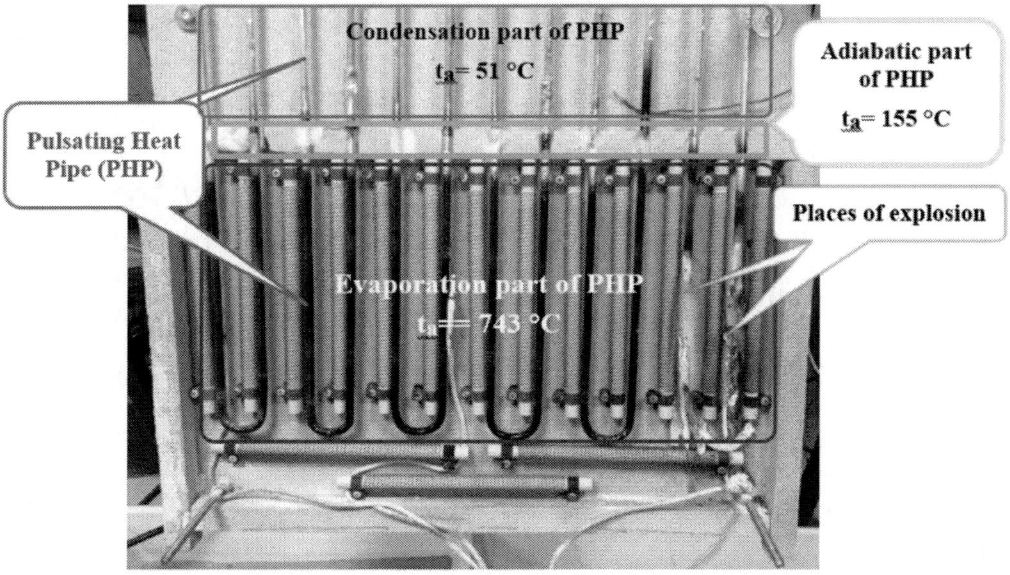

Fig. 1. Detail of the PHP experimental setup and the damage caused by explosion.

It is visible that the character of compounds inside the PHP changes over different parts (condensation – adiabatic transfer – evaporation) is caused by different temperatures. The average temperature (t_a) in each PHP part is shown in Figure 1. For the identification of the black substance covering the pipe walls, the Scanning Electron Microscopy (SEM) equipped with Energy Dispersive X-Ray Spectroscopy (EDS) and X-Ray photoelectron spectroscopy (XPS) methods were used to analyse the factors causing the explosion. The XPS analysis shows the presence of pure carbon in high concentration upon the quartz pipe wall (Table 1.).

Table 1. Chemical composition of the black products on the pipe wall, XPS analysis.

Element	**Weight %**	**Atomic %**
Carbon C	81.11	87.79
Oxygen O	9.9	8.04
Silicone Si	9.0	4.17

Based on the XPS results, the explosion was caused by the high temperature followed by the devitrification of fused silica pipe due to sodium [4] and heterogeneity of reactions inside the experimental system. Increasing of sodium diffusion through the weakened walls and the structural failure were followed by chemical reaction of sodium with components of ambient atmosphere.

Under the thermal load, the devitrified areas tend to weaken the PHP and change its mechanical properties. Also, in such devitrified areas the crystallization of previous amorphous structure was observed. That was probably caused by diffusion channels for transfer of environmental contaminants into the molten sodium [5]. The mechanism of devitrification enhanced by sodium, involves diffusion of sodium into the fused quartz. The sodium is assumed to diffuse quickly through the open structure of the silica via strings of interconnected interstices [6]. The sodium sets up nucleation sites for crystallization. One can imagine a piece of fractured glass pipe from a wet oxidation has failed due to devitrification. It can be observed, that the heavily devitrified region is inside of the PHP wall. It can be caused by increasing rate of devitrification from water [7].

Furthermore, the SEM analysis of the damaged PHP exposed to the molten sodium at four different temperatures were carried out (Figure 2).

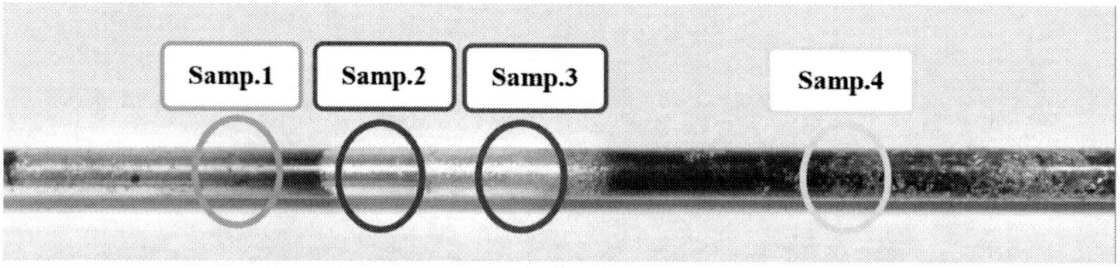

Fig. 2. Labelling of analysed areas of the exposed PHP.

For the SEM analyses, the quartz samples were mechanically cleaned of surface deposits and subsequently washed in water. The samples were mounted using the colloidal silver and sputtered by 20 nm thick carbon film, to ensure charge removal. The results of SEM (EDS) analyses (chemical mapping and point analyses) of all samples are shown in Figure 3-14. The sample areas were imaged in Secondary (SE) and Backscattered (BSE) electrons. For the disambiguation, the spectral lines are vertically shifted.

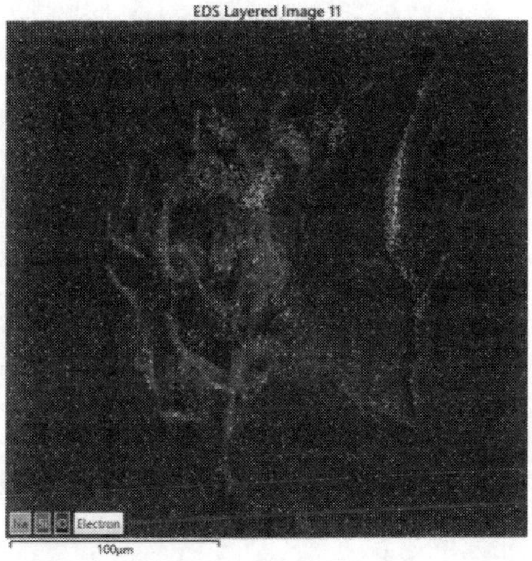

Fig. 3. Sample 1 – EDS mapping of sodium (green channel), silicon (red channel) and oxygen (blue channel) elements.

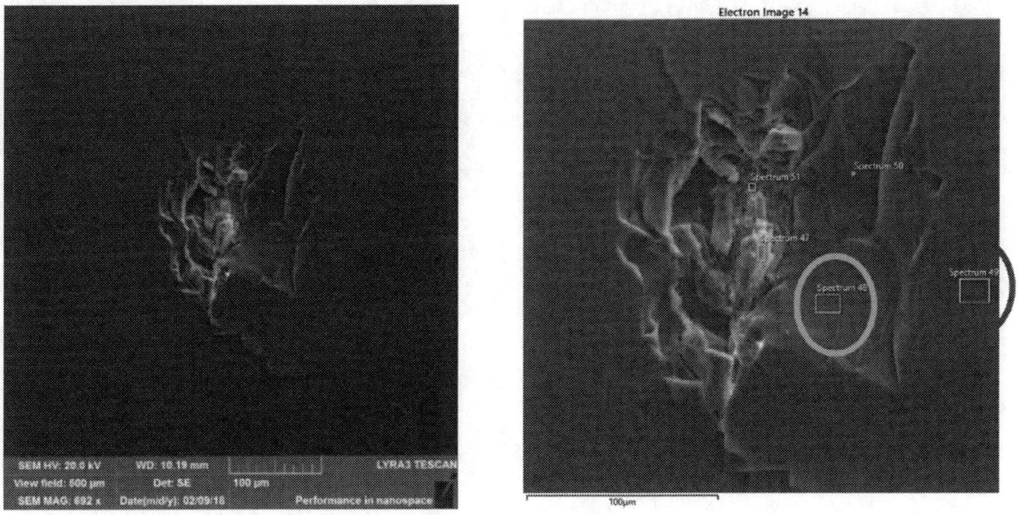

a) b)

Fig. 4. Sample 1 – SEM imaging of the inner surface (a). Detailed SEM image of the crack in the inner surface with analysed regions emphasized (b).

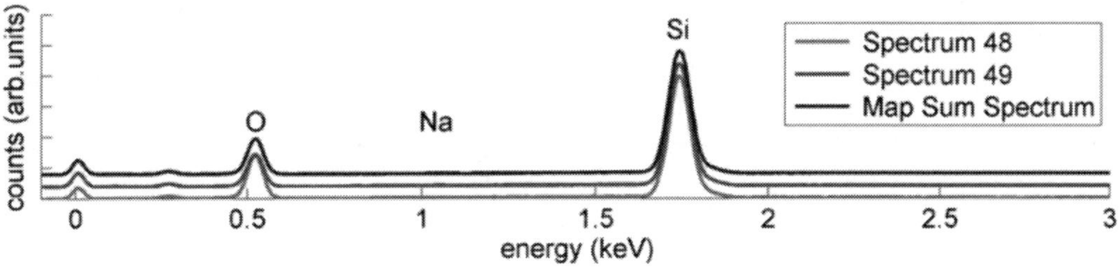

Fig. 5. Sample 1 – EDS spectrum of the region in Fig. 4a (blue line), the sodium-exposed surface (red line) and the uncovered surface (green line).

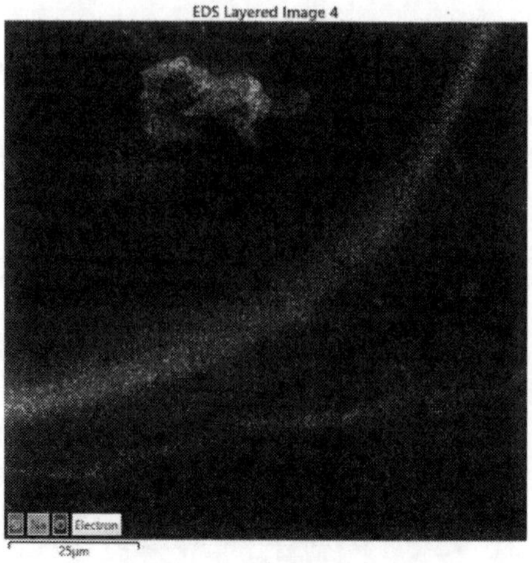

Fig. 6. Sample 3 – EDS element mapping.

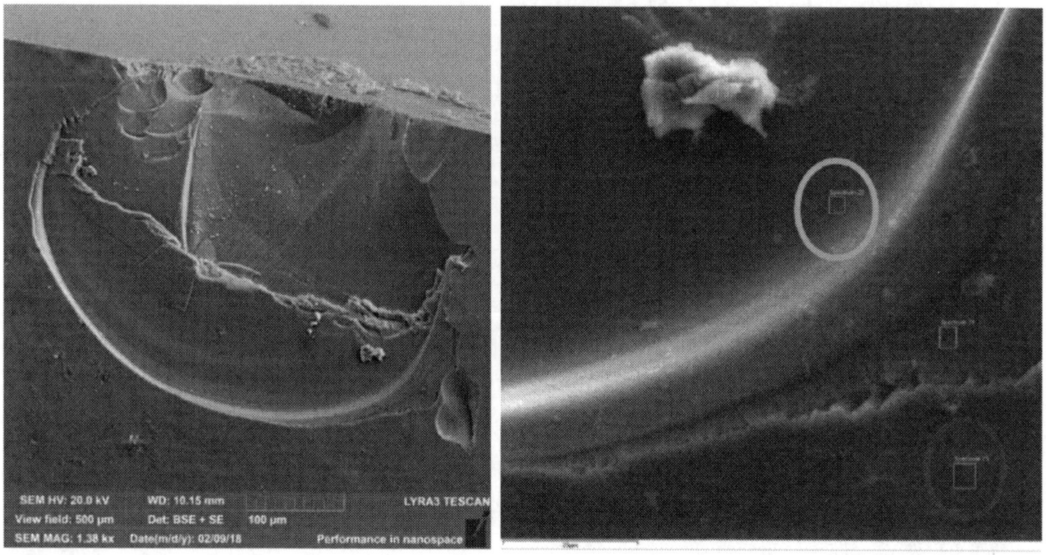

a) b)

Fig. 7. Sample 2 – SEM imaging of sample surface (a) and crack detail with labelled areas of EDS analysis (b).

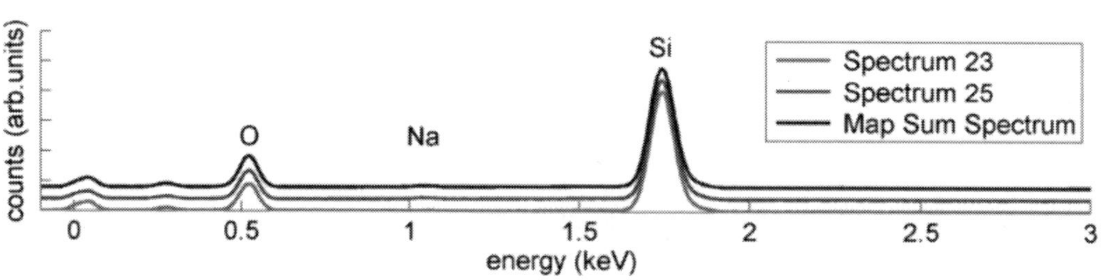

Fig. 8. Sample 2 – EDS spectrum of selected areas of the exposed material (Spectrum 23), the inner surface of the PHP (Spectrum 25) and EDS spectrum from the whole image (Map Sum Spectrum).

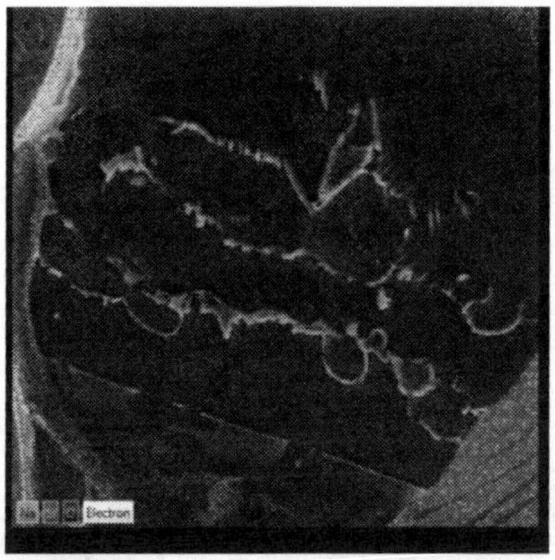

Fig. 9. Sample 2 – EDS element mapping.

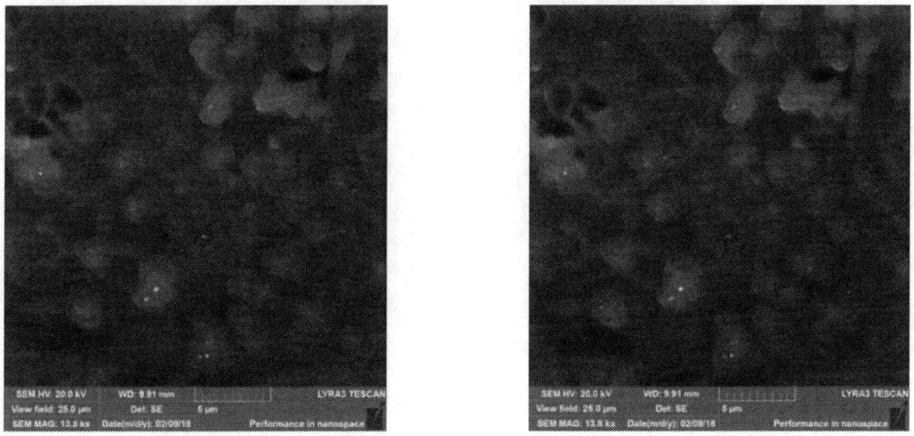

a) b)

Fig. 10. Sample 3 – Detailed image of high sodium concentration.

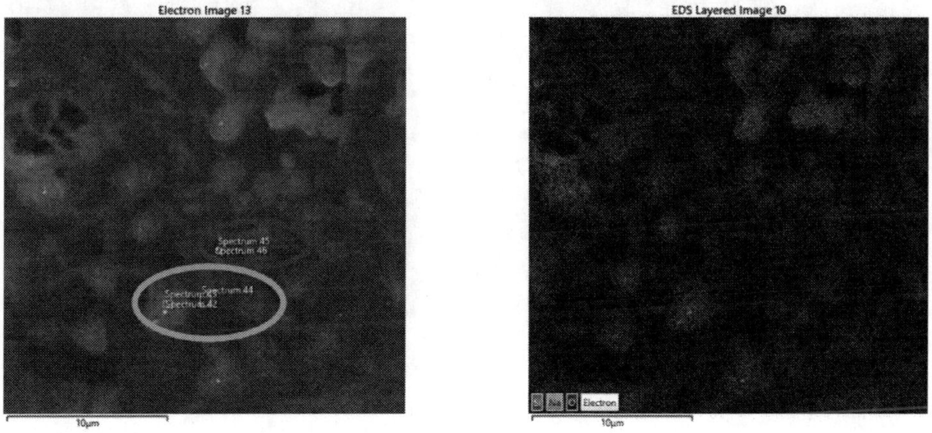

a) b)

Fig. 11. Sample 3 – Detailed image of high sodium concentration with marked areas of EDS analysis.

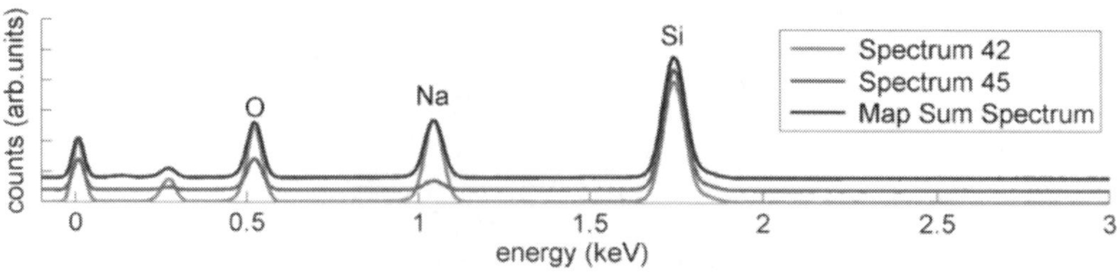

Fig. 12. Sample 3 – EDS spectra of selected areas of exposed material (Spectrum 42), the inner surface of the PHP (Spectrum 45) and spectrum from the whole image (Map Sum Spectrum).

a) b)

Fig. 13. Sample 4 – The inner material of the PHP. a) BSE image, b) SE image.

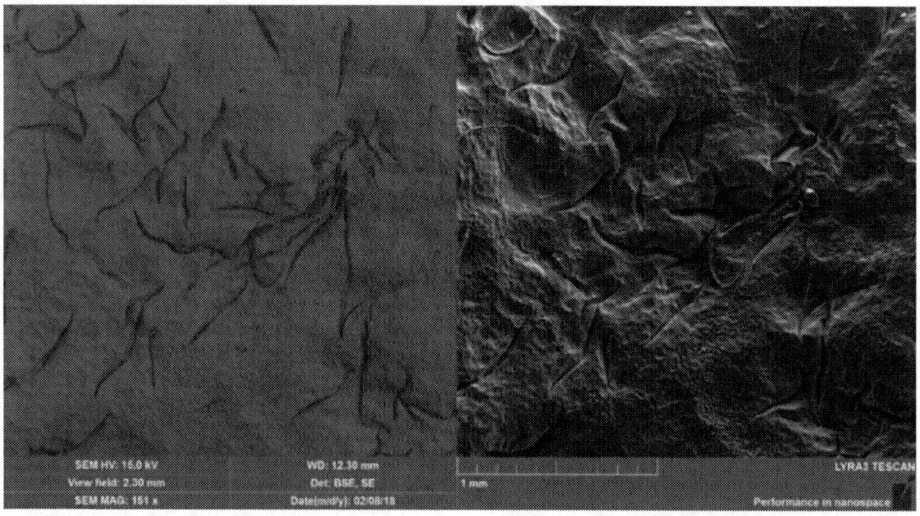

a) b)

Fig. 14. Sample 4 – Crack detail. a) BSE image, b) SE image.

Table 2. Chemical composition of pipe wall, EDS analysis.

	Uncovered surface of PHP			**Sodium-exposed surface of PHP**		
	O (At. %)	Si (At. %)	Na (At. %)	O (At. %)	Si (At. %)	Na (At. %)
Sample 1	64.2	35.6	0.2	68.6	31.4	0.0
Sample 2	64.2	35.0	0.8	64.0	35.7	0.3
Sample 3	54.4	24.5	21.1	60.2	38.6	1.2

Results from XPS and EDS analyses for samples 1 - 3 suggest formation of a sodium-enriched surface layer and broken silicon, oxygen stoichiometry. The structure of PHP is more fragile with increasing amount of sodium in the glass tube.

The sodium contamination of samples 1 and 2 was lower than 1 at.% and the sodium-exposed surface was smooth and free of deep cracks. The cracks observed were probably caused by excessive load at a time of explosion.

The sodium contamination of sample 3 is significantly higher and deeper compared to contamination of samples 1 and 2. The topmost sodium-exposed layer contains 21 at.% of sodium while the deepest region uncovered contains 1.2 at.% of sodium. As displayed in Figures 9-12, the sodium attack is highly localized by forming sodium-rich fibers (17-23 at.% Na) distributed in a silicon-rich (4-5 at.% Na) matrix. Such fibers form diffusion channels for further sodium contamination and act as vitrification and crack initiators.

Sample 4 exhibits the surface morphology completely different from samples 1 – 3. The sample is characterized by a rough surface full of deep cracks. (Note: Chemical analyses were not performed due to inappropriate sample sizes - the raised parts of the tube removed the signal for the EDS).

3 Conclusions

Based on the different temperatures in the PHP, it can be expected that samples 1 and 2 were particularly stressed by sodium vapour, which might explain the low sodium contamination of the pipe material. The colour transition from dim white to dark brown can be probably explained by an increasing contamination of the inner surface, where sodium replaced the silicon in the oxide structure. With an increasing distance from the molten sodium level, a lower temperature and lower sodium vapour pressure can be expected.

The black surface of sample 4 can be explained as a very intensive sodium attack on the amorphous SiO_2 structure by a molten sodium. The thermal effect gradient on the devitrification process shall be considered. Sample 4 was localized in the heated part of the experimental apparatus. In this area, heat is conducted through the wall in the molten sodium at higher temperatures, whereas samples 1 – 3 were localized in cooled part of the experimental apparatus where the thermal gradient is in opposite direction and temperatures are significantly lower.

The pipe wall thinning can be expected due to the chemical reaction of molten sodium with the PHP's wall. The pipe was not resistant to further mechanical stress caused by operating pressure thermal gradients.

The acknowledgement: This article is supported by the financial assistance of the project KEGA No. 042ŽU-4/2016 "Cooling by physical and chemical processes".

References

1. R. Nosek, et al., *AIP Conference Proceedings*, **1917**, 030001 (2017)

2. M. Jobb, L. Kosa, M. Holubcik, R. Nosek, *Applied Mechanics & Materials*, **832**, 184-191, (2016)
3. J. Krätsmár-Šmogrovič et al. *Všeobecná a anorganická chémia* (in Slovak), II. edition, Osveta ISBN 8080632458 (2007)
4. K.B. Richter, *Prevention of sodium diffusion through furnace tube quartz*, Lehigh University-Lehigh Preserve Theses and Dissertations (1984)
5. P.F. Schmidt, *Contamination-Free High Temperature Treatment of Silicon or other materials*, J. Electrochem. Soc., **130** (1983)
6. S.J. Rothman, *Diffusion of Alkali Ions in Vitreous, Silica*, 11 J. Am. Ceram. Soc.T, **65** (1982)
7. W. Heng-yu, W. Jia-fan, *A study on the devitrification of silica glass tube the diffusion furnace*, Silicate Journal, **9** (1981)

One dimensional modelling of flow in a helium loop

František Világi[1,*], *Branislav* Knížat[1], *Marek* Mlkvik[1], *František* Urban[1], *Róbert* Olšiak[1], and *Peter* Mlynár[1]

[1]Institute of Energy Machinery, Faculty of mechanical engineering STU, 812 31 Bratislava 1, SR

Abstract. The natural circulation helium loop is a facility designed for decay heat removal from ALLEGRO fast nuclear reactors. The article deals with the observation of pressure and velocity relations during steady flow of helium. A one-dimensional numerical model of flow capable of determining the velocity with sufficient accuracy is presented in the article. The model describes the flow of highly compressed gaseous medium with variable density in direct pipelines with local resistances. It's a hydraulic model, which means that the temperature distribution along the loop must be known. The article also includes the evaluation of local resistances in DHR and GFR, which significantly affects the resulting accuracy. The results from numerical model are compared with experiments.

1 Introduction

After the Fukushima nuclear disaster on 11. Mach 2011, the role of passive safety of nuclear power plants has increased considerably. The new main objective is to minimize the dependence on electrical supply. In the age of carbon dioxide emission regulations, the nuclear energy is an alternative to green energy sources, which puts passive safety systems on the path of sustainable development. New generation of nuclear power plants are being developed, which are based on the success of the use of nuclear power and the experience from operating facilities. Some new reactor solutions rely on passive systems that can meet safety requirements. In recent years, passive security systems have proven to be elements that can make a significant contribution to simplifying and potentially improving the economy of new nuclear power plants. The use of passive safety systems eliminates the costs associated with installation, maintenance and operation of active safety systems that require multiple pumps with independent and redundant power supplies. Consequently, passive safety systems are considered in new reactor concepts particularly in the IV. generation reactor designs [1]. Another incentive to use passive security systems is the potential for increased security through increased system reliability, because passive systems don't have any moving parts which could fail during operation. Another safety factor is the used medium, helium. Because of suitable chemical and physical properties helium can't get radioactive at used conditions and can't interact with other substances as

* Corresponding author: frantisek.vilagi@stuba.sk

it's chemically inert. It is non-toxic and has high values of thermal conductivity and specific heat.

2 Experimental facility

Helium loop should serve as a safety system for decay heat removal from nuclear reactors of IV. generation. Helium is supposed to be heated directly from the nuclear reactor and then to be cooled down in the heat exchanger (water). The flow of helium should be ensured only by the difference of densities in hot and cold piping branch. The power to drive is thus obtained directly from the heat supplied [2]. It is a vertical heater-horizontal cooler type, single-phase, closed loop. The experimental helium loop facility was built as a model of real helium loop supposed to be used in nuclear power plants and its purpose is the research of helium flow characteristics at various conditions. It is located in the area of Energomont s.r.o. in Trnava. It consists of two main elements, the GFR (reactor substitution) and the DHR (heat sink), along with two piping branches (hot and cold). The GFR is a heater with maximum installed power of 500 kW of which the adjustable power range is 220 kW. The projected heat output of the DHR is also 220 kW with possible water flow regulation. Pressure, pressure difference on the elements, helium temperatures, surface temperatures, water flow, inlet power and velocities are measured during the experiment at several points along the loop. Measurement of helium velocity is performed using a Pitot probe. The compressibility of helium can be neglected along the loop due to relatively small changes in pressure (maximum hundreds of pascals). Helium can achieve the following operating values based on the settings of GFR and DHR devices:

- Helium temperature at the inlet to GFR: 400 °C – 520 °C
- Helium temperature at the outlet from GFR: 150 °C – 250 °C
- Operating pressure: 3 MPa – 7 MPa

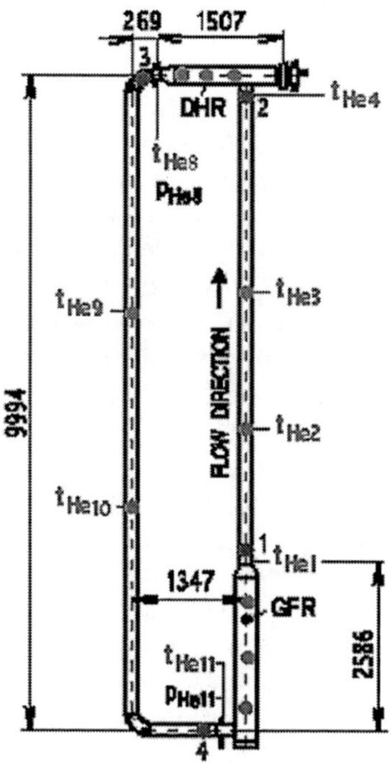

Fig. 1. Diagram of experimental helium loop built within the Allegro project.

The experimental facility provides an opportunity to verify and research the thermodynamic and hydraulic properties of a self-driven, highly compressed, helium flow. The facility enables modelling of changes in input and output power of elements, as well as changes in operating pressure. It is used for the research of steady and unsteady helium flow in a given range of pressures and temperatures [4].

Figure 1 represents the diagram of the experimental facility. Besides the geometry it also shows the positions of measuring elements for temperature, pressure and velocity. The facility encompasses an advanced system of automatized data collecting and evaluating. As the heat conduction through pipes to the surroundings is generally neglected, the most important values from the measurement are temperatures and velocities from the control points 1-4, which are marked with red colour. The measured temperature distribution along the loop is required as an input to the numerical model along with the operating pressure.

The experimental facility further allows: vacuuming of the primary circuit before it is filled with the test medium (He), controlling the primary circuit pressure in the range of 3 MPa to 7 MPa by releasing the test medium into the low-pressure reservoir, refilling the medium back into the high-pressure reservoir, continuous refilling of the test medium to the circuit from an external source in case of leakage of the test medium [4].

3 Hydraulic model of flow

The authors of this article describe a method based on previous works [2], [3]. The basis is that the flow depends on the difference of densities in the cold and hot branch. For a closed loop we can write the balance:

$$\left(\rho_{CB} - \rho_{HB}\right) gh = \sum_{i=1}^{n} p_{li} \tag{1}$$

Where the left side of the equation represents the potential energy available due to different densities in the cold (ρ_{CB}) and the hot (ρ_{HB}) branch. During steady flow, the equation must be valid as the flow can not continuously accelerate or slow down, so the potential energy gained must be spent on the pressure losses p_l. The sum of pressure change across the loop (i.e., 1-2-3-4-1) must be zero (steady flow):

$$\sum_{i=1}^{3}\left(p_i - p_{i+1}\right) + \left(p_4 - p_1\right) = 0 \tag{2}$$

The method uses a relation for calculation of pressure change between two points during steady flow [3].

$$p_{j+1} = \sqrt{\frac{p_j^2}{e^{\alpha}} - \lambda \frac{L}{d} \frac{zRT_{mean}}{A^2} Q_m^2 \left(\frac{e^{\alpha} - 1}{\alpha e^{\alpha}}\right)} \tag{3}$$

$$\alpha = \frac{2g\left(h_{j+1} - h_j\right)}{zRT} \tag{4}$$

where p_j – pressure at the inlet to pipe; h_j, h_{j+1} – height at the inlet and outlet of pipe, respectively; Q_m – mass flow, z – compressibility factor; v – mean velocity; A – cross-sectional surface. As temperatures must be known, this method serves in particular, to observe the effect of pressure loss on the mass flow during the operation. The equation contains two unknowns that are dependent on each other, namely the mass flow Q_m and the pressure at the next point p_{i+1}. Thus, the calculation was possible only with iterative method [3].

We calculate the pressure p_{i+1} with equation (3) from the pressure at the previous point. This way, we can gradually calculate the pressure at each control point on the loop and return to the starting point 1. As the initial pressure, we choose the total operating pressure - $p_{1,est}$ which is either chosen as the condition at which we want to calculate the velocity or in our case, this is the measured pressure at point 1. Then, the calculated pressure $p_{1,calc}$ can be obtained as follows:

$$p_{1,est} \xrightarrow{Hot\ Branch} p_2 \xrightarrow{DHR} p_3 \xrightarrow{Cold\ Branch} p_4 \xrightarrow{GFR} p_{1,calc} \quad (5)$$

When calculating the pressure at points 2, 3, 4, 1, the relationship (3) is always used. At places where the local losses are significant (GFR and DHR) the friction factor λ is expressed in terms of local losses coefficient ξ as described in section 3.2 equation (12). During steady flow, the pressure difference must be zero, so we can write

$$f(Q_m) = p_{1,calc} - p_{1,est} = 0 \quad (6)$$

Due to this condition, which must be kept, we can iteratively calculate the mass flow as well as the flow velocity from measured temperatures and operating pressure. The mean temperature T_{mean} that is in equation (3) is determined from the measurement as the average between two points. The calculation is performed by the iterative Newton method (7), which serves to find the root of the equation (6). The new mass flow rate after each iteration is determined as:

$$Q_m^{j+1} = Q_m^j - \frac{f(Q_m)}{f'(Q_m)} \quad (7)$$

For the first iteration, the initial value of the mass flow is chosen in the order close to predicted value. The derivation of the function is calculated by numerical derivation:

$$f'(Q_m) = \frac{f(Q_{m,h}) - f(Q_m)}{h} \quad (8)$$

The member $f(Q_{m,h})$ expresses the difference between the selected pressure $p_{1,est}$ and the pressure $p_{1,h}$. This pressure ($p_{1,h}$) is calculated analogously to $p_{1,calc}$ with increased mass flow by differential value: $Q_{m,h}^j = Q_m^j + h$ where $h = 10^{-5}$.

$$f(Q_{m,h}) = p_{1,h} - p_{1,est} \quad (9)$$

The stop condition is set to $Q_m^{j+1} - Q_m^j < \varepsilon$, where $\varepsilon = 10^{-5}$. When this condition is reached, the iteration cycle ends and the velocity at any point on the loop is calculated from the continuity equation:

$$v_i = \frac{Q_m}{\rho_i A_i} \quad (10)$$

3.1 Pressure losses due to friction

In order to calculate the losses due to friction a method available from the literature was used. The equation for calculation of friction factor, also known as the McKeon relation was used in the form:

$$\frac{1}{\sqrt{\lambda}} = 1.930\log\left(\mathrm{Re}\sqrt{\lambda}\right) - 0.537 \tag{11}$$

The equation (11) is a regression of measured data and is considered as one of the most accurate. The McKeon relation can be used only for smooth pipes and the literature [6] states the maximum error of 1.25 % for low Reynolds values in range of 0 < Re > 310.10^3, and error of 0.5 % for range 310.10^3 < Re > 18.10^6. In our case, the Reynolds numbers are approximately in the range of 15.000 to 37.000 unless we count the start-up.

3.2 Local pressure losses

The biggest hydraulic loss in the loop occurs in the heater (GFR) and in the cooler (DHR). These losses are introduced into equation (3) by calculating the friction factor related to local loss coefficient.

$$\lambda = \xi_{DHR/GFR}\frac{L}{d} \tag{12}$$

Equation (12) needs the local loss coefficient as input, which had to be calculated separately for both elements by adjusting the well-known equation for pressure losses due to local resistance [2]:

$$\sqrt{p_l} = \sqrt{\xi}\left(\frac{\sqrt{\rho_{ie}}}{2}v_{ie}\right) \tag{13}$$

As the flow in GFR and DHR can't be considered as isothermal, the density and the mean velocity change from the input to the output in both elements. Thus, the local loss coefficient ξ was related to the averaged values of the changing parameters v_{ie} and ρ_{ie} [2]. The pressure loss inside the elements was calculated by the Bell-Delaware method for each measurement, which is considered to be sufficiently accurate. Subsequently, losses from sudden expansion and reduction at inputs and outputs were added. Function (13) is a straight line with $\sqrt{\xi}$ being its slope. By plotting the dependence into a graph and after data regression, it was possible to determine the local loss coefficients ξ_{DHR} and ξ_{GFR}.

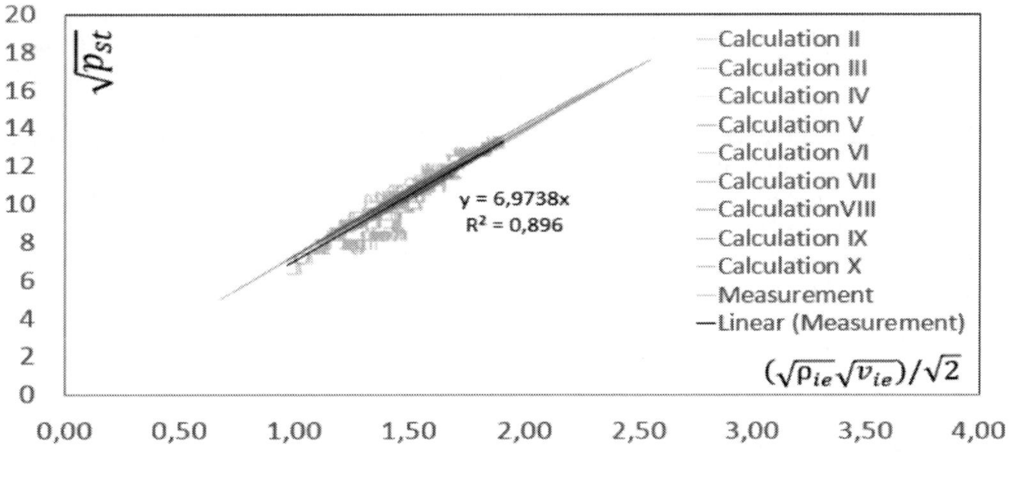

Fig. 2. Determination of local loss coefficient for DHR.

Figure 2 shows nine plotted dependencies in which the pressure loss was calculated by the Bell-Delaware method and one dependency with directly measured pressure loss. The measurement is consistent with calculations. The final value of ξ_{DHR} was thereby determined at: 48.837 and the value for ξ_{GFR} at: 4.875.

4 Comparison of experiments with calculations

The measurement begins with the helium being filled into the system at pressure 2.5 MPa. After the heat input increases, the temperature in the hot branch rises and a flow is induced. By consequent helium cooling in the DHR exchanger, the temperature difference between two branches stays constant at the steady state. In Figure 4, the measured velocities v_H are compared with those calculated using the model described in section 3. The temperature distribution, as well as operating pressure, is a necessary input to the calculation and was obtained from the measurement. The state properties of helium required for the calculation were obtained using the Soave-Redlich-Kwong equation of state, which is proved to be the most accurate from previous work [2] when compared to the reference values from lit. [5].

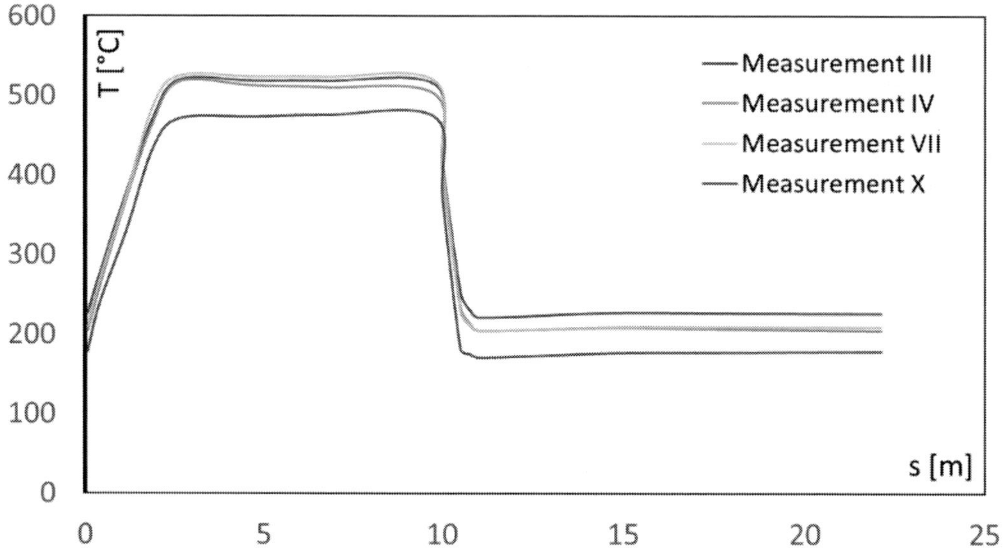

Fig. 3. The distribution of temperatures along the loop for four selected measurements.

Figure 3 shows the distribution of temperatures from measuring III, IV, VII and X during steady flow. The initial temperature is the value from the bottom of the cold branch (control point no. 4), then the values from the temperature sensors in the interior of GFR are displayed, continuing with hot branch and DHR. In the end, the value of cold branch is displayed again (from control point no. 4).

Figure 4 shows a record of mean velocity v_H in the hot branch from four measurements (III, IV, VII and X). The velocity is measured every 1 second and the total duration of measurements vary from 22000 [s] to 28000 [s]. Since the numerical model assumes a steady flow in the loop, it was necessary to select a range for all measurements, where steady state was considered as achieved. For the selected measurements the ranges were: Measurement III. (12500 – 18000) with average operating pressure 6.156 MPa, measurement IV. (15000 – 23000) with average operating pressure 5.730 MPa, measurement VII. (10000 – 17300) with average operating pressure 4.992 MPa, and measurement X. (10000 – 17000) with average operating pressure 3.423 MPa.

When compared, the measured and calculated velocity v_H has a similar course. The averaged deviation between the calculation and the measurements in the considered range of steady-state flow were satisfactory with the values for measurement III. – 4.964 %, measurement IV. – 4.520 %, measurement VII. – 6.311 % and measurement X. – 4.456 %.

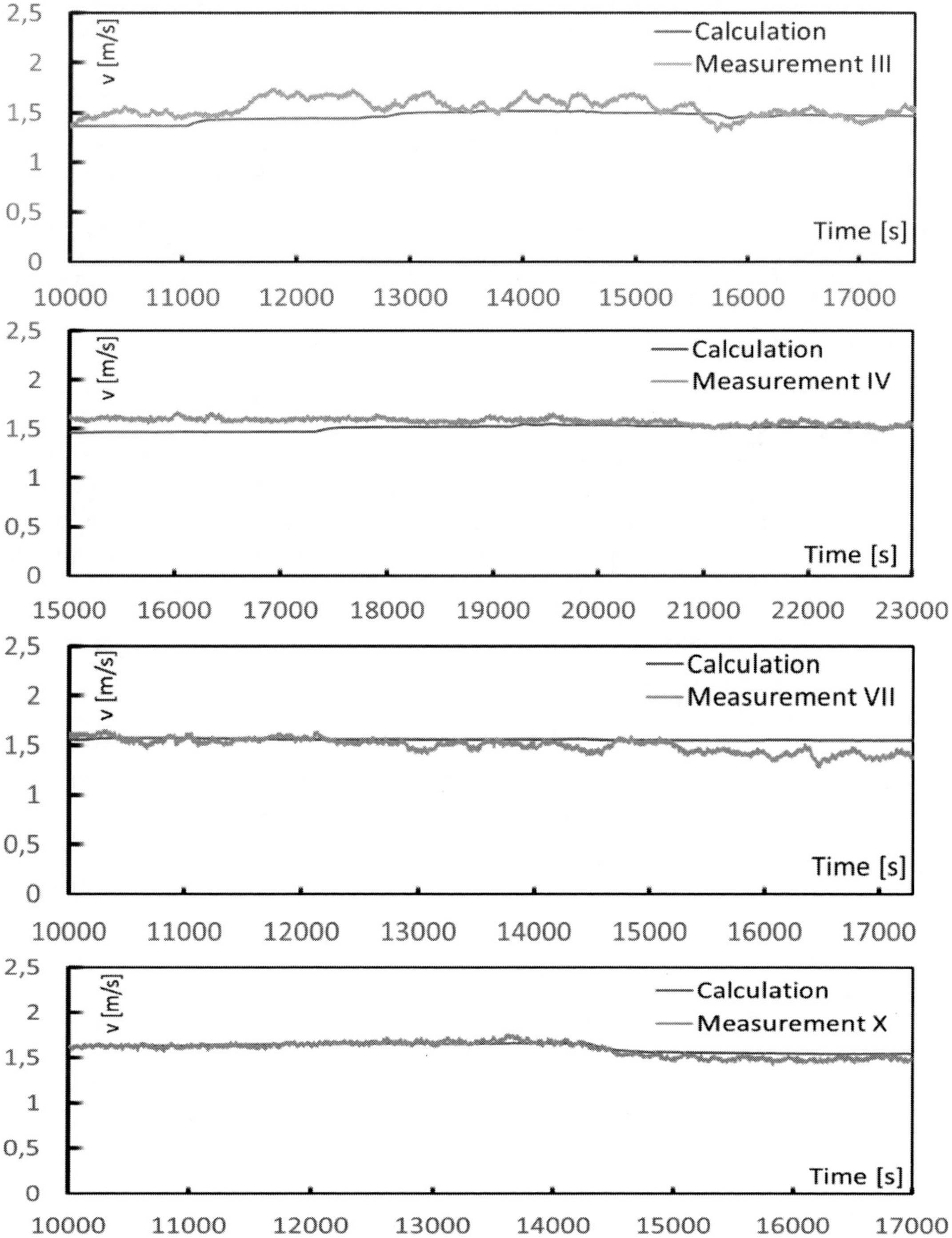

Fig. 4. Comparison of measured velocity (blue) with calculated values using numerical model (red).

5 Conclusion

The article analyses a one-dimensional numerical model of flow in the helium loop with natural circulation. The flow is described as one-dimensional and steady. It turns out that a one-dimensional model can describe the flow in such a facility with sufficient accuracy. The difference between the calculated values and the selected measurements were about 4 % – 7 %. The deviations are apparently because the one-dimensional model does not take into account the flow to the y-axis and in particular, does not consider the backflows and swirls that almost certainly occur when helium enters into the DHR. We want to address the backflow problem in our further research, as well as the enhancement of the current numerical model into a thermodynamic model, capable of calculating the temperature distribution in the loop. The thermodynamic model is expected to have impaired accuracy due to deviations in temperature calculations.

Európska únia
Európsky fond regionálneho rozvoja
Operačný program VÝSKUM a VÝVOJ

This contribution was created on the basis of the project "Research centre ALLEGRO" (ITMS project code: 26220220198), supported by Operational Programme Research and Development funded by the European Regional Development Fund.

References

1. P.K. Vijayan, A.K. Nayak, *IAEA Technical Documents*, **1474**, 89-99, (2005)
2. F. Világi, B. Knížat, F. Urban, R. Olšiak, M. Mlkvik, F. Ridzoň, P. Mlynár, *Proceedings of Engineering Mechanics 2017,* ISBN 978-80-214-5497-2, 1030-1034, (2017)
3. B. Knížat, F. Világi, F. Urban, R. Olšiak, M. Mlkvik, F. Ridzoň, *AIP Conference Proceedings,* **1889**, (2017)
4. F. Urban, P. Élesztos, R. Palenčár, Š. Emmer, K. Jelemenský, B. Knížat, Ľ. Kolláth, *Research report for research center Allegro.* (2016)
5. D.R. McCarthy, D.V. Arp, G.D. Friend: *NIST Technical note,* **1334 (revised)** (1998)
6. B.J. McKeon, M.V. Zaragola, A.J. Smiths: *J. Fluid Mech.*, **538**, 429-443, (2005)

MATEC Web of Conferences **168**, 07008 (2018) https://doi.org/10.1051/matecconf/201816807008
XXI. AEaNMiFMaE-2018

Experimental investigation of the boundary layer transition on the prismatic blade using infrared thermography

Erik Flídr[1,*], and *Tomáš* Jelínek[1]

[1]VZLU - Czech Aerospace Research Centre, Beranových 130, 199 05 Prague, Czech Republic

Abstract. Topic of this contribution is the detection of the laminar-turbulent boundary layer transition on the prismatic blade using infrared thermography. Several Reynolds and Mach numbers were investigated, and their influence on the boundary layer transition was evaluated. Results are in good agreement with experiments that were performed earlier with hot-film anemometry.

1 Introduction

The kinetic energy loss of the flow caused by the presence of the turbomachinery blades in subsonic flow regimes is strongly dependent on the character of the boundary layer (BL) on the blades and also on the separation of the BL before the trailing edge. The character of BL is dependent on the flow conditions which are commonly described by similarity criteria such as Mach and Reynolds numbers. Laminar-turbulent transition of the BL is also influenced by the turbulence level of the free stream, by the pressure gradients, and by the compressibility effect, see Narasimha [1] or Mayle [2]. The kinetic energy loss can be determined experimentally by the pressure measurement behind the trailing edges of the blades. Nowadays the Computational Fluid Dynamics (CFD) is widely used in the turbomachinery design and analysis of the flow field. Fully turbulent computations are used very often and the results are satisfying for high Reynolds number regimes (higher than 10^6), where the transition of the BL starts very early behind the leading edge.

For low Reynolds number (lower than 10^6) flow the BL transition must be taken into account if meaningful results of the CFD computations are to be obtained. Therefore several BL transition models were developed and are widely tested in various examples by Straka et al. [3, 4], Menter et al. [5], Suzen and Huang [6] or Steelant and Dick [7]. To validate the CFD models it is necessary to have the relevant experimental data of the BL transition for the different flow conditions.

BL transition can be investigated using various experimental methods, e.g. surface hot-film anemometry (HFA) or infrared thermography. Many other methods can be found in literature, see e.g. Tropea et al. [8].

Gomes et al. [9] investigated BL transition using HFA. They used an array with 50 sensors for the detection of BL transition. Measurements were performed in the constant

* Corresponding author: flidr@vzlu.cz

temperature anemometry (CTA) regime. Comparison of HFA and hot-wire anemometry (HWA) measurement at the blade cascade was made by Griffin and Davis [10].

Infrared thermography uses differences in the temperature of the body surface to detect the BL transition. These differences are caused by the variation of the heat transfer that is associated with the BL type. Across the turbulent BL, there is higher heat transfer due to the enhanced momentum transfer. Therefore, the body surface will be more heated or cooled, depending on whether the body has lower or higher temperature compared to the oncoming flow. The major advantage of this method compared to the other methods used for the detection of the BL transition is, that the flow field is not influenced by any objects on the surface (e.g. presence of the HFA in the flow field). That is very important, because of a high sensitivity of the BL transition to any disturbances. Howerver, it is only qualitative method how to detect the BL transition and parameters of the flow cannot be determined.

The key elements to obtain meaningful results from the experiment are the differences in temperature between the flow and the body surface and low thermal conductivity of the model body.

Because of reasons mentioned above, many authors use this technique for the detection of the BL transition. This technique was used in the supersonic flow by Saric and Zuccher [11], where for the better results, surface of the model was cooled by the liquid nitrogen or heated by the electrical element during some measurements. Model was also cooled by Yokokava [12], who used cooling blankets for the surface temperature decreasing and experiments were performed in low speed wind tunnel. An electric element for the model heating was used by Liselle at al. [13], De Luca et al. [14] and Patorski et al. [15]. Another technique for the surface heating is to use heat lamp as was shown by Baek and Fuglsang [16] or Raffel et al. [17].

As the initial experiment, the transition on single laminar airfoil at the ambient temperature in the intermittent suck-down wind tunnel was performed at our laboratory by Jelínek [18]. The experiment was designed to study the transition evoked by different thickness of turbulence-generators.

In closed-loop wind tunnels the difference in temperature of the flow and the body surface is an issue. As the infrared cameras can detect roughly 30×10^{-3} °C, the difference of surface temperature of 1 °C seems to be satisfactory to detect the transition. Otherwise the temperature of the model or the flow must be changed by external source. Both approaches are presented and discussed in this paper.

2 Experimental methods, setup and apparatus

2.1 Wind tunnel and flow regime

Experiments were performed in VZLU laboratory of high speed aerodynamics in a closed loop low-pressure wind tunnel (WT). Scheme of the WT is shown in Figure 1.

A linear blade cascade that was assembled of the individual prismatic blades placed between the two side-walls made of plexiglass. Pitch-to-chord ratio of the cascade was $t/c = 0.9$. The flow incidence could be adjusted, with respect to the fixed cascade, by a pair of shaped semi-nozzles placed upstream of the linear cascade. Inlet flow angle was set at constant value $\alpha = 85°$. Configuration of the wind tunnel allows adjusting the Reynolds and Mach number independently. Mach number can be changed by changing the rotational-speed of the compressor, while the change of the Reynolds number can be achieved by reduction of the pressure in the WT by set of vacuum pumps.

Flow regimes with tunnel settings are summarized in Table 1.

Regime of the flow in the WT was set based on pressure measurement. The reference total pressure was measured in the inlet chamber and the static pressure at the wall in front of the cascade. All pressures were converted to the voltage output using pressure transducers with accuracy of 0.1 % rdg. Barometric pressure was measured by means of Druck DPI 145 with accuracy of 0.013 % FS. The measured signals were sampled using A/D card (National Instruments PCI-6259 A/D card: 16-Bit, 1MS/s (Multichannel), 1.25 MS/s (1-Channel), 32 Analogue Inputs). The rate could be set according to frequencies of the observed process. Signal conditioning was provided by DEWETRON system, which was equipped with modules with bandwidth up to 200 kHz. Temperature and humidity of the flow was measured using hygrometer Sensorika Humistar HTP-1.

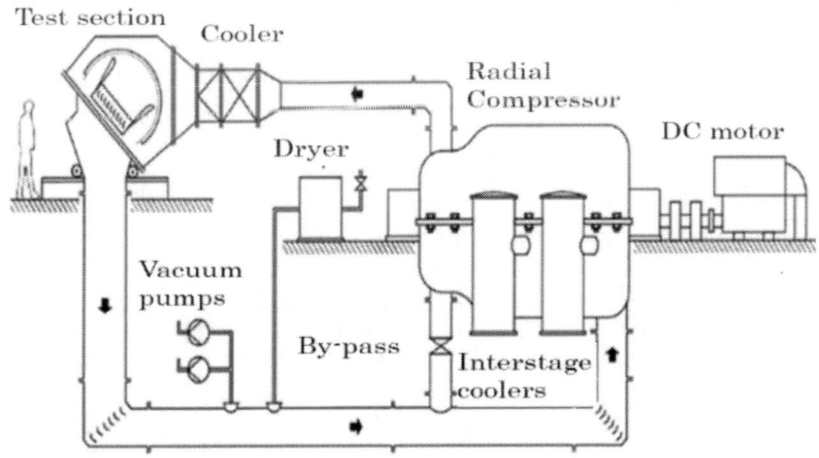

Fig. 1. Scheme of the wind tunnel.

According to the measurement device accuracy, uncertainties of the measured quantities were established according to the GUM document [19]. For the outlet isentropic Mach number M_{2is}, the uncertainty was less than 1 %.

Table 1. Tested flow regimes.

Blade made of Cibatool - blade A				Blade made of PLA natural by 3D printing - blade B				
M_{2is} [1]	Re_{2is} [1]	T_{sref} [°C]	T_a [°C]	M_{2is} [1]	Re_{2is} [1]	T_{sref} [°C]	T_a [°C]	
0.8	5.0×10^5	12.7	14.1	0.8	5.0×10^5	15.7	17.7	Unheated
0.6	7.5×10^5	12.7	13.5	0.6	7.5×10^5	15.5	16.5	
0.6	5.0×10^5	13.2	14.0	0.6	5.0×10^5	16.0	17.0	
0.6	2.5×10^5	13.1	14.0	0.6	2.5×10^5	15.4	16.1	
0.4	5.0×10^5	12.7	13.3	0.4	5.0×10^5	16.0	16.6	
0.4	2.5×10^5	16.1	16.5	0.4	2.5×10^5	19.5	19.7	
-	-	-	-	0.4	2.5×10^5	22.3	20.1	Heated
-	-	-	-	0.4	5.0×10^5	24.2	21.7	

2.2 Infrared camera

2.2.1 Principle of the Infrared thermography

Infrared thermography (IRT) is a method for direct measurement of the emissive power of the body surface. Relation between the temperature and emissive power of the body is given by Stefan-Boltzmann law:

$$E = \varepsilon\sigma T^4, \quad (1)$$

where ε is the emissivity of the body ($\varepsilon = 1$ is valid for black body), $\sigma = 5.68\times10^{-8}$ W/m^{2}K^4 is the Stefan-Boltzmann constant and T is the thermodynamics temperature. Proper camera calibration and the knowledge of the body emissivity is essential for the correct temperature measurements. Here is important to note that the emissivity ε is the function of four variables and can be expressed as function of:

$$\varepsilon = f(\eta, T, \varphi, R), \quad (2)$$

where η is the spectral range, φ is the direction of radiation and R is the roughness of the surface. Because of experimental setup, R and η can be neglected (the blade surface was smooth and spectral range $\eta = 7.5 \div 14$ μm), but the φ influenced the results as will be shown further below in section 3.

There is one more issue with IRT, the reflection from the body surface. This phenomenon will be also discussed in the section 3.

2.2.2 Infrared camera setup

FLIR A655sc infrared camera with resolution of 640 × 480 pixels and maximum frame rate of 50 Hz was used for the experiments. View angle of the camera was 15°. Camera can clearly detect 30×10^{-3} °C differences in temperature. Spectral range of the camera is $\eta = 7.5 \div 14$ μm. Standard temperature range for the FLIR A655sc is $T = (-40 \div 150)$ °C and can operate in the temperature range of $T = (-10 \div 50)$ °C. Best accuracy of the camera is ± 2 °C or ± 2 % rdg.

Camera was placed 260 mm behind the trailing edge of the blades, see Figure 2.

2.3 Hot film anemometry

To identify the transition in the boundary layer the array of hot film sensors was used at first. The sensors were manufactured by Tao systems and were delivered as a set on one polyamide foil with a thickness of about 0.05 μm. The foil was glued to the blade suction side, with asymmetrical distribution of the sensors around the blade centreline (see Figure 3). The sensors were made of two copper leads and the active part was made from nickel (thickness 0.2 μm). Therefore, it was necessary to minimize the thermal conductivity between sensors and the blade, the blade was made of Cibatool (further

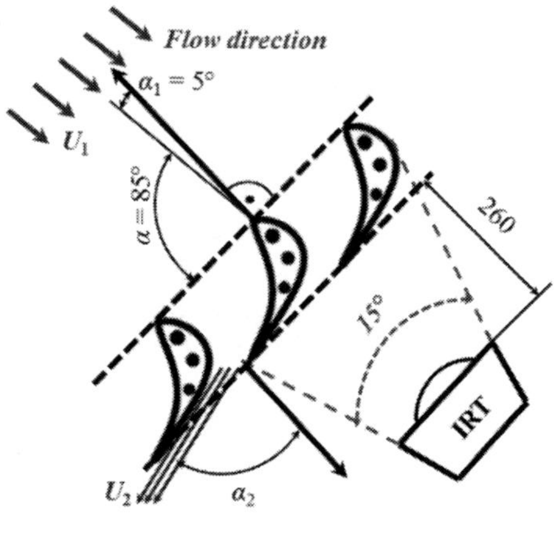

Fig. 2. Blade cascade with IR camera.

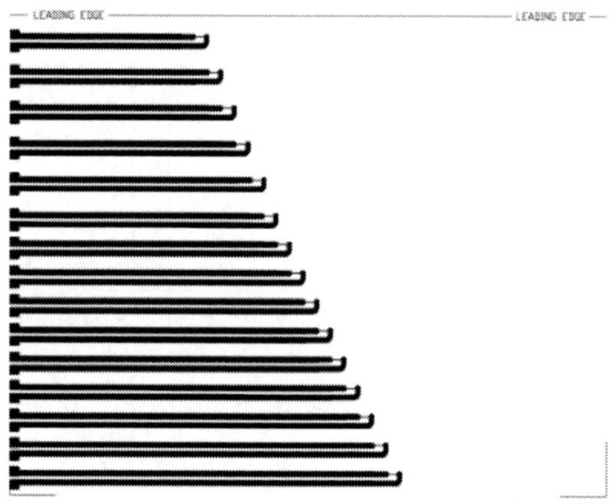

Fig. 3. HF sensors positions.

referred to as blade A).

Constant current mode was used for the measurements. Hot-films signals were acquired for 5 seconds with sampling frequency of $f_{samp.} = 50$ kHz and with low pass filter of $f_{LP} = 10$ kHz. Results from these experiments were summarized in Michálek [20].

2.4 Blades for the visualization of boundary layer transition

Two different blades were used for visualization experiments using IR camera. Prior to use of blade A (made of Cibatool), the hot film sensors were removed and the surface cleaned. A new blade (further referred to as blade B) was manufactured (made of PLA) by the 3D printer in the VZLU laboratory of high speed aerodynamics. The blade was equipped with the cavity and with a thin slit, leading from the cavity to the trailing edge, see Figure 4a. The copper plate was placed to the slit for better heat transfer trough the blade. For better resolution of the camera, blade was heated using hot air. Hot air was blown into the cavity through hole on one side of the blade and was sucked out on the other.

To determine the position of the BL transition detected by IR camera it was necessary to implement another coordinates. Usually there are two coordinates systems for the description of the location on the blade surface. Orthogonal system (x, y) and curvilinear system with coordinate s (see Figure 4b), that is associated with blade surface. However IR camera was rotated by angle $\varphi = 45°$ against system (x, y) and was rotated by angle ε = 15°against tangent t, which leads through the tangent point in trailing edge, see Figure 4b.

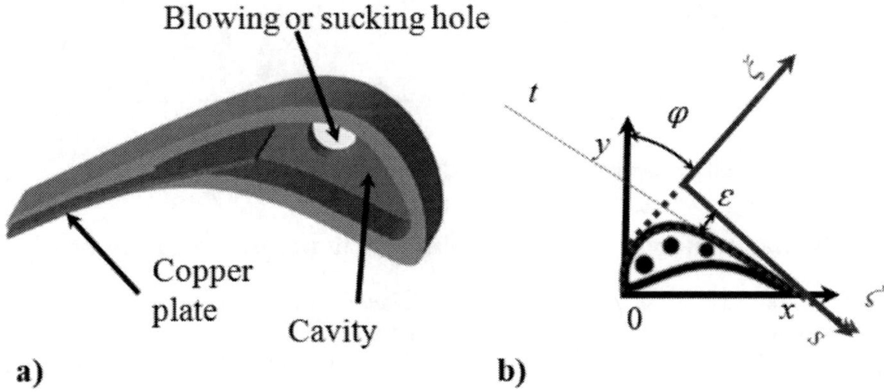

Fig. 4. a) Blade made by 3D print, **b)** coordinate systems.

2.5 Detection of the boundary layer transition

As was mentioned above in section 1, a heat transfer increases significantly when laminar to turbulent BL transition occurs. This is caused by the increase of skin fiction coefficient $C_f = \tau_w/(2\rho u^2)$, that is usually one order higher in the turbulent BL compared to the laminar BL. According to Reynolds analogy, heat transfer is a function of skin friction coefficient as:

$$Nu = \tfrac{1}{2}\, C_f(x)\, Re, \qquad (3)$$

where $Nu = hl/\lambda$ is the Nusselt number and $Re = ul/\nu$ is the Reynolds number. Nomenclature for the Nusselt and Reynolds number is: h is the heat transfer coefficient, l is the characteristic length, λ is the thermal conductivity of the fluid, u is the fluid velocity and ν is the kinematic viscosity. Experiments were performed with air as a medium, therefore the

Prandtl number was approximately 1 and the condition for using Reynolds analogy was satisfied.

3 Results and discussion

3.1 Hot film anemometry

Root mean square (RMS) values of the measured voltages signals from hot-films are shown in Figure 5. Results show, that the BL transition occurs near the trailing edge of the blade ($s/c > 0.8$) where the RMS values of the signals increase for all measured regimes of flow. Free stream turbulence during these measurements was under 2 %, as was shown by Michálek [20].

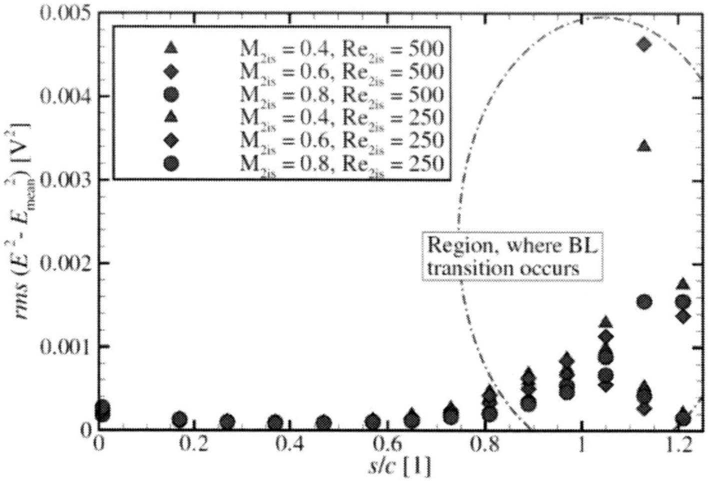

Fig. 5. RMS values of the measured voltage signals from the HFA.

3.2 Effects that influence the IRT measurement and initial conditions

During the initial testing, blades A and Bwere placed in the test section at the same time. This configuration was found to be unsatisfactory for the detection of BL transition; however the images can be used for the demonstration of the artefacts that were observed during the IRT measurement. The images in Figure 6 were captured before the WT was turned on.

Blade B was placed above the blade A. The blade B was heated at the temperature $T = 26$ °C, while the blade A had temperature of the surroundings that was $T_a = 11$ °C.

Regions, where hot air was blown in the blade B and sucked out (places with higher temperature close to the leading edge), can be seen in Figure 6. At the leading edge of the blade there is also the region where the influence of the curved shape can be observed. This conclusion was made because there were no other reasons for such significant decrease of temperature. At the trailing edge of the blade in the upper image in Figure 6 there can be seen a region with the significant decrease in temperature, it is a part of the blade, where the copper plate did not reach.

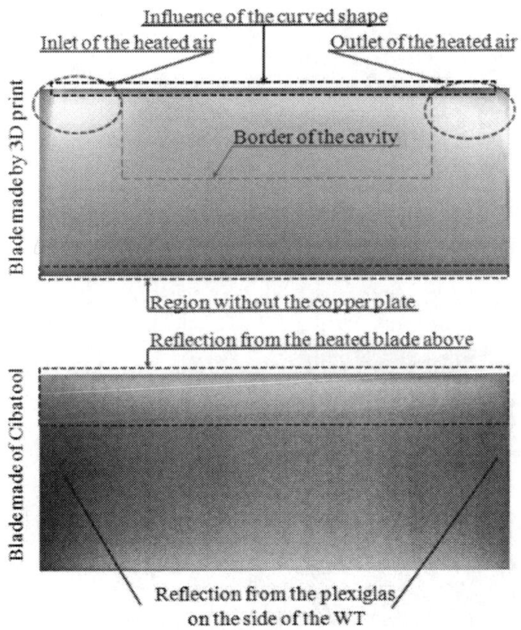

Fig. 6. Effects influences the IRT measurement.

The IRT image of blade A is shown in the lower part of the Figure 6. Region with the reflected radiations, which influence the IRT temperature reading, from both plexiglasses in which the cascade was assembled, are visible. Reflection of the radiation from the heated blade can be seen behind the leading edge in Figure 6. Reflection ends at the point, where the step change of the temperature occurs.

3.3 Boundary layer transition

3.3.1 Transition on blade A and unheated blade B

The BL transition was investigated for several flow regimes described in Table 1.

All regimes were measured with the blade A first and after the measurement was finished the blade A was replaced with the blade B.

Visualization of the BL transition is shown in Figure 7. Surface temperature distribution on the blade centreline is shown on the right hand side of Figure 7. Ratios of the temperatures at each point along the centreline and maximum temperature T_s/T_{max} are plotted in the figure. Mean value from the surface temperature distribution was chosen as a reference surface temperature T_{sref}. Free stream temperature T_a was higher compared to blade surface temperature T_{sref} (see Table) during all tests.

Corner vortices can be seen in every picture in Figure 7. On the right handside of figure, vorticity is enhanced due to the leakage of the fluid through the gap between the blade and the sidewall that was created during the cascade mounting. Because the temperature of the plexiglases was lower than the temperature of the free stream, air that goes through these gaps on both sides of the WT was cooled and as a result of this phenomenon the blades in the region of corner vortex were cooled as well.

Visualization of the BL transition using IRT technique shows good agreement in the location of the transition on the blade A and B for all tested regimes. The transition occurs near the trailing edges of the blade for the $Re_{2is} = 2.5\times10^5$ and for both $M = 0.4$ and 0.6. With increasing Reynolds number, the transition moves upstream, closer to the leading edge. Vortices generated in the transitional area were clearly shorter compared to regimes with Reynolds number $Re_{2is} = 2.5\times10^5$.

Moreover some turbulent wedges generated on the blade B can be seen in Figure 7a, b, c and e. These structures were generated behind the surface defects that were not entirely wiped out from the blade during the surface finishing. It can be seen that smaller defects on the blade surface play more important role in the BL transition if the Reynolds number is higher.

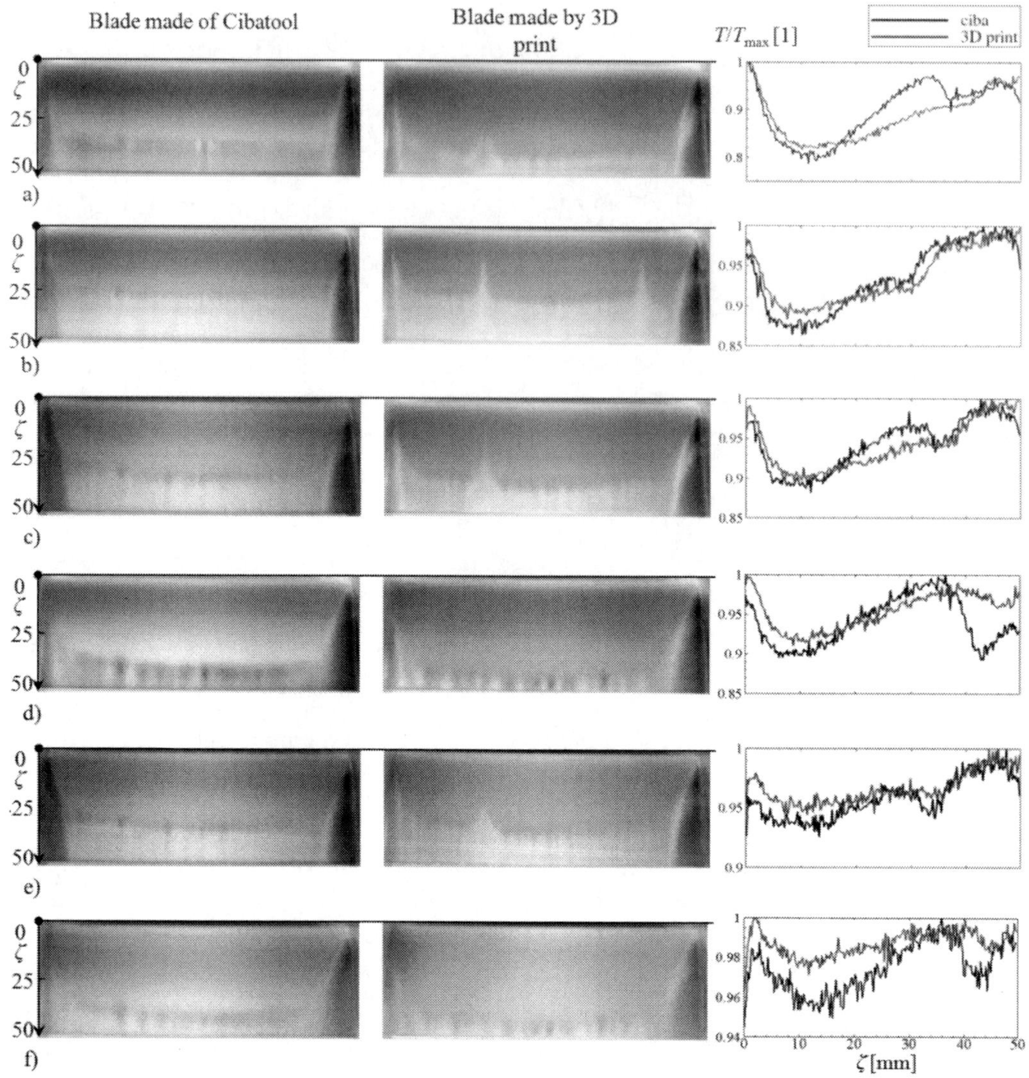

Fig. 7. Visualization of the BL transition using IRT a) $M_{2is} = 0.8$; $Re_{2is} = 5\times10^5$, b) $M_{2is} = 0.6$; $Re_{2is} = 7.5\times10^5$, c) $M_{2is} = 0.6$; $Re_{2is} = 5\times10^5$, d) $M_{2is} = 0.6$; $Re_{2is} = 2.5\times10^5$, e) $M_{2is} = 0.4$; $Re_{2is} = 5\times10^5$, f) $M_{2is} = 0.4$; $Re_{2is} = 2.5\times10^5$.

3.3.2 Transition on heated blade B

In order to get higher differences in temperatures of the medium and the blade surface the experiment with heated blade was performed. For this experiment the blade B (manufactured by 3D printing) (Figure 3) was used.

Comparison of heated and unheated blade B is shown in Figure 8 for Mach number $M_{2is} = 0.4$ and Reynolds numbers $Re_{2is} = (2.5 \text{ and } 5) \times 10^5$. Blade was heated up before the visualization using hot air as was described in 2.4. During the visualization itself, heating was turned off, because the temperature distribution was not uniform across the whole blade and before the picture capturing some time was needed for the temperature compensation. Reference surface temperatures T_{sref} during the measurements are listed in Table 1.

The heating of the blade had not significant effect on the BL transition as can be seen. The heating of the blade increases the temperature resolution; however, the temperature gradients may affect the evaluation of the blade transition. Areas where the blade was heated more than the rest of the blade can be seen in both figures, moreover in Figure 8b, border of the cavity in the blade can be seen, because the air in it was already colder. Vortex structure behind a surface defect was generated when the Reynolds number was $Re_{2is} = 5 \times 10^5$ (Fig. a) in both investigated cases (heated and unheated blade).

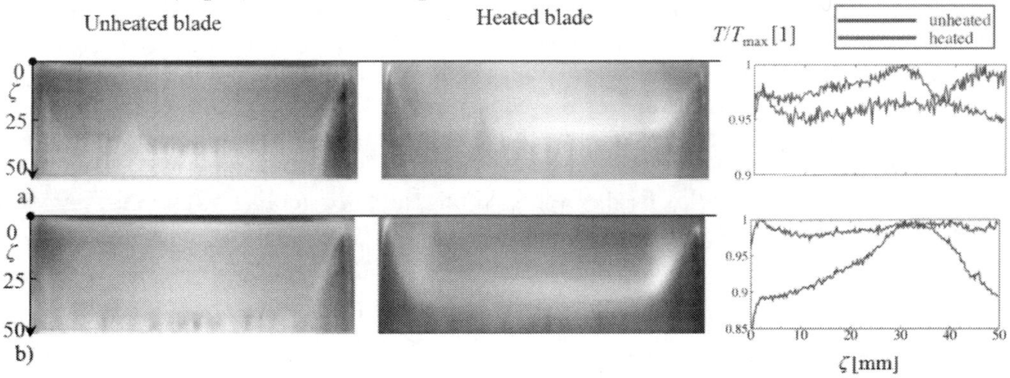

Fig. 8. Comparison of the heated and unheated blade B a) $M_{2is} = 0.4$, $Re_{2is} = 5{\times}10^5$, b) $M_{2is} = 0.4$, $Re_{2is} = 2.5{\times}10^5$.

3.4 Results comparison

Comparison of the results from both used methods is in Table 2, where the positions ζ of transitions are summarized. Coordinates *s/c* used for the description of HFA position was transformed to the system (ζ, ξ) for better clarity. There is a good agreement of the results obtained by the IRT visualization and by the HFA. Position of the transition obtained from the HFA measurements is evaluated with accuracy of $\zeta = \pm$ 3.5 mm because of the distances between individual sensors. Position of the transition obtained from the IRT is evaluated with accuracy of $\zeta = \pm$ 4 mm, because the transitions were evaluated along the one line positioned in the middle of the blade, but as can be seen from the visualization above, the positions of the transitions were not same along the blade height (value of ± 4 millimetres was the worst case from all).

Table 2. Results comparison.

		Blade A		**Blade B unheated**		**Blade B heated**		**HFA**	
M_{2is} [1]	Re_{2is} [1]	ζ [mm]	s/c [1]	ζ [mm]	s/c [1]	ζ [mm]	s/c [1]	ζ [mm]	s/c [1]
0.8	$5.0{\times}10^5$	34	1.011	38	1.088	-	-	38	1.088
0.8	$2.5{\times}10^5$	-	-	-	-	-	-	44	1.209
0.6	$7.5{\times}10^5$	30	0.933	30	0,933	-	-	-	-
0.6	$5.0{\times}10^5$	32	0.972	36	1.050	-	-	35	1.030
0.6	$2.5{\times}10^5$	42	1.165	45	1.223	-	-	42	1.165
0.4	$5.0{\times}10^5$	32	0.972	36	1.050	34	1.011	38	1.088
0.4	$2.5{\times}10^5$	40	1.127	43	1.184	43	1.184	42	1.165

4 Conclusions

The transition of BL on the prismatic blades was determined by IRT for several regimes characterized by Mach and Reynolds numbers. Results were compared to the HFA measurements performed earlier in our laboratory and good agreement was found. The transitions were detected as a rapid change in the temperature of the blade surface, the difference of about 1K in temperature between the medium and blade surface was found sufficient. Locations of the transitions were dependent on the value of the Reynolds number Re_{2is} (with increasing Re_{2is} transition moved toward the leading edge of the blade), no significant influence of Mach number on this phenomenon was observed, what is in accordance with the results made with HFA. The results could be used to verify numerical models. Moreover the corner vortices on the blades can be also detected by IRT. It can be used for the future comparison with measurements performed by multi hole pressure probe behind the cascade, where the kinetic energy losses are usually investigated.

For the future work in this field, some kind of heating or cooling of the bluff body can be recommended, because of better resolution of IRT; however it was shown that technique used during these experiments will need some development.

References

1. R. Narasimha, Prog. Aerospace Sci., **22**, (1985)
2. R.E. Mayle, J. Turbomach. **113**, (1991)
3. P. Straka, J. Příhoda, Proc. Conf., Exp. Fluid Mechanics, TU Liberec, (2010)
4. P. Straka, J. Příhoda, P. Šafařík, Proc. Conf., App. Exp. Num. Meth. Fluid Mech., Žilina, (2012)
5. F.R. Menter, R.B. Langtry, S.R. Likki, Y.B. Suzen, P.G. Huang, S.A. Völker, J. Turbomach. **128,** (2006)
6. Y.B. Suzen, P.G. Huang, J. Fluid Eng. **122**, (2000)
7. J. Steelant, E. Dick, Int. J. Num. Methods Fluids **23**, (1996)
8. C. Tropea, A.L. Yarin, J.F. Foss, *Handbook of experimental fluid mechanics* (2007)
9. R.A. Gomes, et al., J. Turbomachinery **137**, (2015)
10. P. Griffin, M.R. Davies, 16th Symp. Meas. Tech. Trans. Sup. Flow in Casc. Turb., Cambridge, (2002)
11. S. Zuccher, W.S. Saric, Exp. Fluids **44**, 1 (2008)
12. Y. Yokokawa, Trans. Jap. Soc. Aero. Space Sci. **48**, 161 (2005)
13. L.A. Joseph, A. Borgoltz, W. Devenport, Exp. Fluids **57**, 5 (2016)
14. L. De Luca, G.M. Carlomagno, G. Buresti, Exp. Fluids **9**, 3 (1990)
15. J.A. Patorski, G.S. Bauer, J. Theor. App. Mech. **39**, 2 (2001)
16. P. Baek, P. Fuglsang, Eur. Wind Energy Conference, (2009)
17. M. Raffel, Ch.B. Merz, T. Schwermer, K. Richter, Exp. Fluids **56**, 2 (2015)
18. T. Jelínek, Exp. Fluid Mech. conf. proc., (2017)
19. *Guide to the Expression of Uncertainty in Measurement*, JCGM 100:2008, (1995)
20. J. Michálek, VZLU technical report R-6541, (2016)

MATEC Web of Conferences **168**, 07009 (2018) https://doi.org/10.1051/matecconf/201816807009
XXI. AEaNMiFMaE-2018

Importance of the experimental investigation of a concasting technology

Josef Štětina[1], *František* Kavička[1,*], *Jaroslav* Katolický[1], *Tomáš* Mauder[1], and *Lubomír* Klimeš[1]

[1]Brno University of Technology, Faculty of Mechanical Engineering, Czech Republic

Abstract. The solidification and cooling of a continuously cast billet, slab or cylinder, generally of a concasting and the simultaneous heating of the mold is a very complicated problem of three-dimensional (3D) transient heat and mass transfer. The solving of such a problem is impossible without numerical models of the temperature field of the concasting itself which it is being processed through the concasting machine (caster). The application of the numerical model requires systematic experimentation and measurement of operational parameters on a real caster as well as in the laboratory. The measurement results, especially temperatures, serve not only for the verification of the exactness of the model, but mainly for optimization of the process procedure: real process → input data → numerical analyses → optimization → correction of real process. The most important part of the investigation is the measurement of the temperatures in the walls of the mold and the surface of the slab in the zones of secondary and tertiary cooling.

1 Introduction

The production of steels, alloys and metallurgical products in general is constantly developing. Materials with high utility parameters are more in demand and traditional production is being replaced by higher quality steel. More and more sophisticated aggregates using more sophisticated technological procedures are being implemented. In order to maintain competitiveness, diversify production and expand to other markets, it is necessary to monitor technological development.

In the case of concasting, it is not possible to fulfill these requirements without the application of models of all caster processes dependent on thermal-mechanical relationships. The successful application of a numerical model (for example of a steel slab) and its continuous improvement is necessarily conditioned by carrying out demanding experiments and their evaluation [1-3]. The measured parameters serve either as direct input data for the numerical model or they are used for the creation of this input data. This is, for example, the case of determining the heat transfer coefficients beneath the cooling jets that has to be carried out in the laboratory (see below). The range of the measurements carried out and applied was relatively considerable. They can therefore be divided into

* Corresponding author: kavicka@fme.vutbr.cz

experimental research carried out on an operational device (caster) and experimental research carried out in laboratory conditions [4]. Research conducted on the operational device comprises permanent monitoring of the necessary physical and technological quantities during the entire process. Some of them are part of a so-called breakout system. This entails relatively extensive files of values containing individual data as well as time records – for example temperature in the walls of the mould, surface temperatures of the slab in the place of the bend, the pouring temperature, the casting speed, the temperature of the cooling water, etc. Furthermore, other experiments were also conducted for the purposes of research. This includes, for example, measurement of the surface temperatures in other places of the secondary-cooling zone (i.e. the cage), after exiting the cage (in the so-called tertiary-cooling zone) [5] and more.

2 An original off-line model of the temperature field of the concast steel slab 1 530 × 250 mm (width × thickness)

The 3D model had first been designed as an off-line version and later as an on-line version so that it could work in real time. The numerical model takes into account the temperature field of the entire slab (from the meniscus of the level of the melt in the mould to the cutting torch) using a 3D mesh containing more than a million nodal points.

The solidification and cooling of a concast slab 1 530 x 250 mm is a global problem of 3D transient heat and mass transfer. If heat conduction within the heat transfer in this system is decisive, the process is described by the Fourier-Kirchhoff equation. It describes the temperature field of the solidifying slab in all three of its states: at the temperatures above the liquidus (i.e. the melt), within the interval between the liquidus and solidus (i.e. in the mushy zone) and at the temperatures below the solidus (i.e. the solid state). In order to solve these it is convenient to use the explicit numerical method of finite differences. Numerical simulation of the release of latent heats of phase or structural changes is carried out by introducing the enthalpy function dependent on temperature T, preferably in the form of enthalpy related to unit volume H_v. The latent heats are contained here. After the automated generation of the mesh (pre-processing) ties on the entry of the thermophysical material properties of the investigated system, including their dependence on temperature – in the form of tables or using polynomials. They are namely the heat conductivity k, the specific heat capacity c and density ρ of the cast metal.

The temperature distribution in the slabs described by the enthalpy balance equation. The simplified equation, suitable for application on radial-casters with a great radius, where only the speed (of the movement of the slab) component w in the z-direction is considered, is:

$$\frac{\partial H_v}{\partial \tau} = k \cdot \left(\frac{\partial^2 T}{\partial x^2} + \frac{\partial^2 T}{\partial y^2} + \frac{\partial^2 T}{\partial z^2} \right) + w \cdot \frac{\partial H_v}{\partial z} \quad (1)$$

Enthalpy H_v as a thermodynamic function of temperature must be known for each specific steel. It is dependent on the composition of the steel and on the rate of cooling.

The experiment was conducted on a 1530 x 250 mm steel slab, where the pouring conditions were characterised by the temperature in the tundish (1 550 °C), the temperature of the liquid (1 521 °C), and the shift rate of the slab (0.82 m.min^{-1}). The liquidus and solidus temperatures are derived from the composition of the steel.

The exactness of the presented numerical model depends not only on the spatial and temporal discretization, but also on the accuracy with which the thermophysical properties of the materials of all parts of the system are determined. It also depends on the derivation of boundary conditions, i.e. the values of the heat transfer coefficient on all caster boundaries.

3 Experimental measurement for the application of the model of the temperature field of a steel slab 1 530 × 250 mm

3.1 Measuring temperatures in the mould (crystallizer) and in the secondary cooling zone

Figure 1 illustrates the positions of 44 thermo-couples in two lines along all walls of the mould [6] (so called the breakout system). The results of these measurements are indicated in Figure 2, together with the course of the mean value and distribution, and are entered into the numerical model of the mould. Figure 2 shows an example of real immediate temperatures on the mould.

In order to make the model of the temperature field of the mould more accurate and to verify it, it is useful to carry out experimental measurement of the surface temperatures immediately below the mould, i.e. in the area of the retaining rollers. The measurement is carried out using optical pyrometers positioned at the centre of the slab, half-way between the centre and the edge and near the corner (Figure 1). This measurement also makes it possible to assess whether the temperature field is symmetrical along the axis of the cross-section. Figure 3 shows the measurement of the surface temperatures along the large radius in the area of the retaining rollers. The graph illustrates that it is necessary to process the data before it is used, i.e. it is necessary to find a suitable method of filtering. The generally used method of filtering is floating averaging. Its disadvantage is that large deviations are spread along a larger area. The median recursive filter brings the strongest filtration – its output is, for example, usable for the regulation of the secondary cooling. It is possible to consider that interference of the signal from the pyrometer is caused by a quality surface of the slab (scales and solidified casting powder). That is why the maximal values provided by the maximum filter are considered as correct [7].

The experimental temperatures could be used from the breakout system (and not the data from the pyrometers), which, with respect to the densely arranged thermocouples in two horizontal planes (Figure 1) covers the asymmetry better. In order to apply the dynamic model of the temperature field, the investigation of only one half of the slab is sufficient. Figure 3 illustrates the favourable comparison of temperatures measured by three pyrometers with a calculated curve of the surface temperatures of the slab after 4 000 s.

In order to make the model of the temperature field more accurate and to verify it, it was necessary to use two permanently installed pyrometers with both measuring the upper surface of the slab along the small radius. The first was installed by Mannesmann Demag already at the stage of starting up the caster. It is positioned just before the unbending point. This pyrometer was originally used for checking the function of the process secondary cooling. During the integration of the on-line model into the system, it is now used for comparing the temperatures calculated by the model with the real ones. Furthermore, it was joined by another pyrometer of the same type at the exit from the cage. This arrangement of both models also includes a processing unit that filters the measured data. This facilitates their utilisation because the software of the temperature model no longer has to deal with the filtering of the signal. The pyrometers measure within the range from 750 to 1 200 °C and in the case of a measurement failure it shows a temperature of 700 °C. The positioning of the pyrometers in the two mentioned points is necessary for the purposes of checking. The temperature of the slab in the point of the first pyrometer at the beginning of the unbending (point) must be higher than the temperature prescribed for this point (i.e. a slab that is too cold must not be straightened out). On the other hand, the temperature at the exit of the cage should be lower than the prescribed temperature.

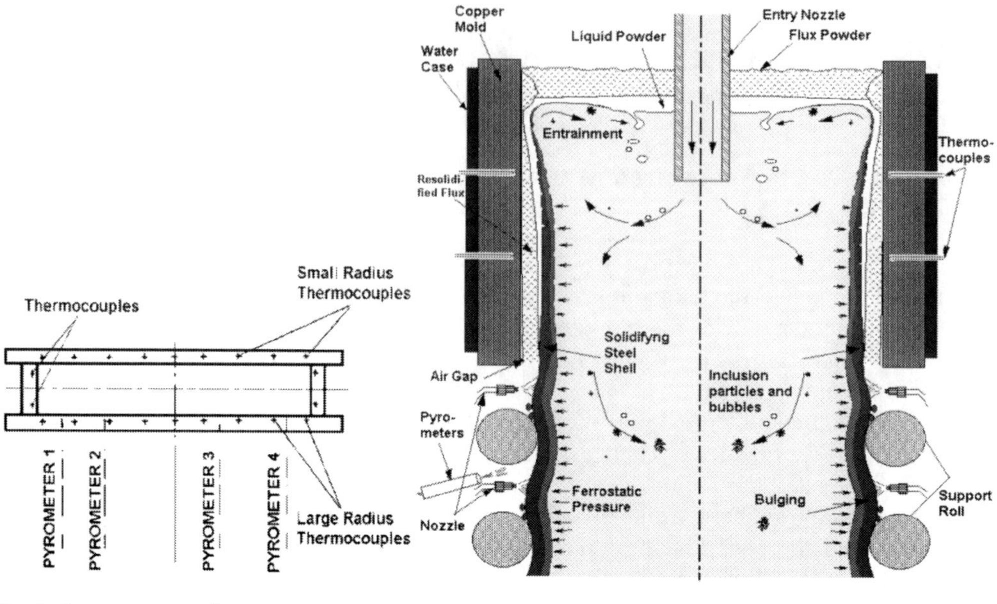

Fig. 1. Arrangement of measurement sensors.

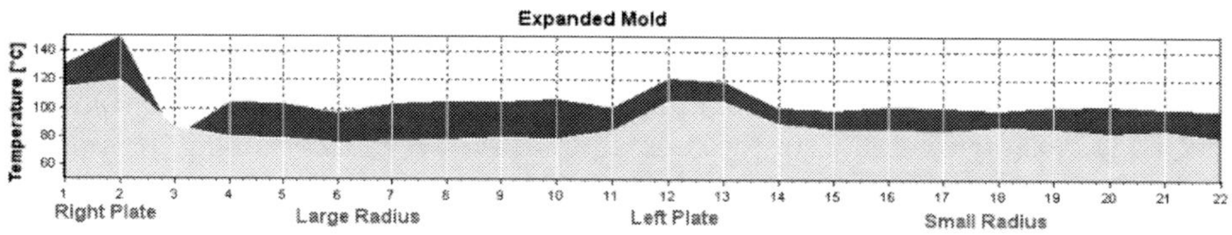

Fig. 2. Immediate temperatures of the breakout system (red – upper line, yellow – bottom line).

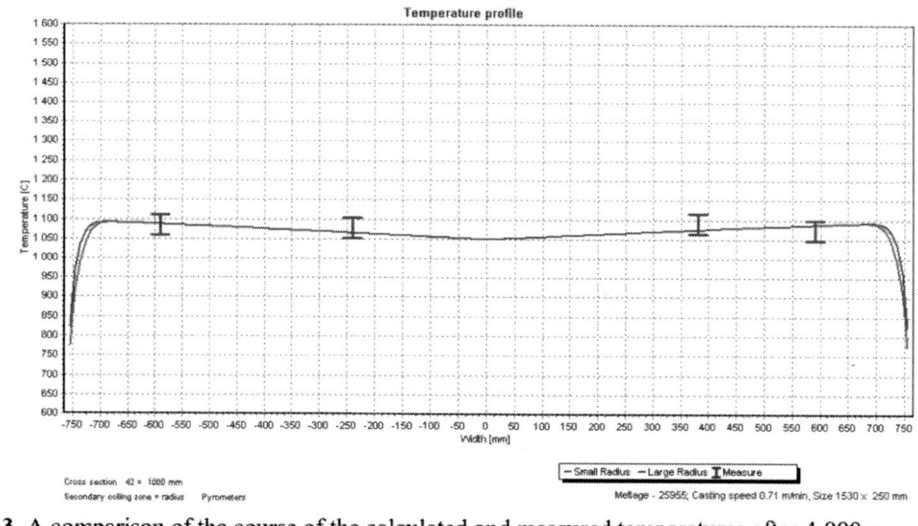

Fig. 3. A comparison of the course of the calculated and measured temperatures after 4 000 s.

3.2 Measuring of cooling effect of nozzles

A real caster contains a total of 8 types of air-jets and geometrical layouts. Since it is not possible to determine the intensity of the air-water jets on an actual caster, it is necessary to transfer the investigation – of each jet individually – to the experimental laboratory device (Figure 4), which is capable of simulating the surface of a concast slab [8-11]. This device also allows the measurement of temperatures beneath the surface within the slab. The temperatures measured are converted to cooling intensities by means of an inverse task, which, in turn, are converted to the courses of the heat transfer coefficients using an expanded numerical model. The numerical model also determines the effect of radiation which is dependent on the surface temperature.

Fig. 4. The laboratory device in use.

3.3 Measuring of slab surface temperatures in the tertiary cooling zone

Three 2 mm insulated thermocouples had been used for the measurement. The hot end of each thermocouple was fixed inside the hollow tip of a stud, carefully positioned on the surface of the slab after cooling (Figure 5) and all were hammered into the same depth during the torch-cutting process [12]. The measurement was initiated on a separated slab, 5-to-6 m from the exit of the cage, and continued until the slab was cold – in the cooling field. The correction of the error of the measurement result, from the heat transfer through the thermocouple insulation and the protruding end, is conducted via comparison measurements. The measured temperatures of three surface points are in Figure 5.

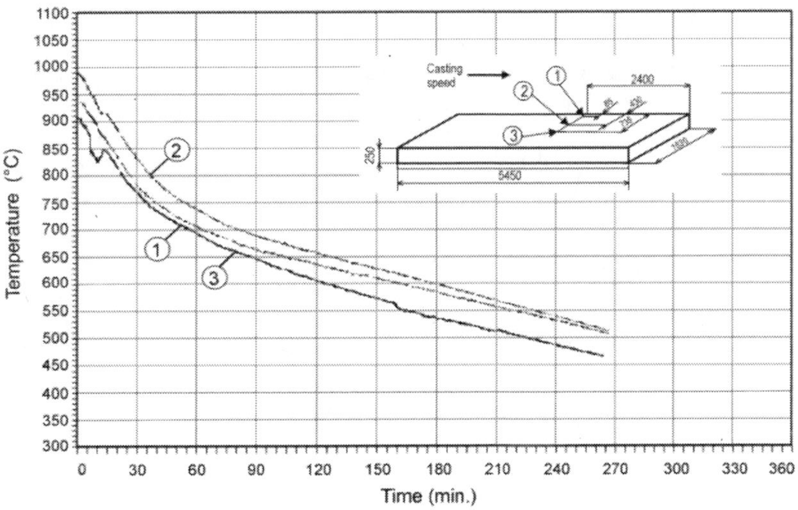

Fig. 5. Temperatures of the three surface points.

4 Conclusion

Research into the thermokinetics of solidification and cooling of concast slabs requires systematic experimental measurement on a real caster. Its results are utilised not only for improving the numerical model of the temperature field but also for assessing the exactness of this model. This dynamic model, which works non-stop in real time, ensures continuous correction of the real process of the caster in question. The main measured quantities are the temperature in the walls of the mould, the surface temperatures of the slab upon exit from the mould, at the unbending point of the slab and upon exit from the cage of the secondary cooling. Furthermore, it is the casting temperature, the casting speed, the temperature of the cooling water, the metallurgical length, etc. A specialised laboratory conducts measurements of the spraying characteristics of individual cooling nozzles. Its output establishes the heat transfer coefficients beneath each of the water or water-air nozzles.

This research was supported through NETME CENTRE PLUS (LO1202) by the Ministry of Education, Youth and Sports of the Czech Republic under the „National Sustainability Programme I“.

References

1. J.K. Brimacombe, Metall. and Mater. Trans. B, **30B**, 553 (1999)
2. J. Miettinen, S. Louhenkilpi, J. Laine, Proc. of Gen. COST 512 Workshop (1996)
3. R.A. Richard, R. Harding, Ch. Beckermann, Metall. & Mater. Ttransactions, **34B**, 297 (2003)
4. J. Štětina, Ph.D. Thesis, TU Ostrava, Czech Rep. (2007)
5. B.G.Thomas, R.J. O'Malley, D.T. Stone, Model. Of Cast. Weld., Solidif. Proc. VIII, San Diego, CA, TMS (1998)
6. F.Kavička, et al., Proc. of 4th Inter. Conf. on Comput. Fluid Dyn. in the Oil, Gas, Metall. and Proc. Ind., Trondheim, Norway (2005)
7. F. Kavička, et al., Communications, 4A2012, **14**, 13 (2012)
8. J. Horský, M. Raudenský, Proc. of 14th Inter. Metal. & Mater. Conf. METAL, Hradec n/M., Czech Rep. (2005)
9. J. Horský, et al., AISTech, Iron&Steel Techn. Conf. & exp., Charlotte, USA (2005)
10. M. Raudenský, J. Horský, Ironm. & Steelm. **32**, 2 (2005)
11. J .Štětina, Assoc. professor. & habilit. thesis, TU Ostrava, Czech Rep. (2008)
12. J. Štětina, et al., Comp. Meth. & Exp. Meas. **XII**, 749, WIT Press (2005)
13. F. Kavička et al., Proc. of 5th Pac. Rim Inter. Conf. on Adv. Mater. & and Proces., Zuerich, Trans Tech Pub. (2004)

Liquefied extinguishing agent discharge to an overpressure-sensitive enclosed volume

Lukáš Hurda[1,*], and *Richard* Matas[1]

[1]University of West Bohemia in Pilsen, New Technologies – Research Centre, Univerzitni 8, Pilsen, Czech Republic

Abstract. The throttling of liquefied substances from high pressure vessels to an enclosed volume starting at atmospheric pressure is described in order to determine thermodynamic state of the extinguished room gaseous contents. Time dependent, 0D mathematical model is implemented describing the state inside the agent container, the isenthalpic throttling in the distribution system, agent vaporization and mixing with air. The agent is modelled as real gas. Other influences on the process including the heat transfer from selected solid parts inside the room and the gas mixture leakage out of the room are taken into account. Main outcome is an MS Excel tool for integrated fire extinguishers design optimization. The optimization balances the two contradictory requirements: Agent volumetric concentration to sustain the fire extinguishing capabilities and tolerable room overpressure. Agent fill weight and discharge time are being adjusted. The discharge time is controlled by the distribution piping and spray nozzles design. System operation is checked concerning various initial and boundary conditions.

1 Introduction

The design apparatus for integrated fire extinguishing system is described. Many aspects of the process are covered by standards, but the need to be worried about the emerging overpressure in the extinguished volume asks for careful analysis and proposal of new techniques.

The thermodynamic system of interest consists of two main parts: an enclosed volume (a room) and extinguishing agent container (a pressure vessel).

The room is initially filled with clean, dry air at atmospheric or near-atmospheric pressure and arbitrary temperature. The heat transfer between walls and its surroundings is omitted completely as the real-life object being modelled is well insulated and relatively low temperature gradients are expected. The model is not completely closed speaking of mass conservation, a simple model of gas mixture leakage based on the pressure rise or fall during extinguishing agent discharge is implemented.

The vessel may have different initial temperature than the room. The heat transfer through the vessel wall is also neglected as the conditions for its evaluations are unclear. The vessel is mounted inside the room, but it is covered in a way that prevents immediate

* Corresponding author: hurda@ntc.zcu.cz

propagation of temperature changes of room gaseous contents to the nearest surroundings of the vessel. Another argument for neglecting the heat transfer is that it should not have a significant effect on the discharge process from the energetic point of view – the energy from air reaches the agent either through the vessel wall or through direct contact after its discharge.

There is a piping system ending with nozzle sets for agent distribution installed in the room. This parts' heat capacity and heat transfer properties are neglected due to similar reasons as for the vessel. The whole distribution system is modelled only in terms of pressure drop.

Three substances' real gas thermodynamic data from the NIST REFPROP database are in use for various calculations of agent state change: CO_2 and two hydrocarbon/halocarbon mixtures. These mixtures are used for approximating a pure compound used in reality that is not available in properties database. The substitutes were chosen based on their interchangeability in refrigeration applications as stated in [3]. The agent is expected to be liquid initially.

The fire extinguishing system proper function is ensured by sufficient volumetric concentration of agent in the extinguished volume, it is checked throughout the time-dependent calculation. The unusual limit of overpressure comes from the demand for integrity of lightweight structural design of the room walls.

2 Thermodynamic state of extinguishing agent in the pressure vessel

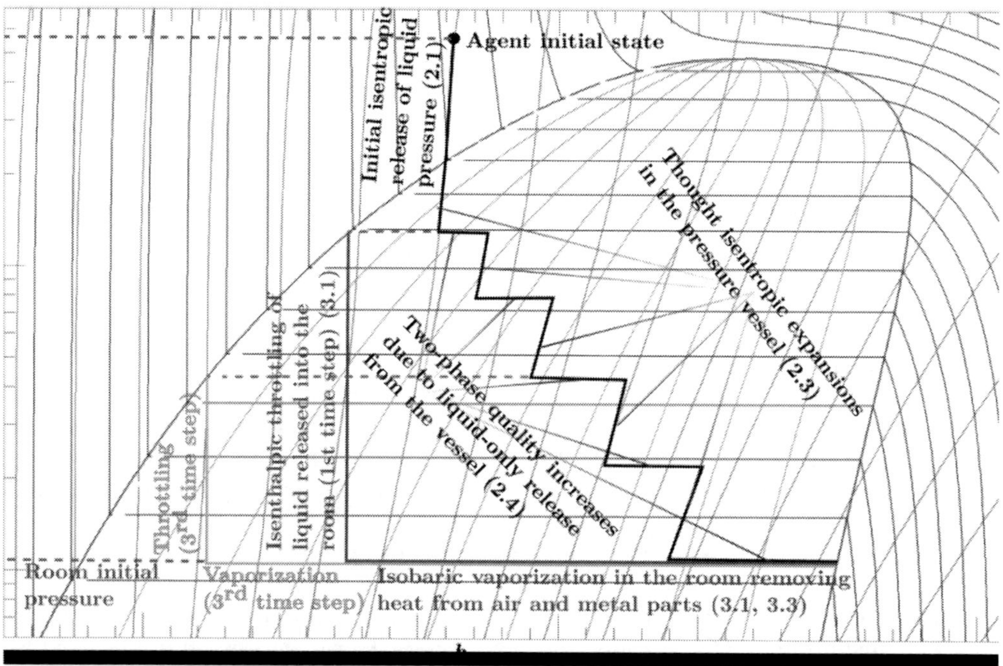

Fig. 1. A schematic pressure-enthalpy diagram of the extinguishing agent with annotated examples of phenomena described in (2, 3). Red – Saturation curves; blue – isotherms; magenta – isentropes, green – curves of constant two-phase quality. Other curves annotated in the figure by text of corresponding colour. (Own work based on NIST REFPROP fluid thermophysical database program).

The main reason why to be concerned with the process going on during emptying of the bottle is to get the starting point of isenthalpic expansion (throttling) which occurs in the piping and mainly nozzles (see (3.1)). Like that, the enthalpy of the released agent is determined.

The vessel is designed to release only the liquid phase of the agent, even though the agent enters the two phase region just after the start of release. This is ensured by siphon tube or bottom release socket. The presence of relatively well separated liquid and vapour phases is expected, despite the ongoing boiling process in the vessel.

This means that the state of the released agent is to be found on the liquid saturation curve. It also needs to be examined how the agent pressure and mass drops with time. After that, the state of agent in the bottle can be determined in discrete time levels.

2.1 Initial release of pressure in fluid

In the first moment after container socket is opened, the pressure drops quickly as the contents of the container are mostly incompressible. In reality, it is necessary for the vessel to be filled with a small fraction of inert gas to compensate for the temperature changes of the vessel contents, but this fraction occupies a small volume to have a significant effect on the pressure release.

According to the statement above, the first time step of the calculation is finding the point on the liquid saturation curve which is reached by isentropic expansion of fluid from the prescribed initial pressure and temperature, see (1). The isentropic nature is expected as the kinetic aspects of the process are minor.

$$p_{sat} = f(s_{sat}); \;\; s_{sat} = s_{init} = f(p_{init}, T_{init}) \tag{1}$$

2.2 Mass flow estimation

An important assumption is made that the mass flow rate is constant during most of the agent discharge process. This conforms with [1, 2]. The value of the mass flow rate is in fact the input quantity for the design of the distribution system, so it is not calculated in any way, but may be a subject of optimization.

Verification of the piping and nozzles design was a subject to experiment. In Figure 2, a constant mass flow rate may be observed for the most of the total mass of agent discharged. The diminishing outflow on the right side of Figure 2 is the part of discharge after "end of liquid" point. This point is reached extremely early in case of CO2 test due to reaching the states under the triple point, where the vessel siphon tube freezes full. This later discharge is neglected in the calculation and the discharge is considered to continue at the prescribed constant flow rate.

An explanation why the two phase, boiling mix discharge rate does not vary with dropping pressure in the container may lie in the subsonic design of the spray nozzles and the fact that pure liquid is entering the distribution system.

Inarguably, the fluid becomes a two phase mixture before reaching the nozzles, but the density of the mixture is probably close to constant. The boiling process in the piping determines the resulting density. It should be dependent only on the time a particle of fluid spends in the pipe and the parameters of the heat transfer through the pipe wall – these are believed to vary few.

The pressure ratio is high enough to choke the flow through the convergent nozzles for the most part of gas release. The critical pressure ratio ranges between 0.55 ÷ 0.58 for all subjected gases.

The end of the discharge is determined by the point where there is only vapour and no liquid in the vessel. The neglected short period of subsonic outflow could possibly start before or after this point.

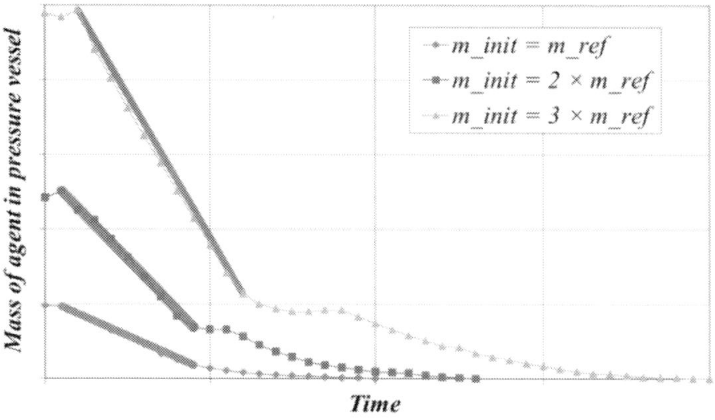

Fig. 2. The discharge process for three experimental tests on a room prototype with CO_2 used for verification of the constant mass flow rate. Thick grey lines indicates the linear approximation of mass, i.e. approximation of constant mass flow rate.

2.3 Isentropic expansion in two-phase region

The first of the two processes modelling the liquid release from the vessel mimics the expansion of the remaining contents of the vessel. The concurrent increase of two-phase quality is described in the next subsection.

The total volume of interest is obtained from total mass of agent and specific volume of saturated liquid in a state after reaching saturation as described in 2.1. The known volume of the container is not taken into account to dismiss the influence of inert gas vessel fill.

As the mass flow rate is known, the released mass per every time step is determined and the related volume too. This released volume is taken as the total volume change for the analysed isentropic expansion. The specific volume after the thought expansion can then be determined as in (2):

$$v_{e\,n} = \frac{V + \Delta m_e \cdot v_{e\,(n-1)}}{m_e} \tag{2}$$

The state is then determined by s_n^e, v_n^e, but as the shapes of isochoric and isentropic curves are not known explicitly in two phase region, a numerical solution is made rather than curve reconstruction. The pressure and temperature after expansion is found from table based on the equality of quality evaluated from entropy and from specific volume (note the indices *s*, *v* for respective quantities) (3), .

$$x_{e\,n\,s} = \frac{s_{e\,n} - s_{liq}}{s_{vap} - s_{liq}} \qquad x_{e\,n\,v} = \frac{v_{e\,n} - v_{liq}}{v_{vap} - v_{liq}} \tag{3}$$

Then, the pressure and temperature is found at which the condition $x_{e\,n\,s} - x_{e\,n\,v} = 0$ is met. And then $x_{e\,n\,s} = x_{e\,n\,v} = x_{e\,n}$.

2.4 State change due to liquid release

As the total volume increase discussed in previous subsection is only thought, the mass decrease in the pressure vessel is further modelled. An increase in two-phase quality at constant pressure and temperature is calculated. Only liquid is released thanks to siphon tube or bottom socket of the vessel.

The relation of the quality after thought expansion (pre-release) to the real quality after the concurrent liquid release and expansion is expressed simply in (4).

$$x_{e\,n} = x_{e\,n\,pre-release} \cdot \frac{m_{n-1}}{m_n} \tag{4}$$

This practice is legit as long as the time step is short enough, so the increase of quality is small. This can be possibly measured by the difference of the achieved fluid state from the presented calculation and if the expansion is calculated after the liquid release – in reverse order than presented. The proposed order was chosen to avoid the singular behaviour of quality increase calculation from saturated liquid state, where $x = 0$.

3 Thermodynamic state of gas mixture in the room

3.1 Released agent temperature and enthalpy

The phenomena occurring during the flow of extinguishing agent from the pressure vessel through piping to nozzles in the room is believed to be an isenthalpic expansion, throttling. Therefore, the initial state for the expansion of liquid agent for each time step of the calculation is found on the left saturation curve at current pressure and temperature in the vessel.

It expands to the pressure present in the room and a temperature bound to this pressure in the two phase region. Gas is always in two phase state of different quality and the temperature is usually lower than the room contents.

The amount of heat for complete evaporation of the agent is expected to be transferred rapidly from the current gaseous contents of the room. The portion of the specific enthalpy of state change can be evaluated as follows (5):

$$q_{evap} = h_{d\,n\,vap} - h_{d\,n} \tag{5}$$

The temperature of the room contents before the mixing calculation is initiated is lowered by the heat consumed by the phase change of the agent as described in (6).

$$T_{n\,r\,premix} = T_{r\,(n-1)} - \frac{q_{evap} \cdot \Delta m}{c_{(n-1)\,r} \cdot m_{(n-1)\,r}} \tag{6}$$

3.2 Mixing process

The mixing of the discharged agent with the gaseous contents of the room is estimated with the theory of ideal gas. In the first time step, pure air is mixed with agent and the successive steps are calculated considering the already existing mixture. The agent is considered to be in saturated vapour state, usually colder than the rest of the room.

The specific heat and specific gas constant of the mixture is a mass weighted average. The temperature is an average weighted by mass and specific heat of the components.

3.3 Heat addition from metal bodies

When the room cools down rapidly, some parts may be able to expel its heat content as well as the gases do. There are a few factors to assess the ability to expel heat rapidly. Bodies worth this analysis need to be made of high conductive materials, have thin walls and large surface area exposed directly to the air. Also the total amount of heat per Kelvin (mass times specific heat) in such bodies must be high enough to have any impact on the whole process of interest. The eligible parts of room interior are sheet metal and thin metal profiles.

The presented method is very simple, but it was verified by more complex apparatus and has shown good agreement. The temperature distribution in the profile is neglected. A single value of bulk temperature is used. For the initialization of an iterative process, the metal part temperature is expected not to change from the previous time step (7).

$$T_{m\,n\,init} = T_{m\,(n-1)} \tag{7}$$

Only two iterations of the process presented in (8), (9) give satisfactory results.

$$Q_{m\,n} = \alpha \cdot A_m \cdot \left(\frac{T_{m\,(n-1)} + T_{m\,n}}{2} - T_{n\,r}\right) \tag{8}$$

$$T_{m\,n} = T_{m\,(n-1)} - \frac{Q_{m\,n}}{c_m \cdot m_m} \tag{9}$$

A qualified guess has to be made for the heat transfer coefficient α estimation. It may lie somewhere between 10÷50 $W/(m^2K)$. It can be adjusted concerning the exposure of the metal surfaces to the nozzle wake areas, a possible contact with liquid particles of the agent and other factors.

3.4 Control volume leakage

Some slots and chinks may be present in the outer bounds of the room structure. When an overpressure between the room and its surroundings occurs during the fire extinguishing process, the gases may leak out of the room. This may have a positive effect of the dropping overpressure.

The amount of leak per time step of calculation is based on current overpressure, the surroundings are considered to stay at constant pressure equal to the initial room pressure. Due to low discharge velocities, incompressible gas dynamics theory is adopted. The discharge velocity is expressed simply as $w = \sqrt{\Delta p \cdot v_r}$. The calculated velocity reaches about 90 m/s. The Mach number of the gas close to regular atmospheric conditions is then reaching a maximum of about 0.3. Under the circumstances of approximations already made, such flow could be considered incompressible.

The equivalent area of the ideal nozzle A is estimated from an experiment carried out at reference conditions. The leaked mass at current time step is then $m_l = \Delta t \cdot w/v_r$.

3.5 Resulting overpressure and volumetric concentration

The overpressure between the room and the surroundings is expressed by perfect gas equation of state (10). The mass here is the mass of agent added to the mass of initial air content in the room and the leaked mass is subtracted. The specific gas constant and mixture temperature is discussed in 3.2. The volume of the room is a predefined constant.

$$p_r = \frac{m_r r_r T_r}{V_r} \tag{10}$$

4 Conclusions

A set of coherent calculations was built together for the purpose of integrated fire extinguishing system design and analysis of various operating conditions. The overpressure in the extinguished room is checked throughout the time dependent calculation.

There are a few key parts of the work besides many trivial calculations.

The first is the assumption of constant mass flow rate of the extinguishing agent into the room, which is inevitable, but it is not causing inaccuracies as it is close to reality. The assumption is based on both external references and an experimental results.

The second is the thorough decomposition of the thermodynamic processes and the decisions on which way to model them. A summary is given in Figure 1. Many issues has been overcome on the side of using REFPROP tables imported as raw real gas data into MS Excel.

Last but not least, the identification of significant influences is made. The heat of metal bodies and control volume leakage (chapters 3.3, 3.4) are incorporated into the calculation.

According to the preliminary experiments made on the room prototype, there may be other influences on the process. The walls are not rigid, so the overpressure causes deformation resulting in enlargement of the total room volume. The gases may also loose significant amount of internal energy due to the work being done on the walls deformation. Analysis of those phenomena may lead to the future enhancements of the design tool.

The results were developed in the specific research (ZCU SGS-2016-045).

Nomenclature

c	- specific heat (isobaric)	$[J/kg/K]$	h	- specific enthalpy	$[J/kg]$
m	- mass	$[kg]$	p	- absolute pressure	$[Pa]$
Q	- total heat	$[J]$	q	- heat per kilogram mass	$[J/kg]$
r	- specific gas constant	$[J/kg/K]$	s	- specific entropy	$[J/kg/K]$
T	- absolute temperature	$[K]$	t	- time	$[s]$
V	- volume	$[m^3]$	v	- specific volume	$[kg/m^3]$
w	- flow velocity	$[m/s]$	x	- two-phase quality	[-]
α	- heat transfer coefficient	$[W/m^2/K]$			

Indices:

d	- agent discharge state	e	- extinguishing agent state
init	- initial	l	- leak
liq	- liquid	m	- metal
n	- time step number	vap	- vapour
r	- room	sat	- saturation state

References

1. M. Ahmad, L. Buit, O. Florisson, C. Hulsbosch-Dam, M. Bögemann-van Osch, M. Spruijt, F. Davolio, *Experimental Investigation of CO2 Outflow from a High-pressure Reservoir,* Energy Procedia, **37**, 3005-3017, ISSN 1876-6102, http://www.sciencedirect.com/science/article/pii/S187661021300430X, (2013)
2. National Fire Protection Association, *NFPA Code 12A: Standard on Halon 1301 fire extinguishing systems* ISBN: 978-145591885-0, https://www.nfpa.org/codes-and-standards/all-codes-and-standards/list-of-codes-and-standards/detail?code=12A, (2018)
3. Linde AG, *Retrofit Guide,* https://www.linde-gas.com/internet.global.lindegas.global/en/images/Refrigerant%20Retrofit%20guides17_109809.pdf?v=5.0, (2017)

Influence of temperature on the production and size distribution of fine particles released from beech wood samples

Ján Poláčik[1,*], *Jiří* Pospíšil[2], *Ladislav* Šnajdárek[1], and *Tomáš* Sitek[1]

[1]Faculty of Mechanical Engineering, Brno University of Technology, Technická 2896/2, 61669 Brno, Czech Republic
[2]NETME Centre, Brno University of Technology, Technická 2896/2, 61669 Brno, Czech Republic

Abstract. Fine particles generated from laboratory biomass combustion are discussed in this study. The approach combines the thermogravimetric analysis during thermal decomposition of beech wood sample with detailed monitoring of the size distribution of fine particles produced. Thermogravimetric analysis (TGA) allows monitoring the exact temperature influence of a small fuel sample (wood) according to the desired schedule. The cool aerosol stream leaving TGA enters a Scanning Mobility Particle Sizer (SMPS) where the particle size fractions are separated. The monodisperse aerosol is counted by the condensation particle counter (CPC). The parametrical study was carried out to assess the influence of composition, size and surface of the wood sample on the production and size distribution of ultrafine particles

1 Introduction

Understanding of particles formation and their behaviour related to biomass burning process are objective of the intense research. Particulate matter (PM) concentration level in cities is increasing. Especially, attention is increasingly drawn to smaller factions. Most of the particles produced by the combustion process are PM_1 (particles with a diameter of less than 1μm).

These fractions have a considerable impact on the health of the population. They are easily inhalable and can penetrate deep into the human respiration system. In addition, compared to larger particles, they have a much greater active surface at the same weight; and therefore, have a high ability to bind other harmful substances [1, 2].

The highest concentrations of particulate matter are generally present in the inner parts of urban areas. Fireplaces and biomass boilers have become the main source of fine particles in these areas.

* Corresponding author: jan.polacik@vutbr.cz

1.1 Particulate formation

As shown in Figure 1, volatile components released from biomass have different values of partial pressure. After reaching the saturation point, the formation of a new phase begins. We call it the nucleation process. The molecules are clustered into ultrafine particles up to 0.1 μm in size. The resulting particles can be further enhanced by coagulation mechanisms (aggregation of colloidal and macromolecular organic particles into larger aggregates), by agglomeration (bonding based on the adhesion of the surfaces), and by oxidation reactions or condensation on the particle surface [3].

Particles formed in the combustion chamber are called primary particles. Secondary particles are formed in the flue gas duck and in the atmosphere. According to volatility, the resulting organic compounds are divided into volatile organic compounds (VOC) and semi-volatile organic compounds (SVOC) [4].

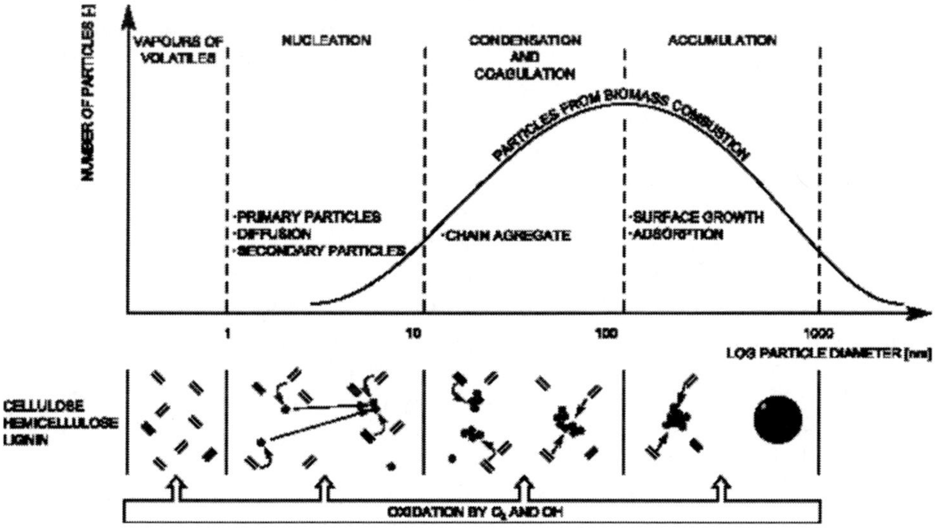

Fig. 1. Nucleation and growth of the particles in combustion process.

Describing particle formation in detail is possible only under laboratory conditions. Real biomass combustion is always associated with the fine particles formation in the flue gas. Their production depends on the properties of the fuel (humidity, granulometry, etc.) and the type of combustion device. However, mainly it depends on the combustion temperature, reaction time and volume of oxygen in the atmosphere.

1.2 Composition of the beechwood

Cellulose is the most common polymer in nature and it can be expressed by the chemical formula $[C_6H_{10}O_5]_n$. Individual molecular chains of cellulose are hold by hydrogen bonds and van der Waals forces creating fibrils. These cellulose fibrils are the main element of the trees and plants [5]. The cell walls components, representing the bulk of wood, are principally the lignin fraction and the total carbohydrate fraction (cellulose and hemicellulose), termed holocellulose. Residues from woody plants, such as bark and sawdust, have differing compositions.

Table 1. Chemical composition of the beech wood.

Density	730 kg/m^3
Humidity	6.1 %
Lignin	20 %
Holocellulose	78 %
α-cellulose	50 %
Extractives	2 %

The beech belongs to the group of *beech trees* with composition: Extractables (2 %), Lignin (20 %) and Holocellulose (78 %) [6].

2 Experimental Setup

This article presents a result of laboratory research of particle production during the burning of redwood sample. The laboratory procedure which was developed combines the advantages of thermogravimetric analysis with the detailed monitoring of the size distribution of produced fine particles. Thermogravimetric analysis (TGA) allows to monitor the exact temperature influence of a small fuel sample according to the desired schedule. TGA also influences the composition of the atmosphere flowing around the sample. During the test, the TGA monitors the weight of the heat-affected sample and identifies its loss. From the analysis of the development of temperature rise, it further identifies the presence of endothermic and exothermic reactions.

The measurement was carried out with utilizing the TGA device *NETZCH – Jupiter F3*. The base component of the STA-449 Jupiter analyzer is a very precise digital weighting system with vertical design. The weighting system is connected to a shielded ceramic module (TG-module) into which the analyzed samples are placed. During the measurement, the entire module is inserted into a gas-tight laboratory furnace with controlled heating rate. The result of the measurement is a TGA curve showing the weight change in dependence on the temperature of the sample.

The gaseous components released from the fuel sample during the TGA are scattered in the test atmosphere and removed from the device. Subsequent flow through the connecting pipeline cools the stream to the ambient temperature. The cool aerosol stream enters a *Scanning Mobility Particle Sizer* (SMPS) where the particle size fractions are separated. Nearly every particle has some level of electric charge. Electrostatic classifier requires aerosol to achieve a steady state of charge distribution. This is achieved by Aerosol Neutralizer device, which provides discharging and charge neutralization process. After this process, particles pass through a bipolar charger and reach equilibrium charge level on the particles. Then, aerosol flows into the electrostatic classifier where sizing occurs (Figure 2).

In a particle sizing system, the *Electrostatic Classifier* separates particles by size for high resolution measurements of particle size distribution. Scanning Mobility Particle Sizer (SMPS) allows classifying particles in the range from 10 to 1 000 nanometers in diameter. *Differential mobility analyzer* (DMA) selects flowing particles of specific size from a polydisperse aerosol. Particles are separated according to their electrical mobility. As a result, there is a highly monodisperse aerosol containing particles of a certain size. Monodisperse particles leaving the electrostatic classifier continue to a *Condensation Particle Counter* (CPC). This device measures particle number concentration. A number of particles in the particle-size fractions are identified by the condensation particle counter (CPC).

In the CPC, particles pass through a heated saturator in which Butanol is vaporized. The aerosol together with evaporated Butanol flow into a cooled condenser. Particles, as

condensation nuclei, get enwrapped by the supersaturated Butanol vapour. They become larger droplets so they are easily counted passing through an optical detector [7].
The production of fine particles during the process of burning the wood sample was measured by TSI-SMPS device (*Model 3080-Series Electrostatic Classifiers including CPC 3775, TSI Inc.*).

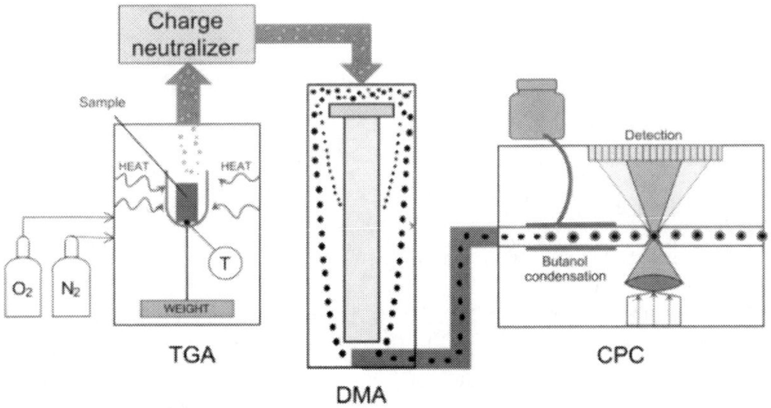

Fig. 2. Scheme of the Measuring equipment and the particle flow: TGA – Thermo-gravimetric analysis; DMA – differential mobility analyzer; CPC – Condensation Particle Counter.

3 Results and discussion

The parametrical study was carried out to assess the influence of temperature rise on the production and size distribution of ultrafine particles released during a heating of beech wood sample. Different weight and form of the beech wood samples was tested as shown in following table. Water content of 6.1 % was measured at 25 °C.

Table 2. List of tested samples.

Material	Sample dimensions	Weight
Beech heartwood	20×13×2 mm	280 mg
Beech heartwood	20×8×2 mm	150 mg
Beech heartwood	20×3×2 mm	80 mg
Beech bark	8×6×3 mm	80 mg
Beech heartwood sawdust	Particle size up to 0.4 mm	80 mg

3.1 Different size of samples

The measuring method described in the previous section was used for series of experiments with beech wood samples. The first study was carried out with the target to identify the influence of the size of the sample on particle production during controlled heating. Three samples of heart beech wood were used for this study, namely 280 mg, 150 mg and 80 mg. The samples were obtained from the same part of the wood log with water content 8 %. All experiments were carried out in oxidation atmosphere mixed in a laboratory from pressurized vessels of pure oxygen and pure nitrogen in volume ratio 21:79 %.

Thermogravimetric analysis was used for continuous increase of sample temperature. The heating sequence started at 20 °C and the temperate rise was set to 10 °C/min. The sample was heated continuously until its temperature reached 630 °C. The Figure 3 shows the experimentally obtained relationships between the temperature of the tested samples and time period (the increasing curve) and weight of the samples and time period (the decreasing curves). With increasing temperature, the weight of the samples decreases due to volatile fraction release and solid material oxidation.

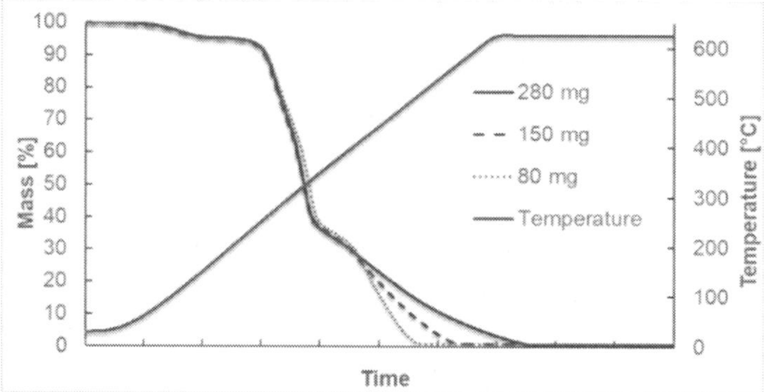

Fig. 3. Results of thermogravimetry analysis for beech wood samples of different mass.

During temperature increase up to 120 °C, water evaporates from the tested samples. In the interval 120 °C to 250 °C, light volatile compounds are released. The interval 250 °C to 300 °C is associated with the rapid loss of samples weight induced by the intensive release of other volatile compounds. Exothermic reactions are commenced at these temperatures. The associated heat generation promotes the release of other volatile compounds. The interval 300 °C to 500 °C is associated with the gradual charcoal decomposition of carbon.

The volatile vapours, particles and gaseous combustion products released during TGA are mixed with the flow of the oxygen-nitrogen mixture entering the thermogravimetric analyzer (flow rate 0.3 l/min). The final aerosol passes through the measuring device capable to identify concentration and size distribution of particulate matter. The Figure 4 shows obtained the size distribution of fine particles identified in the final aerosol generated by TGA. Obtained results are presented for four temperature levels, namely 200 °C, 300 °C, 400 °C and 500 °C.

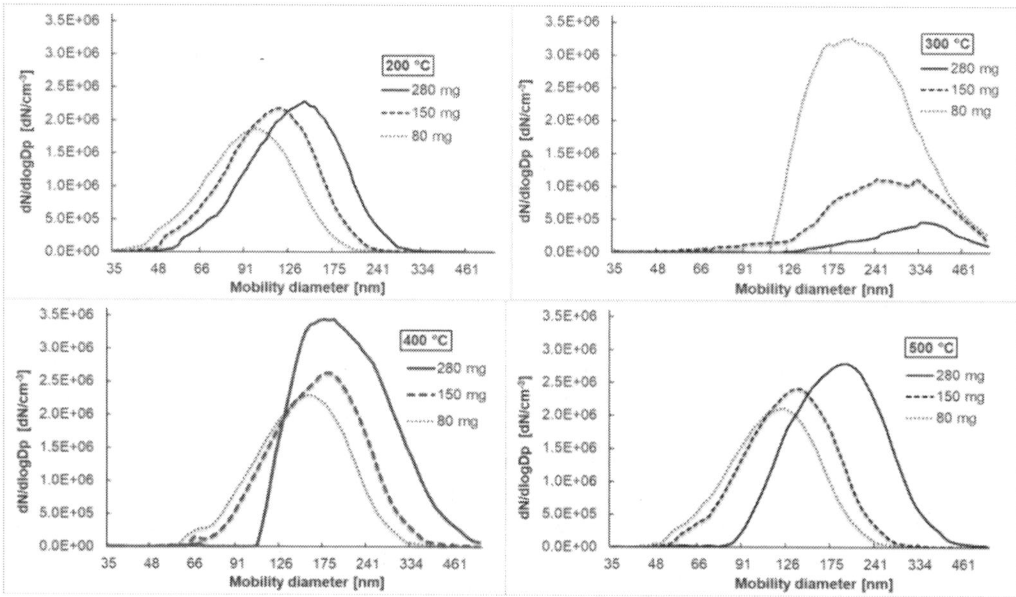

Fig. 4. Size distribution of fine particle (PM1) emitted from Beech wood sample of the weight 280, 150 and 80 mg in certain temperature levels.

The trend visible in Figure 4 shows to increase the particle concentration with increase of the tested sample of wood. The size of particles increases with the size of the sample too. An anomaly in these trends presents the temperature level 300 °C. In this temperature, the combustion of volatiles stars and process becomes to be significantly unstable. The same trend is visible in the relationship between the mass concentration of PM and size of particles expressed for same temperature levels, see Figure 5.

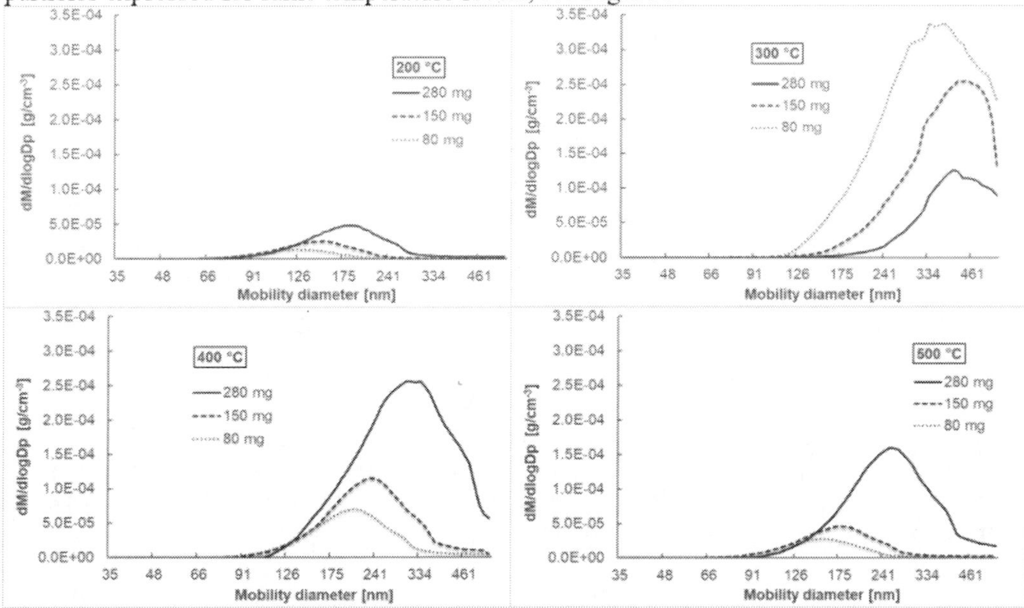

Fig. 5. The Relation between the mass concentration and the particle size for beech wood samples of the weight 280, 150 and 80 mg in certain temperature levels.

For a correct assessment of the combustion processes is necessary to sum particle production during the entire period used for heating of samples from 20 °C to 630 °C. The Figure 6 shows particle concentration and mass concentration of fine particles in form of the size distribution curves for the entire testing period. These results support two prior outcomes. Increasing size of tested sample results in the increase of the mass concentration of emitted particulate matter. Size of particles increases with the increasing size of samples too.

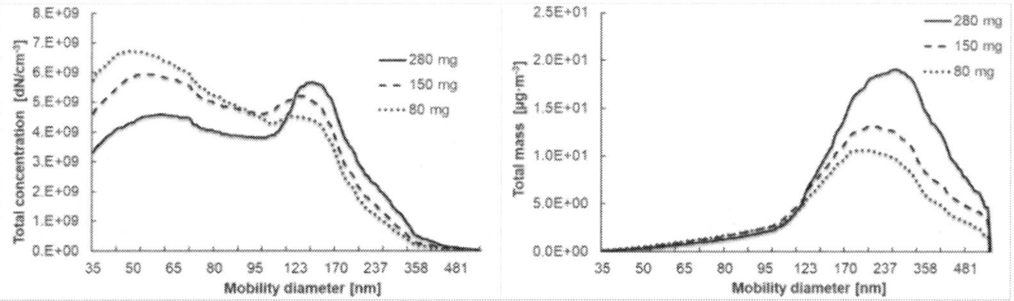

Fig. 6. The size distribution curves comparing particle production for entire testing period.

3.2 Different form of samples

Another part of the study is focused on identification of particle emission during heating of different form of the beech wood. The tests were carried out with three different samples of wood utilized as energy biomass, namely heartwood, bark and sawdust. These samples were evaluated by same laboratory process as described in the previous section. Figure 7 shows the result obtained from TGA. It shows experimentally obtained relationships of i) the temperature of the tested samples versus time period, ii) weight of the samples versus time period. Comments to this Fig. are similar as mentioned before for Figure 3. There is visible faster combustion of sawdust and slower combustion for bark in comparison with the heartwood.

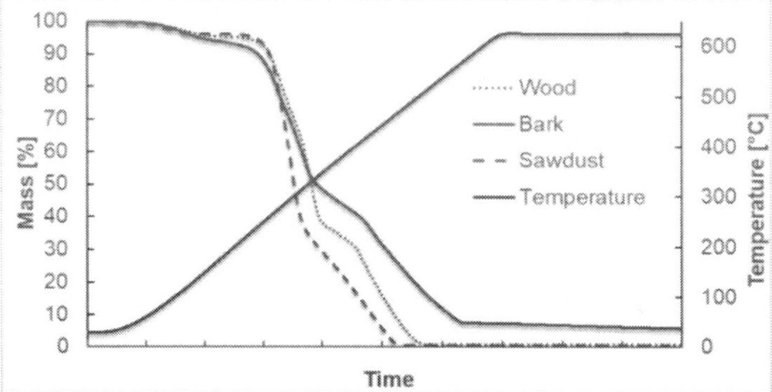

Fig. 7. Results of thermogravimetric analysis for beech wood samples of different form.

There is no clear trend in particle size or particle concentrations obtained for the monitored temperature levels, see Figure 8. Size of the majority of generated particles is in the wide range from 40 nm to 500 nm. The temperature dependent mass concentration of particles is without any general trend too, see Figure 9.

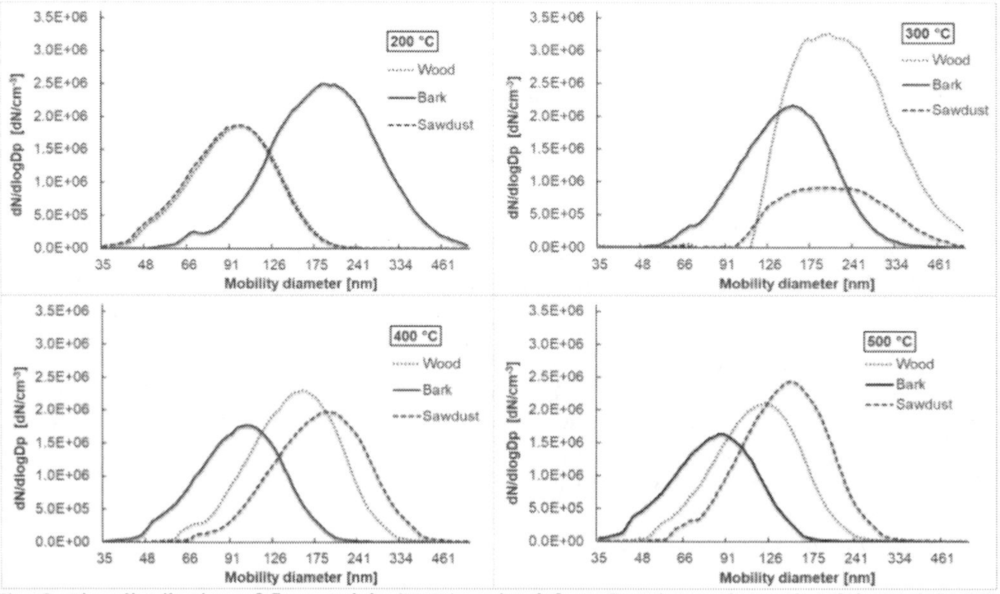

Fig. 8. Size distribution of fine particle (PM1) emitted from Beech wood sample of the weight 280, 150 and 80 mg in certain temperature levels.

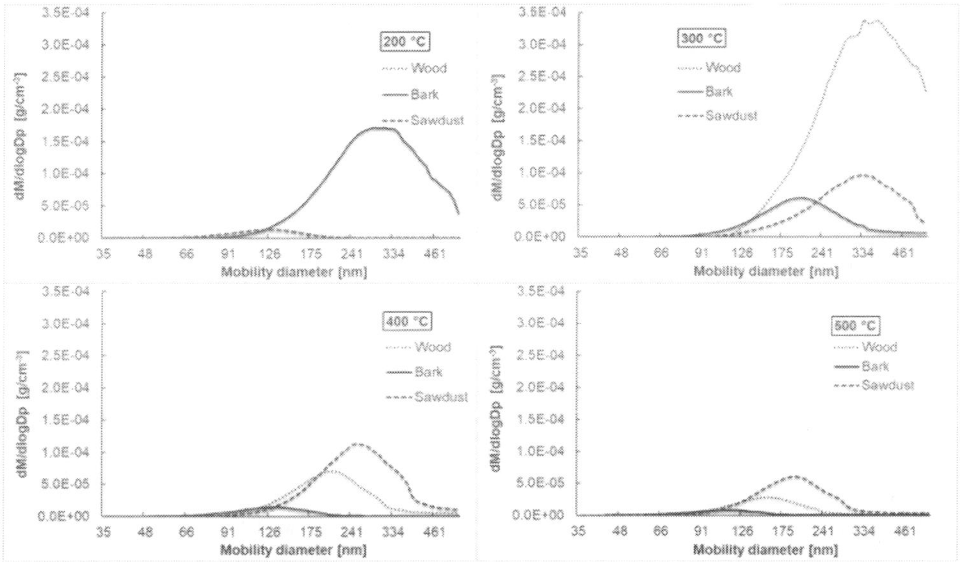

Fig. 9. The Relation between the mass concentration and the particle size for beech wood samples of the weight 280, 150 and 80 mg in certain temperature levels.

The Figure 10 shows particle concentration and mass concentration of fine particles in form of the size distribution curves for the entire testing period. The highest particle concentration was identified for the ultrafine particles with the diameter close to 40 nm. The bark produces the highest number of ultrafine particles from all tested samples. The highest mass concentration was identified close to the particle diameter 200 nm for all samples. The bark produces the lowest mass concentration for dis particle size. The heartwood and sawdust produce significantly more. These outcomes are valid for assessment of the entire testing period.

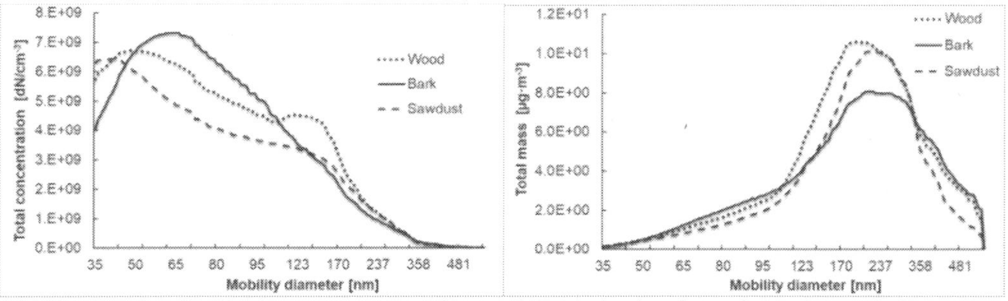

Fig. 10. The size distribution curves comparing particle production for entire testing period.

4 Conclusions

This study deals with laboratory investigation of fine particulate matter production over controlled heating of beech wood samples in the atmosphere with 21 % of oxygen. From carried out measurements follow that majority of emitted fine particles is in the range 40 nm to 500 nm. The peak number of particles is between 100 nm an 300 nm. The mass concentration of produced particulate matter increases proportionally with increasing size of the sample. The number of particles is increasing too, but not proportionally.

In the second step, the samples in form of sawdust, bark and heartwood were analyzed from the view of particulate matter production. Obtained results are strongly influenced by actual combustion conditions that significantly differ between wood samples forms. It is not easy to find any clear trend between temperature and size distribution of produced particles. From the general view, the bark produces higher number concentration of ultrafine particles in comparison with heartwood and sawdust samples. The total weight of produced particulate matter is very close for all tested form of wood samples.

Presented results were obtained within the project NETME CENTRE PLUS (LO1202), created with financial support from the Ministry of Education, Youth and Sports of the Czech Republic under the „National Sustainability Programme I".

This work has been financially supported by TACR under the grant TJ01000331.

References

1. J.A. Baldock, R.J Smernik, *Chemical composition and bioavailability of thermally altered Pinus resinosa (Red pine) wood*. Organic Geochemistry. **33(9)**, 1093-1109 (2002)
2. T. Kaivosoja, P.I. Jalava, H. Lamberg, A. Virén, M. Tapanainen, T. Torvela, U. Tapper, O. Sippula, J. Tissari, R. Hillamo, M.R. Hirvonen A.J. Jokiniemi, *Comparison of emissions and toxicological properties of fine particles from wood and oil boilers in small (20–25 kW) and medium (5–10 MW) scale*. Atmospheric Environment. **77**, p. 193-201 (2013)
3. M. Obaidullah, S. Bram, V.K. Verma, J. De Ruyck, *A Review on Particle Emissions from Small Scale Biomass Combustion*, International Journal of Renewable Energy Research, **2**, 147–159 (2012)
4. O. Sippula, *Fine particle formation and emission in biomass combustion*, PhD Thesis, Department of Environmental Science, University of Eastern Finland, Finland (2010)
5. M. Poletto, V. Pistor, J. Ademir, *Structural Characteristics and Thermal Properties of Native Cellulose*. VAN DE VEN, Theo G.M., ed. Cellulose - Fundamental Aspects, DOI: 10.5772/50452. ISBN 978-953-51-1183-2 (InTech, 2013)
6. Di Blasi, Colomba, et al. *Pyrolytic behavior and products of some wood varieties*. Combustion and Flame, **124.1-2**: 165-177 (2001)
7. TSI Inc. Tsi.com. Available at: http://www.tsi.com/

Volume ratio of two immiscible phases measured by electrical impedance tomography

Michal Malík[1,*], *Jiří* Primas[1], *Michal* Kotek[2], *Václav* Kopecký[2], *Bohuš* Kysela[3] and *Radek* Šulc[3]

[1]Institute of Novel Technologies and Applied Informatics, Faculty of Mechatronics, Informatics and Interdisciplinary Studies, Technical University of Liberec, Czech Republic
[2]Department of Physical Measurements at the Institute for Nanomaterials, Advanced Technology and Innovation at the Technical University of Liberec, Czech Republic
[3]Department of Process Engineering at the Faculty of Mechanical Engineering, Czech Technical University in Prague

Abstract. The mixing of two immiscible phases is a process commonly seen in many branches of industry. Whether it be desirable (e.g. mixing of ingredients in chemical or food industry) or undesirable (e.g. sediments or contaminants in water purification) process, it has to be taken into consideration and a detailed description would be beneficial to any end product. This paper deals with a method for observing the volume ratio of two immiscible phases from the state of total separation, during the mixing process until the state of a homogenous mixture using industrial Electrical Impedance Tomography (EIT) system ITS p2+. The paper also shows a great agreement between the data obtained through measurement using this method and data derived from a theoretical formula. Also using EIT the authors were able to provide additional information describing the mixing process in real time.

1 Introduction

In a vast amount of industrial applications (water treatment, energetics, chemical engineering, etc.) there is some form of mixing of different phases involved. In case of deliberate mixing the end goal of this process is usually reaching a homogeneous mixture with the minimum expenditure of energy and time. In case of unwanted mixing we need to know the volume of the second phase or the state of the mixture. Both cases benefit from observation or measuring the process for the purpose of determining if and how the end goal has been reached.

This paper focuses on the use of Electrical Impedance Tomography (EIT) as a measuring tool for the use in mixing, where it is not possible to use various other methods (e.g. ultrasonic measuring methods, capacitance measurements or optical methods such as image processing, particle image velocimetry, interferometric particle imaging, etc.). Since optical measuring methods are the most commonly used, the main advantage of using EIT becomes evident if the mixed media or the container is optically non-transparent.

* Corresponding author: michal.malik@tul.cz

The principle behind the Electrical Impedance Tomography is measuring the conductivity/impedance of a medium between two sets of electrodes [1]. As with all tomography methods this is done over a whole measuring plane by using several sets of electrodes in the same plane and switching between them. Since unwanted electrochemical effects may occur in direct current measurements, the whole EIT measurement is done using alternating currents in the range of milliampers at tens of kHz. Mathematically, the problem of reconstructing conductivity from surface measurements of current and potential is a non-linear inverse problem and is severely ill-posed. The detailed mathematical formulation of this problem has been stated by Alberto Calderón [2]. This means, the vast amount of measured data have to be processed by a suitable modelling software with the results usually displayed in the form of scalar maps. These show the planar conductivity distribution in the interrogated area.

This principle has several advantages over other methods usable for the same purpose. Mainly, we can see "inside" a lot of processes, but the method does not require us to introduce any kind of ionizing radiation or foreign agents for tracing purposes. This is a huge advantage in medicine, where EIT is widely used – e. g. lung and breast EIT diagnostics. Another very important advantage is that EIT also does not require any probe or other physical object to be inserted into the interrogation area. The only requirement for the electrodes of the tomography system is to have a conductive contact with the medium. They can thus be on the sides of the container/pipe, where they merge with the surface of the container and do not intrude into the medium itself. This also means that the use of EIT does not affect the character of flow in the measured system.

It can be said that nearly all ordinary chemical substances have different conductivities, which mainly depend on free ion content [3]. Thus EIT allows us to observe the course of their mixing accurately and with ease from the initial state, when the substances are separate, to the final state, when a homogenous ideal mixture is achieved.

One such common problem is to describe the mixing of a conductive medium with a non-conductive one. The theory behind the conductivity of such a mixture is well known [4]. However to be able to analyse the mixture using the theoretical description, we need it in its homogenous state. Here is the point, when the EIT system comes into play, since it is a very suitable tool for ascertaining if the mixture has reached homogeneity. This is due to its capability of showing a planar (or even spatial) distribution of conductivity. If during the mixing process the measured conductivities in each point of the measured plane (volume) reach the same value, we can consider the mixture to be homogenous. This allows us to characterise the mixture by this single conductivity value, which we can use to represent it in the theoretical description.

There are many ways of calculating the conductivity of a two-part mixture with known original conductivities. In [4] it has been shown that the volume ratio of two mixed substances, where the first one has a known conductivity and the second one is non-conductive, can be calculated by the means of the following formula:

$$\varepsilon = \frac{2-\frac{2\gamma}{\gamma_0}}{2+\frac{\gamma}{\gamma_0}}, \tag{1}$$

where γ is the conductivity of the final homogenous mixture, γ_0 is the conductivity of the conductive substance and ε is the volume ratio of both mixed substances and can be defined as follows:

$$\varepsilon = \frac{V_1}{V_0}, \tag{2}$$

where V_1 is the volume of non-conductive substance and V_0 is the volume of the conductive substance in the mixture.

Using EIT we are able to monitor the time development of the mixture conductivity and we can easily find the point, when the conductivity remains constant. Then we can be sure that the mixture is homogenous and the measured conductivity value is the right one.

2 Experiment methodology

In our experiments we used the EIT ITS p2+ system for measuring two different mixtures. In both cases the mixtures consisted of a conductive phase (water) and a non-conductive phase (transformer oil, silica sand). Because of the need for a sufficient difference between conductive and non-conductive phases the conductivity of water was increased by adding a small amount of sodium chloride (NaCl). The conductivity of a strongly diluted solution of NaCl depends linearly [5] on the concentration of salt in water. This way its conductivity was raised to 0.8 mS/cm in the first experiment and to 1.5 mS/cm in the second one.

The container used for the experiment was a cylindrical vessel with four sets of sixteen electrodes each (thus giving us four independent EIT measuring planes), though for both following experiments we used only three. The mixing process also took place in this vessel using a Rushton turbine [6] made of plastic on a long plastic-composite shaft (see Figure 1). The vessel is equipped with a lid, which prevents aeration of the mixture during the mixing process. This is very important, because the media measured are truly only the two phases we introduced and the experiment is not invalidated by the generation of air vortex or air bubbles. With the lid closed the total volume of the measuring vessel is 6.36 l.

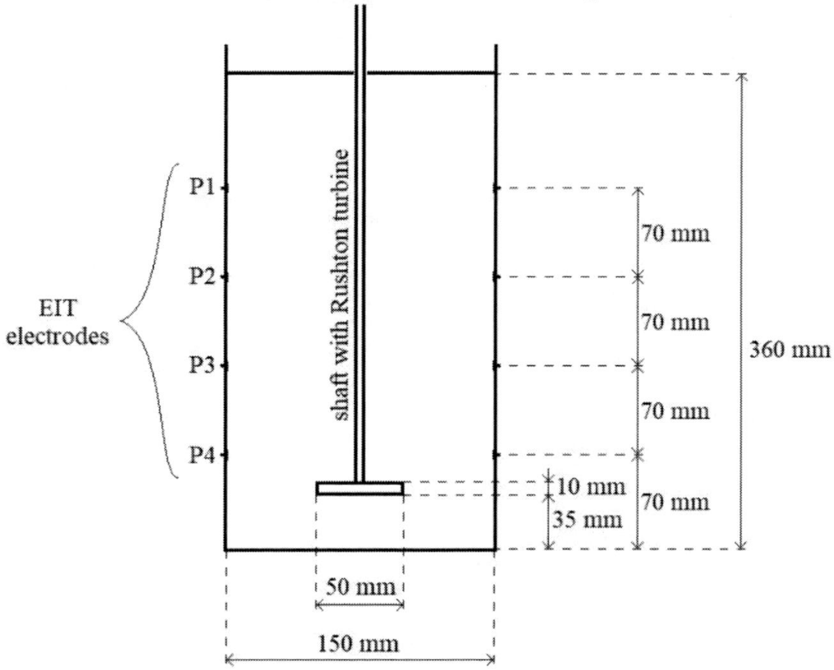

Fig. 1. Diagram of the measuring vessel with the Rushton turbine and EIT electrodes.

The materials of the shaft and turbine were chosen intentionally non-conductive, so the EIT analysis software would be able to mask their presence, thus ensuring that the measurement results would show only the two mixing phases.

3 Experiments

The following two experiments were chosen to verify the theoretical formula (1). They also serve to prove the viability of ITS p2+ tomography system for measuring and analysing of mixing processes.

3.1. Mixing water and oil

For the first experiment we poured 0.7 l of transformer oil into the measuring vessel and the rest of the space was filled by water with its conductivity increased to 0.8 mS/cm. This gave us 11.1 % of transformer oil in the whole volume of the measuring vessel. The EIT measurement was then started synchronously with the turbine. This provided the data from the whole mixing process from the very beginning until the homogenous mixture was reached. According to [7] while using Rushton turbine at speeds of several hundred rotations per minute, the mixed liquid should reach homogenous state after less than a minute. In our case the speed of the motor connected to the turbine was set to 760 RPM. The measurement was carried out through the point when the average conductivity in each measured plane became constant and was continued for another two minutes to assure homogenous conductivity distribution.

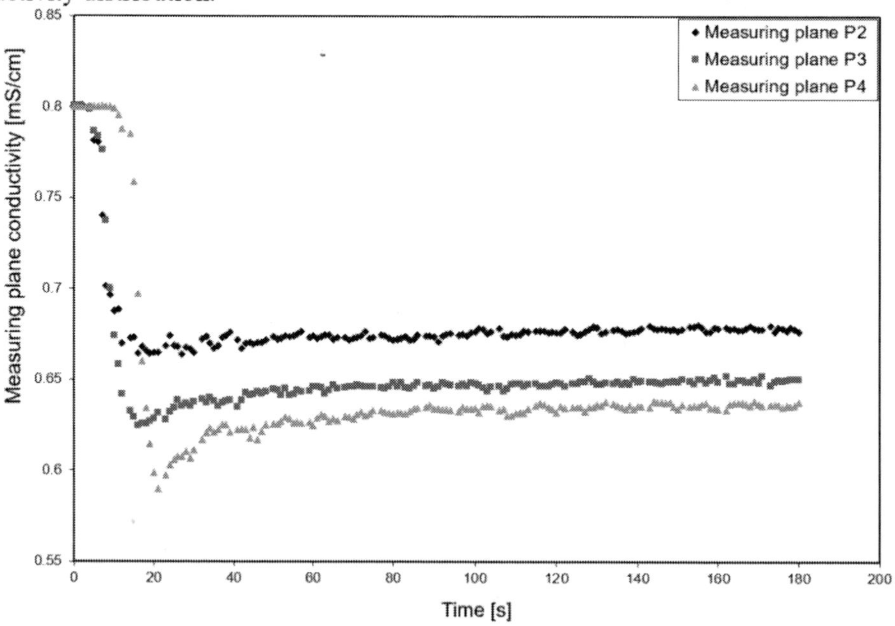

Fig. 2. Graph of time development of average conductivity in each measuring plane for the mixing process of water and transformer oil.

The mixing process was initiated by starting the turbine and simultaneously the measurement was started. This can be seen in Figure 2, which shows the time development of average conductivity in each measuring plane changing from the initial value of unmixed components to a final value representing a homogenous mixture. A series of photos showing the mixing process can be seen in Figure 3.

a) b) c)

Fig. 3. Photos of the measuring vessel a) with unmixed components (water – transformer oil), b) at the start of the mixing and c) at the point of homogenous mixture.

Since the transformer oil has lower density than water and at the beginning of the experiment it was all on top (see Figure 3a). This also affected the time of homogenisation as can be clearly seen from Figure 2, since the conductivity values in plane 2 are the first to become constant and the values in plane 4 are the last.

Second fact seen in Figure 2 is that after 35 seconds conductivity values in all three measuring planes become constant. The difference between them, though slight, is noticeable and it is caused by the dynamic equilibrium of forces affecting the mixed liquids (buoyancy, gravity and forces originating from the turbine caused vortex).

By using the average value of the three conductivities to represent the whole volume and putting it into the formula (1) we get the volume ratio ε = 0.12 (nonconductive phase / conductive phase). From the initial conditions of the experiment we know the real volume ratio of the phases was 0.7 l : 5.66 l, which equates to ε' = 0.124. This shows a perfect agreement.

3.2. Mixing water and sand

In the next experiment we wanted to test the methodology in the case of mixing the conductive liquid phase with a non-conductive solid phase. This time the measuring vessel was first filled with water to a height of 340 mm (volume 6 l). This gave us a space of 20 mm from the water surface to the lid (volume 0.36 l). Next we poured in silica sand (with particle size of 0.6 – 1.2 mm) so that the water level rose to the surface of the lid, thus filling the vessel completely. This way the sand took up 5.5 % of the total volume of the vessel. The reason for using only half the volume for the second phase was the possible strain on the plastic turbine, if we were to use more silica sand. This could lead to either heavy abrasion or even breaking of the turbine. Through the addition of NaCl the conductivity of water was again adjusted to a higher value, this time to 1.5 mS/cm. Because the difference in densities of the first and second phase is in this case much greater than in the first experiment we had to use higher turbine speed. For this purpose the speed of the turbine was set to 1530 RPM. The measuring method used was also slightly modified. This time the EIT was used to capture not only the process of mixing from start until relative homogeneity, but also the process of relaxation after the motor has been turned off (see Figure 4). This way we can watch the conductivity of the mixed medium affected by sand rising from the bottom of the vessel because of the vortex and also by the opposite process when the sand is settling down during the relaxation.

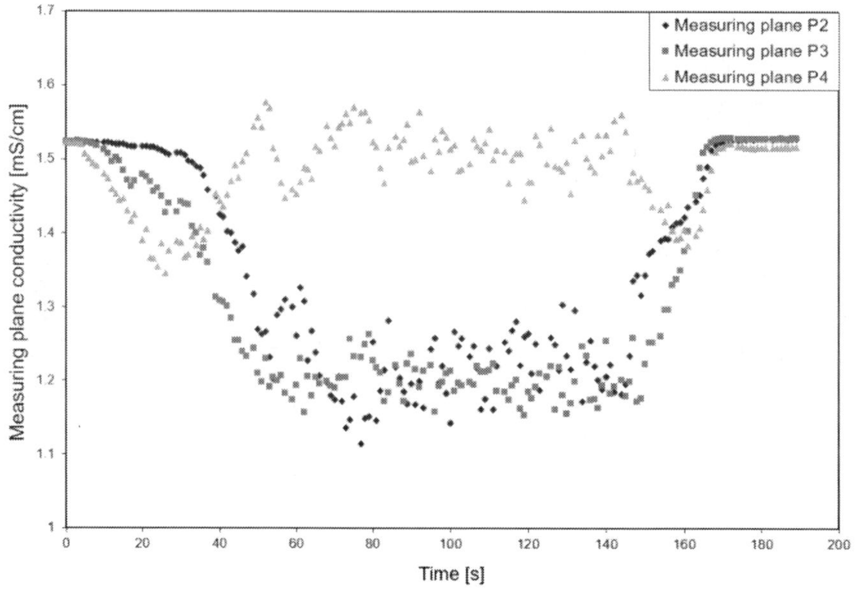

Fig. 4. Graph of time development of average conductivity in each measuring plane for the mixing process of water and silica sand.

A series of photos again shows the mixing of silica sand with water in the measuring vessel – see Figure 5.

a) b)

Fig. 5. Photos of the measuring vessel a) with unmixed components (water – silica sand) and b) at the point of relatively homogenous mixture.

This time the second phase (silica sand) is much denser than water and at the beginning of the experiment it was all at the bottom of the vessel (see Figure 5a). This again affected the homogenisation times (see Figure 4), since the conductivity values in plane 4 are changing sooner and faster than the others. Unlike in the first experiment the conductivity in plane 4 returned very soon after the start of mixing to its initial value (on average). This was caused by the fact the silica sand rose all the way from the bottom of the vessel because of the vortex (in this experiment much stronger due to the higher RPM) affecting it. After the first 25 seconds almost all the sand left the measuring plane 4 and concentrated in between planes 2 and 3 as can be seen in Figure 5b. The overall fluctuations of the conductivity values were caused by the imperfect homogeneity and very strong turbulent flow.

By using the average value of conductivities in plane 2 and 3 to represent the mixture and putting it into the formula (1) we get the volume ratio $\varepsilon = 0.14$. From the initial conditions of the experiment we know the real volume ratio of the phases was 0.36 l : 6 l, which equates to $\varepsilon' = 0.06$. These values appear not to agree at all, but if we consider that the vast majority of the silica sand was concentrated in the area covered by measuring planes 2 and 3, thus only it was present only in half of the total volume of the vessel. This corresponds with the fact the theoretically obtained value is approximately double of the real one. In which case we can consider the formula (1) to be still valid.

4 Conclusion

In this paper it has been shown that EIT system ITS p2+ is a viable tool for measuring mixing processes, where one of the mixed phases is considered non-conductive. It has also been shown, it works for both solids and liquids. This paper also brought experimental proof of an existing theoretical formula (1). The agreement of experimental results with calculated volume ratios is in the first experiment better than 96 % and in the second experiment better than 85%, which can be considered to be very good.

Tomography measuring methods are commonly used to view or analyse processes taking place inside objects. From the results of both described experiments it can be seen, the EIT provides more information and in greater detail about the examined processes than many other methods. Optical methods are usually used with great success for extremely precise measurements and observation of flows and possibly mixing of phases. However in our case they would fail, because the mixed materials used in our experiments are optically non-transparent. At much smaller volume ratios of mixed phases optical measuring methods (like shadowgraph method or particle/speckle image velocimetry [8]) may suddenly become viable, but this paper proves, the tomography measuring methods and namely the used EIT excel in these difficult conditions at great volume ratios of mixed phases.

Research of observing the mixing processes using EIT continues on the Technical university of Liberec in Czech Republic.

The Authors gratefully thank to the support of Grant Agency of the Czech Republic (GA ČR) 16-20175S Local turbulent energy dissipation rate in dispersion systems, and LO1201 co-funding from the Ministry of Education, Youth and Sports as part of targeted support from the „National Programme for Sustainability I".

References

1. F. Dickin, M. Wang, "Electrical resistance tomography for process applications" in *Meas. Sc. and Tech.*, **7**, IOP Publishing, (1996)
2. A. P. Calderón, "On an inverse boundary value problem" in *Comp. and App. Math.*, **25**, Petropolis, (2006)
3. D. S. Holder, *Electrical Impedance Tomography: Methods, History and Applications*. London: Institute of Physics, (2004)
4. M. Liu, T. Wang, W. Yu and J. Wang, "An electrical conductivity probe method for measuring the local solid holdup in a slurry system" in *Chem. Eng. J.*, **132**, Elsevier, (2007)
5. A. Kay Smith, R. A. Gortner, "The Electrical conductivity of mixed salt solutions" in *J. of Phys. Chem.*, **37** (1932)

6. D. Jašíková, M. Kotek, V. Kopecký, "Time Resolved PIV Measurement of Fluid Dynamics in Agitated Vessels" in *Proceedings of SPIE*. USA : SPIE (2015)
7. D. Jašíková, M. Kotek, B. Kysela, V. Kopecký, "Experimental Identification of the Flow Vortex Structures Generated in the Agitated Vessels" in *Rec. Adv. in Mech. Eng.*, **11**, Italy: World Academy of Science, (2014)
8. D. Jašíková, B. Kysela, J Konfršt, R. Šulc, "Estimation of turbulence dissipation rate by Large eddy PIV method in an agitated vessel" in *EPJ Web of Conferences*. France: EDP Sciences, (2015)

Explosion Characteristics of Hydrogen for CFD Modelling and Simulation of Turbulent Gas Flow

Jan Skřínský[1,*], *Jan* Koloničný[1], and *Tadeáš* Ochodek[1]

[1]Energy Research Centre, VSB-TU Ostrava, 708 33 Ostrava - Poruba, Czech Republic

Abstract. Renewable energies became more and more important in the last years. Hydrogen as a promising energy carrier is a perfect candidate to supply the energy demand of the world. The state of the hydrogen gas (turbulences and point concentrations) has a significant impact on the gas explosion indices. A gas cloud is formed by a partial-pressure method in gas explosion experiments in the spherical $20.0{\cdot}10^{-3}$ m^3 chamber. Gas in the chamber reaches an uniform state beyond in hundreds of ms. The absolute pressure for gas dispersion should be higher than 0.01 MPa for the H_2 of concentration larger than 30 vol. % of fuel. The initial temperature also influences turbulent gas flow before ignition, especially in the case of the gases lighter-than-air.

1 Introduction

1.1 Interest

Results of explosion experiments depend on many different parameters of the investigated process, such as the energy and type of ignition source, size and shape of explosion chamber, initial temperature and composition of the flammable mixture [1]. One way is to study these processes by mathematical computational fluid dynamics (CFD) modelling. CFD has sufficient accuracy and a much higher potential for being optimized if based on currently used experimental methods such as standard $20.0{\cdot}10^{-3}$ m^3 [2].

1.2 Previous studies

Explosion indices of individual gases at ambient temperature and atmospheric pressure have been studied systematically. Only limited data are available for hydrogen gas at elevated temperatures and pressures until now. In work [3] has been reported measurement on explosion pressures of hydrogen-air mixtures, at various initial concentrations close to stoichiometric, in closed vessels of different volumes. The results have shown a similar behavior of investigated system. The maximum explosion pressures have been (7.9±0.3) bar for measurements made in the standard $20.0{\cdot}10^{-3}$ m^3 spherical vessel. Higher deviations from these values are observed, however, when using smaller or larger size vessels, where

* Corresponding author: jan.skrinsky@vsb.cz

radiative and convective heat losses to the walls could be neglected. In [4] have been reported different values of maximum explosion pressures for hydrogen-air mixtures, depending on the volume of the explosion vessel: 8.1 bar in a $120.0 \cdot 10^{-3}$ m^3 explosion vessel. The reported results from $6.0 \cdot 10^{-3}$ are identical to those results obtained in the $20.0 \cdot 10^{-3}$ m^3 and $120.0 \cdot 10^{-3}$ m^3 despite the great difference in volume. The present contribution introduce the explosion pressures of stoichiometric hydrogen - air mixtures measured experimentally in $20.0 \cdot 10^{-3}$ m^3 chamber for various initial temperatures. The temperature influence on the maximum explosion pressure was investigated at atmospheric pressure.

2 Experiment

2.1 Experimental device

Experiments were done in $20.0 \cdot 10^{-3}$ m^3 chamber (CA 20L) made by OZM Research in accordance with EN 1839 [5]. It is the stainless steel (1.4435) double wall vessel of spherical shape with an internal diameter of 336 mm. The vessel is provided with an opening of an inside diameter of 148 mm. The whole system is schematically introduced at Figure 1.

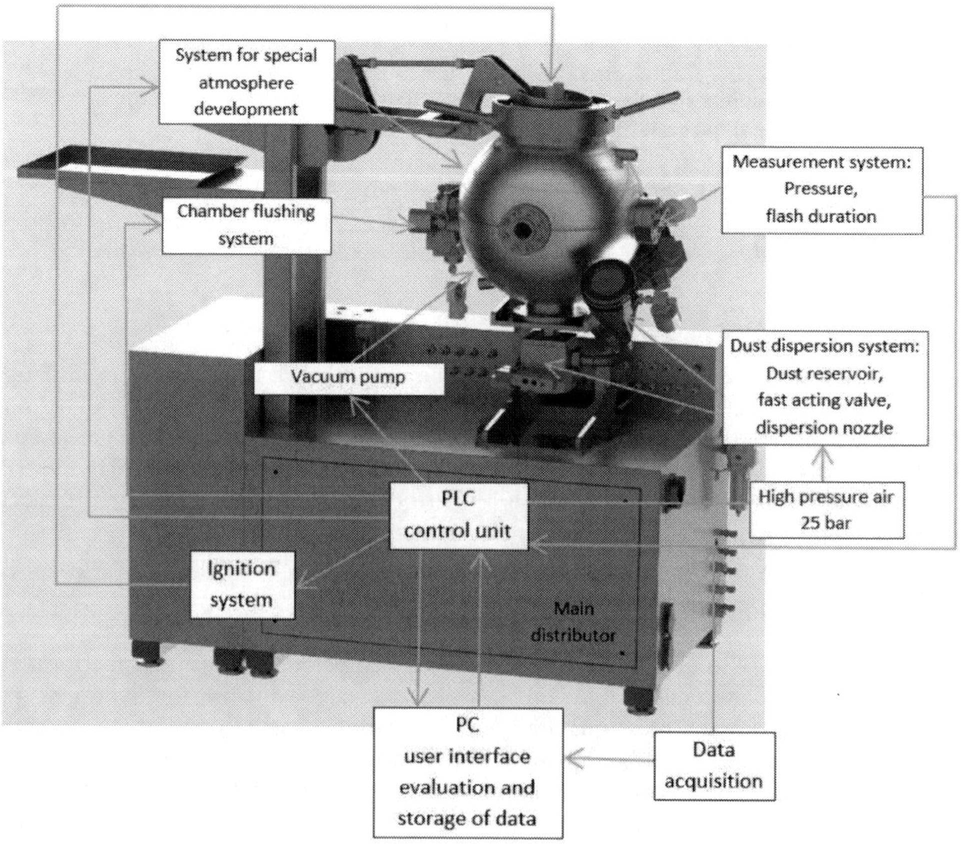

Fig. 1. Schematic introduction of the $20.0 \cdot 10^{-3}$ m^3 chamber system.

Highly dynamic temperature control system Presto A30 by JULABO is used for heating of the chamber. Permanent spark generator was made in accordance with EN 15967 [6] with Tungsten electrodes with a distance of 6 mm. A pair of piezoelectric pressure sensors by Kistler, type 701A. Pressure range set for the sensors is 2.5 MPa and natural frequency 70 kHz. Data are recorded be four-channel data card with a sampling rate 50 kS/s/channel. PLC Siemens Simatic 1215 connected to PC used as a user interface automatically control whole procedure including fast acting valve timing. The chamber is equipped by thermocouple for temperature monitoring, especially for measurement of temperature in time of ignition. Correct hydrogen gas - air mixture composition is assured by partial pressure method. Pressure transducer measures internal pressure. For the experiment 99.995 % hydrogen guaranteed by Siad, s.r.o. has been used [7].

2.2 Experimental procedure

Procedure for gas explosion starts with evacuation to pressure less than 0.4 bar to leave a space for the gas. For example, the use of hydrogen concentrations of 30 vol. % and 40 vol. % require the partial pressures of methane to be added are 0.3 bar and 0.4 bar respectively. Then the evacuation pressures should be 0.7 bar and 0.8 bar respectively. PLC starts the experiment where 60 ms is counted as ignition delay time, and then the gas is mixed by blowing dispersion air into the hydrogen-air mixture inside the chamber. After that, the mixture is ignited by the electric discharge. The explosion indices are measured and calculated [7].

3 Results

3.1 High-speed images

Figure 2 illustrates the real non-adiabatic explosion. The first left foto in the Figure 2 illustrates the nearly laminar spherical flame that have been produced during the first phase (i.e. in the first 10 ms) of the explosion. Beyond in hundreds of ms the spherical flame became turbulent and the flame speed increases monotonically with increasing turbulences and fuel air mixing.

Fig. 2. High-speed images illustrating the evolution of hydrogen gas-air spherical flames.

3.2 Explosion indices

Figure 3 illustrates the normalized peak explosion pressure, P_{ex}/P_0, versus equivalence ratio at atmospheric pressure and four initial temperatures. The P_{ex}/P_0 explosion pressure is the maximum value of explosion pressure, as indicated in Figure 3. With the increase of temperature, the peak explosion pressure is decreased due to the decrease of the total fuel mass in gas/vapour cloud (turbulences). This is because the species in flames at elevated temperatures are more reactive and taking into account the effect of density (point concentrations). As shown in Figure 4, $(dP/dt)_{max}$ and K_G give an approximate value at varied initial temperatures, indicating that $(dP/dt)_{max}$ and K_G are sensitive to the variation of temperature. Rate of pressure rise is affected by both flame speed (turbulences) and heat release (point concentrations). Flame speed increases monotonically with the increase of temperature, but the decrease of total fuel mass results in the decreased heat release and offsets the effect of flame speed on the explosion pressure.

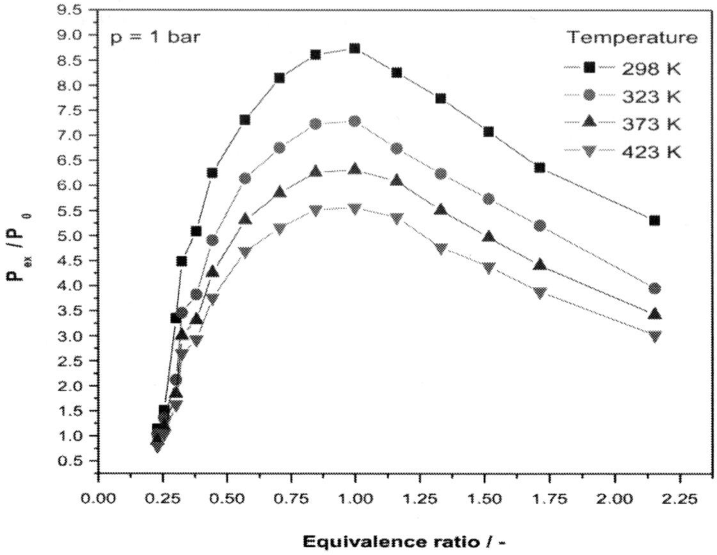

Fig. 3. Normalized peak explosion pressure, P_{ex}/P_0, versus equivalence ratio.

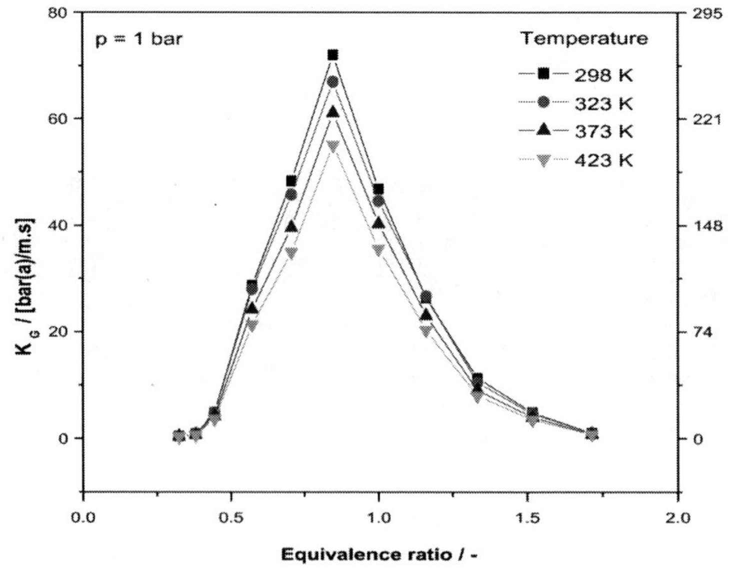

Fig. 4. Rate of pressure rise, $(dP/dt)_{ex}$, and deflagration index, K_G.

4 Conclusion

Explosion indices data are essential for a quantitative risk assessment of explosion hazard associated with the use of hydrogen gas. The present work is the preliminary study to investigate the influence of various initial temperatures on explosion characteristics of hydrogen gas/air mixtures. The explosion pressure, the maximum rate of pressure rise and the deflagration index are the fundamental explosion characteristics. We have described, that they have decreased with the increase of initial temperature. In all cases, the turbulent gas flow and point concentrations play crucial role in the promotion of the whole reactions due to fuel air mixing and hence the faster flame speed, and heat release. In the following experiments, these values will be used as approximate initial values for explosion experiments carried out in heated 1 m^3 explosion apparatus built by OZM Research s.r.o. at Energy Research Centre, VŠB - Technical University of Ostrava.

This work was prepared within the projects: 1) „Innovation for Efficiency and Environment – Growth“, identification code LO1403 with the financial support from the Ministry of Education, Youth and Sports in the framework of the National Sustainability Programme I and 2) Energy use by ORC technology and turbines, identification code SP2018/171.

References

1. J. Skrinsky, J. Veres, V. Peer, P. Friedel, J. Travnickova, A. Dalecka, Chemical Engineering Transactions, **53**, 7-12 (2016)
2. J. Skrinsky, J. Veres, V. Peer, P. Friedel, AIP Conference Proceedings, **1745**, art. no. 020057 (2016)
3. K. Hochtappels, Journal of Loss Prevention in the Process Industries, **19**, 4 (2006)
4. K.L. Cashdollar, I.A. Zlochower, G.M. Green, R.A Thomas, M. Hertzberg, Journal of Loss Prevention in the Process Industries, **13**, 3-5 (2000)
5. EN 1839:2013. Determination of explosion limits of gases and vapors. Berlin: Beuth Verlag (2013)
6. EN 15967:2012. Determination of maximum explosion pressure and the maximum rate of pressure rise of gases and vapours. Berlin: Beuth Verlag (2012)
7. B. Janovsky, J. Skrinsky, J. Cupak, J. Veres, Draft-paper in 12th International Symposium on Hazards, Prevention and Mitigation of Industrial Explosions Kansas City, USA - August 12-17 (2018)

Experimental Modelling of Autoignition Temperature for Alkyl/Alkenyl Products from Fischer-Tropsch Synthesis

Jan Skřínský[1,*], *Ján* Vereš, and *Karel* Borovec

[1]Energy Research Centre, VSB-TU Ostrava, 708 33 Ostrava - Poruba, Czech Republic

Abstract. Interest in Fischer-Tropsch technology is increasing rapidly. Alkyl/alkenyl products from Fischer-Tropsch synthesis are alternative, renewable, environmentally and economically attractive fuels and there are considered one of the most favorable fuels for conventional fossil-based fuels. The chemistry of this gas-to-liquid industry converts synthesis gas containing carbon monoxide and hydrogen to oxygenated hydrocarbons such as alcohols. The fire hazards associated with the use of these liquid hydrocarbons mixtures are obvious. This article aims to explore the fundamental fire and explosion characteristics for main products composition from Fischer-Tropsch synthesis.

1 Introduction

1.1 Interest and challenges

Nowadays there is a worldwide demand to develop energy efficient and economical processes for sustainable production of alternative chemical compounds and fuels as a substitute for those emerging from petroleum [1]. Autoignition temperature (AIT) is one of the most important variables used to assess fire and explosion hazards of flammable liquids.

It is the lowest temperature at which the test substance will ignite when mixed with air under the conditions defined in the test method. In the industry, it is used for convenient and reliable classification of the flammability of liquids. Investigation of autoignition is always a challenge because AIT depends on several physical-chemical properties, which are discussed in [2].

1.2 Previous studies

We realized after a careful bibliographic search autoignition temperature data for methanol, ethanol, propanol, 2-butanol, 1-butanol, and 2-methyl-2,4-pentanediol [3]. These authors soon carried out a re-investigation of the effect of experimental conditions on measuring autoignition temperatures of pure liquid chemicals [4]. In our recent investigation, we identify the effect of experimental conditions on measuring autoignition temperature of

* Corresponding author: jan.skrinsky@vsb.cz

methanol and ethanol binary mixtures [5]. We have extended the measurement of autoignition temperature of 1-pentanol and its binary mixtures with water [6]. Presented results should provide essential information for the chemicals prepared by Fischer-Tropsch micro catalyst bed avaible at Energy Research Centre, VSB-TU Ostrava and include light hydrocarbons (C1 and C2), olefins, LPG (C3-C4), naphtha (C5-C11), diesel (C12-C20) and wax (>C20) [7].

2 Experiment

2.1 Experimental device

Experiments were done in the test vessel made by OZM Research s.r.o. and is described in [2]. It generally consists of: 1) mirror mounted above the flask so that the observer may see into the flask without having to be directly over it; 2) insulated cover; 3) electrically heated crucible furnace; 4) aluminium, to promote temperature uniformity; 5) test temperature Chromel-Alumel thermocouple; 6) borosilicate round-bottom, short-necked boiling flask; 7) external thermocouple (bottom); 8) external thermocouple (middle); 9) external thermocouple (top). The photography of the whole system is at Figure 1a and schema is introduced at Figure 1b.

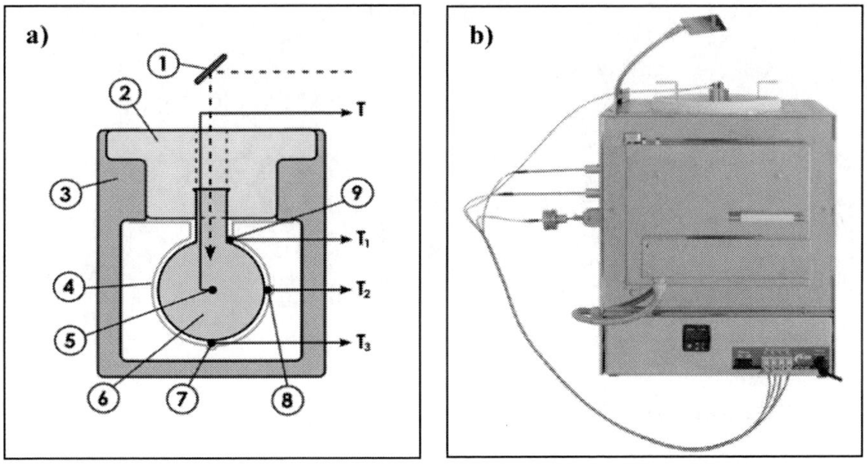

Fig. 1. a) Photography and b) scheme of the autoignition temperature apparatus.

2.2 Experimental procedure

The experimental procedure meet the test requirements described by the test method of ASTM E659-78:2005. The amount of substance and the temperature of the test vessel, which is filled with air are varied to find the lowest temperature (of the hot surface) that causes ignition. This standard test method is designed to determine the autoignition temperature of a flammable vapour in air mixture with air at ambient pressure and temperature up to 650 °C. After the internal flask temperature has reached the desired temperature, we have adjusted the temperature controller to maintain this temperature within the designated limits and allow the system to equilibrate. The sample, approximately 50-250 uL, was injected into the uniformly heated flask containing air at a predetermined temperature. After insertion of the sample, the contents of flask were observed in a dark room for 10 min, or until autoignition occurred.

Autoignition was evidenced by the sudden appearance of a flame inside the flask and by a sharp rise in the temperature of the gas mixture. When the mixture exhibited flames at the preset temperature, the next sample was tested at a lower temperature. These procedures were repeated until the lowest temperature at which the mixture exhibited flames was obtained [5].

3 Results

3.1 Autoignition images

Figure 2 depicted the occurrence of an autoignition by the sudden appearance of a flame outside the flask following by a sharp rise in the temperature of the gas mixture in Figure 3-4. The illustrative sample is taken from [5]. The extent of this temperature rise depends upon the difference between the rate of heat generation within the mixture and the rate of heat loss to the chamber walls. The characteristic thermal effects when a combustible liquid is introduced into the chamber at different initial temperatures are illustrated by the time-temperature curves of Figure 4.

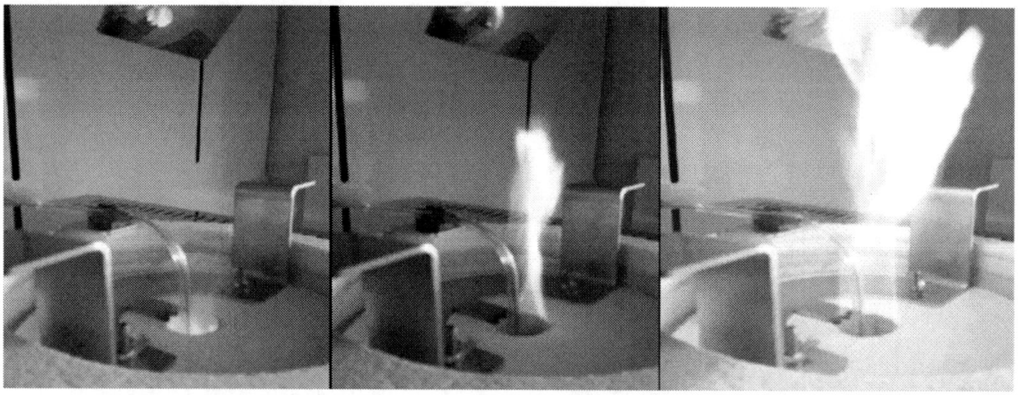

Fig. 2. Development of the flames emitted above the top of the flask (200 ul, 432 °C).

3.2 Combustion plots

According to ASTM E659-78, recorded two values for each test (temperature, ignition delay time). Figure 3 illustrates minimum autoignition temperature for alkyl/alkenyl products from Fischer-Tropsch synthesis that is denoted by the red colour. The measurement of the auto-ignition started according to previous studies (between 440-460 °C) that was in the temperature region where our thermometer has its highest sensitivity and where the lowest alcohol has its AIT. The sample to be tested, approximately 100 uL, was inserted into a uniformly heated 500-ml glass flask containing air at a predetermined temperature (460 °C). This procedure led to detection of more than 350 identified AIT lines, most of them in the 430-480 °C. At relatively low initial temperatures, there may be further thermal effects. The presented AIT record include the following information: hot-flame selfignition temperature, SIT (°C), ignition temperature, IT (°C), minimum ignition temperature, MIT (°C), slow combustion, SC (°C), and reaction threshold temperature, RTT (°C), for preflame reaction. Figure 4 show the combustion plot of experiments for the alkyl/alkenyl products from Fischer-Tropsch synthesis. In this figure, the x-axis is the ignition delay time between application of heat to a material and its ignition, and the y-axis is the preheated temperature of the combustion container.

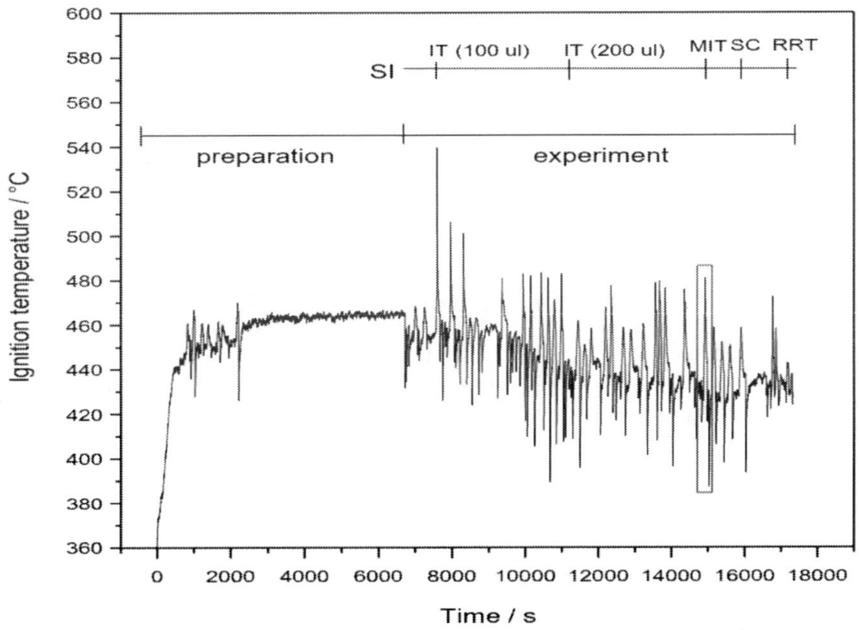

Fig. 3. AIT owing to the alkyl/alkenyl products from Fischer-Tropsch synthesis.

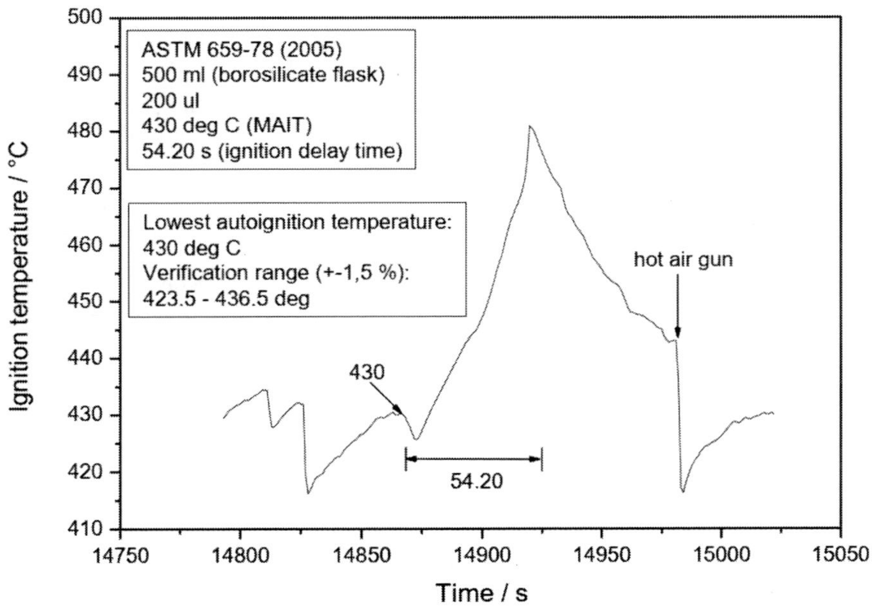

Fig. 4. Example of the AIT time curve with minimum autoginition temperature.

4 Conclusion

In this preliminary study, the AITs of alkyl/alkenyl products has been characterized in the gas phase for the first time. The analysis of the observed records led to identification of more than 350 identified AITs measured in compliance with the ASTM E659-78:2005 test method. An extensive set of autoignition temperatures and ignition delay times have been obtained allowing a systematic comparison between existing standards for open glass flasks to find the reproducible "worst case" value. In the following experiments, these values will be used as approximate initial values for experiments carried out in heated 200 ml flask apparatus to details available information regarding alternative test conditions according to EN 14522:2006.

This work was prepared within the projects: 1) „Innovation for Efficiency and Environment – Growth", identification code LO1403 with the financial support from the Ministry of Education, Youth and Sports in the framework of the National Sustainability Programme I and 2) Purification of gas produced by gasification by various materials, method of continuous analysis of ammonia content in ash, identification code SP2018/174.

References

1. N. Moazami, M.L. Wyszynski, K. Rahbar, A. Tsolakis, H. Mahmoudi, Chemical Engineering Science, **171**, 32-60 (2017)
2. ASTM E659-78:2005. Standard Test Method for Autoignition Temperature of Liquid Chemicals. Washington: ASTM Committee E27 (2005)
3. C.C. Chen, H.J. Liaw, C.M. Shu, Y.C. Hsieh, J. Chem. Eng. Data, **55**, 5059-5064 (2010)
4. C.C. Chen, Y.C. Hsieh, Ind. Eng. Chem. Res., **49**, 5925-5932 (2010)
5. J. Veres, J. Skrinsky, Chemical Engineering Transactions, **52**, 1267-1272 (2016)
6. J. Veres, J. Skrinsky, J. Travnickova, A. Dalecka, S. Sevcikova, Chemical Engineering Transactions, **53**, 43-48 (2016)
7. M. Wnukowski, Report on the Fischer-Tropsch Micro Catalyst Bed (MCB) operation. Ostrava: 29.08.2017, 26

MATEC Web of Conferences **168**, 07015 (2018) https://doi.org/10.1051/matecconf/201816807015
XXI. AEaNMiFMaE-2018

Designing the solar collector position with the Arduino electronic platform

Peter Vician[1,*], *Peter* Ďurčanský[1], and *Jozef* Jandačka[1]

[1]University of Žilina, Department of Power Engineering, 010 26 Žilina

Abstract. The article deals with the utilizing of solar energy by the means of concentric trough collector based on own design. It focuses on achieving optimal position of the trough with respect to incident radiation during its operation. Due to the collector virtual sun tracking system, which was proved to be ineffective and could not achieve the desired results, the optimum position had to be maintained by manual control of rotation. The measurement of the angle of rotation of the concentric trough is realized using the ARDUINO electronic platform, for which the program was created in the software ARDUINO (IDE) 1.8.4.

1 Introduction

The thesis deals with the use of a concentric trough collector located on the building of University of Žilina. The design of the collector was made according to its own design. Due to its later modification and future application it is necessary to test the collector and determine its performance parameters. The collector has an automatic virtual tracking system, which uses mathematical algorithm to determine the position of the sun and subsequent rotation of the collector. Early measurement has shown some performance deviations due to incorrect trough position. Further tests have shown that the actual tracking system is unable to maintain an optimal position throughout the day. For the purposes of the following measurements, functional monitoring should be ensured. Tilt of the trough can be set manually using the control panel located in the laboratory. The optimal values of tilt angle for time interval of specific day was determined according to calculation of solar angles. However, for the manual operation of the collector and the angle of rotation, it is necessary to provide a response that can be recorded in the form of data. This task is accomplished by the Arduino Mega 2560 Rev3 electronic programmable platform with an additional module MPU-6050 filling function of accelerometer and gyroscope. The program for calculating the angle of rotation and recording the measured values into Excel was created in the freely available ARDUINO (IDE) software 1.8.4. The device itself and the attached modules are powered via the USB A/B connecting data cable. Thus, the device can be programmed and simultaneously powered by one cable. Once the device is connected to an external power source, we can also use it in difficult-to-access areas such as the roof of a building and its operation is limited by the capacity of the battery.

* Corresponding author: peter.vician@fstroj.uniza.sk

However, in our case, for measurement purposes, it is not enough to simply put the measured data on the microSD card but also to get a real-time response. For this reason, the created program displays the measured values in the specified time interval via the serial monitor.

Fig. 1. USB Data Extension 1.1 (LAN) cable and RJ-45 network cable.

For real-time data transmission, you must ensure that the device is connected continuously to a computer using a data cable. The USB data cable can fully transfer the data to a distance of approximately 10 m, but this is not enough in our case. For data transmission over longer distances the signal has to be amplified by repeater. A better solution is to use an easy-to-get network cable with RJ-45 and a USB/LAN extension to a distance of 30 m. A short measurement of the collector power parameters was performed to verify the device's functionality under realistic conditions. Measured data using the MPU-6050 chip was compared with data obtained from the control panel.

2 Solar tracking systems

The tracking mechanism must be reliable and capable of monitoring the sun with a certain degree of accuracy, even during alternating cloud, which depends on the collector acceptance angle. Another important task that the mechanism provides is to return the collector to its original position at the end of the day or at night. Some systems, in addition to tracking, also perform a protective function and turn the collector out of focus to protect it from hazardous environmental and working conditions such as severe wind shocks, overheating, and failure of the heat flow mechanism. There are different forms of tracking mechanisms from complex to very simple. They can be divided into two broad categories: mechanical and electronic systems. Electronic systems generally have greater reliability and tracking accuracy. These can be further subdivided into mechanisms using electronically controlled motors through sensors that measure the size of solar illumination and mechanisms using computer controlled engines with feedback provided by solar-flow sensors on the receiver [1, 2, 3].

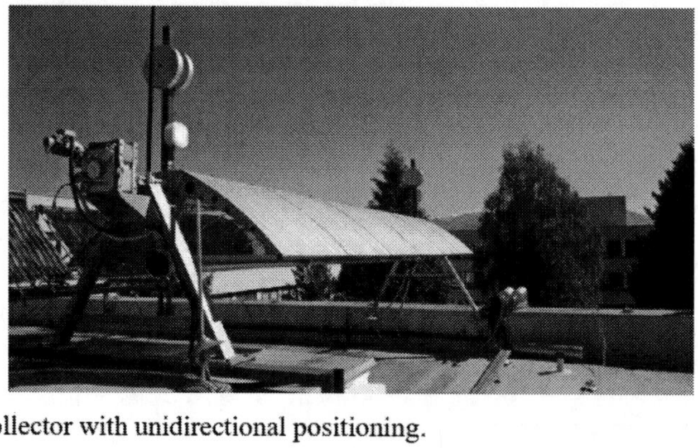

Fig. 2. Parabolic collector with unidirectional positioning.

Another form of tracking system is called "virtual" tracking, where a traditional sun monitoring unit with sensors that detect the position of the sun has been replaced by a system based on calculating the position of the sun using a mathematical algorithm. The unit is implemented with a 13-bit optical angle encoder mechanically coupled to the collector rotation axis. By comparing the position of the sun and the collector axis with the electronic devices, a signal will be sent to the propulsion system to move to the optimal position [4].

3 Design of measurement device

Arduino is a freely available electronic platform based on easy-to-use hardware and software. Arduino boards can read input values using additional modules. The input can be light on the sensor, finger on the button, Twitter message or other. Any input parameter can be changed to output, e.g. engine activation, LED on, online publishing. The function and usage depend on the set of instructions that are sent to the microcontroller on the board. The Arduino programming language, based on the Wiring programming framework, allows you to write multi-layer software to control devices connected to a wide range of microprocessor boards to create all kinds of creative codes, interactive objects and environments. The Arduino platform is used by a wide community of students, musicians, artists and professionals. It has been used for thousands of projects ranging from everyday objects to complex scientific instruments.

There are several Arduino platforms and a number of sensor types. For our project, the Arduino Mega 2560 Rev3 microcontroller model, the MPU-6050 with accelerometer and gyro function, and the Arduino SD card module were used.

Fig. 3. Arduino Mega 2560, module MPU-6050, and Arduino SD card module.

The sensor InvenSense MPU-6050 contains a MEMS accelerometer and a MEMS gyro in one chip. It is very accurate because it contains 16-bit analog digital converter hardware for each channel. The chip captures values in coordinate axes x, y and z simultaneously. The sensor uses an I2C bus to connect with Arduino.

The design of the program was based on a manual for the Arduino programming language and tutorials, which are freely available on the Internet. Program created in ARDUINO (IDE) software 1.8.4 can be theoretically divided into four functional units.

For proper communication of the microcontroller with peripheral devices, it is necessary to address the corresponding pins via bus. This step can be done manually by using physical addresses or automatically by placing the appropriate library in the program.

In the first part of the program we will include the necessary libraries, whose commands we will use. Define the bit length and variable type with the appropriate name. Next, we create a file object to record data to the SD card and define the pin chipset for the SD card module.

In the second part we start one of the main parts of the program, which contains commands executed only once, always when the program is turned on. Commands include initiating communication with the COM serial port and defining the pin of the chipset as the output. Next, initialization of the SD card with the corresponding response in the form of a message on the serial monitor follows. The following is the creation of a text file named Data.txt, in which the beginning of the new measurement is written, and it is simultaneously printed on the serial monitor. If a file is not created, the process failure information is listed. In the last step, we begin communication with the gyro module and read the values recorded by the gyroscope thousands of times every 3 microseconds into the created variables. We divide the recorded values by thousands to get the average value. The last set of commands is important for the next program run, such as the start of integrated time logging, resetting the calculation loop, and the write loop. Since, in our case, no real-time module is used to fill the required I2C pin with the gyro module, time will be measured using an integrated timer that will count down seconds after the measurement starts. Since the number of calculations per time unit is different from the number of entries, different timers have to be loaded.

The third part contains a calculation loop that reads the gyro value and subtracts it from the previous value, calculating the difference. This value is converted to the climb angle. Since the sinus function works with radians, the obtained values have to be converted. The values obtained with the accelerometer are processed similarly and the resulting tilt values are obtained by the sum of the component of the gyroscope and the accelerometer at the given ratios. To avoid excessive variations in output values, damping has been introduced. This means that the output values will be only 10 % of the new value and 90 % of the previous ones. The tilt angle values and the measurement time are displayed on the serial monitor at a specified interval of 2 seconds, with the calculation loop running every 4 microseconds.

The last part of the program is used as a setup and calibration function of the MPU-6050. By sending the specified values to the individual microchip registers, we can adjust the sensitivity of the gyroscope and accelerometer by changing the measuring range.

4 Results comparison

The created program was loaded on the measuring device placed on the counterweight frame of the collector. This part is a solid plane intersecting the axis of the concentric trough and is the ideal place for mounting the measuring device. Since the solar collector is located in an outdoor environment where it is exposed to environmental influences such as rain, snow, wind, sun, or birds, it is essential to protect the surface connections of the measuring device.

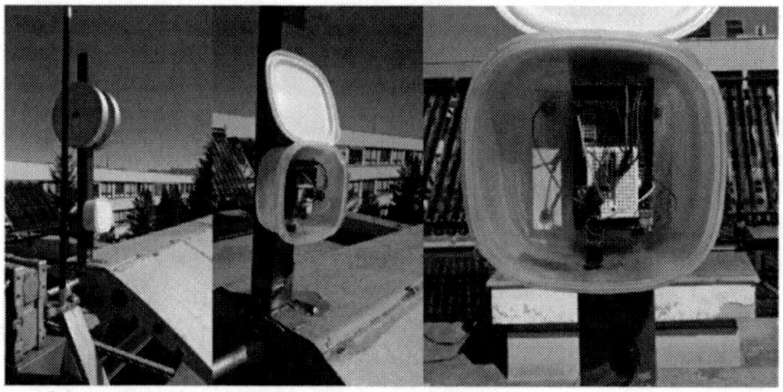

Fig. 4. Placement of the measuring device with protective case.

For proper fitting, protection, and ease of access when replacing chips or removing a microSD card, the device was stored in a modified plastic container with a removable cap (Figure 4). Device communication and initialization is controlled via a combination of USB/LAN cables, as mentioned earlier, through the shaft to the laboratory premises.

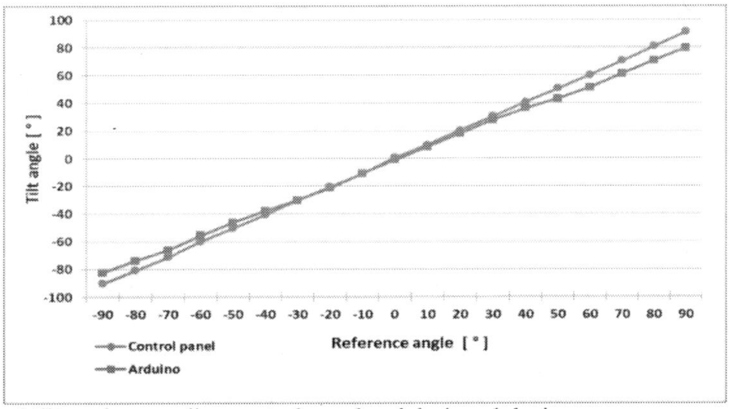

Fig. 5. Measured tilt angle according control panel and designed device.

In order to verify the correct operation of the device, the tilt angle indicated by control panel display and the angle obtained from the designed device located on the collector was compared (Figure 5). The graph was designed for reference angles from -90° to 90° in 10° increments. Based on the geographic position and the location of the collector, it was found that the 180° range is sufficient to provide optimal position throughout the day. Angle -90° is the standby position (trough facing downwards) and the 90° is a fully open collector with the focus in the vertical axis. On the chart we can see that the values measured by Arduino in the range of -90° to 0° are higher than the values obtained from the control panel and vice versa in the range of 0° to 90°.

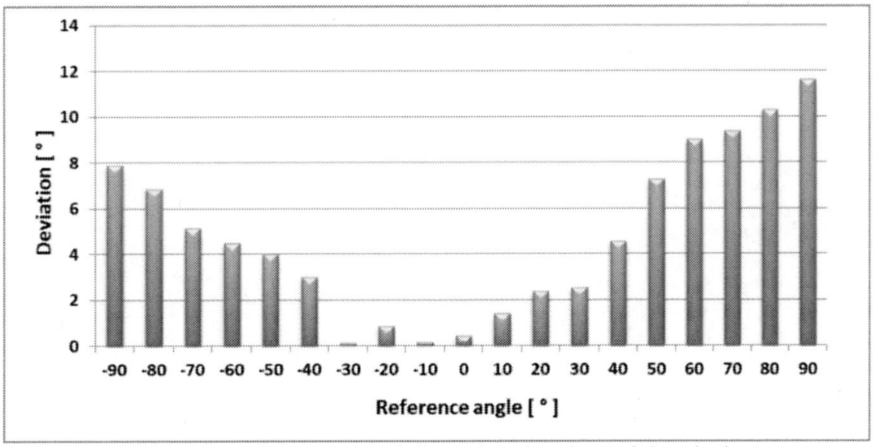

Fig. 6. The values of deviation of the measured tilt angle using the Arduino device.

The measurement results show relatively small deviations in the area around the horizontal position of the gyroscope, maximum of 2.53° at a 30° inclination. Deviation values gradually increase with increasing tilt angle to the vertical position. The highest deviation of 11.62° was measured in a vertical position of 90°. The measurement shows that the deviances range from 0.15° to 11.62°.

Taking into account the effect of the deviation from the optimal position on the concentric solar collector power, the observed deviations represent a significant decrease in collector efficiency. For this reason, the device is not suitable for precise control of the collector operation. The detected deviations may be caused by the incorrect formulation of the microchip program MPU-6050 and must be subjected to a closer analysis.

This article has been released with financial support from the European Union and project support. We support research activities in Slovakia. The project is co-financed by EU funds. Project title: "Research on new methods of heat conversion from RES to electricity using new progressive thermal cycles" ITMS 26220220117 and APVV-15-0790 "Optimization of biomass burning with low melting point of ash".

References

1. S.A. Kalogirou, *Solar energy Engineering* (Academic Press Elsevier, California, 2014)
2. S.A. Kalogirou, Solar Energy **57** (1996)
3. F. Briggs, *Tracking-Refinement Modeling for Solar-Collector Control.* (Sandia National Laboratory, Albuquerque) (1980)
4. K.D. Boultinghouse, *Development of a Solar-Flux Tracker for Parabolic-Trough Collectors* (Sandia National Laboratory, Albuquerque, 1982)
5. G. Brakmannet, *Concentrated solar thermal power - now!* (ESTIA, Brussels, 2005)
6. G.L. Morrison, *Solar collectors* (James and James, London, 2001)
7. K. Chia-Wei, Y. Pei-Shan, Ch. Wen-Chey, Ch. Keh-Chin, Procedia Engineering, **79** (2014)
8. H. Caiqin, X., Procedia Engineering **121** (2015)
9. A. Rabl, Solar Energy **18** (1976)
10. H. M. Guven, R. B. Bannerot, Solar Energy **36** (1986)
11. A. Čaja, P. Nemec, M. Malcho, *AIP Conference Proceedings*, **1608** (2014)
12. P. Nemec, M. Smitka, M. Malcho, Scientific world journal, **2014** (2014)

Impact of oxygen enhanced combustion of natural gas on thermal efficiency of combustion aggregate

Róbert Dzurňák[1,*], *Gustáv* Jablonský[1], *Augustín* Varga[1], *Ján* Kizek[1], *Marcel* Pástor[1], and *Ladislav* Lukáč[1]

[1]Technical university of Kosice, Faculty of Materials, Metallurgy and Recycling, Institute of Metallurgy, Department of thermal technology and gas industry, 042 00 Letná 9/A Kosice, Slovakia

Abstract. The aim of the present article is to present the achieved results on the experimental combustion aggregate. In the experimental measurement, the increased of the oxygen concentration in the oxidizing agent in the combustion of natural gas has been used. The results presented in this article are focused to achieve the thermal efficiency of the device, which is expressed by useful heat of the combustion aggregate. Part of the article is the analysis of the heating heat exchanger based on heat flow by convection and radiation and CFD mode of heat exchanger

1 Introduction

Energy efficiency control in industrial equipment has always been a relevant subject in industrial activity. In this context, there has been a lot of research in new methodologies and new processes of combustion control provided by the scientific community for technological development. Among researches, the industrial application of air combustion with oxygen enrichment has appeared in recent years (oxygen enhanced combustion). The recent technologies of oxygen production have enabled greater competitiveness of oxygen use for application in industrial combustion.

Baukal is the main structured reference on the oxygen enhanced combustion use. It is commented that the lowest levels of the enrichment (volumetric fractions of oxygen (O_2) in the combustion air below 30 %) are normally used in retrofit applications, in which only small modifications are necessary in the existing equipment. Moreover, expressive benefits are obtained with a significant increase in the production rate in processes of heating with low levels of enrichment. In most cases, burners can operate successfully using air of combustion air enriched with up to 28 % of oxygen, without modification in the equipment [1, 2].

Furthermore, Baukal Jr. showed that many industrial heating processes can be improved by substitution of part or total of the air with oxygen of high purity. Typical applications include heating and fusing of metals, fusing of glass and calcification. In a report from Gas

* Corresponding author: robert.dzurnak@tuke.sk

Research Institute, the applications which were identified as possible candidates for the OEC: processes which have high exhaust temperatures and low thermal efficiency, due to constraints in the heat transfer typical in such conditions; processes with loading limitations, which could benefit from an additional heat transfer, without an adverse effect on the quality of the product; processes in which exhaust compounds are both in the gas and particle phases; processes with high NOx emission rates and processes with limitations in the volume of the exhaust gases[1, 3].

2 Combustion aggregate

In order to assess the influence of the mentioned oxidizing agent properties on the combustion process, an experimental device (Figure 1) was constructed on which practical measurements were made. Within the measurements, a whirling burner burning natural gas using oxygen as an oxidizing agent at concentrations $\xi 1$ = 21 %, $\xi 2$ = 25 %, $\xi 3$ = 30 %, $\xi 4$ = 35 % $\xi 5$ = 40 %. In the case of individual measurements, a constant natural gas supply corresponding to a burner input of 9 kW and a surplus of oxidizing agent m = 1.1 was maintained. Enriching of the combustion air with oxygen was accomplished by injecting oxygen into the combustion air supply. The mixing of the natural gas with the oxidizing agent was carried out in the combustion chamber. The heat generated by combustion of natural gas was taken through a heat exchanger in which the cooling water flowed.

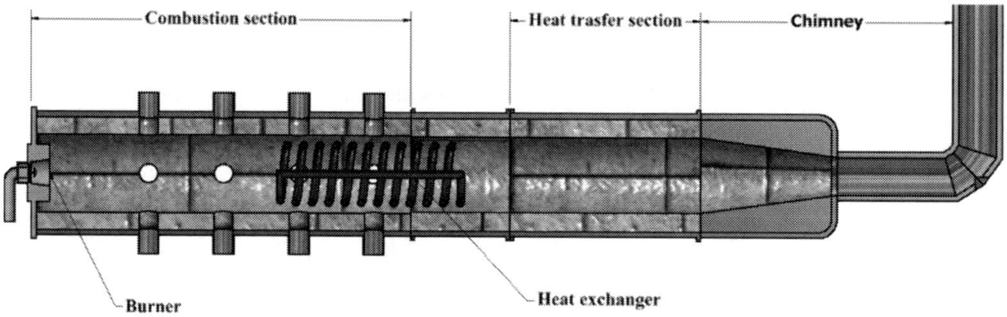

Fig. 1. Combustion aggregate Variant A.

The evaluation was developed into two design models. Variant A had a heat exchanger placed in the combustion chamber (Figure 1). In Variant B (Figure 2), the heat exchanger was moved away from the burner. The objective was to determine the effect of the flue gas temperature on the efficiency of the combustion aggregate. The flue gases in Variant B have a lower temperature than variants A due to thermal losses along the combustion chamber.

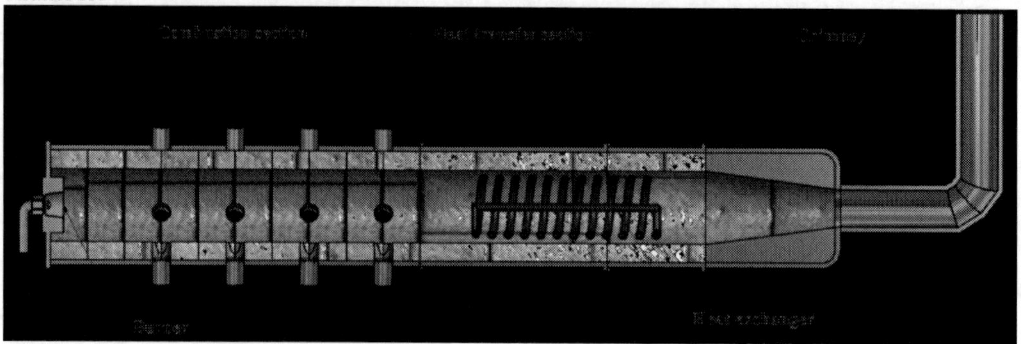

Fig. 2. Combustion aggregate Variant B.

2.1 Heat exchanger

Heat exchanger is a technical device that allows the exchange of heat between two substances. In assessing the effect of increased oxygen concentration in the oxidant to the combustion process, the heat released by the combustion of natural gas has to be transformed into useful heat item [4].

On Figure 3 shows a heat exchanger, this was used during laboratory measurements. This is a regenerative type of heat exchanger in which heat exchange takes place through a heat exchange surface. The heat exchange surface is formed by a steel tube bent in a spiral shape. The heat transfer in the exchanger is provided by a combination of convection heat flow and the heat flow caused by radiation.

Description	Dimension	Size
Length of the exchanger	mm	4 500
Number of turns	ks	11
Diameter of outer tube	mm	20
Diameter of inner tube	mm	16

Fig. 3. Description of heat exchanger.

The heat flux by convection is induced by flowing of the fresh combustion gases from the combustion around the heat exchange surface of the exchanger. However, flue gas as a heat transfer medium is characterized by a relatively small passage through the heat exchange surface of the heated body, therefore a high temperature of the flue gas is required in order to bring the temperature difference and the heat transfer as high as possible. The disadvantage of heating by exhaust gas is also the danger of local overheating of the exchanger and in the presence of oxygen in the exhaust gas also oxidation of the surface of the heat exchange surface. The heat flux is caused by an increased concentration of CO_2 and H_2O components in the flue as well as by radiation from the surface of the refractory material from the combustion and exchange section of the combustion aggregate.

The heat transfer medium in the heat exchanger was water. The water flow rate entering the exchanger was set so that the outlet heat exchanger temperature was below 50 °C. This condition was necessary due to possible thermal degradation of the hose, which served to drain water from the combustion aggregate.

2.2 Results

Based on experimental measurements, we can evaluate the effect of increased oxygen concentration in the oxidizing agent as follows. In Table 1 are plotted the results for variant A. Calculation of useful heat was given by the relationship:

$$Q_{,H2O} = m_{H2O} \, . \, (c_{p,t2,H2O} \, . \, t_2 - c_{p,t1,H2O} \, . \, t_1) \qquad (1)$$

The calculation of the efficiency of the combustion aggregate is given by the relationship:

$$\eta = Q_{,H2O} / Q_{,c} \qquad (2)$$

Relationship (2) expresses the ratio between useful heat and total heat that has been brought to the aggregate.

Table 1. Results of useful heat for Variant A.

	21 %	**25 %**	**30 %**	**35 %**	**40 %**
t_1 (°C)	19.8	18.2	18.4	19.8	17.7
t_2 (°C)	45.8	46.2	47	48.1	47.3
$m_{,H_2O}$ ($kg.h^{-1}$)	106.8	104	103.3	106	106.8
$Q_{,H_2O}$ (W)	**3228.73**	**3385.89**	**3435.16**	**3491.30**	**3675.78**
$q_{,H_2O}$ ($W.m^{-2}$)	9224.94	9673.99	9814.75	9975.15	10502.24
η (%)	**38.47**	**39.96**	**40.81**	**41.71**	**43.72**

Table 2. Results of useful heat for Variant B.

	21 %	**25 %**	**30 %**	**35 %**	**40 %**
t_1 (°C)	17.6	15.2	15.8	15.3	14.8
t_2 (°C)	36.9	35.6	36.7	35.8	36.9
$m_{,H_2O}$ ($kg.h^{-1}$)	106.8	104	103.3	108.2	104
$Q_{,H_2O}$ (W)	**2396.71**	**2466.87**	**2510.31**	**2579.11**	**2672.44**
$q_{,H_2O}$ ($W.m^{-2}$)	6847.74	7048.19	7172.32	7368.89	7635.54
η (%)	**28.72**	**29.57**	**29.77**	**30.38**	**31.60**

The results presented in Table 1 and Table 2 confirm the positive effect of increasing the oxygen concentration of the oxygen in the oxidizing agent to the combustion process and to increase the efficiency of the combustion aggregate. However, the flue gas temperature decreases that the heat exchanger heating efficiency of Variant B is approximately 10 % lower than that of Variant A.

Figure 4 illustrates the effect of individual heat transfer processes on the heat exchanger depending on the concentration of oxygen in the oxidizing agent for both investigated variants. Heat transfer by convection decreases as a result of reduced flue gas production. Heat transfer through radiation is the dominant form of heat exchange, due to an increase in the concentration of the components CO_2 a H_2O.

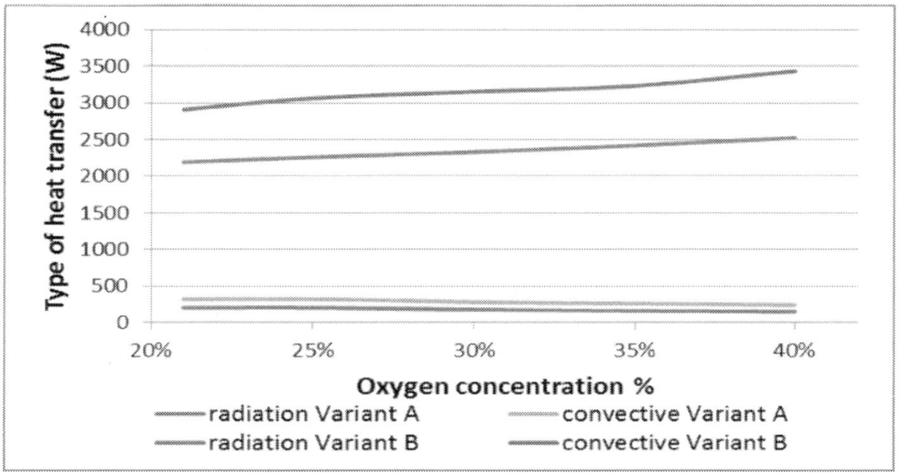

Fig. 4. Type of heat transfer in combustion aggregate for Variant A and Variant B.

In order to visualize the influence of changing the heat exchanger storage in the combustion aggregate, a mathematical model was developed in ANSYS. Figure 5 shows the distribution of achieving temperature field in the outer surface of the heat exchanger.

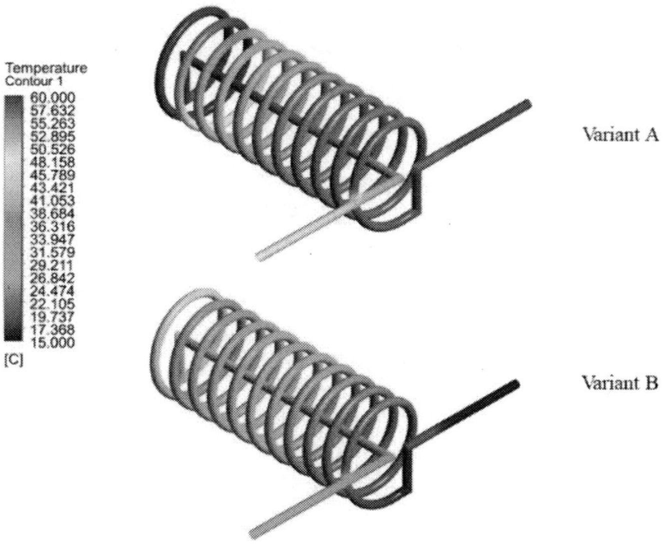

Fig. 5. CFD model of heat exchanger.

3 Conclusion

The results of the experimental measurements confirmed the positive effect of the increased concentration of oxygen in the oxidizing agent on the thermal efficiency of the combustion plant. Increasing combustion temperatures intensifies the heat exchange in the heat aggregate and results in a faster heating of the batch material, thereby increasing furnace efficiency and decreasing the amount of consumed gas per unit amount of batch material. The submitted information is only partial results of the dissertation work.

This publication was supported by the project VEGA 1/0578/16 and obtained results are part of the solution from this grant project.

References

1. C.E. Baukal Jr., *Oxygen-Enhanced Combustion*, 2st ed., CRC Press: New York (2013)
2. A. Santos, A. Torres, P. Pereira, Critical evaluation of the oxygen-enhanced combustion in gas burners for industrial applications and heating systems. J. Braz. Chem. Soc.. **22**, 1841-1849 (2011)
3. S.J. Williams, L.A. Cuervo, M.A. Chapman, *Report GRI-89/0256*, Gas Research Institute: Chicago (1989)
4. D. Holoubek, *Spaľovacie zariadenia, výmenníky tepla a kotly,* Košice: Technická univerzita: Košice (2002)

Comparison of particulate matter in combustion of wood chips and sawdust at nominal and reduced output

Alexander Čaja[1,*], and *Nikola* Kantová[1]

[1]University of Žilina, Department of Power Engineering, Univerzitna 1, 010 26 Žilina, Slovakia

Abstract. Biomass has been still on the rise in recent years in the world. But combustion of biomass led to the increase of emissions in the atmosphere. Mainly in winter time is formed a large amount of particulate matter. They have adverse impact on our health when human breathes them. This is a reason why is important to produce low level of these emissions. However, several aspects impact on their formation. This article examines impact of type of fuel and operation of heat source on PM formation. It compares particulate matter from combustion of wood chips and sawdust at nominal and reduced output.

1 Introduction

As a source of heat and energy, biomass has been still on the rise in the recent years in the world [1]. We distinguish different types of biomass, such as wood chips or wood sawdust. Wood chips may consist of wood, bark, needles, green leaves, twigs and undressed admixture [2]. The main source of sawdust is chipboard production. Both of fuels are mainly suitable for combustion in boilers with higher outputs [3]. During the combustion of these fuels are formed emissions. Under ideal conditions, complete combustion of carbon would result only in generation of carbon dioxide (CO_2) and water vapour. Any products other than CO_2 are often called products of incomplete combustion and include particulate matter (PM) and gases [4]. The most important parameters for complete combustion conditions are a high combustion temperature, a sufficient amount of combustion air supply, and adequate mixing of combustion air and fuel gas [5, 10].

Particulate matter could be a mixture of solid particles and liquid droplets, which vary in concentration, composition (can contain various organic and inorganic compounds) and size distribution [4].

Particulate matter has adverse impact on our health when human breathe them. This is a reason why is important to produce low level of these emissions. Their amount depends on the method of the combustion process, the type, shape and moisture content of fuel, the fuel distribution in the furnace, a dose of fuel, quantity, temperature and speed of the combustion air and so on [6, 7]. This paper examines impact of type of fuel and operation of heat source

* Corresponding author: alexander.caja@fstroj.uniza.sk

on PM formation. It compares particulate matter from combustion of wood chips and sawdust at nominal and reduced output.

2 Material and methods

Fuels, wood chips and sawdust were tested on experimental heat source with the cyclone separator in the chimney tract. This separator was installed because of capture of particulate matter. Combustion of these fuels was realized at nominal output of 450 kW and then at reduced output of 120 kW. Particulate matter was obtained by gravimetric method using by sampling probe Figure 1 placed in the chimney behind the cyclone separator and caught on a membrane filter,

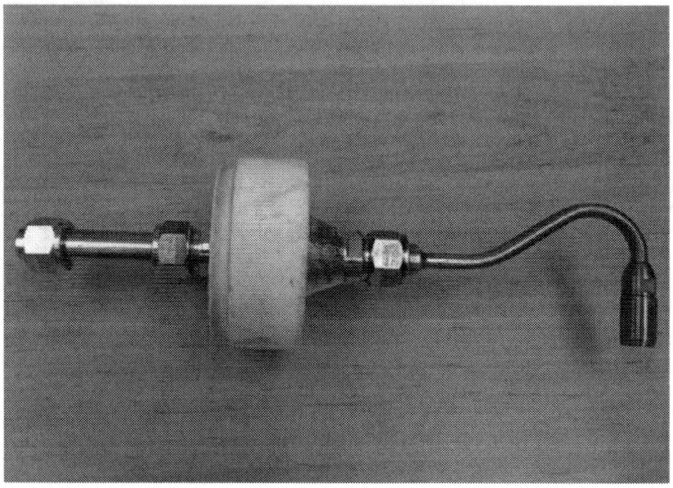

Fig. 1. Sampling probe.

There was determined the amount of particulate matter from the weight of membrane filter before sampling and after sampling. For this evaluation was used the drying and accurate analytical weighing scales. During sampling, there was ensured isokinetic condition Figure 2.

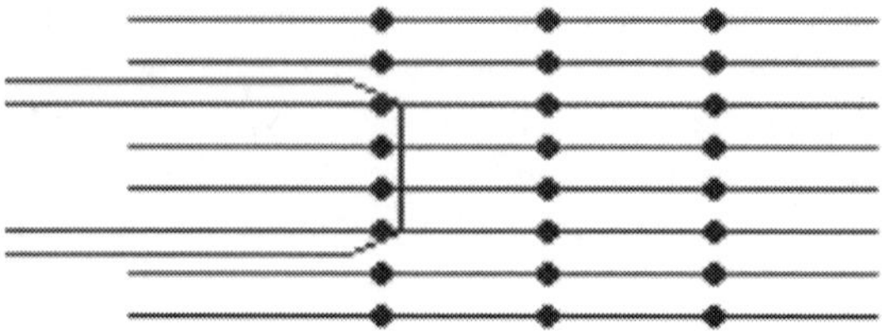

Fig. 2. Isokinetic condition.

It means that velocity of gas entering the sampling probe is exactly equal to the velocity of the gas stream. Then the concentration in the entering of the sampling probe is the same as the concentration in gas stream [8].

3 Results

Analysis of results of amount of particulate matter in combustion of wood chips and sawdust at nominal and reduced output is realized in the Figure 3 and Figure. 4. Values of PM are converted to 10 % reference oxygen in the flue gas according to STN EN 303-5 "Heating boilers. Part 5: Heating boilers for solid fuels, manually and automatically stoked, nominal heat output of up to 500 kW" [9].

During the combustion of wood chips in the first measurement of particulate matter at reduced output was marked by unsatisfactory combustion. The unsatisfactory combustion could be caused by high fuel humidity, which was approximately 50 % at the time of measurement. Based on the results, it can be concluded that, the concentrations of particulate matter in wood chips combustion are significantly higher at the reduced output than at nominal output.

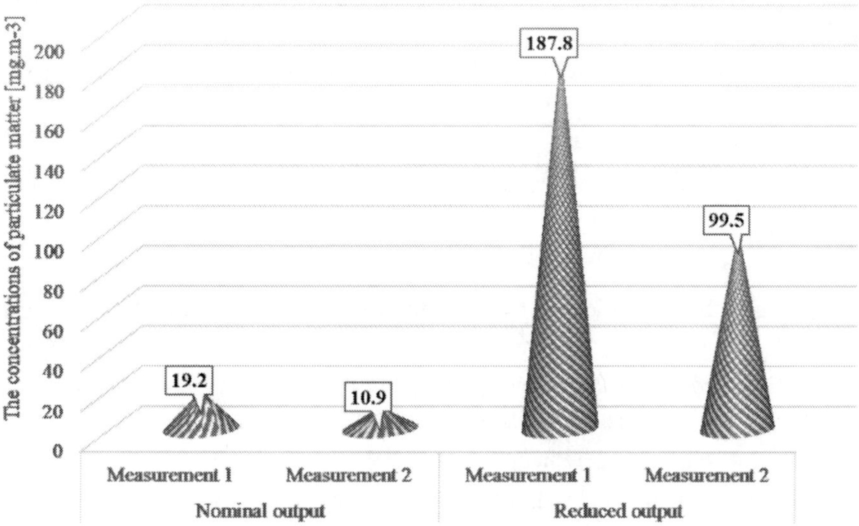

Fig. 3. The graphical analysis of measured concentrations of PM in wood chips combustion.

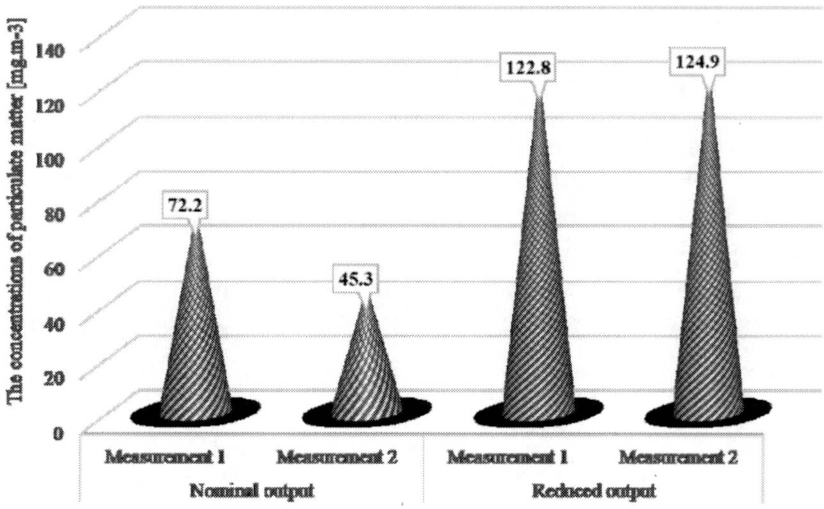

Fig. 4. The graphical analysis of measured concentrations of PM in wood sawdust combustion.

Based on the results, it can be concluded again that, the concentrations of particulate matter in wood sawdust combustion are significantly higher at the reduced output than at nominal output. We can see also a significant increase in PM formation in wood sawdust combustion than wood chips combustion. This increase at the nominal output in the flue gas was on average up to 74.4 % higher than that of fuel wood chips.

4 Conclusion

Examination of combustion process is important for decreasing concentrations of emissions in our environment, such as particulate matter. This paper examines impact of type of fuel and operation of heat source on PM formation. Based on the results, it can be concluded that the least values of particulate matter were obtained during the wood chips combustion at the nominal output on the experimental heat source. The type of fuel, as well as the operation of the heat source has an impact on formation of particulate matter. In terms of PM formation, wood chips are more suitable fuel like wood sawdust and also the operation of the heat source at reduced output has negative influence on their formation.

This work has been supported by the project VEGA 1/0864/16/7 "Analysis and optimization of the factors entering into the process of burning dendromass in small heat sources".

References

1. F. Fachinger, F. Drewnick, R. Gieré, S. Borrmann, Atmospheric Environment 158 (2017)
2. A. Banski, L. Dzurenda, Acta Facultatis Xylologiae Zvolen **56**, 2 (2014)
3. J. Geffertová, A. Geffert, Acta Facultatis Xylologiae **53**, 1 (2011)
4. L. Morawska, J. Zhang, Chemosphere 49 (2002)
5. S. Van Loo, J. Koppejan (eds.) "Handbook of biomass combustion and co-firing". Twenty University Press, Enchede. (2008)
6. M. Holubčík, P. Ďurčanský, P. Nemec, *SGEM 2016 Conference Proceedings*, 1 (2016)
7. J. Tissari, Fine particle emissions from residential wood combustion. [online]. Doctoral dissertation. (2008)
8. R. Nosek, M. Holubčík, J. Jandačka, Acta Facultatis Ecologiae: journal of Faculty of Ecology and Environmental Sciences Technical University in Zvolen (2012)
9. STN EN 303-5: Vykurovacie kotly. Časť 5: Vykurovacie kotly na tuhé palivá s ručným a automatickým prikladaním paliva s menovitým výkonom do 500 kW. Terminológia, požiadavky, skúšanie a označovanie. December (2012)
10. A. Zrak, J. Meško, J. Moravec, R. Nigrovič, D. Kadáš, Journal for Science, Research and Production **16**, 1 (2016)
11. P. Nemec, M. Smitka, M. Malcho, Scientific world journal, **2014** (2014)
12. P. Nemec, M. Malcho, J. Jandačka, *AIP Conference Proceedings*, **1558**, (2013)

Alternative solid fuels combustion in small heat source

Michal Holubčík[1,*], *Nikola* Kantová[1], *Jozef* Jandačka[1], and *Zuzana* Kolková[2]

[1]University of Zilina, Faculty of Mechanical Engineering, Department of Power Engineering, Univerzitna 1, 010 26 Zilina, Slovakia
[2]University of Zilina, Research centre, Univerzitna 1, 010 26 Zilina, Slovakia

Abstract. Air quality is related to the using of solid fuel based heat sources in which the human factor has a major influence on the quality of combustion, which can lead to higher emissions into the air. One of the negative factors is the use of alternative fuels in heat sources. The article deals with the combustion of various alternative fuels, on a waste basis, in small heat sources. There were tested 4 types of fuels: beech wood pieces, 2 types of solid alternative fuel on the base of municipal waste and wood waste. In the experiment, it was tested the influence of used fuel in the fireplace on the heat output, efficiency, production of gaseous emissions and particulate matter. The results confirmed that combustion of fuels not recommended by the heat source manufacturer reduces the efficiency of combustion and significantly increases all monitored emissions.

1 Introduction

Small heat sources are mostly used to provide thermal comfort in family houses, especially outside cities [1, 2]. Paradoxically, in this out-of-town environment, air quality is rapidly deteriorating with the onset of cold weather. The media are increasingly reporting on the occurrence of smog situations and deteriorating air quality. This is related with the using of heat sources based on solid fuels where the human factor has a major influence on the quality of combustion, which can lead to higher emissions into the air [3, 4]. One of the negative factors is the use of alternative fuels in heat sources or often co - combustion of waste with standard fuels. These types of fuels are completely out as determined by the heat source manufacturer [5-7]. Incorrect operation of incomplete combustion heat sources leads to production of enormous emissions.

In 2014, according to the emission inventories in Slovakia, household heating accounted for up to 81 % of total emissions of dust particles (PM2.5) and about 84 % of pollutants of polyaromatic hydrocarbons (PAH) [8]. Dust particles of micro dimensions penetrate into lung alveoli and cause many respiratory and cardiovascular diseases. These particles are carriers of other harmful substances, benzo (a) pyrene (belongs to the group of substances -

* Corresponding author: michal.holubcik@fstroj.uniza.sk

PAH), which causes lung cancer, damages the liver, kidney and genetic information of cells [9-11].

The article deals about the combustion of various alternative fuels, namely beech wood pieces, 2 types of municipal waste and wood waste in the local fireplace – fireplace insert, testing the impact of the used fuel on heat output, efficiency, production of gaseous emissions and solid pollutants.

2 Materials and methods

2.1 Used materials

The following types of samples were used for combustion tests (Figure 1):

• Sample 1: Beech wood - air-chopped split wood in a length of about 45 cm, moisture 12.24 %, combustion heat 17.168 MJ/kg, heating 15.559 MJ/kg.

• Sample 2: Solid alternative fuel with higher carton content - municipal waste - residual after separation of waste, humidity 10.78 %, combustion heat 16.077 MJ/kg, calorific value 14.422 MJ/kg.

• Sample 3: Solid alternative fuel with higher content of plastic - municipal waste - residual after separation of waste, humidity 3.2 %, combustion heat 17.307 MJ/kg, calorific value 15.427 MJ/kg.

• Sample 4: Wood waste - waste from forest dendromass processing in the form of chips, moisture content 9.8 %, combustion heat 17.582 MJ/kg, calorific value 15.875 MJ/kg.

Fig. 1. Tested fuel type samples.

2.2 Methodology of experiments

Experimental measurements to determine the influence of fuel type were made on a local heat source - fireplace insert with a rated thermal output of 13 kW (Figure 2) in accordance with STN EN 13 229 "Inset appliances including open fires fired by solid fuels. Requirements and test methods" [12].

Fig. 2. Fireplace insert used as heat source for experiment.

Each sample was burnt three times in succession, and the results below are the arithmetic mean of these measurements. The combustion of each sample was carried out under the same conditions, the same setting of the combustion air supply and distribution, the same chimney draft (12 ± 2 Pa), approximately the same temperature (21 ± 2 °C) and the relative humidity (40 ± 2 %) of the ambient air.

The heat output (P) and the efficiency (ɳ) of the fireplace insert were determined by calculation of indirect method in accordance with EN 13 229. Measurement of gaseous emissions, namely carbon monoxide (CO), nitrogen oxides (NOx), organic gaseous carbon (OGC), sulphur dioxide (SO_2), carbon dioxide (CO_2) and oxygen (O_2) were measured using an NDIR photometric sensor. Emission production values were converted to standard conditions (0 °C, 101 325 Pa and 13 % oxygen content in flue gas). The production of solid pollutants was determined by a gravimetric method in accordance with STN ISO 9096, provided that the isokinetic flue gas was observed. The chimney temperature and ambient temperature were determined using of K-type thermocouple (NiCr-Ni). The constant chimney draft was provided with a flue gas fan whose speeds were controlled by the frequency converter. These conditions are similarly to methods of other works [13, 14].

3 Results and discussion

The average values of measured parameters during experiments are shown in Table 1.

Table 1. Average values of measured parameters during experiments.

Sample	Test	P [kW]	η [%]	O_2 [%]	CO_2 [%]	$CO_{13\%}$ [mg.m^{-3}]	$NO_{x,13\%}$ [mg.m^{-3}]	$OGC_{13\%}$ [mg.m^{-3}]	$SO2_{13\%}$ [mg.m^{-3}]	$TZL_{13\%}$ [mg.m^{-3}]
1	1	13.8	80.3	11.1	10.4	770.9	57.3	9.1	0	22.06
	2	13.8	80	11.3	10.1	626.8	59.1	5.4	0.1	28.89
	3	13.7	80	11.3	10.2	780.8	50.3	6.4	0	47.02
	Aver.	**13.8**	**80.1**	**11.2**	**10.2**	**726.2**	**55.6**	**7.0**	**0.0**	**32.7**
2	1	8.7	50.6	18.1	2.5	1778	179.4	37.4	60.99	468.8
	2	7.5	46.8	18.3	2.4	2148	156.8	47.9	57.4	397.6
	3	8.3	49.6	17.9	2.8	1589	168.3	31.1	68.4	356.7
	Aver.	**8.2**	**49.0**	**18.1**	**2.6**	**1838.3**	**168.2**	**38.8**	**62.3**	**407.7**
3	1	11.4	66.1	16.3	4.2	1664	233	28	41.91	241.3
	2	11.2	65.8	16.5	4.1	1921	197	31.2	49.6	327.8
	3	10.8	63.4	16.9	3.8	2119	166.7	38.1	55.3	297.3
	Aver.	**11.1**	**65.1**	**16.6**	**4.0**	**1901.3**	**198.9**	**32.4**	**48.9**	**288.8**
4	1	9.8	56.1	16.9	3.9	2750	71.6	48.8	21.4	296.3
	2	8.5	51.4	17.6	3.4	2846	59.3	53.3	18.1	181.3
	3	9.6	57.2	16.5	5.1	1988	85.9	39.1	19.6	147.9
	Aver.	**9.3**	**54.9**	**17.0**	**4.1**	**2528.0**	**72.3**	**47.1**	**19.7**	**208.5**

The highest average heat output of the fireplace insert (Table 1) was as expected during burning the sample 1 - beech wood pieces which are the right fuel according to the manufacturer's recommendation of fireplace insert and reached an average of 13.8 kW. All other fuel samples achieved lower heat output (8.2 ± 11.1 kW). The heat output results corresponded to the efficiency values when the lowest value was recorded in the combustion of the solid alternative fuel with a higher carton content, which was, on average, more than 30 % lower than the combustion efficiency during wood pieces burning. The percentage of oxygen (O_2) and carbon dioxide (CO_2) content in flue gas corresponded in accordance with EN 13 229 with the efficiency of combustion of individual samples - the higher the efficiency of combustion, the lower the O_2 concentration and the higher the CO_2 concentration in the flue gas.

The production of all gaseous emissions was the lowest during combustion of beech wood pieces. The CO concentration during the combustion of solid alternative fuels was approximately 2.5 times higher in comparison with sample 1. The highest production of CO was achieved during wood waste burning. The increased CO concentration was mainly caused by local incomplete combustion of samples 2, 3 and 4 were significantly worse than the wood pieces.

Production of nitrogen oxides (NOx), after conversion to 13 % oxygen content in the flue gas, was the lowest in combustion of beech and the highest during waste with higher content of plastic burning. Similar results were recorded in the production of organic gaseous carbon (OGC) and sulphur dioxide (SO_2) which were significantly higher than during beech wood burning.

The highest difference was recorded in the production of solid pollutants – particulate matter (PM) where alternative fuels had an approximately 6 to 12.5 times higher concentration of TZL in flue gases compared to the manufacturer's recommended fuel, lumber. It was also observed the colour difference of PM filters (Figure 3) during combustion of beech wood pieces (A) and alternative fuel with higher plastic content (C).

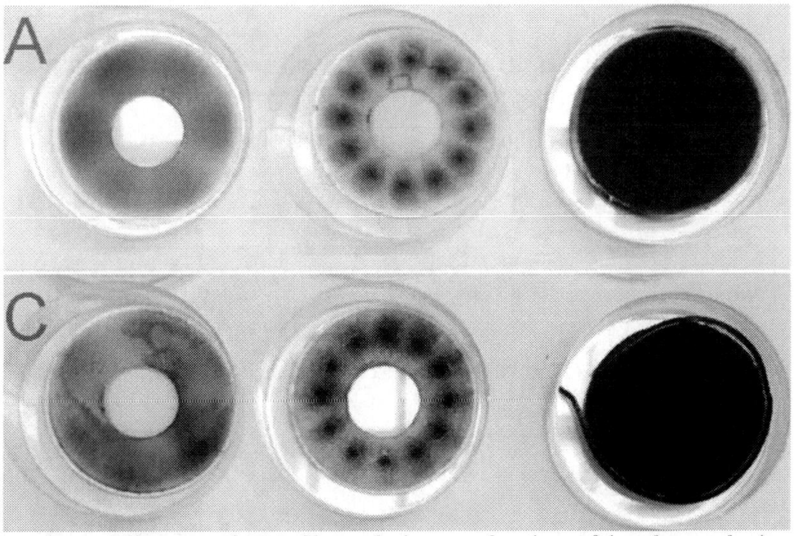

Fig. 3. The colour difference of PM filters during combustion of beech wood pieces (A) and alternative fuel with higher plastic content (C).

4 Conclusions

Due to the fact that small heat sources are among the largest air polluters this issue needs to be addressed to a sufficient degree. The most vulnerable from this point of view are the manually operated heat sources which also allow users to change their operation beyond the fuel changes.

The results have shown that combustion of alternative fuels, especially on the basis of waste, significantly increases the production of emissions, mainly particular solid pollutants. Also the results of thermal performance and efficiency showed that users of heat sources should only use the permitted fuels recommended by the producer of the heat source.

This work was supported by the projects APVV-15-0790 "Optimization of biomass combustion with low ash melting temperature", KEGA 046ŽU-4/2016 "Unconventional systems using renewable energy" and KEGA 033ŽU-4/2018 Heat sources and environmental pollution.

References

1. L. Soos, M. Kolejak, F. Urban, *Biomasa – obnoviteľný zdroj energie*, Vert, (2012)
2. P. Chudíková, M. Taušová, K. Erdélyiová, P. Tauš, Potenciál dendromasy SR a jeho aktuálne využitie v tepelnom hospodárstve, Acta Mont., **15**, 2, 139-145 (2011)
3. P. Križan, *The densification process of wood waste*, IMSETQM (2015)
4. L. Dzurenda, Model of heat load on the atmosphere by flue fases, Man. Tech., **15**, 5, 804-808 (2015)

5. J. Chabadová, Š. Papučík, R. Nosek, *AIP Conf. Proc.*, **1608**, 67-70 (2014)
6. M. Balas, M. Lisý, P. Kracik, J. Pospíšil, Municipal solid waste gasification within waste-to-energy processing, MM Sc. J., **2017**, 1783-1788 (2017)
7. J. Pospíšil, M. Lisý, M. Spilacek, Adv. Mech. Eng., **8**, 6 (2016)
8. L. Dzurenda, E. Hroncová, J. Ladomerský, Forests, **8** (1): 1-9 (2017)
9. J. Moravec, Increase operating life of active parts of Cold-Moulding Tools, Techn. Gaz., **24**, 143-146 (2017)
10. M. Vantúch, K. Kaduchová, R. Lenhard, *AIP Conf. Proc.*, **1745**, (2016)
11. URL: www.vykurovanie.enviroportal.sk
12. STN EN 13 229 Inset appliances including open fires fired by solid fuels. Requirements and test methods (2002)
13. T. Tomsej, J. Horák, S. Tomsejová, K. Krpec, J. Klanova, M. Dej, F. Hopan, Chemosphere, **196**, 18-24 (2018)
14. Š. Papučík, R. Lenhard, K. Kaduchová, J. Jandačka, J. Koloničný, j. Horák., *AIP Conference Proceedings* **1608** (2014)
15. P. Nemec, M. Smitka, M. Malcho, Scientific world journal, **2014** (2014)
16. P. Nemec, M. Malcho, J. Jandačka, *AIP Conference Proceedings*, **1558**, (2013)

Characterization of syngas produced from biomass gasification

Jan Najser[1,*], *Ondřej* Němček[1], and *Marcel* Mikeska[1]

[1]ENET Centre, VŠB -Technical University of Ostrava, 17. listopadu 15, Ostrava-Poruba CZ-708 33, Czech Republic

Abstract. This conference paper presents biomass (wood chips from the production process of sawmills) as a renewable energy source to produce electric energy by gasification. Executed measurements from the gasifier with a fixed bed and properties of produced syngas are described in this paper. The goal is to use the produced syngas at co-generation units with piston combustion engines to produce electric energy and heat. The syngas composition is the final result of these measurements. The syngas from gasification contains pollutants which cause technical difficulties during its use in engines with internal combustion and for this reason it is necessary to choose an adequate system of syngas cleaning.

1 Introduction

World energy consumption trend is rising and surely it will not be limited by more efficient activities and technologies. The reason for this state of affairs is population growth but also the needs of industry and households. This is another reason of huge interests of new and cheaper sources of energy, alternative fuels, energy recovery and finally renewable energy sources which include biomass [1–3].

Current efforts to reduce CO_2 emissions and the amounts used of fossil fuels are the cause of wider use of renewable fuels as well as waste [4]. This leads to the arising of new installation using local fuel sources as well as with various types of process waste [5]. Wastes are increasingly being used to produce electricity and heat. A high level of pollution in waste requires applying more efficient and resistant to different substances devices.

The current situation in the energy industry emphasize the desirability of implementation the small cogeneration systems to national grid. The interests in renewable energy sources as well as the use of waste generated in various processes are increasing [6]. The cogeneration units must meet the both criteria – economic ones as well as demand for high reliability and low environmental impact.

* Corresponding author: jan.najser@vsb.cz

2 Experimental

During gasification, generator gas is produced in the presence of a gasification medium within adequately selected conditions. The low heating value of the generator gas is usually about 4-7 $MJ.m_n^{-3}$, during gasification with oxygen is possible to obtain syngas with LHV from 14 to 18 $MJ.m_n^{-3}$. The produced gas contains primarily CO, CO_2, H_2, CH_4 and N_2 if air is used as a gasification medium [7]. Besides these, the gas also contains other components, mainly undesirables such as dust, tars, alkali, sulphur and nitrogen compounds, hydrogen chloride, hydrofluoride etc [8]. The greater content of volatile inflammable matter gives rise to the greater amount of hydrocarbons in a form of permanent gases such as ethylene, acetylene, benzene, toluene and xylene. As well, high water content results in a formation of tars [9].

In terms of combustion in engines or turbines, ethylene, benzene and toluene are desirable substances – they increase the total low heating value of syngas and improve its combustion properties. However, these substances are undesirable for fuel cells.

Before the syngas can be used in combustion engines or turbines, it needs to be cleaned [10]. Requirements for the cleanliness of produced syngas rise in the case of combustion engines and turbines to extremely low values that are required for trouble-free operation of fuel cells.

By comparing these requirements with a composition of common syngas from the biomass gasification, it can be concluded that without the effective cleaning it is not possible achieve the required syngas quality [11]. To be able to use the syngas in a combustion engine, dust and tar contents must be monitored. The amount of these substances in syngas can be effectively influenced upon their formation by selecting an adequate gasifying reactor.

2.1 Gasification unit description

The unit consists of gas engine adapted for propulsion by the generator gas and synchronous electric current generator. The unit supplies a rated electrical output of 300 kWe and a rated electrical output of about 510 kWt. The gasification generator is downdraft reactor which operates in vacuum mode. As a fuel, cut pieces of waste from the production process of sawmills are used. The combustion medium is air and temperature in the reactor ranged from 795 – 815 °C.

Main parts of the gasification unit:

1. Gasification generator
2. Cyclone separator
3. Series coolers
4. Gas washer
5. Second cooler
6. Centrifuge
7. Filter

Fig. 1. Gasification unit.

2.2 Fuel used in gasification process

Fuel used in this study was waste from the production process of sawmills. The fuel is collected from several processes and stored in open containers for better drying. The resulting waste has a different size. Pieces of wood are from 2 to 10 cm thick, 5 cm wide and 5 to 20 cm in length (see in Figure 2). Such a particle size of fuel is appropriate for the reactor and allows it to contact the gasification agent throughout the reactor. Composition of the fuel was obtained using elemental analysis by the X-ray fluorescence spectroscopy (SPECTRO XEPOS new energy dispersive X-ray fluorescence spectrometer).

Fig. 2. Wood chips used as fuel.

2.3 Methods used for syngas characterization

A detailed analysis of the generated gas was performed by ON-Line and OFF-Line method. The ON-Line method included measurements of basic components (O2, CO, CH_4, CO_2) with GAS 3100P SYNGAS analyzer. The OFF-Line method included later detailed analysis of proced gas which was colected to the tedlar bags. The composition of colected gas was consequently determined in the laboratory by gas chromatography (Chrom 5) with MS 5A, Chromosorb 102, and Porapak QS columns with TCD and FID detectors.

3 Results and discussion

3.1 Characterization of the fuel used in gasification process

The following Table 1. presents the physical and chemical composition of the fuel used in the installation. Moisture content (7 % mass) in the fuel is at a low level, within the limits of tolerance for a reactor. The high calorific value of 18.86 $MJ.kg^{-1}$ positively effects stability of the gasification process. Results of composition of fuel flammables are shown in Table 2.

Table 1. Characterization of the fuel used in gasification process.

Properties	**value**
Fuel type	wood chips
High heating value HHV	18.86 $MJ.kg^{-1}$ d.b.
Moisture W_r	7.0 % mass r.b.
Ash A_d	0.7 % mass d.b.
Volatile flammable matter V_d	77.2 % mass d.b.
Fixed carbon FC_d	22.1 % mass d.b.
Bulk density BD	180 $kg.m^{-3}$ d.b.

Table 2. Composition of fuel flammables.

Component unit	**value**
C	47.87 % mass
H	6.41 % mass
O	45.62 % mass
N	0.1 % mass
S	0.01 % mass

3.2 Characterization of produced gas

Table 3 presents the results of the analysis of syngas made by ON-Line measurements and also analysis of samples taken to the tedlar bags. In presented summary of the main components occurs oxygen which is the result of a leak in the system for gasification or sampling installation. Its content is so small and does not cause danger to the whole devices. More than 50 percentage nitrogen is caused by using air as an agent in the gasification process. This is the simplest form of fuel gasification in which we obtain a gas with low calorific value 5.2 $MJ.m^{-3}$. Percentage part of flammable substances in gas is about 40% and depends on the temperature of the gasification process, equitable access medium, time of reaction, calorific value and fuel humidity.

Table 3. The composition of syngas.

Substances	Content % vol.
O_2 in sample	1.095
CO_2	10.5175
H_2	16.4225
CO	20.5175
CH_4	1.03
N_2	50.71
Ar	0.62
ethane	0.00225
ethylene	0.08925
acetylene	0.07375
benzene	0.01425
toluene	0.001
Q_s, MJ.m^{-3}	5.225

3 Conclusions

The fixed bed gasification generator (down-draft), was able to steadily operates within various gasification conditions. The stability of syngas composition was very good. The low calorific value was 5.2 MJ.m^{-3} and percentage part of flammable substances in gas was about 40 %. In addition, purifying from the dust reached more than 99.3 % efficiency. The increasing requirements of devices such as internal combustion engines, gas turbines and fuel cells requires lower dust limits in developing technologies with higher efficiency.

Experience gained in the operation of the presented technology of biomass gasification will help to create a fully commercial installation, permitting to supply the needs of heat and electricity from local sources.

This paper was conducted by project LO1404: Sustainable development of Centre ENET, within the framework of the project CZ.1.05/2.1.00/19.0389: Research Infrastructure Development of the CENET, by project of VŠB- Technical university of Ostrava, ENET Centre, Specific research SP2018/54- Measuring stand for testing of water-ring pumps, was supported by the Ministry of Education, Youth and Sports of the Czech Republic under OP RDE grant number CZ.02.1.01/0.0/0.0/16_019/0000753 "Research centre for low-carbon energy technologies", project Thermal processing of residues after dry fermentation TH02030120 and project Development of technology for purification and treatment of synthesis gases TH02020032 .

References

1. A. Demirbas, *Prog. Energy Combust. Sci.* 30, 219 (2004).
2. S. Honus, S. Kumagai, and T. Yoshioka, *Energy Convers. Manag.* 126, 1128 (2016).
3. S. Honus, S. Kumagai, O. Němček, and T. Yoshioka, *Energy Convers. Manag.* 126, 1118 (2016).
4. J. Kotowicz, A. Sobolewski, and T. Iluk, *Energy* 52, 265 (2013).
5. H. Raclavská, A. Corsaro, D. Juchelková, V. Sassmanová, and J. Frantík, *Fuel Process. Technol.* 131, 330 (2015).
6. J. Najser, V. Peer, and M. Vantuch, Biomass gasification for liquid fuel production, (2014), pp. 71–75.
7. I. Janajreh and M. Al Shrah, *Energy Convers. Manag.* 65, 783 (2013).

8. J. Amaro, A. Z. Mendiburu, and I. Ávila, *Energy* 152, 371 (2018).
9. H. Knoef, *Handbook biomass gasification*, BTG Biomass Technology Group (2012).
10. N. Abdoulmoumine, S. Adhikari, A. Kulkarni, and S. Chattanathan, *Appl. Energy* 155, 294 (2015).
11. P. J. Woolcock and R. C. Brown, *Biomass and Bioenergy* 52, 54 (2013).

Substrates for slow pyrolysis

Václav Peer[1,*], *Jaroslav* Frantík[1], *Jan* Kielar[1], and *Drahomír* Mašek[2]

[1]VSB - Technical university of Ostrava, Centre ENET, 17. listopadu 15/2172, 708 33 Ostrava-Poruba, Czech Republic
[2]VSB - Technical university of Ostrava, 17. listopadu 15/2172, 708 33 Ostrava-Poruba, Czech Republic

Abstract. Slow pyrolysis of solid materials can produce new materials usable for energy or chemical industry. The advantage of pyrolysis devices is the simple construction and process control and the ability to utilize materials with different properties (composition, ash content). Produced gaseous, liquid and solid materials could be used as a sources of energy, raw materials in chemical industry or substances for improving of soil properties. At article are described products of slow pyrolysis of biomass (wood chips), agrifuels (hay, wheat straw) and sewage sludge.

1 Introduction

Slow pyrolysis is perspective method for transformation of solid fuel into three products – pyrolysis gas, pyrolysis oil (containing main part of pyrolysis water) and solid residue. Pyrolysis is physical-chemical process occurs at temperatures above the heat resistance of the substances (400 – 900 °C). Pyrolysis is carried out without an oxidizer (oxygen from air, water vapour, carbon dioxide etc.).

Processes that occur during pyrolysis can be divided by temperatures up to two temperature intervals. Temperatures up to about 200 °C result in material drying. From 200 to 500 °C, a process of so-called dry distillation takes place, where the macromolecular structures are split into shorter chains due to thermal stability.

Important parameters influencing the pyrolysis process are: pyrolysis temperature, retention time in the reactor, heating rate of the fuel.

Subsequent use of the products depends on their properties. They can be used directly in the process to reduce energy inputs (reactor heating by pyrolysis gas), other equipment (cogeneration unit), or possibly converted to better fuel (liquid refinement).

When compared to other methods of thermal use of materials, it offers the possibility of producing raw materials for transportation liquid fuels, sorption materials or chemical specialties.

Gasification or combustion can produce heat or electricity [1, 2], but there may be problems with melting of ash or incrustation on heat exchangers.

* Corresponding author: vaclav.peer@vsb.cz

2 Experimental part

2.1 Substrates

2.1.1 White wood chips

This fuel was made by a knife cutter with dimensions of approx. 30 x 30 x 5 mm and comes mostly with spruce wood. The white wood chips are obtained from debarked wood, usually cuttings for sawmill production. It is mainly used for the production of chipboard or as a high quality fuel.

For our experiments it was chosen as a standard sample of biomass fuel.

2.1.2 Hay and wheat straw

If these materials can´t be used at agriculture (e.g. contamination, mildew), it is possible to use it as a fuel. Dry material has to be dried, shredded and pelletized or briquetted [3]. We used 8 mm agripellets from hay and wheat straw.

Fig. 1. Samples of white wood chips and hay pellets.

2.1.3 Sewage sludge

The sewage sludge is generated as solid waste from Waste water treatment plants. The dry sludge consists mainly of 60 – 70 % organic matter and the rest is inorganic. 80 % of suspended particles have a size above 0.1 mm and have a large surface area. Our sample was drained, solidified and dried to the actual humidity.

This material was chosen because it would be appropriate to use it as a source of energy [4].

Fig. 2. Samples of wheat straw pellets and sewage sludge.

Table 1. Fuel properties.

	Proximate analysis (wt.%)			Ultimate analysis (wt.%)					HHV (MJ/kg)
	VM	Water	Ash	C	H	S	N	O	
White wood chips	73.4	10.6	0.3	44.3	5.28	0.00	0.45	39.1	19.497
Hay	65.7	9.5	7.6	40.0	5.52	0.01	1.33	36.1	16.999
Wheat straw	67.8	8.0	6.9	41.9	5.65	0.61	0.61	36.9	17.190
Sewage sludge	52.6	9.6	42.7	29.8	3.97	3.2	2.57	8.1	13.753

2.2 Laboratory pyrolysis unit

Laboratory pyrolysis unit is equipped with batch reactor (maximal process temperature up to 900 °C). It consists of four parts: reactor, cooling system, cleaning of pyrolysis gas and gas analysis.

Batch pyrolysis reactor has volume of 2.3 l for 100 – 500 g of fuel (due to it´s density and particle size). It is heated by 3 kW electrical heating elements which can be independently controlled. Hot pyrolysis gas is conducted by heated piping into the tube water cooler. From the bottom part of the cooler is collected pyrolysis liquid (oil and water). Gaseous phase is cleaned in three washers with water, oil and mineral wool. Amount of pyrolysis gas is measured by flowmeter and it is conducted into gas analysis section. Excess gas is combusted in the burner.

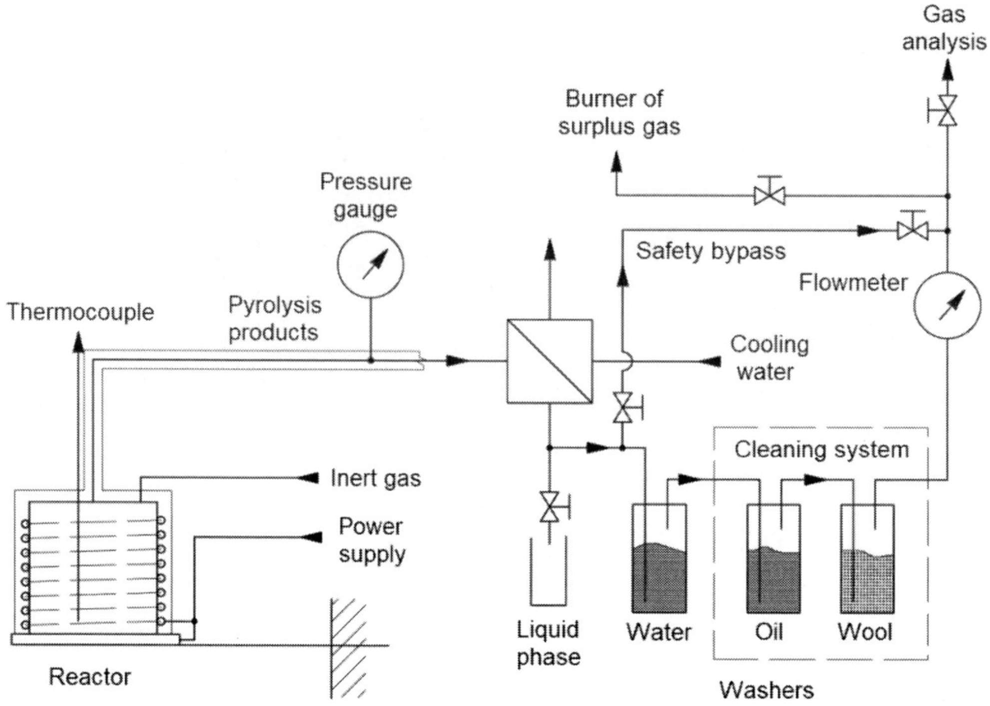

Fig. 3. Scheme of laboratory pyrolysis unit.

Gas analysis is provided by extractive analyser ULTRAMAT 6 (for determination of carbon monooxide, carbon dioxide, methane). Hydrogen amount is measured by CALOMAT 6 (based on thermal conductivity of gas). Total organic carbon TOC is measured by analyser FIDAMAT 6 with flame-ionisation detector. Concentration of higher hydrocarbons are detected by gas chromatograph MASTER GC with flame-ionisation detector.

3 Results and discussion

3.1 Parameters of pyrolysis

For correct pyrolysis, the conditions obtained from the TGA analysis were set. For wood was determined lowest temperature and shortest retention time. Both agrimaterials have similar conditions and highest pyrolysis temperature. This is caused by amount and structure of volatile matter and amount of ash in input fuels.

Table 2. Parameters of pyrolysis.

	Pyrolysis temperature (°C)	Retention time (min)
White wood chips	360	100
Hay	430	130
Wheat straw	430	150
Sewage sludge	480	130

3.2 Pyrolysis products ratios

After pyrolysis were obtained three products – pyrolysis gas, pyrolysis oil and solid residue. Product ratios are mainly influenced by type and structure of input substrates. Pyrolysis of woody biomass produced highest portion of solid residue and lowest amount of gas, sewage sludge produces higher amount of gas and high amount of solid residue (due to high ash content in substrate). Agrifuels produced higher amount of liquid phase and similar amount of gaseous and solid phase.

Table 3. Pyrolysis products ratios.

	Pyrolysis gas (wt.%)	Pyrolysis liquid (wt.%)	Solid residue (wt.%)
White wood chips	1	26	73
Hay	24	45	31
Wheat straw	31	38	31
Sewage sludge	10	27	63

3.3 Pyrolysis gas composition

Compositions of pyrolysis gases were measured during pyrolysis experiments with different solid fuels. Gas from wood have good heating value, but it has very low volume. Highest portion of hydrocarbons contains gases from agrisubstrates – hay and wheat straw.

Table 4. Pyrolysis gas composition.

	CO (vol.%)	CO_2 (vol.%)	H_2 (vol.%)	CH_4 (vol.%)	Hydrocarbons C2 – C4 (vol.%)	Other (vol.%)
White wood chips	34	36	1	9	7	13
Hay	13	36	1	8	35	7
Wheat straw	14	36	2	8	35	5
Sewage sludge	30	40	1	10	6	13

4 Conclusions

Slow pyrolysis of four substrates was done in laboratory pyrolysis unit. Pyrolysis temperature and retention time was established by TGA. Lowest temperature and retention time was for woody biomass, highest was for sewage sludge. This was caused by amount and structure of volatile matter and amount of ash in substrates.

Product ratios are mainly influenced by type and structure of input substrates. Pyrolysis of woody biomass produced highest portion of solid residue and lowest amount of gas, sewage sludge produces higher amount of gas and high amount of solid residue (due to high ash content in substrate).

For highest amount of solid residue is the best substrate wood chips. For highest amount of liquid and gaseous phases are the best agrimaterials – hay and wheat straw.

This work was supported by project LO1404: Sustainable development of Centre ENET, within the framework of the project CZ.1.05/2.1.00/19.0389: Research Infrastructure Development of the CENET, by project of VŠB- Technical university of Ostrava, ENET Centre, Specific research SP2018/54- Measuring stand for testing of water-ring pumps, was supported by the Ministry of Education, Youth and Sports of the Czech Republic under OP RDE grant number CZ.02.1.01/0.0/0.0/16_019/0000753 "Research centre for low-carbon energy technologies", project Torrefaction CZ.01.1.02/0.0/0.0/15_019/0004771 and project Pyrolysis unit with induction heating CZ.01.1.02/0.0/0.0/15_019/0004681.

References

1. M. Balas, M. Lisy, P. Kracik, J. Pospisil, MM Sci. J., **2017**, 1783-1788 (2017)
2. J. Kotowicz, A. Sobolewski, T. Iluk, Energy, **52**, 265-278 (2013)
3. P. Križan, M. Matúš, J. Beniak, M. Kovácŏvá, Acta Polytechnica, **54**, (1), 35-41 (2014)
4. Agar, D.A., Kwapinska, M., Leahy, J.J. Environ. Sci. Pollut. R. (to be published) (2018)

Economy of heat recovery from mining water in Ostrava region conditions

Petra Petričko[1], *David* Kupka[1], *Jan* Koloničný[1*], *and Aleš* Richter[1]
[1]VŠB – Technical University of Ostrava, Energy Research Center, 17. listopadu 15, 708 33 Ostrava, Czech Republic

Abstract. The aim of the article is to assess the possibilities of using low-potential heat by means of industrial heat pumps for heating purposes. The article focused on a relatively specific area, namely the geothermal energy of mine water.

1 Introduction

The article deals with the use of waste heat in mine water. It solves the technical solution of the actual realization, which is specific in connection with the chemical composition of the mine water, but also deals with the economic and balance solution of the connection of the proposed heat pumps to the central heat supply system. The technical solution was also calculated from the economic indicators in comparison with the installation of the classic gas boiler room.

2 Mining water issues

A suitable source of low-potential heat is mine water in flooded areas of deep mines. The water is located deep enough to ensure both the qualitative stability given by the chemical composition and the temperature of the mine water, as well as the quantitative stability expressed by the retention volume of the mine workings, which provides sufficient reserve in the required yield for the economical operation of the current types of heat pumps. The amount of mine water accumulated in the underground at the selected mining sites is shown in the Table 1.

Table 1. The amount of mine water accumulated in the underground at the selected mining sites [1].

Location		Mineral	Reservoir volume [mil.m^3]	Discharged water quantity [l/s]	Water temperatur e [°C]	Geothermal gradient [°C/m]
	Ostrava sub-basin	black coal	23.87	160	26	30-33

* Corresponding author: jan.kolonicny@vsb.cz

Ostrava-Karvina coal basin	Karvina sub-basin	black coal	3.63	38	25	30-33
Rosicko-oslavanská basin		black coal	10	55	14	52
Zlate Hory		non-ferrous metals	0.14	60	8-10	67
Pribram		uranium	22	80-90	21	57
Olsí		uranium	1.84	7	10-12	55

With regard to the magnitude of the energy potential, the possibilities of utilization of the geothermal energy of mine water in the Ostrava-Karviná basin (OKB) were evaluated. It is divided into three sub-basins - Ostrava (OSB), Petřvald (PSB) and Karviná (KSB) sub-basin. In order to prevent mining waters from flowing from the OSB as well as PPB into functional mines in the Karviná area, and at the same time to eliminate risk of poisoning and explosions due to emissions of coalbed gas on the surface, the level of the mine waters must be maintained at the specified level. In the OPB this is performed by means of central pumping from a single mine - Jeremenko water pit (JWP). Mining water is currently discharged without any benefit to the Ostravice River, which causes the thermal and chemical pollution of the river.

An analysis of the mine environment in the OKB was carried out in terms of the yield of the source, the chemical composition and the temperature of the mine waters. The most promising from the point of view of the use of geothermal energy by heat pumps is the mining water from JWP.

The main qualitative parameters of the mine water are $SO4^{2-}$ and Cl^{-}. These indicators have been identified from the outset as indicative and are therefore regularly monitored. Over time, the Fe content has come into the sphere of interest, especially due to the visual effect in the area of the discharging object. In other hydrochemical parameters, mine water does not appear to be problematic. However, mine water is highly mineralized and therefore very aggressive to conventional materials. The key parameter in terms of energy balance is the water temperature, which is around 25 °C throughout the year.

3 Technical solution for the use of mine water

For the mining of mineral water used as a source of thermal energy, two basic technical solutions and their possible combinations according to the specific conditions of the given site are possible:

- mining water extraction at a location where mining water is controlled from underground,
- create a separate sampling point for these purposes only.

The disadvantage of the first solution is usually the lower temperature of the mine water and usually the location outside the area of energy consumption. These disadvantages can be eliminated by realizing the second solution, but at the cost of building two separate wells (pumping and overflowing). The technological arrangement of the geothermal heating plant for the first variant was proposed, since JWP is located directly in Ostrava near potential consumers.

Drainage of mining water with the help of a submersible high-pressure aggregate takes place all year round with a flow rate of 160 to 170 l/s. The available energy potential from this heat source through heat pumps is about 10 MW.

The proposed technology for the use of geothermal power of mine water consists of a set of 7 water/water heat pumps with partial output of 1.45 MW connected in a cascade. The circulation of the refrigerant is provided by semi-hermetic piston compressors to achieve a high output heating water temperature (65 to 80 °C). The heating system of the heating water by means of a heat pump is designed as a two-exchanger because of the extraordinary aggressiveness of the mine water. The primary water/water heat exchanger passes the thermal energy of pumped mineral water to the heat transfer medium that conveys it to the heat pump. The water heated in the heat pump at about 65 °C then passes the heat energy into the secondary heat exchanger to the heating water. The water parameters at the heat pump input are given by the temperature of the mine water reduced by the temperature gradient in the separating exchangers.

Fig. 1. Industrial heat pump with a capacity of 1.45 MW.

4 Economic evaluation of implementation

The installed technology was assessed in terms of investment demand and a comparison of the production costs of GJ heat from the geothermal heating plant against the current sources of heat, i.e. mainly the heat produced in the natural gas boiler and the heat extracted from the central supply system. On the basis of the heat pump installed capacity and the expected annual operation, the energy consumption balance and the amount of heat produced were calculated.

Table 2. Basic parameters of installed heat pumps.

Parameter	Unit	Value
Number of heat pumps installed	-	7
Maximum heat output of one heat pump	MW	1.5

Total installed heat pump heat output	MW	10.2
Installed electrical input of one heat pump	MW	0.5
Total installed heat input of the heat pump	MW	3.8

The Figure 2 illustrates the schematic connection and the heat pump principle. The design is considered with R134a refrigerant due to its use for higher output temperatures.

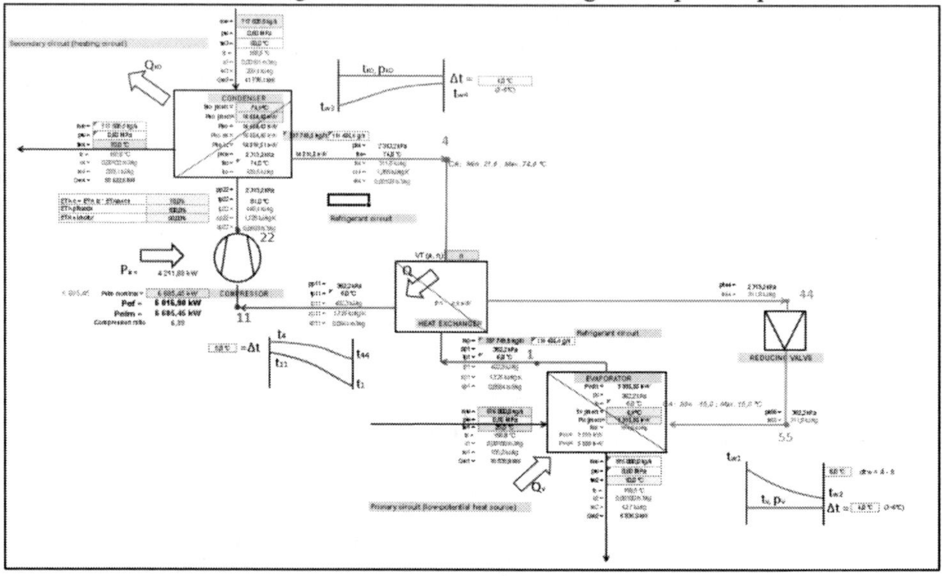

Fig. 2. Basic diagram of the principle of the heat pump connection.

The heat of mine water is fed to an evaporator with mass flow set at the value from 160 to 170 l/s at temperature 25 °C. According to the input parameters, the thermal output that can be obtained from the mine water is 10 MW, if it is considered with the cooling of the mine water at 10 °C. Optimum temperature is required to deliver compression work. The design of the compressor performance is based on the required performance and characteristics of the refrigerant R134a. In order to evaluate the advantage of installing heat pumps, it is necessary to know the balance of the heat produced, including the actual heating factor of the proposed system. Table 3 summarizes the most important installation parameters.

Table 3. Balance of design of heat pumps using mining water with determination of heating factor.

Parameter	**Unit**	**Value**
Thermal output (heating capacity)	kW	9 999
Mass flow rate of refrigerant (working fluid)	kg/h	397 750
Mass flow of heating water (output)	kg/h	717 507
Mass flow rate of mine water (heat source)	kg/h	576 000
Evaporation pressure, evaporator temperature	kPa, °C	362; 6
Condensing pressure, condenser temperature	kPa, °C	2 313; 74

Compression ratio	-	6.39
Isentropic efficiency	%	70
Isentropic compressor power	kW	4 212
Real compressor power	kW	6 017
Power consumption of the compressor motor	kW	6 686
Condenser power output	kW	14 211
Coefficient of performance - no auxiliary energy	-	3.37

Investment costs in Table 4 are determined on the basis of indicative offers from technology manufacturers.

Table 4. Estimation of investment costs for the implementation of a new heat source.

Item	Unit	Value
Installed heat sources and other boiler room technologies	thous. CZK	62 000
- heat pumps		
- accessories for heat pumps		
- storage tank		
- special exchanger (mineralized mine water)		
- circulating pumps		
Other	thous. CZK	10 500
- measurement and control		
- installation and commissioning		
- other (PDs, related technologies, building modifications, etc.)		
Total cost of investment (excluding VAT)	**thous. CZK**	**72 500**

Heat consumption in central heat supply systems is characterized by peak take-offs in the morning and evening hours and, in the summer months, there is a significant decrease due to heat consumption only for the preparation of hot water. For this reason, it is assumed that the installed thermal output is 60 %. The price of electricity (fixed cost + reserved capacity) that is included in the calculations is based on average rates for 2017 in the Czech Republic.

Table 5. Determination of energy costs for heat production from the system.

Parameter	Unit	Value
Hours of operation	-	5 840.0
Capacity factor	%	60.0
Heat production	GJ/year	126 144
Electricity consumption by heat pump compressor	MWh/year	10 398
Auxiliary power consumption	MWh/year	175
Total electricity consumption for heat production	MWh/year	10 573

Average price of electricity - fixed costs	CZK/MWh	902.0
Fixed cost of electricity consumed	thous. CZK/year	11 720
Electricity payment for heat production	thous. CZK/year	21 257

If we want to evaluate the benefits of installing a new heat source that uses geothermal power of mine water, it is necessary to compare the production with another fuel source. A comparison was made with a typical gas boiler with a heat output of 89 %. The price of natural gas (fixed cost + reserved capacity) is also based on average rates in the Czech Republic in 2017.

Table 6. Cost of natural gas for the production of heat from a gas boiler room.

Parameter	Unit	Value
Heat production in gas boiler room	GJ/year	126 144
Average efficiency	-	0.89
Consumption of natural gas for the production of heat	GJ/year	141 735
Average natural gas price - fixed costs	CZK/MWh	167
Fixed cost of collected electricity	thous. CZK/year	1 925
New payment for natural gas for heat production	thous. CZK/year	25 589

Table 7. Cost savings in heat production from geothermal heating plant.

Parameter	Unit	Value
Cost of heat produced in a gas boiler	thous. CZK/year	25 589
Costs of heat produced in heat pumps	thous. CZK/year	21 257
Cost saving on heat generation from geothermal energy	thous. CZK/year	4 332

Note: Only energy costs are evaluated, the evaluation does not include overheads, wages, insurance, etc. It is assumed that the geothermal heating plant is identical to the cost of operating the gas boiler.

Table 8. Economic evaluation of the new geothermal heating plant.

Parameter	Unit	Value
Investment costs	thous. CZK	72 500
Operating revenue	thous. CZK/year	25 589
Operating costs	thous. CZK/year	2125
Cash flow (CF)	thous. CZK/year	4 332
Discount Rate	%	4.00
Net Present Value NPV	thous. CZK	-11 272
Internal Revenue Rate IRR	%	2.0
Payback period	years	16.7
Real time of return	years	>EP
Evaluation period (EP)	years	20
Accumulated CF at the end of the EP	thous. CZK	14 140

5 Conclusion

Within the activity, research and development work was carried out, aiming at a theoretical assessment of the possibility of using heat pumps for the use of geothermal energy contained in mine waters. Due to the significant energy potential, the proximity of consumers and the favorable parameters of the mine waters, the area of the Ostrava sub-basin (OSB), which is part of the Ostrava-Karviná district, was analyzed in detail. From the OSB, the mining water is currently discharged to the surface through a central pit and is discharged into the Ostravice River. A technical solution in the form of a 10 MW geothermal power plant based on industrial heat pump technology was proposed to use the low-potential heat of these waters. Both the technical requirements resulting from the problematic composition of the mine water and the economy of the operation of this source were taken into account compared to the traditional technologies.

Mining water is highly aggressive due to high mineralization, which means it cannot be fed directly to the heat exchanger that forms the evaporator of the heat pump. For this reason, the low-potential heat from the mine water must be indirectly discharged with the aid of an inlet circuit in which the intermediate heat transfer medium flows. This fact exacerbates the realization of the geothermal heating plant. A technical - economic evaluation of this implementation has been carried out. The operation economy of the new source was compared with the operation of the classic gas boiler room with an efficiency of 89 %. According to economic appreciation, the payback time of the new heating plant is 16.7 years. When considering a 4 % discounted rate and a 20-year evaluation period, the return on investment is beyond the lifetime.

The optimization of the technical solution for the realization of a geothermal heating plant of this magnitude depends mainly on the development in the field of heat pumps. More detailed design and evaluation is a matter of next year's research.

This article was published with the support Student Grant Competition project entitled Technology for Waste Energy Utilization, no. SP2017/179, and project „Innovation for Efficiency and Environment – Growth“, identification code LO1403 with the financial support from the Ministry of Education, Youth and Sports in the framework of the National Sustainability Programme I.

References

1. B. Michálek, D. Holéczy, P. Jelínek, A. Grmela, Acta Montanistica Slovaca, **12**, 92 (2007)

Transformation of Waste Heat into Electricity

Aleš Richter[1,*], *Jan* Koloničný[1], and *Jan* Janša[1]

[1]VŠB – Technical University of Ostrava, Energy Research Center, 17. listopadu 15/2172, 708 33 Ostrava, Czech Republic

Abstract. The aim of this article is to introduce the transformation of waste heat into electricity using conventional Rankine steam cycle. Uncommon part of this cycle is a special designed turbine generator, whose impeller is fitted with rows of steel wires instead of turbine blades. For this reason wet steam can be fed into the turbine. The article describes the principle of connection of research unit together with measured data from the operational measurement and graphical and table outputs.

1 Introduction

Currently, two basic principles of transformation of waste heat into electricity are being used. It is either transformation using the thermoelectric effect, or one of the thermodynamic cycles. Electric generators based on the thermoelectric effect can reach nominal electric power up to tens of watts. The most widely used thermal cycles include Rankine cycle, Organic Rankine cycle, Stirling cycle and Kalina cycle.

To use waste heat at low temperatures (200 °C and lower), the units based on the Organic Rankine cycle or eventually the Kalina cycle can be used effectively. Waste heat with a higher potential can be utilized in the conventional Rankine cycle or Stirling cycle. In this case, the waste heat from flue gas of cogeneration units with a temperature above 400 °C is available, therefore Rankine cycle was chosen.

2 Description of the principle of the research technology

The research technology, a special designed turbine generator, has been installed in an animal and plant farm in Želatovice village. This location was chosen because of large amounts of high-potential waste heat in the form of flue gases from cogeneration units processing biogas from near biogas station. The cogeneration units are operated year-round. Electricity is supplied to the distribution network, while a large part of produced heat is wasted into the atmosphere. The flue gases from the cogeneration units have temperature about 420 °C. A bypass was built at exhaust duct of the cogeneration units in order to divert the flue gases to a steam generator. The steam generator is a classic tube heat exchanger, which transfers flue gases heat into water to generate steam driving the special designed turbine generator. The

* Corresponding author: ales.richter@vsb.cz

whole technology is placed in a container, steam generator situated on its "roof" (see Figure 1).

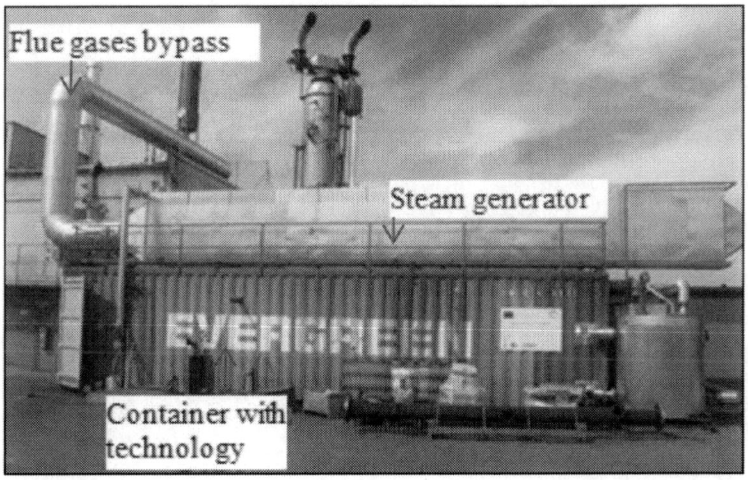

Fig. 1. Research technology.

A simplified scheme of the technology depicted in Figure 3. Water as a working fluid is pumped by a high-pressure pump into the steam generator from a feed water tank with a volume of 1.2 m^3. By passing through the primary side of the steam generator, the flue gases are cooled to 170 °C. The heat transferred from the flue gases generates saturated steam with temperature about 220 °C by boiling the feed water. The saturated steam is fed into the special designed turbine generator through 9 pressure hoses to 9 nozzles, where it expands to barometric pressure. The outputs of turbine were brought out into atmosphere to simplify the tests. The turbine has 2 outputs – axial and radial. The connection between the generator and the turbine rotor is created with belt clutch equipped by tensioner. The generator operates at nominal parameters at 3 000 rpm. The nominal electric power of the generator is 18.5 kW. The turbine rotor is fitted with rows of steel wires instead of turbine blades and therefore it is not susceptible to steam quality. Individual parts of the turbine generator are described in the Figure 2.

Fig. 2. Turbine generator.

3 Operational measurement

Construction modifications of the turbine generator were carried out during 2017. At the end of the year, several operational measurements were carried out, of which a 26-minute interval was selected for analysis. Data from this interval are evaluated in detail below. In order to evaluate the efficiency of the steam turbine generator, the system was equipped with measuring elements. The inlet mass flow of steam was measured on the water side of the generator by an ultrasonic flowmeter. All measured data are summarized in Table 1. For each parameter there is a mark, which corresponds with the schema in Figure 3. In the last three columns of the table, the minimum, maximum and average values for each parameter in the evaluated interval are determined.

Table 1. Summary of measured values in evaluated time interval.

Parameter	Mark	Unit	Min	Max	Average
Generator Speed	o	rpm	3000.0	3011.3	3006.8
Feed Water Pressure	p_3	bar (abs)	31.7	43.7	39.3
Feed Water Temperature	t_3	°C	31.0	56.6	43.8
Steam Temperature	t_4	°C	209.6	226.6	221.1
Flue Gas Temperature	t_1	°C	402.8	423.8	421.3
Flue Gas Temperature	t_2	°C	146.1	189.1	171.3
Electrical Power	P_e	kW	1.1	4.8	3.5
Relative Humidity of Air	Rh	%	30.6	47.3	37.1
Barometric Pressure	p_b	bar (abs)	1.0	1.0	1.0
Water/Steam Mass Flow	m	$t.h^{-1}$	0.6	0.8	0.7

It is clear from the previous table that saturated steam parameters are relatively low – mass flow 0.7 $t.h^{-1}$, temperature 221.1 °C. At these parameters, the turbine achieved average electrical power 3.5 kW.

Summarized data needed for determination of the net efficiency of the turbine generator are presented in Table 2. The calculated value is shown in the last row of the table. In this case, it is very difficult to determine the wet steam enthalpy at the turbine outlet because wet steam quality is not known.

Therefore, the net efficiency of the turbine generator was determined as a ratio of generated electrical power to the theoretical power, which is given by the steam mass flow and the ideal isentropic enthalpy gradient from the inlet pressure to barometric pressure. This efficiency includes all losses, both losses within the turbine itself, and mechanical losses and loss of the generator.

Table 2. Summary of the resulting values in evaluated time interval.

Parameter	Mark	Unit	Min	Max	Average
Feed Water Enthalpy	h_3	$kJ.kg^{-1}$	132.7	239.9	186.9
Steam Pressure	p_4	bar (abs)	18.9	26.3	23.8
Steam Enthalpy	h_4	$kJ.kg^{-1}$	2797.2	2802.5	2801.1
Steam Entropy	s_4	$kJ.kg^{-1}.K^{-1}$	6.2	6.4	6.3
Steam Enthalpy after Isentropic Exp.	$h_{5\ ise}$	$kJ.kg^{-1}$	2256.5	2302.2	2271.2
Enthalpy Drop after Isentropic Exp.	Δh_{ise}	$kJ.kg^{-1}$	494.9	546.0	529.8
Theoretical Power	P_{theor}	kW	92.7	119.7	104.0
Turbogenerator Efficiency	η	%	1.2	4.4	3.3

Average measured and calculated values were also plotted for better interpretation in the simplified functional scheme in Figure 3.

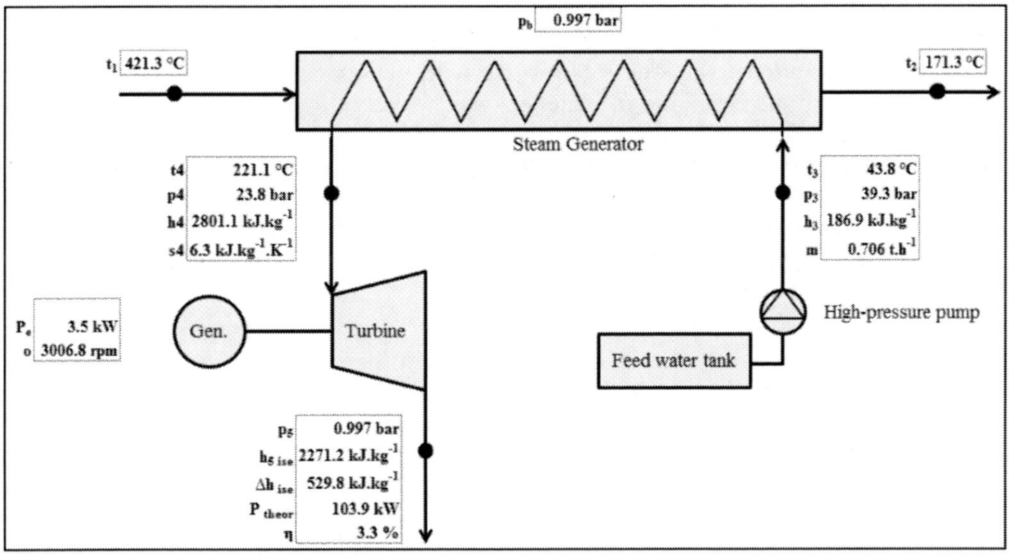

Fig. 3. Simplified scheme of the technology.

4 Comparison with traditional turbine generator

In the framework of the project, a company that produces classic single-stage blade steam turbines was addressed in order to compare the achieved efficiency of the presented turbine with that of conventional technology. The same parameters in which the rated turbine was run were sent to this company for evaluation. This manufacturer commonly produces turbines for the same pressure and temperature, but for higher mass flows. Even so, their smallest machine is capable of operating at such reduced conditions and can be compared to the rated turbine. The net efficiency of the competing machine was determined in the same way - as a ratio of produced electrical power to the theoretical power. The comparison results are shown in Table 3.

Table 3. Comparison results with the traditional turbine generator.

Parameter	Unit	Evaluated Turbogen.	Traditional Turbogen.
Mnufacturer	-	GWRD Inc.	G-Team Ltd.
Type	-	NTR	TR 100
Steam Mass Flow	$t.h^{-1}$	0.7	0.7
Inlet Steam Temperature	°C	222.0	222.0
Inlet Steam Pressure	MPa (abs)	2.4	2.4
Outlet Steam Pressure	MPa (abs)	0.1	0.1
Theoretical Power	kW	111.0	111.0
Real Power	kW	3.5	14.0
Turbogen. Efficiency	%	3.2	12.6

5 Conclusion

Processes for transformation of waste heat into electricity are very briefly described in the introduction of the article. This is followed by a practical description of one of these principles - conventional Rankine steam cycle with incorporated special designed turbine generator. The turbine is unique in impeller type, which is fitted with rows of steel wires instead of turbine blades. Functionality of this system was proved by series of measurement. At this time, the maximal net efficiency of the presented turbine is about 4.4 %, while classic blades turbine in these low steam parameters achieve about 12.6 %. Despite the lower efficiency compared with classic blades turbine, the analyzed turbine may be suitable for operations that have problems with steam purity, since even small droplets can seriously damage blades of the conventional turbine.

This article was published with the support Student Grant Competition project entitled Technology for Waste Energy Utilization, no. SP2017/179, and project „Innovation for Efficiency and Environment – Growth", identification code LO1403 with the financial support from the Ministry of Education, Youth and Sports in the framework of the National Sustainability Programme I.

References

1. Š. Rozsypal, *Brush Steam Turbine*, Brno University of Technology, Brno, (2013)
2. California Institute of Technology, *Thermoelectrics*, see: http://thermoelectrics.matsci.northwestern.edu/thermoelectrics/index.html

Impact of the production of wood pellets on their properties

Matej Palacka[1,*], *Michal* Holubčík[1], and *Jozef* Jandačka[1]

[1]University of Žilina, Department of Power Engineering, Univerzitná 8215/1, 010 26 Žilina

Abstract. Process of production wood pellets is difficult and this process has an impact on the final properties of pellets. The article deals with the comparison of the energy and mechanical properties of experimental pellets that have been made under different conditions. Sample A was manufactured in laboratory conditions and sample B was made in factory production. At the end of the work are displayed the results of the experiments.

1 Introduction

Wood pellets are relatively used as a source of heat energy with great potential to replace fossil fuels. These pellets are made from waste material such as sawdust and shavings. Important properties of the pellets include size, shape, mechanical resistance, moisture content, volatile content, ash melting temperature and caloric value. These properties are influenced by the quality of the input material as well as by the ways and technologies of their production. The production of wood pellets is dealt with by some companies, which provide input materials from woodworking and sawmill industry. Nowadays, the cost of purchasing pelleting equipment is decreasing and for this reason more and more people and households are considering buying these instruments for producing their own domestic wood pellets.

This article discusses the energy and mechanical properties of clean sawdust pellets produced under laboratory conditions and pellets produced in a specialized plant that has been manufacturing pellets for many years.

2 Production of samples

Samples of wood pellets were produced in two ways:

- Sample A is from laboratory production,
- Sample B is from factory production.

Laboratory production was to simulate the production of wood pellets in domestic conditions, which is mostly used by smaller woodworking companies and carpenters, who have waste sawdust and shavings. A pellet press (Figure 1) with a theoretical production of 200 kg pellets per hour was used to produce the experimental samples, but under realistic

* Corresponding author: matej.palacka@fstroj.uniza.sk

conditions, only about 80-100 kg pellets per hour were obtained during experimental production. The sample A (Figure 1) was made from pure sawdust on this machine. Factory production of wood pellets was carried out at a specialized facility in an external laboratory (Figure 2, illustrative photo). The production line used has a production capacity of approximately 1 000 kg pellets per hour. Sample B (Figure 2) was formed on this machine from the same input material as for sample A. The input material came from spruce sawdust from a private sawmill located in northern Slovakia. The preparation of the base material consisted in adjusting the moisture to the required parameters. As the material delivered reached a moisture content of 8.11 %, in the case of a small laboratory production of wood pellets, the sawdust needed to be dampened to about 15-17 %. In the case of the production of wood pellets on the production line, the humidity was controlled by the operator to approximately 12-15 % depending on the output quality of the wood pellets.

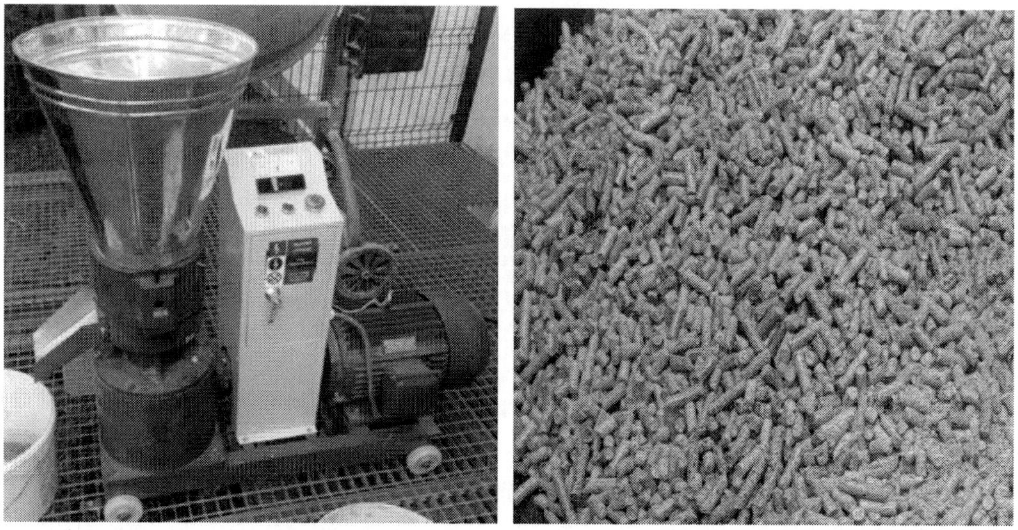

Fig. 1. Laboratory device and sample A.

Fig. 2. Specialized device and sample B.

3 Experiments

During the experiments were determined moisture, volatiles, fixed carbon content, ash content, elemental analysis, gross calorific value and calorific value obtained using these devices:

• First thermogravimetric analyzer (TGA, Figure 3) – is used to determine the composition of the organic, inorganic or synthetic materials. Measured by weight loss as a function of temperature in a controlled environment in a predefined atmosphere. The device consists of a computer and a muffle furnace that allows monitoring of up to 19 samples to be analyzed simultaneously. Empty cups on the samples are placed in carousel the furnace after selecting the method of analysis. The analysis method controls the carousel, furnace, and operating weights that gradually weigh all samples were placed on the carousel. After weighing all the crucibles, is each a crucible is gradually adjusted for the operator of device for the sample application. Individual weights of samples are measured and saved automatically. When all the crucibles are loaded begins analysis. The loss in weight of each sample was monitored and the temperature of the oven is controlled according to the chosen analytical method. The percentage weight loss of each sample is known at the end of every step of the analysis. The device is easy to use with a software program Windows that allows analytical methods that can be designed to meet the most analytic applications according to current norms. The temperature, speed of rising temperatures and atmosphere are optional for each step. Thermogravimetric analyzer can determine the moisture, volatiles, fixed carbon and ash but also combination of all these parameters. In this work thermogravimetric analyzer measured the moisture, the proportion of solid carbon and volatile matter and ash content.

Fig. 3. Termogravimetric analyser.

• Elemental analyzer (Figure 4) – is a device for elemental analysis of organic samples. The device in the combustion chamber burns weighed sample in the absence of atmospheric gases and after burning comes detecting the resulting gases. The device has a separate module for the determination of sulfur. Sample is also combusted, but at a higher temperature (1 350 °C) to detect the presence of sulfur oxides.

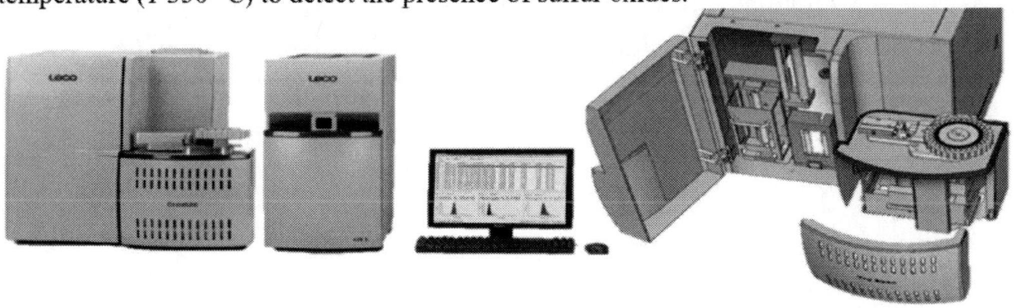

Fig. 4. Elemental analyser.

• Calorimeter (Figure 5) – calorific value of biofuel samples was determined according to STN EN 14918 through the calorimeter LECO AC 500. From the measured gross calorific value, measured relative moisture and the contribution of hydrogen is determined calorific value spruce wood from the following equation:

$$Q_i = Q_s - 2.453\,(M_{ar,} + 9H) \qquad (1)$$

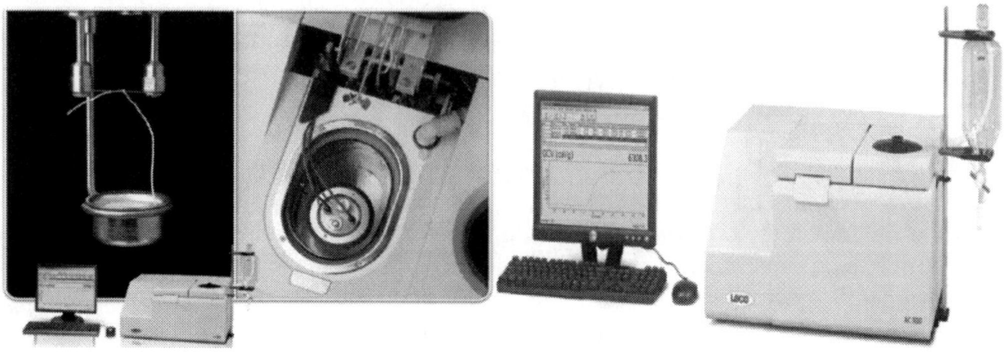

Fig. 5. Calorimeter.

4 Experiment results

The results of the experiments are shown in the table (Table 1). Moisture content, volatile matter, solid carbon and ash in wet and dry conditions.

Table 1. Results of experiments.

		Sample A	**Sample B**
Moisture	[%]	7.33	6.50
Volatile	[%]	81.66	81.91
Fixed carbon	[%]	17.89	17.58
Ash	[%]	0.48	0.51
Volatile (dry)	[%]	75.56	76.45
Fixed carbon(dry)	[%]	16.55	16.41
Ash(dry)	[%]	0.45	0.48

The table (Table 2) shows the results of the fine particle fraction and the abrasion resistance of samples A and B, these values influencing the mechanical properties of the pellets.

Table 2. Results of experiments.

	F test (g)		The proportion of fine particles (g)	DU test (g)		The abrasion resistance (%)
	Before	After		Before	After	
Sample A	100.23	98.90	1.33	100.20	96.65	96.46
Sample B	107.892	107.865	0.027	107.865	107.335	99.509

The table (Table 3) shows the results of moisture, gross calorific value, calorific value of samples A and B.

Table 3. Results of experiments.

	Moisture (%)	Gross calorific value (MJ/kg)		Gross calorific value (MJ/kg)	Calorific value (MJ/kg)
Sample A	7.33	18.8363	18.7560	18.7962	17.3121
Sample B	6.50	19.9657	19.7683	19.8670	18.3626

Different moisture values have been caused by a different production method, in particular the need for greater moisture demand in the case of domestic (laboratory) production of wood pellets. Volatile and solid carbon values are roughly identical due to the same input material - spruce sawdust. Ash content for samples A and B are almost identical. Experiments to determine the fine particle size and abrasion resistance proved that under domestic conditions, on a conventional small pelletizing press, it is not possible to produce sufficiently solid wood pellets in view of the fact that high pressing pressures and pressures can't be achieved as in the case of the large pelletizing press in the production line. Sample A achieved lower calorific value and gross calorific value than sample B, due to higher moisture content in sample A by other production methods. Therefore, it results from the measured values that sample A was of lower production quality than sample B. In the case of combustion in a conventional heat source for pellet burning, this causes wood pellets produced in the small pelletizer to be decomposed more rapidly during the feeding and during the combustion and thus can't achieve the same heat output and efficiency of the heat source as the wood pellets produced in a large production line. During the operation of the heat source, this would result in higher pellet consumption and higher emissions.

5 Conclusion

The difference between commercial and domestic wood pellet production is quite clear from the results obtained. In domestic conditions on small pelletizing presses, high pressing pressures and temperatures can't be achieved, as in the case of the production of wood pellets on a production line with high production. The produced wood pellets differ mainly in qualitative mechanical parameters, when domestic wood pellets are rather decaying, thereby reducing the bulk density, which is negative in their combustion. However, unless the commercial use of wood pellets is anticipated and their domestic consumption is assumed, it is important to produce wood pellets in small pellet presses.

This work was supported by the projects APVV-15-0790 "Optimization of biomass combustion with low ash melting temperature" and 046ŽU-4/2016 "Unconventional systems using renewable energy ".

References

1. L. Dzurenda, A. Banski, Archives of Thermodynamics, **38**, 1 (2017)
2. L. Dzurenda, Ľ. Pňakovič, Annals of Warsaw University of Life Sciences, (2014)
3. J. Beniak, P. Križan, Ľ. Sooš, M. Matúš, Waste Forum, 5 (2017)
4. L. Soos, M. Kolejak, F. Urban, Vert: Bratislava, (2012)
5. L. Martiník, V. Drastichová, J. Horák, Z. Jankovská, K. Krpec, P. Kubesa, F. Hopan, Z. Kaličáková, Chemicke Listy, **108**, 2 (2014)
6. M. Carnogurska, M. Prihoda, M. Koško, R. Pyszko, J. of Mech. S. and Tech., (2012)
7. J. Geffertova, A. Geffert, Acta Facultatis Xylologiae, 53/1 (2011)
8. J. Chabadová, Š. Papučík, R. Nosek, AIP Proceedings, **1608**, (2014)
9. A. Čaja, P. Nemec, M. Malcho, AIP Proceedings, **1608,** (2014)
10. P. Taus, M. Tausova, Acta Montanistica Slovaca, **14**, 1 (2009)
11. P. Nemec, M. Smitka, M. Malcho, Scientific world journal, **2014** (2014)

Solutions for the size & concentration measurement of aerosols from combustion

Lucia Bustin[1,*], *Florian* Dahlkötter[1], *Carsten* Kykal[1], *Torsten* Tritscher[1], and *Oliver* F. Bischof [1]

[1]TSI GmbH, Neuköllner Str. 4, 52068 Aachen, Germany

Abstract. Currently the main pollution comes from the anthropogenic sources related to residential heating and combustion. Most of the emitted particles from this sources are in the mode below 1 µm. Therefore, only by using methods based on the electric mobility separation and size enhancement we can measure the size and concentration of this particles from nano to ultrafine sizes. This contribution describes the most recent methods how to study particle size and concentration of the particles coming from combustion sources.

1 Introduction

Many aerosol sources of interest, like combustion, cigarette smoke, emission form biomass burning and any other anthropogenic production consist of particles in the size range from few nanometers to 1 µm [1]. Current mass limits of PM 2.5 and PM 10 demand measurement of particle size up to 2.5 µm or 10 µm, which do not represent particles coming from anthropogenic sources mainly from combustion. A key parameter for the combustion source characterization is the number size distribution measurement from few nano meters up to 1µm by either high resolution Scanning Mobility Particle Sizer™ (SMPS™, TSI model 3938) spectrometer or with fast sizer Engine Exhaust Particle Sizer (EEPS™, TSI model 3090) working at 0.1 Hz sampling rate. The size range of spectrometers can be extended by an Optical Particle Sizer (OPS, TSI model 3330) that covers larger sizes from 300 nm to 10 µm. Or Aerodynamic Particle Sizer (APS®, TSI model 3321).

2 Methods

Electrical mobility technique has increased use in the in-situ near real-time sizing of aerosol-based processes like diffusion flame synthesis, spray pyrolysis, thermal plasma, and combustion processes etc. With the commercialization of nanotechnology, and increasing interest in exposure monitoring and ambient air monitoring interests, the demand on precise sizing of nano particles, ultrafine particles and micro particles has grown. This contribution provides a brief overview of the electrical mobility technology as integrated in TSI

* Corresponding author: lucia.bustin@tsi.com

SMPS™ and EEPS™ spectrometers followed by the use of electrostatic classification [2, 3] as the only method for on-line measurements of size and number concentration of nano and ultrafine particles that are mainly present in the combustion sources.

TSI spectrometers consist of a sample preconditioner, a bipolar charger, a nanoparticle size classifier and a particle detector. The charger (also referred as neutralizer) establishes charge equilibrium on the particles. This defined charge condition is necessary for the size classification using electrical mobility. Particles are size classified in a Differential Mobility Analyzer (DMA). The electric field inside DMA separates the particles according to their electrical mobility, which is inversely related to the particle size. The electrical mobility of these selected particles is a function of flow rates, geometric parameters and the voltage of the center electrode. The size selected particle stream exiting the DMA is counted either by a Condensation Particle Counter (CPC) in SMPS or by electrometers placed around the center rod to collect the signal from electrons repealed from the center electrode in EEPS. This technique has a considerable reputation for its reliability and the fact that it is based on basic physical principles.

Typical experimental arrangements to study particle number-sized distribution (PN) of aerosols coming from combustion is depicted in Figure 1. The exhaust emissions are diluted and recently sampled by TSI spectrometers.

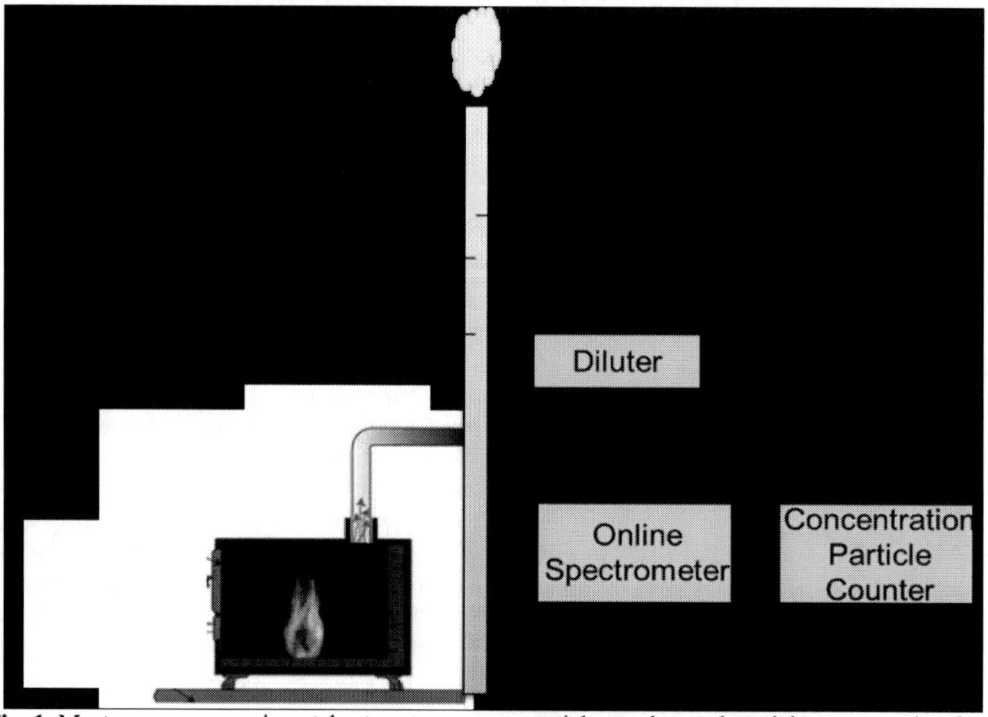

Fig. 1. Most common experimental set up to measure particle number and particle mass coming from combustion source.

3 Results and Discussion

The emission during combustion of wood in a small batch reactor were investigated by one of TSI spectrometers sizing from 10 nm up to 10 μm to measure number size distribution. However, we pay attention only to results below 1 μm.

The different phases of wood combustion are distinguishable by the particle number size distribution. The Figure 2 shows the differences in particle concentration present in nucleation mode so called Aitken mode (up to 40 nm) [4, 5]. The increase of particles concentration in nucleation mode is visible by adding the compressed wood. The nucleation mode is created by particles freshly emitted from the combustion source. It is shown that the combustion conditions (e.g. load, air supply, temperature…) strongly influence the particle size and concentration.

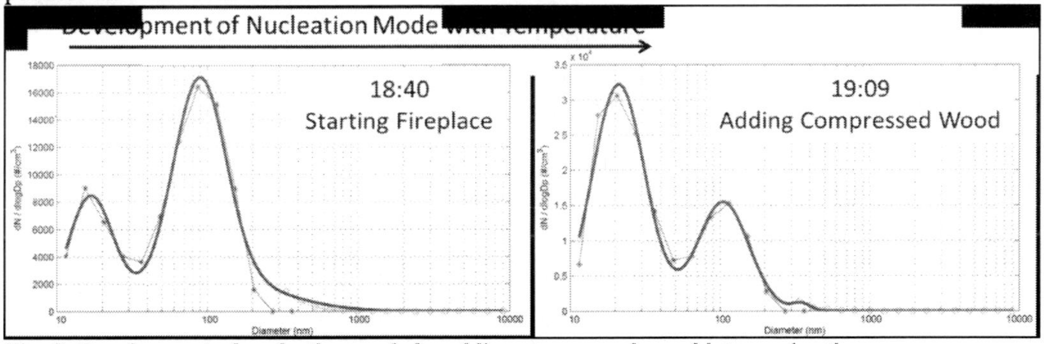

Fig. 2. Development of nucleation mode by adding compressed wood into combustion process.

Besides, the importance of the particle number size distribution determination is its relation the NOx emission. These days' scientists focus on NOx reduction. There are strategies to reduce NOx impact by using PN / PM control. As PM depends on combustion mode and NOx control strategy [6].

References

1. M.L. Souza, A.G. Allen, A.A Cardoso, *Understanding aerosol formation mechanisms in a subtropical atmosphere impacted by biomass burning and agroindustry. Atmospheric Research*, Vol. 183, 1, p.94-103 (2017)
2. E.O. Knutson, K.T Whitby, *Aerosol Classification by Electric Mobility: Apparatus Theory and Applications*, J. Aerosol Sci. Vol 6. p. 443-451, (1975)
3. S.C. Wang, R.C Flagan., *Scanning Electrical Mobility Spectrometer. Aerosol Science and Technology*, Vol. 13, 2 (1990)
4. J.H. Seinfeld, S.N Pandis, *Atmospheric chemistry and Physics*. John Willey and Sons, NewYork. (2006)
5. W.C Hinds, *Aerosol Technology: properties, behavior and measurements of airborne particles*. 2nd ed., A Wiley Interscience publication. ISBN 0-471-19410-7(1999)
6. 2016 Annual Merit Review, Vehicle Technologies Office Results Report December 2016 Prepared by Allegheny Science & Technology for the U.S. Department of Energy, Vehicle Technologies Office. ORNL, SAE 2016-01-0934, DOE AMR(2016)

Ways to heat hot water via the heat pipes

Peter Hrabovský[1,*], *Zuzana* Kolková[1], *Jozef* Matušov[1] and *Patrik* Nemec[2]

[1]University of Zilina – Reasearch Centre, Univerzitná 8215/1, 010 26 Zilina, Slovakia
[2]University of Zilina – Department of Power Energy, Univerzitná 8215/1, 010 26 Zilina, Slovakia

Abstract. The article deals various ways of heating hot water where heat is transferred by the basic phenomena of heat transfer theories, which take place on an innovative basis with the advantage of the absence of mechanical elements. The heat transfer in this case ensures a change in the phase-in phase of the working substance – the fluid – from which the thermal output and the efficiency of the device are derived. The devices described in this article work on the same principle of heat transfer. Each device is characterized by own construction and the principle of heat transfer. Heat pipes are classified according to the way of operation and the place of use. Subdivision of the heat pipe, in terms of its actual, use in the desired temperature range. At present, heating plants use cooling technologies (nuclear power, space stations, IT).

1 Introduction

Currently, the European Union's energy policy, as well as the Slovak Republic, pays great attention to the more efficient use of different forms of energy. Increasing emphasis is placed on energy self-sufficiency, increased efficiency of energy facilities and the use of renewable energy sources. Such a source of energy is also used in small heat sources, mostly to ensure thermal comfort in households.

The article describes various designs of heat pipes, which can be used in the heating of hot water by various methods. The basic distribution of heat transfer to the DHW cylinder is two ways – direct and indirect. The direct heat transfer method is defined by the direct contact of the heated substance with the heat transfer material. Indirect method of heat transfer is not directly in contact with the heated substance, but the heat is transferred through the heat transfer medium. The article describes hot water heating devices, collectively called heat pipes.

2 Gravitational heat pipe

Gravitational heat pipe is a device defined in the literature as a heat transfer device with a high thermal conductivity higher than the thermal conductivity of copper or other heat conductive materials used in thermal engineering [1, 2]. The Gravity Heat Pipe (GHP) works on the principle of phase change, from vapour to heat, and from steam to liquid when

* Corresponding author: peter.hrabovsky@rc.uniza.sk

the heat is transferred [3, 4]. The functional GHP contains a predetermined amount of a working substance capable of group transfers, which is in liquid and gaseous form. GHP consists of 3 parts: evaporation, adiabatic and condensation [5]. With regard to the material and purpose of use, a working substance is selected for the optimal functioning of the GHP [6].

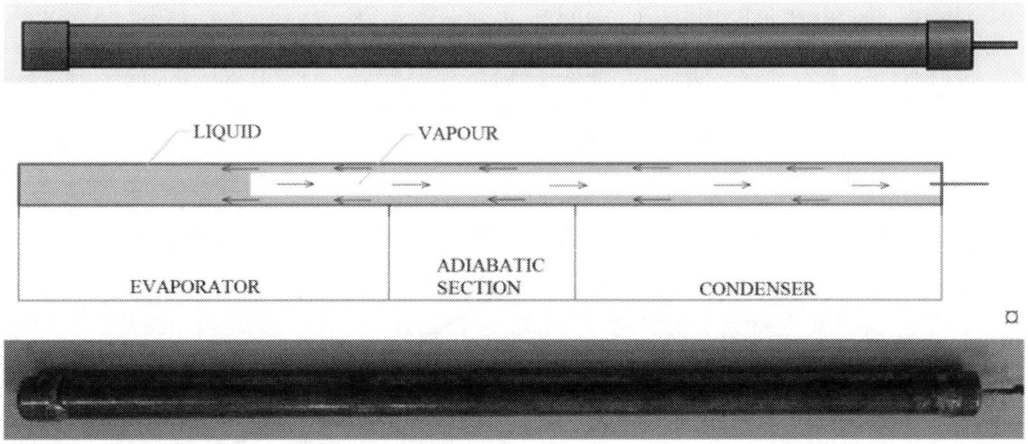

Fig. 1. Gravity heat pipe.

3 Loop heat pipe

The loop heat pipe works on the same principle as the gravity heat pipe. The circulation of the working substance occurs in the loop heat pipe. This type of heat pipe is used for long-distance applications where longer piping needs to be used. An important aspect of the operation of this heat pipe is the actual position of the individual components. As mentioned in the previous chapter (2), the heat pipe consists of the evaporation, condensation and transmission lines. In this type, the position of the evaporator and the condenser must be ensured from the construction point of view. The evaporator must always be under the condenser. Other aspects of the design of this type of heat pipe are based on the application in question, where the type of heat source and the place of consumption of heat energy are decided.

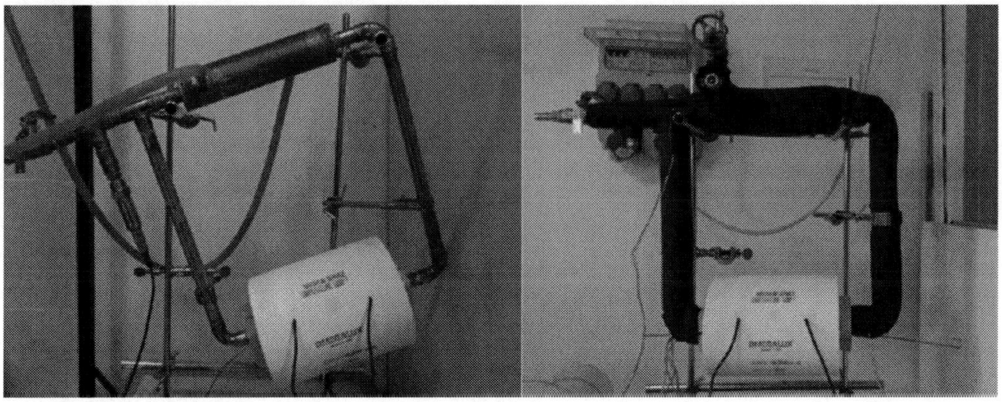

Fig. 2. Real state of loop heat pipe model.

The heat transfer media, respectively the working fluid of the loop heat pipe is distilled water. At the inlet and outlet sections of the loop heat pipe and heat exchanger, temperature sensors are located. At the top there is a pressure sensor and a valve that serves to fill and empty the volume of the loop heat pipe with the working fluid.

4 Proposal of loop heat pipe for water heating

In the design of the plant, requirements have been laid down for removing heat from combustion gases generated by the combustion process in a heat source. From the point of view of the construction and arrangement of the individual parts of the device, it is necessary to ensure the location of the evaporator part of the heat pipe in the heat source so that the combustion occurs with the surface of the evaporator.

Suggested variants show the possibility of removing heat from the flue gas. The first variant is structurally based on the removal of heat from the chimney section. In the chimney pipe, a vaporizer is located, which is a separate block with the possibility of assembly and dismantling from the chimney pipe.

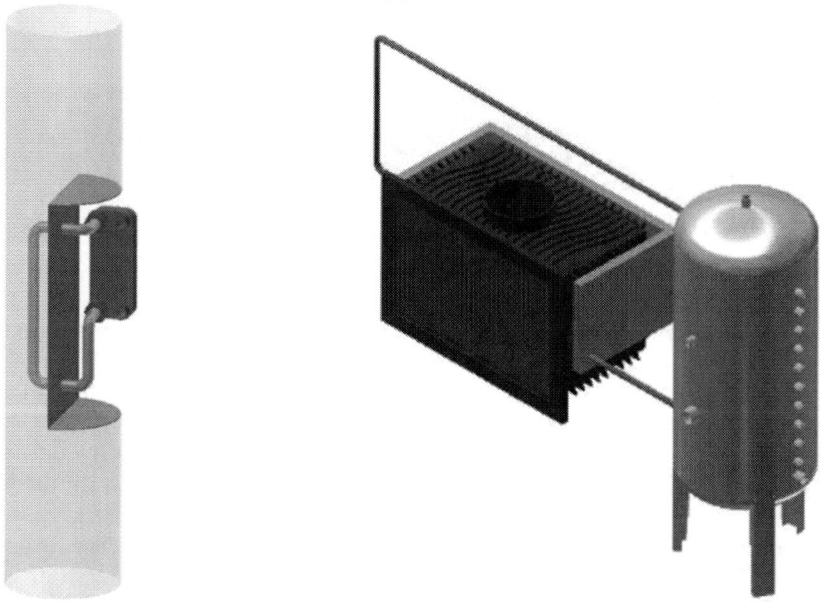

Fig. 3. Left variant evaporator with reflow exchanger, right variant with evaporator around combustion.

The second variant is based on heat removal by the evaporator, which is located on the outside of the combustion chamber shell. Its design consists of a block copying the shape of the combustion chamber.

A third variant is based on the removal of heat from the combustion gases directly in the combustion chamber. Its structure consists of a block located in the back of the heat source combustion chamber.

The second and third variants are connected to a heat exchanger in a storage tank of 110 litres.

Fig. 4. Third variant of evaporator located in rear part of combustion.

In the production of the heat pipe, emphasis was placed on the suitability of the material with respect to the location of the individual parts of the device. Selection of the material has resulted in the selection of a suitable working substance to avoid chemical reactions between the material used and the working substance, which would lead to non-condensable gases and to reduce the efficiency of the thermal equipment (heat pipe).

From a design point of view, a second variant was chosen to extract the heat from the flue gases directly from the combustion chamber. The heat source is a double-shell of evaporator inserted in fireplace of heat performance 11 kW.

The evaporator part of the heat pipe is located in the rear upper part of the combustion chamber. It is made of anti-corrosion material plate, AISI 306 steel, with a thickness of 4 mm. The outside dimensions are 500 x 200 x 20 mm. To prevent deformation of the evaporation part, which can be caused by pressure in the device, the inner space is reinforced with ribs. For the supply and removal of the working substance in the steam and liquid state, a steel pipe with diameter DN15 is located on the rear wall in the diagonal corners.

Fig. 5. Evaporator of heat pipe.

The DHW cylinder, resp. The storage tank is located at the heat source due to the compactness and control of the individual parts of the heating device. The container is a staggered oval shape with a capacity of 110 litres. The storage tank is made of anti-corrosion material AISI 304. The upper part of the container is a welded flange which serves to control and exchange the condensation part of the heat pipe, which is secured inside the container by means of a screw connection with a seal.

In the DHW cylinder or in the DHW cylinder. Of the storage vessel is a heat exchanger in the form of a threaded coil forming the condensing part of the heat pipe. The condensation part has been designed based on the water heating condition, which depends on the surface of the exchanger itself. For this reason, according to design, the largest possible area of the heat exchanger, resp. Condensing portion. The diameter of the spiral is 340 mm, length 26 m, pipe diameter 18 mm, heat exchange area 1.45 m^2 and volume 5.5 litres. The heat exchanger is made of copper, which is a very good heat conductor.

Fig. 6. Heat exchanger for hot water heating – condenser of heat pipe.

5 Conclusion

This article is an introduction to a series of exploration experiments using heat pipes for hot water heating. The next stage of the research will be designed and constructed experimental equipment. This device will be measured based on different requirements for different purposes. One of the objectives of the research will be to test the safety of the equipment for practical use and trouble-free operation of the equipment under all conditions. Hot water has become an integral part of our life. For its preparation, the technology in the area of heating and hot water production is being modernized and their design proposals help us to reduce the consumption of energy sources. Heat pipes can be used in the future to extract heat from the flue gas for heating or heating the water.

The research is supported by the European Structural Fund - project ITMS 26220220121 (50%).
The research is supported by project "Limits of Radiative and Convective Cooling through the Phase Changes of Working Fluid in Loop Thermosyphon", APVV-15-0778 (50%).

References

1. R. Nosek, S. Gavlas, R. Lenhard, M. Malcho, V. Sedlak, S. Teie, AIP conference proceedings, 1917, (2017)
2. P. Ďurčansky, Š. Papučík, J. Jandačka, M. Holubčík, R. Nosek, The scientific world journal 2014, (2014)
3. J. Jandačka, J. Mičieta, M. Holubčík, R. Nosek, Energy & fuels. 31, 3 (2017)
4. R. Lenhard, K. Kaduchová, Š. Papučík, J. Jandačka, EPJ Web of Conferences 67, (2014)
5. P. Ďurčanský, M. Patsch, J. Jandačka, AIP conference proceedings, 1745. (2016)
6. P. Pilát, M. Patsch, Š. Papučík, M. Vantúch, AIP Conference Proceedings, 1608, (2014)

AUTHOR INDEX

AUTHOR INDEX